Ocean Law and Policy

Ocean Law and Policy

20 Years under UNCLOS

Edited by

Carlos Espósito
James Kraska
Harry N. Scheiber
Moon-Sang Kwon

BRILL
NIJHOFF

LEIDEN | BOSTON

Cover illustrations: By Marina Caron. © 2016.

Library of Congress Cataloging-in-Publication Data

Names: Espósito, Carlos, editor. | Kraska, James, editor. | Scheiber, Harry N., editor. | Kwon, Moon-Sang, editor.
Title: Ocean law and policy : 20 years under UNCLOS / edited by Carlos Espósito, James Kraska, Harry N. Scheiber, Moon-Sang Kwon.
Description: Leiden ; Boston : Brill/Nijhoff, [2016] | Includes index.
Identifiers: LCCN 2016034418 (print) | LCCN 2016034498 (ebook) | ISBN 9789004311435 (hardback : alk. paper) | ISBN 9789004311442 (E-book)
Subjects: LCSH: Law of the sea. | United Nations Convention on the Law of the Sea (1982 December 10) | Maritime law. | Marine biodiversity. | Seawater—Fertilization.
Classification: LCC KZA1120.3 .O24 2016 (print) | LCC KZA1120.3 (ebook) | DDC 341.4/5—dc23
LC record available at https://lccn.loc.gov/2016034418

Typeface for the Latin, Greek, and Cyrillic scripts: "Brill". See and download: brill.com/brill-typeface.

ISBN 978-90-04-31143-5 (hardback)
ISBN 978-90-04-31144-2 (e-book)

This book is printed on acid-free paper and produced in a sustainable manner.

Contents

Acknowledgments

Since 1965 the conferences and publications of the Law of the Sea Institute—originally a chartered body founded at the University of Rhode Island, later long headquartered at the University of Hawai'i, and since 1992 a research unit of the University of California, Berkeley, School of Law—has provided one of the ocean community's most important venues for the debate of ocean law and policy development. The present volume continues in this splendid tradition, being the latest in our LOSI publication series issued by the Nijhoff/Brill house. In the course of the book's planning, editing, and development for publication, many debts have been incurred by the Berkeley LOSI staff and myself as LOSI director. It is a pleasure to acknowledge them here.

All the papers published in this volume were presented in preliminary form at an international conference held in Madrid, Spain, in late 2014; subsequently, all have been fully vetted and edited. For the support of the Madrid conference, as a collaboratively funded LOSI-Korea Institute for Ocean Science and Technology (KIOST) project, we gratefully acknowledge the important co-planning and financial roles of KIOST and especially of our volume coeditor Dr. Moon-Sang Kwon. The participation of presenters from fifteen nations was made possible in part by additional funding contributions from the Jon Van Dyke Institute of the University of Hawai'i; the law firm Briscoe, Ivester & Bazel (San Francisco); Istanbul Bilgi University (Turkey); the Fridtjof Nansen Institute (Norway); ANCORS, in Wollongong University (Australia); the Law Faculty, University of British Columbia (Canada); National University of Ireland, Galway; and the Dickson Poon School of Law, Kings College, London.

For his carrying with grace and extraordinary effectiveness the responsibilities of local negotiations for the Madrid conference's local arrangements, our thanks go out to Professor Carlos Espósito, a LOSI Distinguished Fellow, who has also played a lead role in the vetting and editorial processes. Valued advisers to the conference organizers included Judge/Professor Tullio Treves, and Professors Nilufer Oral and Seokwoo Lee. On scene, the conference staff of the Hotel Westin Palace extended every possible courtesy.

Staff support for the conference was capably provided in the LOSI headquarters at UC Berkeley by Ms. Karen Chin and Ms. Toni Mendicino, and financing was extended by the Office of the Law Dean through the School's Institute for Legal Research. At the KIOST offices in Seoul, Ms. Charity Lee provided essential liaison between our two research organizations in both planning and air travel logistics and funding. Ms. Jordan Diamond, serving as Academic

Coordinator of the LOSI, provided indispensable support (if both the initiating of ideas and the implementation on every front in regard to the Madrid event may be termed merely "support") of all my own efforts as Institute director and conference co-director. Thanks are also owed to Professor James Kraska, Distinguished Fellow of LOSI and also a coeditor of this book.

The most important acknowledgement, from myself and the co-editors, must be to the roster of distinguished scholars who have so generously devoted their time and effort to their roles in this enterprise for reflecting on twenty years of change and on the further advancement of oceans law and policy analysis. Their association with the project has been truly an honor for LOSI and for the co-editors, and we deeply appreciate their contributions to what we confidently expect will be seen as an important contribution to the continuing debates that animate the oceans community globally.

Harry N. Scheiber, Director, LOSI
Stefan Riesenfeld Professor of Law & History,
and Chancellor's Professor (Emeritus)
School of Law, University of California, Berkeley

List of Illustrations

List of Contributors

Kamal-Deen Ali
Captain in the Ghana Navy; Director of Research at the Ghana Armed Forces Command and Staff College and the Navy's Legal Director; Executive Director of the Centre for Maritime Law and Security (CEMLAWS) Africa; Adjunct Lecturer in Law and other departments at the University of Ghana.

Mariano J. Aznar
Professor of Public International Law, University Jaume I, Spain.

Vasco Becker-Weinberg
Professor at the Faculty of Law of the Universidade NOVA de Lisboa and a qualified lawyer at the Portuguese Bar Association. Dr. Becker-Weinberg is the co-coordinator of the LLM program on "Law of the Sea and the Sea-Economy."

Sherry P. Broder
Lecturer in Law at the William S. Richardson School of Law at the University of Hawai'i, Adjunct Research Fellow at the East-West Center, and long-time Hawai'i attorney with expertise in complex civil litigation.

Chang Soo Choe
Professor of Public Administration at the Cyber Hankuk University of Foreign Studies (CUFS) in Seoul, Korea, and adjunct professor at Central Officials Training Institute of Korea (COTI).

Assunção Cristas
Portuguese lawyer, professor, politician and current president of the Centro Democrático e Social—Partido Popular/CDS—People's Party. Served from 2011 to 2015 as Minister of Agriculture, Sea, Environment and Territorial Planning, Portugal.

Carlos Espósito
Professor of Public International Law, University Autónoma of Madrid (UAM), and Distinguished Fellow, Law of the Sea Institute.

James Kraska
Howard S. Levie Professor of International Law, Stockton Center for the Study of International Law, U.S. Naval War College; Distinguished Fellow, Law of the Sea Institute.

Moon-Sang Kwon
Principal Research Scientist, Ocean Policy Institute, in the Korea Institute of Ocean Science and Technology (KIOST) and Chair, Korea Ocean Policy Society.

Seokwoo Lee
Professor, Inha University Law School; Co-Editor-in-Chief, Asia-Pacific Journal of Ocean Law and Policy (APJOLP).

Hee Eun Lee
Professor, Handong International Law School.

Ronán Long
Jean Monnet Chair of European Union Law, National University of Ireland Galway, and Managing Director, Marine Law and Ocean Policy Research Services, Ltd.

Said Mahmoudi
Professor of International Law, Stockholm University.

Nilufer Oral
Member of the Istanbul Bilgi University Law Faculty; Distinguished Fellow, Law of the Sea Institute.

Seoung Wook Park
Division Leader, Policy Research Section, Ocean Policy Institute, Korea Institute of Ocean Science and Technology (KIOST).

Bimal N. Patel
Director and Professor of International Law, Gujarat National Law University, Gandhinagar (Gujarat); Member, 21st Law Commission of India.

Jin-Hyun Paik
Judge, International Tribunal for the Law of the Sea; Distinguished Fellow, Law of the Sea Institute and Professor of International Law, Seoul National University.

Kerrylee Rogers
Australian Research Council (ARC) Future Fellow in the School of Earth and Environmental Science and GeoQuest Research Centre at the University of Wollongong.

Harry N. Scheiber
Director, Law of the Sea Institute; Stefan Riesenfeld Professor, and Chancellor's Professor of Law and History (Emeritus), School of Law, University of California, Berkeley; Elected Fellow of the American Academy of Arts and Sciences.

Clive Schofield
Director of Research, Australian National Centre for Ocean Resources and Security (ANCORS) and Leader of the University of Wollongong's Global Challenge on Sustaining Coastal and Marine Zones.

Katherine Seto
Ph.D. candidate at UC Berkeley in the Environmental Science, Policy, and Management Department.

Tullio Treves
University of Milan, former Judge of the International Tribunal for the Law of the Sea, Senior International Law Consultant, Curtis, Mallet-Prevost, Colt & Mosle LL.P; Distinguished Fellow, Law of the Sea Institute.

Yasuko Tsuru
Professor of International Politics at Faculty of Global Studies, Sophia University, Tokyo, Japan.

Davor Vidas
Research Professor in International Law and Director of the Law of the Sea Programme at the Fridtjof Nansen Institute, Norway. He is the Chair of the ILA Committee on International Law and Sea Level Rise, and a member of the Anthropocene Working Group of the International Commission on Stratigraphy.

Introduction

*Harry N. Scheiber**

The Law of the Sea Institute is privileged to bring to publication this collection of important studies of ocean law and policy, providing searching analyses of developments under the UN Convention on the Law of the Sea (UNCLOS) during the two decades since it went into force. As shown in several chapters here, in many respects the success of UNCLOS has been greater than hoped; its provisions have entered into, or form, customary international law. In other regards, the dramatic advances of science, technology, and initiatives (private and public sector alike) have generated new issues and previously unappreciated possibilities for expanding the scope of UNCLOS and the regulatory regimes that were envisioned when the Convention was originally opened for ratifications in 1982.

This volume is divided into four parts. Part 1 deals with a set of prominent challenges and responses to ocean resources management. The approach is openly ambitious, as it covers analysis of regional actors' policies and practices regarding marine living resources and underwater cultural heritage, particularly the experience of EU and East Asian nations, but it also embraces more general views on global issues, such as the effects of climate change and sea level rise.

In Chapter 1, Ronán Long addresses the question of whether stakeholder participation in European Union regulatory regimes are legitimately seen as "new wine in new wineskins." Long indicates how the EU's common fisheries policy (CFP) is rooted in European treaties and springs from regulations promulgated by secondary legislation to conserve, manage and develop marine living resources. In recent years the CFP has shifted toward an emphasis upon achieving an ecosystem-based approach to the management of fisheries. Long explores the original scope and ongoing problems of ecosystem management, giving close analysis of the prominent emerging issue of recent years as to whether the CFP reform successfully decentralizes management and empowers stakeholders without introducing excessive barriers to the realization of overall goals. His deeply researched explanations of the institutional fragmentations, divergent objectives of the numerous interests identified as relevant

* Harry N. Scheiber is Director, Law of the Sea Institute; and Stefan Riesenfeld Professor, and Chancellor's Professor of Law and History (Emeritus), School of Law, University of California, Berkeley.

 | DOI 10.1163/9789004311442_002

stakeholders (fishing industries, regional and national regime structures and actors, scientists and managers, and environmental NGOs) is unique in the literature and one may fairly term it a "book-within-a-book"! Long concludes that the reformed CFP has the potential to strengthen the effective role and defined responsibilities of industry in the management and conservation of marine living resources.

Part 1 advances further the theme of coastal State resource management with a chapter by Katy Seto on West Africa and the New European Common Fisheries Policy. Her chapter complements Long's examination of the EU CFP by considering how it has historically affected fisheries in West Africa. The West African states are some of Europe's closest neighbors, and the waters comprising their Exclusive Economic Zones (EEZs) are some of the most productive fishing grounds being worked by European fleets. Combined CFP policies have led to a range of consequences, from degraded habitat and diminished stocks to conflicts between domestic and foreign fishers, and thus have stymied local economies. Seto seeks to sort how the recently enacted reforms to the CFP may alter or perpetuate trends in this complex relationship.

In Chapter 3, Davor Vidas examines what climate change means for international law of the sea and global geopolitics in an age of geographic instability along the coasts. He bases his analysis on the concept of "the Anthropocene" that is now gaining currency in the field of earth science. Coastlines now form the key objective circumstance for generation of rights of states to maritime zones. With variable (and shrinking) coastlines that are being affected by global climate change, however, profound re-examination of international law may be required to recast such concepts as the legitimacy of existing territorial claims and even basic notions of human rights. In 2009, the International Commission on Stratigraphy established the Anthropocene Working Group (WG), tasked with examining the stratigraphic basis for the Anthropocene as a distinctive, new epoch in geological history. Findings of this Working Group could raise awareness of the human impact on the Earth System, and thereby prompt new reflections on today's social structures, including the basic premises as to coastal geography and, more generally, as to international law and the law of the sea.

Mariano Aznar, author of Chapter 4, assesses the 2001 Convention for the Protection of Underwater Cultural Heritage (UCH) against the backdrop of UNCLOS. Some ambiguities that reside in the Convention text, taken together with strong differences of interest in the global community as to the allocation of rights to culturally important—often of high monetary values—vessels and artistic artifacts lost at sea, have impeded the progress of implementation. The author provides a finely textured comparison of UNCLOS and UCH, indicating the high stakes that are involved in finding a path to resolution of outstanding national and interest-group differences that are obstructing the path to more universal ratifications. In addition to assessing the full dimensions of the of the juxtaposition of UNCLOS and UCH, Professor Anzar raises a set of questions in depth regarding the legal status of sunken State vessels and the phenomenon of creeping coastal State jurisdiction.

The theme of underwater cultural heritage is further developed in Chapter 5, by Seong Wook Park and Chang Soo Choe, who are concerned with the topic from the standpoint of Korea's current laws and outstanding points of contention regarding new directions for policy. Park and Choe consider Korea's experience with UCH, which entered into force in 2009. Korea still lacks a legal framework for the management and protection of its underwater cultural heritage, even though it is surrounded by water on three sides and has a rich underwater cultural heritage. Korea nonetheless has made progress to protect this cultural heritage through the 2010 Law on the Protection and Investigation of Buried Cultural Property. The authors suggest the new law misses the opportunity to better define the concept of underwater cultural heritage, and it fails to address issues needing resolution, such as warships and other government ships and military aircraft found underwater, the scope of temporal limits, and ownership of, and jurisdiction for, underwater cultural heritage. The authors call for a new Korean legal framework that would include standardization of recovery criteria and the specification of procedures and measures to prevent illicit recovery or export.

Part 2 consists of six chapters that address new concepts of maritime security law throughout Europe, Africa, America, Asia and Australia. This Part also offers substantive analysis of a broad series of topics, such as how to address civil disobedience in the maritime domain, piracy at sea, and liability for wrongful acts committed by private military and security companies on board ships.

Chapter 6, by co-editor James Kraska, analyzes civil disobedience in light of established ocean law. The author considers the legal treatment of the Sea Shepherd Conservation Society in the aftermath of its protests against Japanese whaling ships and Greenpeace's campaign against offshore oil development in the Arctic. These types of aggressive civil disobedience raise questions about appropriate boundaries between two bodies of international law—viz., international human rights norms that reflect the rights of all people to exercise free expression, and the principles of good order and safety at sea embodied in the international law of the sea. Kraska suggests that there is a key difference between laws regulating protest on land and protest at sea. Efforts to address civil disobedience on land unfold within clear boundaries of sovereignty and jurisdiction, and they typically involve the timely intervention of police authorities to separate protesters from other citizens in order to preserve public safety and order, while also protecting the right of free expression to make demands on the polity. In contrast, civil disobedience at sea occurs under conditions and areas of authority shared by more than one state, so competence to prescribe and enforce rules are shared. Frustration on both sides is palpable and can lead to violence.

Kamal-Deen Ali contributes Chapter 7, on the response to piracy in the Gulf of Guinea. Ali, a naval officer and ocean law scholar, recounts the growing set of general international rules to address piracy in West Africa, including United Nations Security Council Resolutions 2018 (2011) and 2039 (2012) and the Code of Conduct for the Repression of Illicit Activities at Sea in June 2013 at Yaoundé, Cameroon. These instruments build on the basic regime for suppression of maritime piracy in UNCLOS, and supplement the classic law enforcement approach by also addressing underlying politico-economic drivers of maritime crime in West Africa.

Chapter 8, by Vasco Becker-Weinberg, explores some of the issues associated with one response to maritime piracy—the employment of privately contracted armed security personnel. His chapter, focused on liability for wrongful acts by private military and security companies on board ships, lays out a set of important questions regarding the use of force by such companies, with regard to corporate civil liability for wrongful acts. Various PMSCs have been successful in suppressing piracy in the Western Indian Ocean and are used in the Gulf of Guinea as well, but not without controversy. Flag state, coastal, and port state jurisdiction are all in play against a background of—and often in tension with—basic UNCLOS provisions and a set of larger international law concepts. Moreover, even where carrying weapons on board and where the use of force is authorized by international law and guidelines, there remains the difficulty in harmonizing human rights standards and penal regimes. Becker-Weinberg concludes that victims of wrongful action can seek both criminal and civil penalties against PMSCs, although there is not a comprehensive system for doing so.

The maritime security laws and policies of Australia, India, Singapore, and South Africa are evaluated in Chapter 9, contributed by Bimal N. Patel. Ever since the attacks of 9/11, maritime security has become a central focus in the development of maritime law more generally. The 2008 UN Secretary-General Report on Oceans and the Law of the Sea identified seven specific maritime security threats: piracy, terrorism, illicit trafficking in arms and WMD, narcotic drugs, smuggling and trafficking of persons, IUU fishing and damage to the marine environment. Like the earlier chapters in this volume on the African coastal fishing states, and on Korean law and its adequacy in meeting the need for underwater cultural protection, Professor Patel's chapter narrows the focus of analysis to provide a detailed example of how the issues are being confronted (or substantially neglected) in a local or regional context. His study is particularly valuable, therefore, for its comparison of municipal legal provisions and varying practices among four nations of the Indian Ocean region; and he identifies lapses and lacunae in regard to both national policies and also regional cooperation. Yet he also finds common threads that reflect global and regional norms.

Regional maritime security requirements and initiatives have expanded globally along with the expansion of international trade. In Chapter 10, Nilufer Oral examines key dimensions of the maritime security regime in the Black Sea area. The navigational passages to the Black Sea are regulated under the 1936 Montreux Convention Regarding the Regime of the Straits, which has remained in force since its adoption. The Convention created a unique regime of passage for the Turkish Straits and entry into the Black Sea by imposing limits (enforced by Turkey) on the total number and tonnage of military vessels passing through the Straits, and on the duration of operations conducted by military vessels. Historically, the Convention has safeguarded the security of Turkey as well as that of the Black Sea more generally, as expressly stated in the Preamble of the Convention. The Montreux Convention persisted through World War II and the Cold War, under often-intensive pressure from other states to reduce Turkey's authority to impose and enforce controls; and the determination of the Turkish governments to avoid any strong legal challenge to their authority has also motivated that nation's refusal to ratify UNCLOS. Now the Montreux regime faces new challenges with the redefinition of NATO, the emergence of reconfigures regional and global conflicts, and new energy corridors.

Seokwoo Lee and Hee Eun Lee are co-authors of Chapter 11, on Northeast Asian maritime security. The authors search in European diplomacy for a security model that might reduce tension in Northeast Asia—a region at risk from North Korean belligerency, nationalism, the economic deflation and domestic political instability of Japan, the rising strength and assertiveness of China, the reduced U.S. military budget, and subsequent rebalance of forces to Asia. Northeast Asia does not have a collective security mechanism such as NATO, which links Europe to the United States and offers a security umbrella but also mutual obligations for cooperation. Lee and Lee conclude that the recent history of cooperation between NATO and Japan, and especially the role of Japan in Afghanistan, provide a glimpse of how the western model of regional security might be exported to Asia through Japan.

Part 3 contains two chapters concerning the interrelated challenges to currently accepted doctrines in international law (and, equally, challenges to the status quo in international relations more generally) that arise in relation to the applications of new technologies throughout the entire area of the global oceans. New geo-engineering techniques to address climate change and methods to protect biodiversity in areas beyond national jurisdiction pose new challenges for UNCLOS and the progressive development of international law.

Opening Part 3, Sherry Broder considers in Chapter 12, in the intersecting contexts of technical change and emerging global responses in law and regime definition, how geoengineering alternatives such as ocean iron fertilization (OIF) may benefit the marine food chain and increase the ability of the ocean biological pump to draw carbon dioxide from the atmosphere. OIF experiments in the ocean have precipitated action by international bodies to adapt international regulations, such as the London Protocol to the London Convention. In 2010, States parties adopted an Assessment Framework to provide criteria for scientific OIF research based on the precautionary principle. The Convention on Biological Diversity also may play a role. Given the consensus on the dire future of climate change, it is evident that geoengineering will continue to evolve as a technology, and further changes in the law will become imperative as this development continues.

Yasuko Tsuru writes in Chapter 13 about the relationship between UNCLOS and changes in our understanding of the marine environment as background to consideration of a possible new international agreement on marine diversity in the area beyond national jurisdiction. Specifically, when UNCLOS was negotiated, States aimed for commercial exploitation of seabed mining in Part 11. Today, however, the most promising resources of the seabed may be living creatures and their genetic material, rather than mineral nodules. Debate is unfolding over whether or to what extent the framework in Part 11 should capture marine genetic resources within the concept of the "common heritage of humankind." The Convention on Biodiversity (CBD) seeks to conserve biological diversity, use its components sustainably, and effect a fair and equitable sharing of benefits derived from genetic resources. Tsuru describes the omission of genetic resources in Part 11 as an "institutional deficit" that has opened a structural conflict among developing countries, the EU and other developed States. She then analyzes whether a new implementation agreement proposed by the Rio+20 (UNCSD) meeting, could fill this gap. Meeting new challenges will require the continued evolution of international law.

In Part 4, Chapters 14 and 15 assess the trajectory of ITLOS jurisprudence, casting light upon leading juridical challenges faced by UNCLOS after twenty years of rich and extensive practice. Specific case law analysis of some controversial issues in international law of the sea portends to even greater development, such as the place of human rights under the jurisdiction of the ITLOS, the scope of the standard of review, and the application of UNCLOS to Non-Party States. Importantly, the two authors represented in this Part are a sitting and a former ITLOS judge, both of them eminent jurists but also of high reputation as university professors who have commented with great vision on the present condition and future problems of ocean law.

In Chapter 14, Judge Jin-Hyun Paik of ITLOS offers "reflections from the bench" on the subject of standard of review under UNCLOS. Judge Paik follows the arc of international case law and the various standards for judicial decision. How have courts applied standards, such as "necessary and appropriate measures" or determined what might be considered "reasonable" under particular circumstances? In such cases, international courts may invoke lower or more deferential standards, or higher and stricter standards. Paik suggests that the standard of review can be a sensitive issue, "as it touches upon the fundamental balance of power between States and an international judicial institution." The standard of review also affects the efficiency and legitimacy of the international court. Courts that apply stringent standards may be criticized for being too intrusive into State sovereignty, for example.

In Chapter 15, former ITLOS Judge Tullio Treves examines the problem of how non-party states relate to UNCLOS. The non-parties to UNCLOS include United States, Colombia, Israel, Peru, Turkey and Venezuela. These States are bound by customary international law, much of which is reflected in UNCLOS. Such rules are binding on all States. Treves analyzes the operation of this rule of construction in international law as a feature in two cases decided in judgments of the International Court of Justice involving a State party to UNCLOS and a non-party: the 2012 *Nicaragua v. Colombia* judgment and the 2014 *Peru v. Chile* judgment. Both judgments consider provisions of UNCLOS that correspond to customary international law—the first concerning the regime of islands and the continental shelf, and the second concerning delimitation of the exclusive economic zone and continental shelf.

The other papers in Part 4 deal with the challenges of ocean rise, proposed policy imperatives for a "blue economy," and, finally, the unique issues of a navigation regime for the Arctic. Chapter 16, by Assunção Cristas, former minister of agriculture and the sea in the Portuguese government, is an intriguing report on policy objectives of her government. She contends that development of the blue economy has become an economic and political imperative, with maritime spatial planning (MSP) as the essential dimension. Development of a legal framework for MSP, as envisioned for Portugal and other States, will advance the blue economy agenda so that economic benefits are realized throughout the maritime sectors, within conditions of transparency and environmental sustainability.

Clive Schofield is author of Chapter 17, in which he explores how marine spatial issues are becoming ever more important to effective oceans governance—creating "ever more 'lines in the sea.'" Rapid geospatial changes are set against the evolving international law context, and Schofield suggests that the definition of such maritime limits and boundaries is essential to the exercise of jurisdiction and effective marine management.

With coauthor Kerrylee Rogers, Schofield also contributes Chapter 18, on planning for coastal ecosystem services in an era of climate change. This chapter considers how the range of coastal fauna and flora are involved in sequestering carbon from the atmosphere, known as "blue carbon," as they do not become carbon saturated and decomposition is hindered by saline conditions. Coastal ecosystems also have some capacity to adapt to sea-level rise by maintaining their position within the tidal frame by vertically accreting sediments or migrating to suitable intertidal elevations at rates equivalent to rates of sea-level rise. This phenomenon, however, is not captured in coastal ecosystem protection policy and legislation. Governments should plan for sea-level rise and the maintenance of ecosystem services by protecting land that will accommodate the migration of natural coastal ecosystems in the future. Without appropriate planning, ecosystem services may be significantly reduced, and our capacity to mitigate climate change by maximizing 'blue carbon' opportunities will be diminished.

Our volume closes with a study of international navigation in the Arctic, by Said Mahmoudi, who considers in Chapter 19 the new standards for navigation in this region that have been instituted at national, regional and international levels. Article 234 of UNCLOS provides special legislative and enforcement powers for the Arctic coastal States as regards measures against vessel-source pollution in ice-covered areas; both Canada and Russia have leveraged the provision to strengthen their rules. These States act against the backdrop of entry into force of the IMO Polar Code. But the relationship between stringent national requirements under Article 234 and the Polar Code is uncertain. The chapter also explores the legal status of Arctic waters adjacent to Canada and the Russian Federation enclosed by straight baselines. The two countries consider these waters as historic internal waters that are subject to national sovereignty. The United States and the EU have challenged these claims. The increased accessibility of the Northern Sea Route to international navigation may tempt vessels of other nations to test the seriousness of the Russian claim, something that the United States unsuccessfully attempted during the 1960s. Professor Mahmoudi includes in his analysis a suggestion of how a pragmatic arrangement like the one that exists in the Baltic Straits might be able to accommodate the concerns of all States.

It is fitting that we have been able to include in this volume reflections on twenty years of ocean law development under UNCLOS and, also, analyses of major ongoing issues related to climate change, science and engineering, implementation of UN initiatives, institutional and doctrinal innovation, regime-management reform, and basic concepts in ocean law jurisprudence. The Law of the Sea Institute and the authors represented in this book are dedicated to the tenet that a robust rule of law in the oceans is an essential prerequisite if the global community is to fulfill the universal hopes for an optimistic future—a future of resources robustly sustained, of healthier and richer lives for the world's peoples, with the maintenance of peace and security of the global oceans realized.

PART 1

Managing Ocean Resources: Rising Challenges & New Responses

CHAPTER 1

Stakeholder Participation in the European Common Fisheries Policy: Shifting the Legal Paradigm toward Rights and Responsibilities

*Ronán Long**

Introduction

The general theme of the 2014 conference convened by the Law of the Sea Institute (LOSI) is twenty years of development of oceans law and policy under the 1982 United Nations Convention on the Law of the Sea (UNCLOS). With this in mind, if one takes a quick look at the proceedings of the 1985 LOSI conference,[1] it is evident that much of the academic discussion on marine resource management at that particular time was focused on who would reap the economic benefits of fisheries over the coming decades under the then relatively new jurisdictional framework advanced by the UNCLOS.[2] Some of the contributors to the 1985 LOSI conference were optimistic about the establishment of 200-mile fishing zones and the vesting of sovereign rights and management powers in a single unified entity, the coastal State, away from the global commons, on the grounds that it would resolve many of the over-exploitation issues stemming from the insuperable common property approach that

* Jean Monnet Chair of European Union Law, National University of Ireland Galway, and Managing Director, Marine Law and Ocean Policy Research Services, Ltd. The chapter is an updated account of the Author's contribution to the EU project: SOCIOEC FP7-KBBE-2011-5 Project No. 289192. The author wishes to acknowledge the assistance of Margaret Armstrong (MSc) in compiling the research material for this chapter and comments on an earlier draft of the SOCIOEC report from Professor Jesper Raakjær (Aalborg University) and Dr. Ralph Doering (Johann Heinrich von Thünen Institute—Institute of Sea Fisheries). The publication reflects the views of the author, and the European Union is not responsible for the opinions or information contained therein.

1 E. Brown, R. Churchill, *The UN Convention on the Law of the Sea: Impact and Implementation* (Honolulu: Law of the Sea Institute and William S. Richardson School of Law, University of Hawaii, 1987) (cited as *L. Sea Inst. Proc.* 1987).

2 See inter alia: L. Anderson "Introduction," *L. Sea Inst. Proc.* 1987 at 137; G. Pontecorvo "The Impact of the Law of the Sea Treaty on the Organization of World Fisheries: some Preliminary Observations on Production," *ibid.*, at 138–156.

 | DOI 10.1163/9789004311442_003

had plagued the management of fisheries for much of the 20th century.[3] Others argued that there were going to be winners and losers under the UNCLOS regime.[4] Indeed, this conclusion was borne out by the fact that states were beginning to take many different management approaches to fishery resource allocation entitlements, with New Zealand setting an international precedent by adopting a system of individual transferrable quotas in order to provide a stable framework for decision-making by companies.[5] As a result, one contributor to the 1985 LOSI conference was able to report that as a small coastal fishing state, New Zealand had "developed a significant fishing industry and still had room for expansion based on sound biological and economic policy,"[6] The distinguished scholar, Professor Rögnvaldur Hannesson, in his insightful and prescient contribution to the 1985 LOSI conference noted that the strength of this market-based system is that the economics prospects of the fishing industry "prevail or perish according to its own profitability."[7]

If we now look back at the progress that has been made in fisheries management and governance in the Europe Union (EU) over the past three decades, what a different story! Perhaps it is best to start by pointing out that the EU has adopted, since its foundation in the late 1950s, a diverse range of policies touching many areas of economic, political and social endeavour that impinge upon fisheries including international trade, the environment, competition, employment, social rights, regional development, foreign affairs, human rights and state aid.[8] Among these, the common fisheries policy (CFP) regulates and manages the activities of the European fishing industry from the Arctic to tropical waters. The policy has extensive geographical, material and personal scope, and applies to all third country vessels operating in Union waters, as well as to the vessels and nationals of EU Member States both within and beyond the EU.[9] The conservation of marine biological resources is an exclusive EU competence and the law-making powers are thus firmly vested in the

3 *Ibid.*

4 G. Munroe *L. Sea Inst. Proc.* 1987 at 204–26.

5 P. Major, "Fisheries Development in New Zealand Since UNCLOS: A Case Study," *L. Sea Inst. Proc.* 1987, 178–203.

6 *Ibid.* at 200.

7 R. Hannesson *L. Sea Inst. Proc.* 1987 at 208. For a provocative review of the growth of ITQs and other private property schemes, see, by the same author, *The Privatization of the Oceans*, (Boston: MIT Press, 2004).

8 H. Wallace, M. Pollack, A. Young, 6th Ed., *Policy-Making in the European Union* (Oxford: OUP, 2010) *passim.*

9 Article 1, Regulation (EU) No. 1380/2013, OJ L 354/22, 28 December 2013.

European institutions.[10] Briefly stated, commercial sea-fisheries in sea areas under the sovereignty and jurisdiction of the 23 coastal Member States are a common pool resource over which the EU is charged with a broad swathe of stewardship and management functions.[11] Over the past three decades, the EU has exercised its extensive law-making powers to adopt conservation and management measures that promote the objective of optimum utilization of the living resources in the sea areas under the jurisdiction of its Member States, as well as on the high seas, in accordance with UNCLOS and related international agreements including the 1995 Fish Stocks Agreement.[12] The CFP is also unique in that it seeks to manage fish stocks in an integrated manner, irrespective of the maritime boundaries of the Member States, traditionally on the basis of maximum sustainable yield (MSY) and more recently by using new normative principles including the ecosystems-based approach and the precautionary principle.[13]

Catastrophically, the traditional approach has not been successful, and 88 percent of EU fish stocks had been fished beyond MSY by 2009.[14] As a result, few would dispute that the CFP remains one of the EU's most controversial policies. Indeed, many of its key features and essential elements can appear at times incomprehensible and perhaps even contradictory.[15] Such perceptions

---

10 Articles 3(d), TFEU. For legal commentary on the CFP, see, R.R. Churchill and D. Owen, *The EU Common Fisheries Policy* (Oxford: OUP, 2010), *passim*.

11 Belgium, Croatia, Cyprus, Denmark, France, Finland, Estonia, Germany, Greece, Latvia, Lithuania, Malta, Ireland, Italy, the Netherlands, Spain, Poland, Portugal, Bulgaria, Romania, Sweden, Slovenia, and the United Kingdom.

12 The EU is party to both UNCLOS and the United Nations Agreement for the Implementation of the Provisions of the United Nations Convention on the Law of the Sea of 10 December 1982 relating to the Conservation and Management of Straddling Fish Stocks and Highly Migratory Fish Stocks of 4 August 1995 (UN Fish Stocks Agreement). Decision 98/392/EC *Concerning the conclusion by the European Community of the LOSC and the 1994 Implementation Agreement*, OJ L 179 (23 June 1998); Decision 98/414/EC *On the ratification by the European Community of the FSA*, OJ L 189/14 (3 July 1998); Decision 96/428/EC, OJ L 177/24 (16 July 1996). The EU has also accepted the FAO Compliance Agreement. Council Decision on acceptance by the Community of the Agreement to promote compliance with international conservation and management measures by fishing vessels on the high seas of 25 June 1996, OJ L 177, 16.7.1996, p. 24.

13 R. Long, "Legal Aspects of Ecosystem-Based Marine Management in Europe," in A. Chircop, M.L. McConnell, S. Coffen-Smout (eds.), *Ocean Yearbook* Vol. 26, (Boston/Leiden: Brill Academic Publishers, 2012), 417–484.

14 COM(2009)163 final, Brussels, 22.4.2009, at 7.

15 R. Long, P. Curran, *Enforcing The Common Fisheries Policy* (Oxford: Blackwell Science, 2000), at 1.

are understandable in light of the failure of the policy to deliver on its core objectives of sustainable fisheries and the consolidation of the European fishing industry as a viable and dynamic economic sector. These failures are all the more critical in light of the deplorable status of world fisheries.[16]

One of the strengths of the CFP, however, is that the policy is subject to a continuous process of review and reform, most notably in 1992, 2002, and from 2009 through to 2013. Leading up to the most recent reform of the policy, the European Commission brought together several multidisciplinary research teams to work on solutions on how to improve fisheries management over the coming decade. One such project, the SOCIOEC project,[17] was tasked with evaluating the socio-economic effects of management measures and with reviewing, among other matters, how EU law influences the possibilities for strengthening governance structures and stakeholder involvement in fishery regulation and management under a reformed CFP. This chapter is an abridged and updated version of my unpublished contribution to the SOCIOEC project. The chapter starts by describing the EU fisheries stakeholders, then shifts to the nature of their participation in the CFP and the rationale for enhancing their role in governance and management. Also considered is the approach taken by the European institutions to improving participatory mechanisms and industry responsibility in resource utilization under the new regulatory framework, as well as stakeholder empowerment more generally.[18] The chapter concludes with comments on some of the lessons that may be derived from the European reform experience in light of 20 years of UNCLOS.

A Common Property Resource: The Source of Intractable Problems

The importance of the CFP can be gauged from the size of the EU fishing fleet, which was comprised of 87,000 vessels in 2015.[19] The majority of these vessels are small (less than 15 meters in length) and they seldom range beyond costal fisheries.[20] Nonetheless, the EU has a sizeable fleet of larger vessels. The principal flag States of the EU are Spain, Denmark, United Kingdom and France,

16 FAO, *The State of World Fisheries and Aquaculture 2014* (Rome: FAO, 2015).

17 <http://www.socioec.eu>.

18 *Op. cit.* note 2.

19 Volume in tonnes live weight and percentage of total. European Commission, Facts and Figures on the Common Fisheries Policy: Basic Statistical Data, (Luxembourg: Publications Office of the European Union, 2014), at 10.

20 *Ibid.*

whose vessels operate primarily in the northeast Atlantic, the Mediterranean Sea and in third country waters under the terms of fisheries partnership agreements concluded by the EU with such countries.[21] The EU fleet catches about 5 percent of the world's catch, which is well behind China at 17 percent but nonetheless makes Europe a global force in terms of resource utilization and environmental impact.[22] In addition, Europe is one of the world's most important markets for fishery products. Consumer demand drives the unrelenting quest for access to new resources and this in turn increases the economic output of fishers. As a direct consequence, the FAO estimates that the average output per fisher ranges "from a high of just above $19,000 in Europe to about $2,231 in Africa and $1,720 in Asia, about a 10-fold difference," which reflects global inequality and the growing concentration of income in the hands of the EU fishing industry.[23]

Despite its economic significance and global footprint, fisheries in Europe have been in a constant cycle of decline and crises.[24] The main problems that have beset the CFP since its inception are the lack of environmental, economic and social sustainability, as well as the complexity of the legislative framework along with a very poor public image.[25] As regards governance, the setting of short-term economic and social objectives, at the expense of long-term environmental sustainability, has resulted in many policy failings.[26] Similar to other world fisheries, the juridical nature of marine living resources as a common property natural resource has also contributed to over-fishing, economic inefficiency, competition and conflict, in sea areas both within and beyond coastal State jurisdiction.[27] Moreover, many of the intractable problems that have beset the management of European fisheries over the past three decades can be attributed to a complex web of inter-related factors, including poor regulatory implementation by the Member States, and in some instances non-compliance by the industry with their legislative obligations.[28] This is

---

21 On third country agreements, see Chapter 2 in this volume by K. Seto, "West Africa & the New European Common Fisheries Policy: Impacts & Implications."

22 European Commission, *op. cit.* n. 12.

23 FAO, The State of World Fisheries and Aquaculture 2006 (Rome: FAO, 2007).

24 J. Raakjaer, *A Fisheries Management System in Crisis: The EU Common Fisheries Policy* (Aalborg: Aalborg University Press, 2009).

25 SEC(2011) 891, Brussels, 13 July 2011, p. 6.

26 SEC(2011) 891, Brussels, 13 July 2011, pp. 10–11.

27 R. Churchill, A.V. Lowe, *The Law of the Sea* (3rd edition, Manchester: Manchester University Press, 1999) 281–283.

28 See R. Long, P. Curran, *Enforcing the Common Fisheries Policy* (Oxford: Blackwell Science, 2000), (Malden MA: Blackwell Science Inc., 2000), (Carleton: Australia, 2000) *passim*.

borne out by the findings of the Court of Auditors, as well as a number of comprehensive specialist studies on this topic.[29]

When considering the scale of the predicaments in European fisheries, one should not forget that the CFP, like many other EU policies, is largely reflective of Member State preferences regarding the form and content of secondary legislation and policy measures. Furthermore, the fisheries policy is unique because Member States, on the grounds of political expediency, frequently ignore scientific advice and disregard the precautionary principle when making decisions and voting in the Council of Fisheries Ministers, a practice that has undoubtedly undermined the policy in achieving many of its conservation objectives.[30] One other long-standing weakness is that the 28 Member States are responsible for giving full effect to EU legislation, which remains a battleground for legal confrontation with the supra-national EU institutions. The Court of Justice of the European Union in turn has frequently sanctioned Member States for non-compliance with their EU obligations, including most notably imposing a penalty payment of €20 million on France, with the potential threat of an additional six-monthly penalty of €57.8 million,[31] for its failure to comply with a previous Court ruling on serious shortcomings in fishery law

29 See, *inter alia*, European Court of Auditors' Special Report n°7/2007, on the control, inspection and sanction systems relating to the rules on conservation of Community fisheries resources together with the Commission's replies. OJ C317/1 of 28.12.2007; A. Berg, *Implementing and Enforcing European Fishery Law: The Implementation and Enforcement of the Common Fisheries Policy in the Netherlands and the United Kingdom* (The Hague: Kluwer Law, 1999); C. Johnson, "Fisheries Enforcement in European Community Waters Since 2002—Developments in Non-Flag Enforcement," *International Journal of Marine and Coastal Law* 23, no. 2 (2008): 249–270.

30 According to the Commission, this can be shown by two elements: "First, the average percentage deviation of Council TACs decisions from scientific advice (for stocks with such advice), for the years 2003–2010, was 47%. Since 2008 this deviation has been reduced to 34% in 2010. Nevertheless, TACs are still set well above what scientist recommend. Second, the number of stocks for which scientific advice was a zero TAC, and for which the Council adopted a positive TAC, was 17 on average for the period 2003–2010. It has been decreasing since 2007 (20) to 2010 (14) but still is high. The CFP establishes the precautionary approach to ensure sustainable exploitation of stocks and to ensure that the impact of fishing on marine ecosystems is kept at a sustainable level. *De facto*, the principle could amount to giving environmental sustainability some degree of precedence. However, it has been rarely used in reality, and in most situations, the general trend has been to roll-over TAC levels." See SEC(2011)891, Brussels, 13 July 2011, 10–11.

31 Case C-304/02 *Commission v France* [2005] ECR. I-6263; Case T-139/06 *France v Commission* [2011] E.C.R. II-7315.

enforcement at a national level.[32] Crucially, all of these practices and proceedings have undermined public and stakeholder confidence in the CFP.

Law Reform Objectives

In light of the interminable crises in EU fisheries, a central objective of the 2009–2013 law reform process was to vest stakeholders, and the fishing industry in particular, with additional rights and responsibilities. There is nothing new or radical in this development as stakeholder involvement in the CFP is a central pillar of the policy since as far back as 2002.[33] In theory, this ought to mean giving stakeholders ownership of the policy and ensuring that there are appropriate structures in place that facilitate, where appropriate, the devolution of management decisions to fishers.

Such an approach is fully consistent with best practice internationally, where there is increased awareness by the FAO and other specialist bodies of the importance of stakeholder participation in key fisheries management functions.[34] Significantly, the World Bank and the FAO in their seismic report *The Sunken Billions: The Economic Case for Fisheries Reform* called for the implementation of secure marine tenure and property rights systems at local, national and regional levels, worldwide.[35] As noted by the World Bank, common reform elements "could include effective stakeholder consultation processes; sound social and economic justifications for change; and an array of

---

32 Case C-64/88 *Commission v France* [1991] ECR I-2727.

33 Art. 2(2)(c) of Council Regulation (EC) No. 2371/2002.

34 See, *inter alia*: FAO, *Law and sustainable development since Rio—Legal issues and trends in agriculture and natural resources management* (Rome, FAO Legislative Study 73, 2002); J. Lindsay "Designing Legal Space: Law as an enabling tool in community-based management," International Workshop on Community-Based Natural Resource Management, Washington D.C., May 10–14, 1998; R.S. Pomeroy et al., "Conditions Affecting the Success of Fisheries Co-management: Lessons from Asia," 25 (2001) *Marine Policy*, 197–208; E. Schlager, E. Ostrom, "Property-rights Regimes and Natural Resources: a conceptual analysis. (1992) 68(3) Land Economics, 249–262; B. Kuemlangan, H. Teigenem "An Overview of Legal Issues and Broad Legislative Considerations for Community-Based Fisheries Management" in R.L. Welcomme and T. Petr (eds.) Proceedings of the Second International Proceedings of the Second International Symposium of the Management of Large Rivers for Fisheries Volume II. *Sustaining Livelihoods and Biodiversity in the New Millennium, 11–14 February 2000*, (Phnom Penh: FAO & The Mekong River Commission, 2004).

35 World Bank and FAO, *The Sunken Billions: The Economic Case for Fisheries Reform* (Washington D.C.: World Bank, 2009), p. xxi.

social and technical options, including decentralization and co-management initiatives to create more manageable fishery units."[36] There is also considerable support in the specialist literature for the adoption of a rights-based approach characterized by individual transferable quotas and property rights, as a pathway leading to more effective natural resource stewardship in the management of discrete fisheries.[37]

The most recent European efforts to reform the CFP should not therefore be viewed in isolation in so far as they are indicative of developments at an international level to resolve the global crisis in marine fisheries. The shift away from centralised "command-and control" structures to a governance model that reflects the interests of stakeholders is ultimately aimed at making EU fisheries management more efficient, effective, accountable and transparent. Indeed, in light of such considerations it is entirely understandable why the European Commission, when presenting its draft legislative proposals for the reform of the policy, sought to strengthen the structures and procedures for stakeholder engagement and to shift the legal framework towards a rights-based approach for the first time in EU fisheries management.

Who Are the Stakeholders?

There appears to be no simple or definitive answer to this question from a legal perspective, as the term "stakeholder" remains open textured and undefined in EU fisheries law. That said, the regulatory impact assessment published by the Commission on the reform of the CFP sheds some light on this issue as it lists the principal stakeholders concerned with the policy, along with their primary areas of interest. For the sake of completeness, this information is reproduced in Table 1.1.[38]

---

36 *Ibid.*, at 56.

37 See *inter alia*: Committee to Review Individual Fishing Quotas, Ocean Studies Board, Commission on Geosciences, Environment, and Resources, and National Research Council, *Sharing the Fish: Toward a National Policy on Individual Fishing Quotas* (Washington, DC: National Academy Press, 1999); R. Shotton, (ed.) Use of Property Rights in Fisheries Management, Vol. 2, FAO Fisheries Technical Paper 404/2, (Rome: FAO, 1999); J.E. Wilen, "Property Rights and the Texture of Rents in Fisheries" in ed. D.R. Leal, (ed.) *Evolving Property Rights in Marine Fisheries* (Lanham: Rowman and Littlefield, 2004) 49–67; R. Grafton et al. "Positioning Fisheries in a Changing World," *Marine Policy* 32 (2008):630–34; R. Barnes, *Property Rights and Natural Resources* (Oxford: Hart Publishing, 2009) *passim.*

38 SEC(2011) 891, Brussels, 13 July 2011, p. 25.

TABLE 1.1 *Stakeholders and their Key Interests*

Stakeholder	Description of stakeholder	Key interests
Catching sector in the EU	EC vessel owners, operators and crew.	Maintaining profitability and livelihoods.
Dependent businesses & communities	Business and communities dependent upon fisheries for their livelihoods.	Maintaining profitability and livelihoods.
Processing Sector	Those processing raw material both imported and caught within EC waters	Maintaining profitability and livelihoods, stable supplies.
Sector Regulators	National, regional and local bodies regulating fishing	Ensuring an efficient, effective and practical management framework that balances a wide range of stakeholder needs.
Sector research	Scientific research bodies contributing to the conservation and management of stocks	Contribution to an effective fisheries management regime through the timely access to high quality, robust data from fishery dependent and independent sources.
Consumers	Those consuming fisheries products	Availability, cost, quality and nutritional values of fisheries products with varying degrees of environmental scrutiny.
Third countries	Fishing sector in competition with EU fleets. Aquaculture producers, exporters to the EU. Authorities in third countries receiving payments under FPA or private agreements	Conflicting interest between those who see the EU as a very important export market and as a source of revenue and small local fishing communities that face competition from external fleets of the EU on access to local resources.

TABLE 1.1 *Stakeholders and their Key Interests* (cont.)

Stakeholder	Description of stakeholder	Key interests
NGOs, the civil society and EU citizens	Non-governmental organizations advocating sustainable management of fisheries. The wider public with an interest in and concern for fisheries and the marine Environment	To maintain fish populations, marine biodiversity, and the amenity value of oceans, rivers and lakes.

From the information presented in Table 1.1, it is evident that there are two broad categories of stakeholders concerned with the CFP. First, there are the parties linked directly and indirectly with the fishing industry. These include the catching-sector itself, which for obvious reasons must be viewed as one of the dominant stakeholder interest in any discussion concerning reform measures. The second category of stakeholder is much broader and extends to NGOs, civil society and the citizens of the Union. This division is also replicated in the structure and composition of the European Fisheries Advisory Councils [ACs],[39] which are characterized by two very wide-ranging

39 The ACs came into operation by means of Commission Decision with the first regional body established for the North Sea in 2004. See Commission Decision 2007/222/EC of 4 April 2007 declaring operational the Regional Advisory Council for the south-western waters under the common fisheries policy. OJ L 95, 5.4.2007; Commission Decision 2007/206/EC of 29 March 2007 declaring operational the Regional Advisory Council for the High-Seas/Long-Distance Fleet under the common fisheries policy, OJ L 91 of 31.3.2007]; Commission Decision 2006/191/EC of 1 March 2006 declaring operational the Regional Advisory Council for the Baltic Sea under the common fisheries policy OJ L 66, 8.3.2006; Commission Decision 2005/668/EC of 22 September 2005 declaring operational the Regional Advisory Council for the north-western waters under the common fisheries policy, OJ L 249, 24.9.2005; Commission Decision 2005/606/EC of 5 August 2005 declaring operational the Regional Advisory Council for pelagic stocks under the common fisheries policy, OJ L 206, 9.8.2005; Commission Decision 2004/774/EC of 9 Nov 2004 declaring operational the Regional Advisory Council for the North Sea under the common fisheries policy, OJ L 342, 18.11.2004; Commission Decision of 29 August 2008 declaring operational the Regional Advisory Council for Mediterranean Sea under the Common Fisheries

classifications of stakeholder-led organisations, namely: the "fisheries sector" and "other interest groups."[40] Again, the former is defined as meaning "the catching sub-sector, including ship-owners, small-scale fishermen, employed fishermen, producer organisations as well as, amongst others, processors, traders and other market organisations and women's networks."[41] The latter includes: "amongst others, environmental organisations and groups, aquaculture producers, consumers and recreational or sport fishermen."[42] The term "among others" suggests that this list is neither inclusive nor exclusive but remains a representative list of stakeholders. Although not specifically mentioned, the marine scientific community could come within the scope of this definition and are a significant stakeholder group in their own right.[43] The division of stakeholders into two distinctive categories is also reflected in the rules applicable to the functioning of the constituent bodies of the ACs, where certain sessions are open to the public but voting rights and decision-making remain the sole prerogative of fishers and other interest groups.[44] In sum, the term "stakeholder" is not a term of art in EU fisheries law with specific legal meaning. For that reason, it can have several meanings in different management and regulatory contexts. Furthermore, governance structures and stakeholder involvement in management can be viewed from many standpoints ranging from public consultation on broad policy issues to the active participation of fishers in core decisions concerning the practical aspects of fishery management. The main point is that EU law appears to make a clear demarcation between the rights and interests of fishers *vis-á-vis* those afforded to civil society and the public more generally.

Byzantine Architecture for Participation

For those unfamiliar with the functioning of the EU, it is often difficult to identify the precise nexus between stakeholders and the various policy and

Policy, OJ L 232/12, 30.8.2008. See R. Long, "The Role of Regional Advisory Councils in the European Common Fisheries Policy: Legal Constraints and Future Options," *The International Journal of Marine and Coastal Law*, 25 (2010): 289–346.

40 Article 1, Council Decision of 19 July 2004 establishing Regional Advisory Councils under the Common Fisheries Policy, OJ L 256/17, 3.8.2004.

41 Article 1(2), 2004/585/EC: Council Decision of 19 July 2004.

42 Article 1(3), 2004/585/EC: Council Decision of 19 July 2004.

43 Article 6, 2004/585/EC: Council Decision of 19 July 2004.

44 Article 6(5), 2004/585/EC: Council Decision of 19 July 2004.

law-making processes that operate within the complex architecture of the European institutions. At first blush, there is extensive scope for all stakeholders including representatives of the fishing industry to express their concerns regarding the CFP by means of the democratic processes that operate through the good offices and political dynamics of the supranational European institutions. This process is achieved by means of the structures and procedures established under the European treaties and secondary legislation and that apply in the Council (of Fisheries Ministers), the European Parliament, the Court of Auditors to a lesser degree, the advisory bodies of the Union (the Economic and Social Committee and the Committee of the Regions), and in exceptional circumstances through judicial proceedings in the Court of Justice European Union. Several of these institutions played a noteworthy part in the reform of the CFP. That said, the principal actors were the Council and the Parliament, who share joint law-making powers in the EU in relation to adoption of secondary legislation. As will be seen below, the European institutions have very different views on how to improve stakeholder participation in the reformed CFP.

Apart from the Member States and the specialist working groups in the EU institutions, there are several other entities that play an advisory role in policy and law-making processes including specialist fisheries consultative and representative bodies. Indeed, stakeholder involvement in European fisheries can be traced back to the work undertaken by the Advisory Committee for Fisheries since the early 1970s.[45] As mentioned previously, one of the most important developments in the reform of the policy in 2002 was the creation of the ACs. This step, in turn, was an essential step in the regionalization of the CFP and to ensure wider stakeholder engagement in the policy process.[46] Operating in parallel with the ACs, there are several Europe-wide bodies that are consulted by the Commission from time to time on draft legislation and these include: Europêche, the EU Fish Processors' Association, the EU Federation of National Organizations of Importers and Exporters of Fish, the European Association of Fish Producers' Organisations, and the Federation of European Aquaculture Producers.[47]

Good quality scientific and expert advice is pre-requisite for success in fisheries management. In this regard, there are a number of scientific bodies that

45 Commission Decision 71/128/EEC, OJ L 68, 22.3.1971, p. 18.

46 Arts. 31 and 32 of Council Regulation (EC) No. 2371/2002. Council Decision 2007/409/EC of 11 June 2007 amending Decision 2004/585/EC.

47 Churchill and Owen, *op. cit.*, note 3 at 28.

play a crucial advisory role in policy formulation within the EU institutions including the Scientific, Technical and Economic Committee for Fisheries (STECF).[48] The latter is an independent body made up of experts in the fields of marine biology, marine ecology, fisheries science, nature conservation, population dynamics, statistics, fishing gear technology, aquaculture, and the economics of fisheries and aquaculture.[49] They are consulted by the Commission at regular intervals on matters pertaining to the conservation and management of living aquatic resources, including biological, economic, environmental, social and technical considerations.[50] Members of STECF are appointed in their personal capacity and they must act independently of Member States and the interests of stakeholder groups.[51]

All of the aforementioned bodies need to be distinguished very carefully from producers' organisations (POs), which remain the backbone of the CFP and facilitate the decentralization of many management decisions at an operational level in the Member States. More exactly, the Commission maintains a list of over 200 recognised POs in the fishery and aquaculture sector and they play a significant role in fishery management in several Member States.[52] They are commercial associations concerned in the main with the marketing of fishery products.[53] As will be seen below, however, their role extends to promoting methods that encourage sustainable fishing and in some Member States this includes the collective administration and management of quotas among their members.[54] The potential of the POs playing an enhanced part in fishery management arrangements under the revised CFP was flagged by the Commission during the course of the reform debate 2011–2013.[55]

48 Art. 31(4) of Council Regulation (EC) No. 2371/2002.

49 Art. 33 of Council Regulation (EC) No. 2371/2002; and Commission Decision 2005/629/EC, OJ L 225/18, 31.8.2005.

50 Art. 33(1) of Council Regulation (EC) No. 2371/2002.

51 Art. 2(2) of Commission Decision 2005/629/EC.

52 OJ C225/9, 30.7.2011.

53 Article 9 of Council Regulation (EC) No. 104/2000, OJ L 17/22, 21.1.2000.

54 See, for example, Marine Management Organisation, Rules for the management of UK fish quotas for ICES areas I, II, IV, VI, VII and associated areas and vb (Faroese waters) in 2013. Available at: http://www.marinemanagement.org.uk/fisheries/management/documents/quotas/ukrules2013.pdf.

55 Green Paper, *Reform of the Common Fisheries Policy*, COM(2009)163 final, 22.04. 2009, pp. 11–12.

Rationale Underpinning Further Stakeholder Empowerment

The case supporting greater stakeholder participation in fisheries governance, regulation and management is based upon economic, political, social, environmental and legal considerations. The European institutions view dialogue with stakeholders as "essential" for the achievement of the objectives of the CFP and to improve legitimacy, transparency, and accountability.[56] This is underpinned by arguments associated with improving economic efficiency through the establishment of a system of private property rights and the implementation of responsible fishing practices by means of delegating responsibility to users through market-based incentives and goal-based management approaches.[57]

Greater stakeholder involvement in the CFP is also predicated on the view that the CFP will benefit from the knowledge and experience of fishers.[58] The fishing industry is often best placed to implement management measures as they have a clear understanding of the tasks and challenges encountered in the practical aspects of conservation and resource utilization. There is also support for the view that self and co-management will result in higher levels of compliance and accountability by the sector, as the industry has greater interests at stake in protecting private property rights.[59] The economic case supporting the introduction of private property rights is of course strengthened if the costs associated with self-regulatory measures are borne by the industry and this in turn leads to greater efficiencies regarding the utilization of public monies for the purpose of management, administration and oversight. We will return to the controversial issue of cost allocation at the end of the chapter.[60]

From the standpoint of implementation, the traditional centralized, top-down approach to fisheries management makes it difficult for the CFP to respond to regional variations at sea-basin levels in the EU. Devolving rights and powers down to local levels thus accords with the ecosystem approach as articulated by a number of regional seas bodies including the OSPAR Commission, which has called for greater involvement of stakeholders in decision-making

56 Recital 65, Regulation (EU) No. 1380/2013.

57 See, K. Nolde Nielsen, P. Holm, M. Aschan, "Results Based Management in Fisheries: Delegating Responsibility to Resource Users," *Marine Policy* 51 (2015): 442–451.

58 Recital 27, Council Regulation (EC) No. 2371/2002 of 20 December 2002 on the conservation and sustainable exploitation of fisheries resources under the Common Fisheries Policy, L 358/59, 31.12.2002.

59 Communication from the Commission on rights-based management tools in fisheries, COM(2007)73, Brussels, 26.2.2007.

60 See discussion "portends of the future: sharing enforcement costs," *infra*.

concerning sustainable uses of the marine environment and the resources it supports.[61] Indeed, greater stakeholder involvement in decision-making concerning oceanic resources is identified in a whole raft of environmental instruments since the early 1990's, including: Principle 10 of the 1992 Rio Declaration on Environment and Development;[62] Agenda 21 of the 1992 United Nations Conference on Environment and Development;[63] and the IUCN principles on ocean governance, which stress that decision-making ought to be inclusive.[64] In a similar vein, the FAO Code of Conduct for Responsible Fisheries calls for the "effective participation of industry, fishworkers, environmental and other interested organisations in decision making with respect to the development of laws and policies related to fisheries management, development, international lending and aid."[65] All of the aforementioned instruments had a bearing on the outcome of the CFP reform process.

The case for strengthening stakeholder structures thus appears to be a strong one. Nonetheless, when considering the raison d'être for further stakeholder empowerment under the reformed CFP, one must take into account that the interests of fishers frequently diverge from those of other interest groups. This is borne out by the findings of an EU funded inter-disciplinary social science and economics study, which demonstrates that industry representatives have significantly different preferences for the ecological, economic, and social pillars of sustainable fisheries when compared to the more mainstream ecological interests of the representatives of three other stakeholder groups, namely: public authorities, scientists, and NGOs.[66]

---

61 Strategy of the OSPAR Commission for the Protection of the Marine Environment of the North-East Atlantic 2010–2020. Available online: <http://www.ospar.org/documents/dbase/decrecs/agreements/10-03e_NEA%20Environment%20.Strategy.doc#biodiversity>.

62 United Nations Conference on Environment and Development: Rio Declaration on Environment and Development, 14 June 1992 (1992) 31 *International Legal Materials* 874.

63 See Agenda 21, Chapter 17, paragraph 23.2, in UN Doc. A/CONF.151/26/REV.1, Vol. 1, 12 August 1992.

64 Adopted at the IUCN World Conservation Congress, Barcelona, in 2008. Available at: www.iucn.org. For commentary, see D. Freestone, "Principles Applicable to Modern Oceans Governance," *IJMCL* 23 (2008): 385–391; and by the same author, "The Modern Principles of High Seas Governance: The Legal Underpinnings" *International Environmental Policy and Law* 39 (2009): 44–49.

65 Art. 6.13 of the FAO Code of Conduct for Responsible Fisheries.

66 M. Aanesen, C. Armstrong, H. Bloomfield, and C. Röckmann, "What Does Stakeholder Involvement Mean for Fisheries Management?" *Ecology and Society* 19 (2014): 35–44 at 41.

Typology: Industry and Stakeholder Participation

Instructively, one academic study has identified three types of industry and public participation in shaping the environmental policies and laws made by government and regulatory bodies.[67] According to Bell, McGillivray and Pedersen, they are as follows: first, "pluralistic participation, within which industry and other representative bodies speak and act on behalf of their constituent members; second, 'stakeholder participation, within which proposals that have already been formulated are transmitted to interested parties to comment upon and refine', and; third, deliberative participation, which consists of 'agreeing the ground rules'—that is involving the stakeholder in determining what general policies and strategies should be adopted—before moving to the stage of specific proposals for implementation."[68]

Using this framework, we can see that there is a clear division between the entitlements of NGOs, civil society and EU citizens, when compared to more substantive and procedural rights enjoyed by professional associations and commercial enterprises representing the interests of fishers under the CFP. Perhaps a useful starting point is to draw some parallels with the legal requirements and best practices that arise in environmental decision-making under the UN/ECE Convention on Access to Information, Public Participation in Decision-Making and Access to Justice in Environmental Matters (the Århus Convention).[69] At one level, stakeholder participation in the CFP under the Åarhus Convention may just entail notification and consultation in respect of the drawing up of certain plans and programmes relating to the environment including those pertaining to fisheries.[70] Similarly, they may be consulted in relation to the environmental impact of fisheries projects, plans and programmes under the EIA and the SEA Directives.[71] Indeed, the Court

67 See S. Bell, D. McGillivray, O. Pedersen, *Environmental Law, 8th ed.* (Oxford, OUP, 2013) p. 335.

68 *Ibid.*

69 The EC signed the Århus Convention in 1998. The EU approved the Convention on 17 February 2005 and many of its provisions are now reflected in many European legal instruments adopted pursuant to Council Decision of 17 February 2005 on the conclusion, on behalf of the European Community, of the Convention on access to information, public participation in decision-making and access to justice in environmental matters. OJ L 124, 17.5.2005.

70 See M. Pallemaerts, *The Aarhus Convention at Ten: Interactions and Tensions Between Conventional International Law and EU Environmental Law* (Groningen: Europa Law Publishing, 2011).

71 Directive 2011/92/EU OJ L 26/1, 28.1.2012; Directive 2001/42/EC, OJ L 197/30, 21.7.2001.

of Justice has held that the harvesting of cockles constituted a "project" and thus required environmental impact assessment under EU law, along with stakeholder notification and consultation regarding on-going harvesting operations.[72] In marked contrast, stakeholder involvement in the CFP does not appear to tally exactly with the various schemes foreseen for public engagement under the general rubric of EU environmental law. For instance, under the CFP, it consists predominantly of the various processes that are in place for keeping the industry and other stakeholders informed of policy decision, as well as the content and shape of draft legislation at the EU and national levels, mainly through the good office of representative bodies such as the ACs.[73] As will be seen below, however, it goes well beyond this in several EU Member States and could very well be described as "pluralistic and deliberative participation" as it entails vesting industry through their constituent POs with a direct role in quota allocation and utilization by means of a system of property rights or other user entitlements.

The general trend in EU law is to widen public participation in environmental decision-making and many initiatives now stem from EU and national legislation that give effect to the obligations that arise under the Åarhus Convention, as well as by means of the procedural mechanisms that apply under various EU Directives including those concerning EIA/SEA,[74] as well as the Marine Strategy Framework Directive (MSFD).[75] Notably, the European Parliament tabled an amendment to draft Basic Regulation during the reform process, which stated that all actions taken by the Union and by Member

72 Case C-127/02 *Waddenvereniging and Vogelbeschermingsvereniging* [2004] ECR I-7405.

73 See R. Long, *op. cit.* note 31.

74 Directive 2011/92/EU on the assessment of the effects of certain public and private projects on the environment, L26/1, 28.1.2012; Directive 2001/42/EC on the assessment of the effects of certain plans and programmes on the environment, OJ L 197/30, 21.7.2001.

75 Directive 2008/56/EC of the European Parliament and of the Council of 17 June 2008 establishing a framework for community action in the field of marine environmental policy (Marine Strategy Framework Directive) OJ L 164/19, 25 June 2008. For commentary on this Directive see R. Long, "The EU Marine Strategy Framework Directive: A New European Approach to the Regulation of the Marine Environment, Marine Natural Resources and Marine Ecological Services," *Journal of Energy and Natural Resources Law* 29 (1) (2011) 1–45; A. Trouwborst, H. Harm, "Comparing European Instruments for Marine Nature Conservation: The OSPAR Convention, the Bern Convention, the Birds and Habitats Directives, and the Added Value of the Marine Strategy Framework Directive," *European Energy and Environmental Law Review* 20 (2011): 129–149; L Juda "The European Union and the Marine Strategy Framework Directive: Continuing the Development of Ocean Use Management," *Ocean Development and International Law* 41 (2010): 34–54.

States under the CFP must comply fully with the both the tenor and letter of the Aarhus Convention.[76] This proposal was aimed at ensuring that the public are guaranteed as a very minimum certain procedural rights regarding the CFP under the three limbs of the Åarhus Convention to strengthen the participation of citizens in environmental matters and to improve the implementation of marine environmental law. Such an approach brings reduces the risk of conflict in environmental-decision-making, thereby increases transparency and legitimacy.[77]

In light of the disparate nature of stakeholder participation in the CFP, there are two aspects of EU law that merit further consideration here: first, the use of regulatory tools in policy and law-making processes to engender stakeholder engagement; and second, the restrictive nature of the rules pertaining to judicial review proceedings in the European courts.

Regulatory Tools that Foster Participation and Legitimacy

EU laws and policies on fisheries do not evolve in a vacuum. On the contrary, there is a sophisticated suite of regulatory tools available to the EU law-making institutions to ensure stakeholder participation at all stages of the policy cycle from formulation to the subsequent implementation of the rules in the Member States including the use of impact assessment, stakeholder consultation and evaluation. In both theory and practice, the application of these tools ought to strengthen the legitimacy of the adopted measures.[78]

Perhaps it is best to start by pointing out that as part of its legal mandate under the EU treaties the Commission plays a central role in bringing forward draft legislation within the EU institutions for the purpose of formulating and implementing the CFP.[79] As part of this task, it regularly evaluates EU policies, legislation, international agreements and non-binding measures pertaining to the management of fisheries. The process usually begins with the Commission publishing policy blueprints, usually in the form of roadmaps and green papers that outline the general thrust, policy options and scheme of planned

76 Amendment 109, European Parliament First Reading, 6 February 2013.

77 J. Steele, "Participation and Deliberation in Environmental Law: Exploring a Problem-Solving Approach," *Oxford Journal of Legal Studies* 21 (2001): 415.

78 This is particularly the case in discrete fields such as social policy where stakeholders have a role in some instances as co-regulators. See P. Verbruggen, "Does Co-Regulation Strengthen EU Legitimacy?" *European Law Journal* 15 (2009): 425–441.

79 Article 43, TFEU.

regulatory initiatives. Subsequently, all proposed legislation is subject to an *ex-ante* and *post-ante* evaluation of the potential economic, social and environmental consequences of EU measures. The aim of this exercise is to identify both the positive and negative impacts of the proposed measures, as well as their conformity with the principles of proportionality and subsidiarity. Once adopted, EU secondary legislation is periodically subject to a Regulatory Fitness and Performance Program, which are aimed at reducing the regulatory burdens placed on stakeholders. In the words of the Commission, the "programme aims to cut red tape, remove regulatory burdens, simplify and improve the design and quality of legislation,"[80]

In line with this general scheme for EU law-making, the proposed EU reform measures were subject to comprehensive assessment, consultation and evaluation, during the period 2009–2013.[81] As part of this process, the Commission published an Impact Assessment Report (IAR), which reflects the views of the various stakeholders on the reform options advanced in the draft legislation.[82] The IAR followed a well-established methodology, which identified the structure, problem, performance and problems of the current CFP, the need for EU involvement, the justification for reform, an assessment of the objectives of the policy, as well as an assessment of the effectiveness of the reform options and their environmental, economic and social impact.[83] A central element in this process is risk assessment and an examination of potential risk mitigation measures that are capable of ameliorating policy shortcomings. On the whole, the IA undertaken on the draft legislation was a rigorous exercise and provided a clear insight into the rationale underpinning the EU reform proposals. The whole cycle of review and reform revealed that there was an urgent need for further stakeholder empowerment and a radical overhaul of fisheries governance and management structures at both EU and Member State levels. The importance of this exercise should not be underestimated as the Court of Justice has referred to the results of IA in judicial proceedings when reviewing

80 COM(2014)368 final, Brussels, 18.6.2014, at 2.

81 http://ec.europa.eu/fisheries/reform/index_en.htm.

82 Impact Assessment Accompanying Commission proposal for a Regulation of the European Parliament and of the Council on the Common Fisheries Policy [repealing Regulation (EC) No. 2371/2002]. SEC(2011)891, Brussels, 13 July 2011.

83 European Commission, Impact Assessment Guidelines, SEC(2009)92, 1.1. Available at: http://ec.europa.eu/smart-regulation/impact/commission_guidelines/docs/iag_2009_en.pdf. See also P. Craig, G. De Búrca, *EU Law: Text, Cases and Materials*, 5th ed., (Oxford: OUP, 2011) at 93.

the legality of EU secondary legislation.[84] Again, however, we can see that EU law makes a clear demarcation between the rights afforded to industry and other stakeholders in judicial review proceedings in the Court of Justice.

Paradoxes in Participation by Means of Judicial Review Proceedings

The European institutions elaborate the policy framework underpinning the CFP by means of secondary legislation in the form of regulations, decisions and directives. The process of judicial review by the European Court of Justice ought then to be one of the most effective ways of making the EU institutions accountable and to ensure that they act within the confines of what is permissible under the European treaties, secondary legislation and in conformity with international agreements that are binding on the EU including UNCLOS.[85] Broadly speaking, review procedures in the European courts serve two purposes, annulment proceedings to strike down a contested measure, or to compel the EU to act where it has not done so.[86]

Therefore, one would expect that judicial review proceedings should be a viable option for stakeholders to engage with the policy and law-making supranational institutions. As a general matter of practice, however, it is extremely difficult for individuals to gain standing in the EU courts with a view to making the EU institutions more accountable regarding fisheries management decisions. On this subject, three brief points illustrate how EU law draws a clear distinction between the rights of private parties and those of environmental NGOs in judicial review proceedings.

First, provided that they satisfy strict admissibility requirements, natural or legal persons, including professional associations or organisations representing the interests of the fishing industry, are considered as semi-privileged applicants in such proceedings before the European courts.[87] As such, a legal

---

84 See Case C-58/08 *The Queen, on the application of Vodafone Ltd. v Secretary of State for Business, Enterprise and Regulatory Reform*, [2010] ECR I-04999.

85 A. Ward, *Judicial Review and the Rights of Private Parties in EU Law*, 2nd ed. (Oxford: OUP, 2007) *passim*.

86 Articles 263, 265 and 277, TFEU.

87 Joined Cases C-182/03 and C-217/03, *Kingdom of Belgium and Forum 187 ASBL v. Commission of the European Communities* [2006] ECR I-5479; Joined Cases 67/85, 68/85, and 70/85 *Van der Koy and Others v. Commission* [1988] ECR 219; Joined Cases T-481 and T-484/93 *Exporteurs in Levende Varkens and Others v. Commission* [1995] E.C.R. II-2941; Joined Cases T-227/01 to T-229/01 and T-265/01, T-266/01 and T-270/01 *Territorio Histórico*

act of the EU institutions that impinges upon their private property rights or business interests is reviewable by the Courts, provided that the decision is of "individual concern" to an applicant.[88] Under this procedure, for example, a Spanish company established in France with vessels that operated on a regular basis in the waters south of Ireland, was eligible to challenge EU fisheries conservation measures on the basis that they affected their business interests.[89] In stark contrast, environmental groups such as *Greenpeace* or other entities that act in the public interest more generally, lack *locus standi* to initiate such actions, thus depriving them of access to justice.[90] This issue is particularly relevant to the issue of stakeholder engagement with the CFP, because the absence of standing effectively closes the door on environmental groups seeking the intervention of the Court to review the probity of EU legislative and policy measures.

Second, as noted in the introduction to this chapter, international agreements concluded by the EU bind its institutions and the Member States.[91] The Court of Justice, however, will only review the legality of EU acts against the obligations set down in international agreements if the latter creates directly effective rules that confer rights and freedoms on natural or legal persons.[92] In addition, the contested provision at issue in the international agreement must be sufficiently precise and unconditional.[93] Following on from this and in the context of the law of the sea, the Court is unwilling to review the legality of EU measures against the benchmarks set down by UNCLOS and international marine environmental agreements, even where such instruments are binding on the EU. This is because the nature and broad logic of UNCLOS does not grant independent rights and freedoms to individuals and thereby prevents the Court from assessing the legality of EU acts in the light of its provisions.[94]

---

de Álava—Diputación Foral de Álava and Comunidad autónoma del País Vasco—Gobierno Vasco and Others v. Commission [1995] ECR II-2941.

88 Case 25/62, *Plaumann & Co v Commission* [1963] ECR 95.

89 Case C-263/02, *Commission v Jégo-Quéré* [2004] ECR I-03425.

90 Case C-321/95 P, *Greenpeace v. Council* [1998] ECR I-1651 (affirming Case T-585/93 *Greenpeace and others v Commission* [1995] ECR II-2205).

91 *Op. cit.* note 5.

92 Case C-2008/312, *Intertanko and Others*, [2008] ECR I-4057, para. 45; Case C-2008/476, *FIAMM and Others v Council and Commission*, [2005] E.C.R. II-5393, paras. 110 and 120; and Case C-366/10, Air Transport Association of America and others v. Secretary of State for Energy and Climate Change [2011] ECR I-1133, para. 54.

93 *Ibid.*

94 *Ibid.*

Third, the Court of Justice has further restricted the ability of European citizens and environmental NGOs to challenge decisions of EU institutions under legislation giving effect to the Åarhus Convention. The latter, it may be recalled, is partly aimed at providing individuals and environmental groups with wide access to justice on environmental matters.[95] In January 2015, however, in cases concerning EU legislation on European air quality and pesticides, a Grand Chamber of the General Court denied environmental NGOs the right to make a 'request for internal review' of decisions taken by the Commission on the grounds that the challenged acts could not be considered of "individual concern" to the Applicants in the proceedings at hand.[96] An outcome that was subsequently described by the European Environmental Bureau (an organisation that represents the environmental voice of European citizens) as a "step backwards for environmental democracy, public accountability and access to justice with respect to the EU institutions at a time when the EU needs to be proving its democratic credentials."[97] For that reason, the judgment appears to be a retrograde step marking a shift away from stakeholder empowerment on foot of the Åarhus Convention.

Hand-in-Hand: Rights and Responsibilities

In contrast to the impediments faced by certain categories of stakeholders in judicial review proceedings, the Commission pushed during the reform negotiations 2009–2013 for the adoption by the EU of a rights-based fishery management system as a means to allocate "individual fishing rights to fishermen, fishing vessels, enterprises, cooperatives or fishing communities."[98] One of the key passages in the 2009 *Green Paper*[99] on this subject is worth quoting in full:

---

95 *Op. cit.* note 58.

96 Cases C-401 to 403/12, *Council e.a. v. Vereniging Milieudefensie* and C-404 and 405/12, *Council v. Stichting Natuur en en Milieu and Pesticide Action Network Europe v Commission* 13 January 2015. See documents as included in http://curia.europa.eu/.

97 European Environmental Bureau Press Release 13 January 2015: http://www.eeb.org/index.cfm/news-events/news/ecj-rulings-a-setback-for-environmental-democracy/.

98 Communication from the Commission on rights-based management tools in fisheries, COM(2007)73, Brussels, 26.2.2007, p. 3.

99 Green Paper, Reform of the Common Fisheries Policy, COM(2009)163 final, 22.04.2009.

Very little can be achieved if the forthcoming reform fails to motivate the catching sector, the processing and seafood chain as well as consumers to support the objectives of the policy and take responsibility for implementing them effectively. It is critical to the success of reform that industry should understand the need for it, support it and have a genuine stake in its successful outcome. In a mostly top-down approach, which has been the case under the CFP so far, the fishing industry has been given few incentives to behave as a responsible actor accountable for the sustainable use of a public resource. Co-management arrangements could be developed to reverse this situation.

There are two closely linked aspects to involving the industry more closely: responsibilities and rights.

The industry can be given more *responsibility* through self-management. Results based management could be a move in this direction: instead of establishing rules about how to fish, the rules focus on the outcome and the more detailed implementation decisions would be left to the industry. Public authorities would set the limits within which the industry must operate, such as a maximum catch or maximum by-catch of young fish, and then give industry the authority to develop the best solutions economically and technically. Results-based management would relieve both the industry and policy-makers of part of the burden of detailed management of technical issues. It would have to be linked to a reversal of the burden of proof: it would be up to the industry to demonstrate that it operates responsibly in return for access to fishing. This would contribute to better management by making the policy considerably simpler and removing the current incentives for providing false or incomplete information. Proportionality should be observed and the impact on preserving a competitive industry when implementing results-based management.

There are already many examples of such self-management through bottom-up initiatives in the European catching sector. Some Producer Organisations (POs) manage the quota uptake of their members and provide for private penalties against those who overshoot their individual quota at the expense of others. There are examples of groups of vessels that have taken on the burden of proof by providing full documentation of their catches, often as a response to processors' and retailers' pressure to improve traceability. These initiatives could be generalised by turning the POs into bodies through which the industry takes responsibility for documentation and quota/effort management.

> Giving the industry more responsibility requires that safeguard mechanisms are in place and implemented by the Community. There are cases where the catching sector shoulders greater responsibility to adapt the size of the fleet and their management costs. In some cases, this has been achieved by giving *rights* to the industry, which has encouraged them to use their investments more efficiently and eliminate surplus capacity. In the context of a CFP, which gives more rights to the catching sector and relieves the industry the burden of the micro-management it will be relevant to raise the issue of sharing the costs of fisheries management. So far the fishing industry has been given free access to a public resource and management costs have been largely incurred by taxpayers.[100]

The 2009 *Green Paper* also draws attention to a number of related issues that are relevant to the enhancement of stakeholder involvement in the CFP. More specifically, it focuses on the hierarchy for decision-making and the failure of the policy to distinguish decisions concerning the principles of fisheries management from the practical aspects of policy implementation. As a result of this failure, the Commission noted that all decisions tend to be taken at the highest political level by the Council of Fisheries Ministers and implemented subsequently by means of regulations, which are binding on the Member States and on fishers. As a matter of practice, this regulatory approach allowed little scope for a more flexible approach to the technical task of implementation and gave the industry little say in key management decisions. In order to address this shortcoming, the Commission advocated the drawing of a clear hierarchical distinction between decisions concerning the fundamental principles underpinning the policy and the technical task of implementing policy detail in the Member States.

Two reform options to address this challenge are mentioned in the *Green Paper*, namely: greater use of delegated powers within the EU institutions, or "specific regional management solutions implemented by Member States, subject to Community standards and control," Essentially, this latter option would entail the taking of decisions on fundamental principles at the EU level (such as those pertaining to MSY) and the delegation of implementation decisions at Member State level subject to the caveat that Member States would work at the level of the marine region in relation to management of shared stocks and ecosystems.

100 Green Paper, Reform of the Common Fisheries Policy, COM(2009)163 final, 22.04.2009, pp. 11–12.

The underlying rationale for both options (delegation to the Commission or co-management at a regional level) was that it would allow "governments and the industry to adapt the implementation of the policy to their needs and to find the best solutions both technically and economically."[101] Furthermore, regarding the overall hierarchy for decision-making under the CFP, the *Green Paper* was very clear that decisions on principles would remain entirely within the remit of the EU and the Member States and that decisions pertaining to implementation ought to be devolved or decentralised to national and subsidiary bodies in the Member States.

Stakeholders Support a Clearer Chain of Responsibilities

Disappointingly, the interests of civil society in fisheries management were not canvassed in any great detail in the 2009 *Green Paper*. The Commission nevertheless received written contributions from well over 350 public and private parties regarding their reform proposals. In addition, there were in excess of 200 meetings with representatives of the Member States, ACFA, the ACs, trade unions, NGOs, as well as representatives of the fishing sector. In general, this dialogue indicated strong support for the adoption of a decision-making structure where the European institutions agreed on the policy and principles underpinning the CFP and where there is scope for policy implementation by means of devolved structures that are closer to industry and to stakeholders in the Member States.[102]

Indicatively, the industry was broadly, if somewhat cautiously, supportive of the concept of self-management subject to an assessment of the risks and benefits, and provided that it empowers fishers to decide the best technical solutions and marks a shift away from micro-management.[103] In contrast, the majority of environmental NGOs were reluctant to embrace self-management on the grounds that it would require greater oversight as well as the implementation of more robust control and enforcement mechanisms.[104] On the other hand, many NGOs were supportive of the concepts of participatory governance or co-management.[105] Overall, the majority of stakeholders (which

101 *Ibid.*

102 European Commission Staff Working Document, Synthesis of the Consultation on the Reform of the Common Fisheries Policy, SEC(2010)428 final, Brussels, 16.4.2010, p. 6.

103 *Ibid.*

104 *Id.*

105 *Id.*

includes the European Parliament and some Member States) were of the view that industry responsibility in management is closely allied with their active involvement in the decision-making processes regarding resource utilization.

From the perspective of marine scientific research provisions in UNCLOS,[106] it is also interesting to note that there was general agreement among stakeholders that the policy should be based on science and better cooperation between the fishing sector and scientists.[107] Furthermore, the consultative structures established under the revised policy should provide appropriate channels for the fishing industry and other interested parties to engage in the collection and delivery of scientific information for use in the policy process that informs fishery management decisions.

European Parliament Pushes for a Rights-Based Approach

During the period 2009–2013, the European Parliament pushed vigorously for a move away from the traditional hierarchical approach to fisheries management and sought greater engagement with a wider spectrum of stakeholders in the policy process. In particular, the Parliament sought to empower the ACs and other actors in the formulation and implementation of the CFP.[108] Outside of the domain of commercial sea fisheries, the Parliament has been a long-standing proponent of new or alternative governance models in many sectors including employment, social and environmental policies.[109]

The call for a new governance model for fisheries is clearly evident in Parliament Resolution of 25 February 2010, which addressed the topics of management models, decentralization, the need for greater industry responsibility and supervision of fisheries management and implementation measures.[110] In supporting a major shift away from the hierarchal nature of decision-making under the CFP, the Parliament recommended that the responsibilities for developing and implementing fisheries management plans should be devolved

106 Part XIII, UNCLOS.

107 SEC(2011)891, Brussels, 13 July 2011.

108 See, for example, European Parliament Resolution of 24 April 2009 on "Governance within the CFP: the European Parliament, the Regional Advisory Councils and other actors" 2008/2223(INI).

109 See J. Scott, D. Trubek, "Mind the Gap: Law and New Approaches to Governance in the European Union," *European Law Journal* 8 (2002): 1.

110 European Parliament Resolution 2009/2106(INI), 25 February 2010, adopted by 456 votes to 50, with 65 abstentions.

to Member States and regional bodies.[111] In order to achieve this objective, the Parliament called for the abandonment of the traditional top-down approach to fisheries management and its replacement with a bottom-up decision-making process under the reformed CFP.[112]

Instructively, the overlapping concepts of decentralization and regionalization are perceived as closely aligned if not symbiotic by the Parliament. The underlying rationale for greater stakeholder participation in decision-making is to deliver a more effective policy in the longer-term and to further integrate fisheries management with the decisions made in other maritime policy areas. One key point is the unequivocal rejection by the Parliament of any attempt to adopt a single EU fisheries management model for the CFP. Following on from this, the Parliament requested the Commission to explore the options for the adoption of new fisheries management models that are complementary to the TAC / quota system.[113] This includes models that are connected with the management of fishing effort and the use of a rights-based approach to resource utilization. For obvious political reasons, the Parliament considered that any change in the management model underpinning the CFP should be introduced initially on a transitional basis in the Member States with a view to ensuring that long-standing policy features, including the principle of relative stability are fully respected.[114] There was general agreement in the Parliament that robust participatory structures will result in better understanding, acceptance and implementation of management measures by the fishing industry.[115] One central thread running through the parliamentary resolutions concerns the enhanced role that ought to be played by the ACs, as well as other stakeholders such as the European Fisheries Control Agency in the fisheries decision-making processes.[116] Sensibly enough, the model proposed by the Parliament places emphasis on the diversity of ecosystems and the natural production conditions that apply in the different regional seas basins in Europe.[117] The Parliamentary deliberations were largely silent on the allocation of the costs associated with fishery management apart from stipulating that EU monies should only be used to fund "ecologically,

111 Para. 71 of European Parliament Resolution 2009/2106(INI).

112 Paras. A and 91 para. 71 of European Parliament Resolution 2009/2106(INI).

113 Para. 82 of European Parliament Resolution 2009/2106(INI).

114 *Ibid.*

115 Para. 87, European Parliament resolution of 25 February 2010.

116 European Parliament resolution of 25 February 2010.

117 *Ibid.*

economically and socially sustainable" resource utilization practices by the fishing industry and the Member States.[118]

At the first reading of the draft legislation, reflecting its environmental credentials, the European Parliament supported the codification of a requirement for mandatory environmental impact assessment (EIA) prior to EU vessels gaining access to new fisheries.[119] Of course, the assessment process under the applicable EU directives provides a further route for environmental stakeholders to have their say in the decision-making process before the authorization of new fisheries or new fishing practices.[120] This legislation requires national bodies in the Member States to take the views of the public and other stakeholders into consideration *ex-ante* and before a particular development, project, plan or programme is approved or authorised by the licensing or regulatory authority. In general, EIA has improved the quality, accountability and legitimacy of environmental and planning decisions in both the EU and the Member States.[121] Furthermore, the assessment of the environmental impacts of fisheries is fully consistent with the provisions on the controlling and monitoring of pollution under UNCLOS.[122]

Council Protects Member State Prerogatives

The views of the Council of Ministers evolved considerably over the lifetime of the reform debate 2009–2013. At the outset, the Council supported the general thrust of the Commission's proposals on the need for greater stakeholder involvement in decision-making under a revised CFP. They endorsed the need for regionalization and close consultations with all stakeholders to ensure that they later accepted the outcome of the reform process.[123] In 2010, several Member States identified the devolution of decision-making to regional and national levels as a key element for the delivery of more sustainable fisheries.[124]

---

118 *Id.*

119 European Parliament: 2011/0195(COD)-06/02/2013 Text adopted by Parliament, 1st reading/single reading, 06.02.2013.

120 *Op. cit.* note 67.

121 S. Bell, D. McGillvray and O. Pedersen, *Environmental Law*, 8th ed. (Oxford: OUP, 2013), 452–497.

122 Articles 204–206, UNCLOS.

123 See Press Release 2944th Council meeting Agriculture and Fisheries Brussels, 25 May 2009, p. 8.

124 See press Release on the 3025th Council meeting, Agriculture and Fisheries, Luxembourg, 29 June 2010, at pp. 8–9.

On a number of occasions in 2012, the Council debated several governance issues including an enhanced role for the ACs by providing them with a legal basis in submitting recommendations on how to simplify rules on fisheries management.[125] With passage of time, however, the Council did not pursue the concepts of self and co-management of fisheries with the same vigour as the Parliament or in line with the initial thought-provoking views articulated by the Commission in the 2009 *Green Paper*. Towards the end of the reform process, the Council held that the practical aspects of rights-based fisheries management ought to remain within the exclusive prerogative of each Member State.[126] This outcome is perhaps understandable in light of the diverse range of political, economic, environmental and social considerations that were at play in the 22 coastal Member States of the Union.

A Fresh Start: New Basic Regulation

After completing wide consultation with interested parties and circulating the draft legislative act to national parliaments in the Member States, the Commission presented a new proposal for a new basic fisheries management regulation to the Council and Parliament in July 2011.[127] The Basic Regulation, which came into force on the 1 January 2014,[128] prescribes the general framework that applies to fisheries under the CFP, sets down the basic principles and standards, as well as the timeframe for the adoption of management and implementation measures.[129] In addition, it addresses the role of industry, the Member States, the European Commission, and the various bodies and competent authorities that operate at national and EU levels. Moreover, if one studies the text of the Basic Regulation, it is evident that the provisions on decentralization, regional flexibility and the simplification of the regulatory framework, all have the potential to bring about some significant changes to the traditional *modus operandi* of fisheries management pursuant to the CFP.

As a start point, the rules that restrict access to resources within the 12 nautical mile zones of Member States are retained on the grounds that these measures help conservation by restricting fishing effort in the most sensitive

125 Partial general approach on the CFP basic regulation adopted at Council meeting 12 June 2012.

126 See discussion *infra*.

127 COM(2011)425 final. Brussels, 13.7.2011.

128 Article 51, Regulation 1380/2013.

129 Articles 1–3, 5–17, 48, Regulation 1380/2013.

zone of Union waters, that is to say coastal waters.[130] As noted in the Preamble, these rules have also preserved traditional fishing activities on which the social and economic development of certain coastal communities is highly dependent.[131] Significantly, the European Parliament sought, during the course of the reform debate 2011–2013, to strengthen these rules to give preferential access for small scale, artisanal or coastal fishermen. Importantly for Member States on the periphery of the Union, the Council has codified the Hague Resolution in the Basic Regulation.[132] This Resolution, which was first adopted in 1976, is aimed at safeguarding the needs of regions where local communities are especially dependent on fisheries and related activities.[133]

Another significant feature is that the new regime allows Member States to adopt conservation and management measures for stocks in Union waters applicable solely to vessels flying their flag.[134] Notably, the legislation retains the principle of relative stability as a guiding axiom for the policy. The latter principle aims to ensure the relative stability of fishing activities by allocating fishing opportunities among Member States based upon a predictable share of the stocks for each Member State.[135] The strict application of the principle could potentially curtail the scope for the development of a comprehensive system of private property fishery rights at a pan-European level, as it requires a degree of balance in the opportunities afforded to vessels based upon their flag or nationality of registration. In other words, it is an impediment to the application of free market principles to allocation and utilization of resource entitlements among fishers.

On more specific management measures that mitigate the effects of the principle of relative stability, the Member States are requested to "distribute quotas between vessels in a mix reflecting as much as possible the expected composition of species in the fisheries. Mismatch between available quotas and actual fishing pattern could be adjusted through quota swaps with other Member States, including on a permanent basis."[136] They should also consider pooling quotas in POs or in groups of vessel owners.[137] Following on from

130 Recital 19, Article 5(2) and Annex I, Regulation 1380/2013.

131 *Ibid.*

132 Annex VII of Council resolution of 3 November 1976 on certain external aspects of the creation of a 200-mile fishing zone in the Community with effect from 1 January 1977, OJ C 105, 7.5.1981.

133 *Ibid.*

134 *Ibid.*

135 Article 16(1), Regulation 1380/2013.

136 Recital 29, Regulation 1380/2013.

137 *Ibid.*

this, the principal means advanced by the European institutions to enhance governance structures and to expand the scope for stakeholder involvement in the management of fisheries follows a three strand approach that entails the following elements: (1) the voluntary establishment of a system of transferable fishing concessions; (2) the further codification of the role of the ACs in the policy and decision-making processes; and (3) the regionalization and decentralization of aspects of the policy. Each of these elements merits further consideration here.

Means to an End: Transferable Fishing Concessions

The introduction of a system of private property rights was the subject matter of considerable debate in the reform negotiations within the European institutions and in the Member States. Many concerns were voiced about the mandatory introduction of a system that entailed the transfer of ownership of a public resource into private property. A compromise solution to this conundrum was ultimately agreed in the form of transferable fishing concessions (TFCs). The term is defined to mean:

> the revocable user entitlements to a specific part of fishing opportunities allocated to a Member State or established in management plans adopted by a Member State…, which the holder may transfer to other eligible holders.[138]

The system of TFCs as devised originally by the Commission has three fundamental features, namely: (1) the vesting of clearly identifiable economic entitlements or user rights to fishers as natural persons or corporate entities; (2) the alienability of such rights; and (3) a market based approach to their utilization and transfer. In addition, the residual property rights in relation to ownership of the resource remain vested with the Member States. The Commission initially envisaged the introduction of this system for all vessels for a period of at least 15 years.[139] The structure of the proposed scheme was informed by the national management schemes in place for a number of commercial-sea fisheries in Denmark and elsewhere in the EU including Estonia, Finland, Germany, Italy, Latvia, Lithuania, Malta, the Netherlands,

138 Article 4(1)(23), Regulation 1380/2013.

139 Article 28(4), Draft Regulation, COM(2011)425, 13.7.2011. There was however an exception of vessels under 12 meters with passive gear.

Poland, Portugal, Sweden and the UK.[140] There are also many examples of the success of Individual Transferable Quota (ITQ) schemes in the United States, Namibia, Iceland, New Zealand and Australia. The rationale for the introduction of TFCs is based upon international best practice and is supported by expert studies, which demonstrate that private property rights "halts, and even reverses widespread [fishery] collapse", as well as being a useful instrument to drive economic growth within the sector.[141] Furthermore, a number of European fisheries experts have explored the utility of applying such a system more widely within the CFP in light of its success in achieving fleet reductions and improved economic performance of segments of the fishing industry in Denmark, Spain and Estonia.[142] On the other hand, there is also some support for the view that the implementations of management systems based upon ITQs do not always deliver ecological, economic, and social sustainability in fisheries governance.[143] The Commission sought to limit scope for failure of

140 See, *inter alia*: L. Van Hoof, "Design or Pragmatic Evolution: Applying ITQs in EU fisheries management," *ICES Journal of Marine Science* 70 (2): 462–470 (published online January 2013). Also, B.I. de Vos, and J. van Tatenhove, "Trust Relationships between Fishers and Government: New Challenges for the Co-management Arrangements in the Dutch Flatfish Industry," *Marine Policy* 35 (2011): 218–225.

141 C. Costello *et al.*, "Can Catch Shares Prevent Fisheries Collapse?" *Science* Vol. 321 (No. 5896) (2008): 1678–1681.

142 The empirical evidence cited by the Commission is equally impressive in so far as it notes that the introduction of TFCs in Denmark in 2003 resulted in 50% decrease in the pelagic fleet size, and a 30% reduction in the number of demersal vessels, while at the same time both fleets increased their profitability. Similarly, the Commission instance the case of Estonia, which introduced a TFC system in 2001 and by 2009 the fleet is reported as decreasing by 40% in size. In Spain, there appears to be a similar success in so far as the so-called Gran *Sol* fleet decreased by 30% between 1992 and 1997. See SEC(2011) 891, Brussels, 13 July 2011. Also see contribution from M. Schou, Minister's adviser for fisheries and aquaculture, Danish Ministry of Food, Agriculture and Fisheries. Available at: http://cfp-reformwatch.eu/2011/11/transferable-fishing-concessions-the-danish-experience/.

143 See, for example: U. Sumaila, "A Cautionary Note on Individual Transferable Quotas," *Ecology and Society* 15(3) (2010): 36. See generally the papers in Gisli Paálsson and G. Pétursdóttir (eds.), *Social Implications of Quota Systems in Fisherie* (Copenhagen: Nordic Council of Ministers, 1997), including especially Harry N. Scheiber and Chris Carr, "The Limited Entry Concept and the Pre-History of the ITQ Concept in Fisheries Management," at 235–60, and Parzival Copes, "Social Impacts of Fisheries Management Regimes Based on Individual Quotas," at 61–91.

the proposed measures by identifying five principles governing the application of TFCs, as shown in Table 1.2.[144]

Under the proposed scheme, management is based on the allocation of a fixed percentage of the national quota for a specific fish stock. Central to the application of TFCs is the notion of greater stakeholder engagement and the vesting of additional responsibilities with the industry. Again in the words of the Commission:

> If designed correctly, TFCs can be an effective tool for vessel owners to plan their fishing activity along market developments, land all catches and plan their investments. They also offer the possibility to fishermen to leave the industry in exchange for financial compensation. Experience shows that TFC like systems also increase operators responsibility and facilitate discard reduction.[145]

TABLE 1.2 *Five Principles Governing Transferable Fishing Concessions*

1	Marine resources are and must remain a public good. TFCs cannot confer property rights over marine resources, but only user rights to exploit them for a limited time. After the time is up the TFC has to fall back to the Member State, who is free to allocate it again using the same allocation criteria or different ones.
2	Selling, leasing or swapping of TFCs can only happen under strict conditions as only owners of registered and active vessels with the purpose to use them on a licensed and active vessel, can buy TFCs.
3	Relative stability must be respected.
4	Member States have to withdraw TFCs in case of a serious infringement by the vessel owner.
5	Member States have to reserve quotas and TFCs for new fishermen who are looking to enter the fishery.

144 European Commission, CFP reform—Transferable Fishing Concessions. Available at: http://ec.europa.eu/fisheries/reform/docs/tfc_en.pdf.

145 *Ibid.*

The POs were to play a central role in the management of TFCs under the proposed scheme, including the selling and leasing of TFCs among their members, thus further enhancing industry stakeholder empowerment. Moreover, this could extend to collective management by a PO of TFCs of several members, which would "allow for real-time swapping and proper production planning."[146] The design of the national system of TFCs should "favour coastal communities dependant on fishing, more environmentally friendly fishing practices and the artisanal fleets."[147] In order to curtail the practical aspects of the scheme, the Commission published a toolbox of management parameters, many of which were aimed at protecting the rights of coastal communities and smaller participants from encroachment by large commercial enterprises, as shown on Table 1.3.

The debate in the EU institutions and the Member States on the introduction of TFCs was hotly contested during the reform period 2009–2013. In contrast to the original Commission proposal, the Parliament sought to ensure that Member States retained discretion in adopting TFCs should they wish to do so. Moreover, the Parliament recommended that various factors ought to be taken into account when allocating TFCs among fishers including the environmental impact of certain types of fishing gears, the record of compliance by the industry with their legislative obligations, as well as their contribution

TABLE 1.3 *Commission's Proposal: Transferable Fishing Concessions Toolbox*

1	The exclusion of small scale fishing (<12m with passive gear), in order to ensure that fishing rights of this important segment will not be sold to larger vessels;
2	The prevention of excessive concentration by avoiding that too many fishing rights end up in the hands of a few vessel owners [this the Commission suggest must be done by setting maximum percentages of a given resource that can be held by any given vessel owner];
3	Reserving a part of national quotas for coastal communities that depend on small-scale fleets;
4	Limiting the transferability to inside specific fisheries (e.g. whitefish concessions can only be traded with other whitefish concession holders, not to a pelagic concession holder).

146 *Ibid.*

147 *Id.*

to the local economy in coastal areas.[148] The position taken by the Council, articulated during the course of the Irish Presidency, called for the voluntary introduction of TFCs by the Member States with a view to adapting fishing capacity to available fishing opportunities.[149]

Surprisingly, a number of environmental NGOs were strongly opposed to the introduction of a mandatory EU system of TFCs on various legal and public policy grounds. *ClientEarth*, for example, argued that the EU is prohibited, pursuant to one of its own treaty provisions (Article 345 of the TFEU), from prejudicing the rules governing property ownership in the Member States.[150] That is to say, the regulation and privatization of property rights is a matter for the Member States.[151] The Court of Justice, however, has narrowly interpreted this provision in case law concerning private property rights.[152] The Court has also affirmed Article 345 as an expression of the principle of EU neutrality in relation to the rules in Member States governing the system of property ownership.[153] Thus, the Court showed a reluctance to interfere with the national rules that curtailed the transfer of shares held in an electricity or gas distribution system in the Netherlands to private ownership.[154] In this case, the Court pointed out that Article 345 TFEU does not mean that the "rules governing the system of property ownership current in the Member States are not subject to the fundamental rules of the Treaty, which rules include, *inter alia*, the prohibition of discrimination, freedom of establishment and the

148 The Parliament's Fisheries Committee held a hearing on TFCs on 24 April 2012.

149 See Outcome of Proceedings Council Meeting, Brussels, 28 February 2013. Agreement on a general approach, Proposal for a Regulation of the European Parliament and of the Council on the Common Fisheries Policy, Interinstitutional File: 2011/0195 (COD), 11322/1/12. REV 1. Available at: http://register.consilium.europa.eu/doc/srv?l=EN&f=ST%2011322%20 2012%20REV%201.

150 Article 345, TFEU. See ClientEarth, Legislative briefing: The legality of mandatory transferable fishing concessions, August 2011. Available at: http://www.clientearth.org/reports/the-legality-of-mandatory-transferable-fishing-concessions.pdf.

151 *Ibid.*

152 Case 16/74 *Centrafarm BV v. Winthrop BV* [1974] ECR 1183. Some scholars have argued that Article 345 TFEU only concerns the private or public ownership of undertakings, with which the Union shall not concern itself and which can thus be regulated by the Member States themselves. See B. Akkermans, E. Ramaekers, Article 345 (ex. Article 295 EC), its meaning and interpretations, 16(3) *European Law Journal* (2010): 292–314. Also see, F. Losada Fraga *et al.*, "Property and European Integration: Dimensions of Article 345 TFEU," (2012) Helsinki Legal Studies Research Paper No. 17.

153 Para. 29, Joined Cases C-105/12 to C-107/12 *Staat der Nederlanden v Essent* [2013] ECR I-0000.

154 Para. 31, *ibid.*

free movement of capital,"[155] but that they could in some instances be contingent upon overriding public interest objectives that justify restrictions on the application of EU rules.[156] Although the precise meaning and scope of application of Article 345 remains unclear, it appears that it requires EU neutrality and could therefore prove to be a legal impediment to any EU scheme that entails the mandatory transfer of the ownership of fishery resources to private parties.

The agreed TFCs scheme as ultimately jointly agreed by the Council and the Parliament and as set out in the Basic Regulation reflects a compromise of sorts in so far as it entails the introduction of rights-based management at a national level. The provisions in the regulation are enabling in so far as they provide a legal basis for Member States to introduce a system of TFCs on a voluntary basis.[157] Member States are however obliged to establish a registry of any such concessions.[158] The latter requirement, it must be assumed, is aimed at ensuring that the method of allocation remains transparent and susceptible to both public and industry scrutiny. The criteria that must be taken into account by Member States in allocating TFC or other fishing opportunities include: the impact of fishing on the environment; the history of compliance; the contribution to the local economy and the historical record/catches of the vessel in the fishery.[159] With a view to advancing the green agenda in EU fisheries law and policies,[160] Member States must endeavour to provide incentives to fishing vessels deploying selective fishing gear or using fishing techniques with reduced environmental impact, such as reduced energy consumption or habitat damage.[161]

One of the strengths of TFCs is that they strengthen governance and the role of industry in taking management decisions pertaining to the reduction of fishing effort or the economic performance of particular vessels or segments of the fleet. It is difficult to see how industry can be empowered, however, without resorting to such a system in one form or another. Indeed, experience

---

155 *Para. 36*, citing Case 182/83 Fearon [1984] ECR 3677, paragraph 7; Case C-302/97 Konle [1999] ECR I-3099, paragraph 38; Case C-452/01 Ospelt and Schlössle Weissenberg [2003] ECR I-9743, paragraph 24; Case C-171/08 Commission v Portugal [2010] ECR I-6817, paragraph 64; Case C-271/09 Commission v Poland [2011] ECR I-13613, paragraph 44; and Commission v Greece, paragraph 16.

156 Para. 68, Joined Cases C-105/12 to C-107/12 *Staat der Nederlanden v Essent* [2013] ECR I-0000.

157 Article 21, Regulation 1380/2013.

158 *Ibid.*

159 Article 17(1), Regulation 1380/2013.

160 Article 11, TFEU.

161 Article 17(1), Regulation 1380/2013.

in the Netherlands has shown that the negative effects associated with TFCs can be addressed through the design of the system under the CFP.[162]

What is significant is that the EU appears to have passed the ball firmly back to the Member States in so far as the legal measures on TFCs in the Basic Regulation are enabling or permissive and contingent upon action at a national level for specific fisheries. The establishment of property or user rights over marine resources under the CFP will thus vary in the 23 coastal jurisdictions that are members of the EU.

Role of Advisory Bodies: *Peu de Changement*

One of the principles of good fishery governance under the Basic Regulation is the involvement of stakeholders, in particular Advisory Councils, at all stages of the policy cycle, from the formation to the subsequent implementation of regulatory measures.[163] The Basic Regulation also provides a legal basis for the establishment of four new ACs, covering: the outermost regions (West Atlantic, East Atlantic and Indian Ocean); aquaculture; markets; and the Black Sea.[164] Notably, the reformed regulatory regime does not vest the ACs with any powers in relation to the management of fisheries over and beyond their consultative role in the formulation of new measures for the sustainable exploitation of marine biological resources.[165] This role extends to instances where there is "evidence of a serious threat to the conservation of marine biological resources, or to the marine eco-system," and this situation requires immediate action and the adoption of special legislative measures.[166] In such cases, the ACs may submit their written comments on the adoption of temporary measures or emergency measures by the Commission or the Member States.[167] Likewise, they have an advisory role regarding the design and implementation of pilot projects into the "practicable methods for the avoidance, minimisation and elimination of unwanted catches in a fishery."[168] Similarly,

162 See, L. Van Hoof, "Design or Pragmatic Evolution: Applying ITQs in EU fisheries management," *ICES Journal of Marine Science* 70(2) (published online January 2013): 462–470.

163 Article 3, Regulation 1380/2013.

164 Article 43(2) and Annex III, Regulation 1380/2013.

165 Article 6(2), Regulation 1380/2013.

166 Article 12, Regulation 1380/2013.

167 Article 12(2) and 13(2), Regulation 1380/2013.

168 Article 14, Regulation 1380/2013.

multiannual plans should be adopted in consultation with the ACs, operators in the fishing industry, scientists and other stakeholders having an interest in fisheries management at a regional level.[169]

There are several substantive provisions in the Basic Regulation on the establishment and functioning of the ACs.[170] The Basic Regulation states that the AC's are established to promote a balanced representation of *all* stakeholders and to contribute to the achievement of the objectives of the CFP (emphasis added).[171] Furthermore, it opens-up a number of possibilities regarding the future amendment of the mandate of the ACs in so far as the Commission is empowered to make amendments to Annex III by means of delegated acts in relation to the areas of competence for ACs, as well as regarding their composition and function. Of relevance to stakeholders is the general legislative obligation placed on the Commission to carry out appropriate consultations during its preparatory work for the adoption of delegated acts.[172]

The three principal tasks of the AC are as follows: (1) to make recommendations to the Commission or to the relevant Member State(s) on fisheries management and the socio-economic and conservation aspects of fisheries and aquaculture, particularly in relation to the simplification of regulatory measures; (2) to inform the Commission and Member States of problems relating to fisheries management and aquaculture in their area of competence and to propose solutions to these problems; and (3) to contribute to the collection and supply of scientific information to facilitate development of conservation measures.[173] Moreover, one crucial change to their role under the Basic Regulation is that their advice must be taken into account regarding regional cooperation on conservation measures including multiannual plans and on the adoption of joint recommendations pursuant to Article 18 of the Regulation.[174] Apart from setting down a requirement to reply within two months, the Basic Regulation does not place any express obligation on the Commission or the Member States to adopt any recommendation, suggestion and information received from an AC on fishery management or any other matter. There is, however, an obligation placed on the Commission and the Member State concerned to provide detailed reasons or an explanation where the final measures adopted diverge from the contribution received from

169 Article 18(2), Regulation 1380/2013.
170 Articles 43–45, Regulation 1380/2013.
171 Article 43(1), Regulation 1380/2013.
172 Article 46, Regulation 1380/2013. See discussion of other game-changing outcomes below.
173 Article 44(2), Regulation 1380/2013.
174 Article 44(3), Regulation 1380/2013.

an AC.[175] In principle, this provision leaves open a possibility for the ACs to make recommendations regarding the establishment of appropriate schemes of self and co-management of fisheries, including as noted previously on the use and allocation of TFCs for this purpose.

In relation to the proposed structures, three comments may be made about the role and mandate of ACs. First, the structures must consist of a general assembly and an executive committee. Apart from meeting the requirements set down by Annex III of the Basic Regulation, each AC appears to have a degree of discretion regarding the adoption of more precise organisational measures. Second, there are clear limits on this discretion in so far as the Commission is in charge of much of the detail and is vested with power to regulate on the composition and the functioning of ACs by means of a delegated legislative act.[176] Moreover, in line with the Treaty, the Commission can only act within the limits of the authority delegated by the Council and Parliament. The Commission must notify the European Parliament and the Council simultaneously as soon as it adopts a delegated act.[177] Third, ACs are compelled to work with scientists and other stakeholders having an interest in fisheries management with a view to fostering regional cooperation on conservation measures.[178]

Twin Pillars: Regionalization and Decentralization

The reformed policy is founded on the twin pillars of regionalization and decentralization. Thus, for example, there is a clear link between regionalization and the geographical footprint of the ACs, which are structured up with a specific focus on the Baltic Sea, the Black Sea, the Mediterranean Sea, the North Sea, North Western waters and South Western waters in the Atlantic, the outermost regions of the Union, as well as all non-Union waters.[179] The latter refers to sea areas under the sovereignty and jurisdiction of third countries (that is to say, non EU Member States), as well as the high seas. There is also reference to the "regionalization process" in the provisions dealing with the obligations that are placed on fishers to land their catches and on the introduction of discard plans, which require all catches to be kept on board fishing vessels,

175 Article 44(4), Regulation 1380/2013.

176 Article 45(4), Regulation 1380/2013.

177 Article 46(4), Regulation 1380/2013.

178 Article 18(2), Regulation 1380/2013.

179 Annex III, Regulation 1380/2013.

landed and counted against fish quotas.[180] Moreover, echoing UNCLOS, the Preamble of the Basic Regulation notes that all matters in relation to Europe's oceans and seas are interlinked and provides that the policy should aim to be coherent with other EU sector policies that apply in the Baltic Sea, North Sea, Celtic Seas, Bay of Biscay and the Iberian Coast, the Mediterranean and Black Sea.[181] The introduction of a sea-basin approach is entirely consistent with the regional approach to management of all maritime activities under the MSFD.[182] There are also provisions in the Basic Regulation that allow for a regionalised approach to the management of fisheries in the EU's outermost regions, which include Guadeloupe, French Guiana, Martinique, Réunion, Saint-Barthélemy, Saint-Martin, the Azores, Madeira and the Canary Islands.[183] Although the Basic Regulation is largely silent on the matter, it is also foreseeable that a regionalized approach could be adopted by the relevant Member States for the areas where the special needs of regions and the populations are particularly dependent on fishing and related activities in line with the 1976 Hague Resolution.[184]

The main provisions on regionalization are embedded in Article 18 of the Basic Regulation, which deals exclusively with this subject and establishes a framework for regional cooperation by Member States on the adoption of conservation measures that apply to one or more geographical regions. Although these provisions are slightly attenuated and bearing in mind that the Commission has the residual powers to adopt legislation by means of delegated or implementing acts, it is nonetheless very significant that this provision vests Member States with the power to submit joint recommendations aimed at achieving the objectives of the relevant Union conservation measures. This includes making a joint-recommendation on the adoption of measures aimed at achieving the objectives of EU environmental legislation including the MSFD, the Birds and Habitats Directives.[185] Thus, the fisheries policy is brought into line with the more specific obligations in EU law on the protection and preservation of the marine environment.

---

180 Recital 67 and Article 18, Regulation 1380/2013. By February 2015, the Commission had adopted five discard plans pursuant to these obligations: Commission Delegated Regulations (EU) No. 1392/2014 (Mediterranean Sea); No. 1393/2014 (north-western waters); No. 1394/2014 (south-western waters); No. 1395/2014 (North Sea); No. 1396/2014 (Baltic Sea).

181 Recital 17, Regulation 1380/2013.

182 *Op. cit.* note 68.

183 See, for example, provisions on capacity ceilings in Annex II, Regulation 1380/2013.

184 On the Hague Resolution see discussion *infra*.

185 Articles 11(3), 18(1)–(4) and (6), Regulation 1380/2013.

The provisions on joint-recommendations are tangible evidence of efforts under the revised policy to vest Member States with decision-making powers at a national level, albeit subject to oversight by the EU institutions. The Commission, for instance, must not adopt any such delegated or implementing acts before the expiry of the deadline for submission of joint recommendations by the Member States.[186] The relevant Member States in turn are compelled to cooperate with one another in formulating joint recommendations and must also consult the pertinent AC(s). The Commission must facilitate any such cooperation between Member States, including, where necessary, obtaining assistance from scientific bodies. Again, however, it needs to be noted that the residual enabling mechanism or regulatory powers in bringing these measures into force remain vested in the Commission. The latter is empowered to adopt measures based upon the joint recommendations, by means of delegated or implementing acts, provided that such recommendations are compatible with the relevant conservation measure and multiannual plan.[187] Member States in turn must ensure that all joint recommendations on conservation measures are based on the best available scientific advice. In addition, joint-recommendations must be compatible with a number of prescribed criteria, namely: (a) the objectives for the CFP as set out in Article 2 of the Basic Regulation such as the precautionary principle and the ecosystem-based approach to fisheries management; (b) the objectives of the Union conservation measure(s); (c) the quantifiable targets set out in a relevant multiannual plan; and most importantly, (d) must be no less stringent than Union legislation.[188] The default position, should this procedure not work as anticipated, is that the Commission can rely on its law-making powers under Article 190 of the Treaty to adopt appropriate conservation measures.[189]

The case for regionalization and decentralization of decision-making is also evident in the provisions on the application of ecosystem-based management under the CFP. This approach is defined in the Basic Regulation as meaning:

> …an integrated approach to managing fisheries within ecologically meaningful boundaries which seeks to manage the use of natural resources, taking account of fishing and other human activities, while preserving both the biological wealth and the biological processes necessary to safeguard the composition, structure and functioning of the

186 Article 18(1), Regulation 1380/2013.

187 *Ibid.*

188 Article 18(5), Regulation 1380/2013.

189 Article 18(6), Regulation 1380/2013.

> habitats of the ecosystem affected, by taking into account the knowledge and uncertainties regarding biotic, abiotic and human components of ecosystems.[190]

The reformed CFP aims to minimize the negative impacts of fishing activities on marine ecosystems, which is one of the most welcome and progressive developments in EU fisheries law. In order to give teeth to this commitment, the Basic Regulation provides for the use of incentives, including economic incentives, to reduce the impact of fishing activities on marine ecosystems, as well as the adoption of technical conservation measures for this purpose.[191] Moreover, multiannual plans and emergency measures are to be founded in the ecosystem based approach to fisheries management.[192] Within this setting, it should not be forgotten that the ecosystem approach is based on ecological regions and not on the maritime boundaries established by the Member States.[193]

Ecosystem-based management also entails taking management decisions at the most appropriate governance level and provides a useful normative justification for the further empowerment of stakeholders. As noted by a United States specialist advisory panel in their report to Congress, ecosystem-based fishery management can "contribute to the stability of employment and economic activity in the fishing industry and to the protection of marine biodiversity on which fisheries depend."[194] The report to Congress goes on to point out that "systems organized to promote the maximum involvement of stakeholders, including the interests of future generations, and to emphasize the maximum appropriate delegation of responsibility and authority to the lowest possible levels of the management system (e.g., the local or regional level), tend to have the highest credibility among fishery constituents (Pinkerton 1989)."[195] Similar to the U.S. experience, engagement with

190 Article 4(9), Regulation 1380/2013.

191 Article 7(1)(d), Regulation 1380/2013.

192 Articles 10 and 13(1), Regulation 1380/2013.

193 See R. Long, "Legal Aspects of Ecosystem-Based Marine Management in Europe," in Aldo Chircop, Moira L. McConnell, Scott Coffen-Smout (eds.), *Ocean Yearbook*, Volume 26 (Boston/Leiden: Brill Academic Publishers, 2012), 417–484.

194 National Marine Fisheries Service Ecosystem Principles Advisory Panel, *Ecosystem-Based Fishery Management, A Report to Congress by the Ecosystem Principles Advisory Panel*. Available at: http://www.nmfs.noaa.gov/sfa/EPAPrpt.pdf.

195 *Ibid.* at 21. Citing E. Pinkerton, ed., *Co-operative Management of Local Fisheries: New Directions for Improved Management and Community Development* (Vancouver B.C.: University of British Columbia Press, 1989).

stakeholders will facilitate the implementation of ecosystem-based management in Europe, and thus improve the credibility of the CFP with fishers and the public more generally. Regionalization and decentralization can therefore be viewed as critical components of the revised European fishery policy that promote stakeholder participation in management and regulatory decision-making at local and national level in the Member States.

Portents of the Future: Sharing Enforcement Costs

One of the objectives of the revised CFP is the development of a culture of compliance among "operators" in the fishing industry. The term "operators" is given a broad definition in the Basic Regulation and means:

> ... the natural or legal person who operates or holds any enterprise carrying out any of the activities related to any stage of production, processing, marketing, distribution and retail chains of fisheries and aquaculture products.

This definition extends to a wide number of stakeholders including anyone associated with the catching and marketing of fishery products. Apart from nurturing a greater culture of compliance, the EU system of fisheries control and enforcement is founded on a number of key principles including: a global and integrated approach; the use of modern control technologies such as vessel monitoring systems and remote sensing; a risk-based strategy focused on systematic and automated cross-checks of all available relevant data; and the establishment of effective, proportionate and dissuasive sanctions.[196]

In a fundamental shift away from the traditional paradigm for enforcement, the Basic Regulation provides that Member States may require their operators to contribute proportionally to the costs of implementing the EU's fisheries control system.[197] Furthermore, in both a carrot and stick approach to fiscal incentives, financial assistance from the EU to the Member States to cover some of the costs of fishery law enforcement is conditional upon CFP compliance.[198] A similar approach is evident with respect to the fishing industry in so far as the EU's financial assistance to operators is also contingent upon

196 Article 36, Regulation 1380/2013.

197 Article 39, Regulation 1380/2013.

198 Article 41(1), Regulation 1380/2013.

their compliance with the rules underpinning the CFP.[199] Broadly speaking, the enforcement and compliance model as advanced by the Basic Regulation places an emphasis on the role and powers of the Member States, rather than creating incentive structures facilitating voluntary compliance by the industry.[200] An example in point is the expert group on compliance, which is only composed of representatives of the Commission and the Member States, with no representation from industry and other stakeholders.[201] This is a disappointing oversight, as this particular group is tasked with reviewing issues of compliance and fisheries law enforcement including matters concerning the fight against IUU fishing.[202]

Other Potential Game-Changing Outcomes

Despite the constraints imposed by the Treaties, there are several other important and potentially game-changing outcomes of the reform process that have the potential to enhance the governance structures that will apply to fisheries into the future. Some of these have real potential to improve the scope for stakeholder involvement in fisheries management and to expedite the move towards decentralization or devolution of the CFP away from the Council and Parliament. In particular, evidence of this trend includes the extensive range of conservation and management measures that can now be adopted by the Commission by means of implementing act or by means of delegated legislation.[203] This clearly suggests that many key decisions on fisheries conservation and management measures will be taken in the future by the Member States, at the behest of the Commission, without recourse to the Council and Parliament.[204] The Commission as the legislator is compelled to systematically consult experts from the national bodies in the Member States, who are responsible for the implementation of delegated acts. The European Parliament

199 Article 42, Regulation 1380/2013.

200 See R. Döring, G. Leyre, SOCIOEC- Socio-economic effects of management measures of the Future CFP—Overview on a new European FP 7 Project. 2012. IIFET 2012 Tanzania Proceedings, at p. 6. Available at: http://www.socioec.eu/images/SOCIOEC/Media_Centre/Scientific/IIFET%20Tanzania%20Doering%20_%20July2012.pdf.

201 Article 37, Regulation 1380/2013.

202 Article 37(3)(a)–(c), Regulation 1380/2013.

203 Article 46, Regulation 1380/2013.

204 Delegation powers are vested in the Commission pursuant to Articles 11(2), 15(2), (3),(6),(7), and 45(4), Regulation 1380/2013. In relation to implementing acts, Articles 12, 23, 24, 47, Regulation 1380/2013.

and the Council retain oversight roles pursuant to their powers under the EU treaties.[205] The arrangements places the Member States firmly in the driving seat in shaping the policy going forward through the role that is assigned to the Committee for Fisheries and Aquaculture.[206] Under the legislation that governs the functioning of this particular committee, national interests are protected and taken into consideration every time the Commission exercises its implementation powers under the Basic Regulation.[207]

There is other evidence of the efforts that have been made to vest Member States with additional responsibilities under the Basic Regulation in accordance with the general scheme of what is permissible under the Treaties. Notable features of these new obligations include the general duty placed on Member States to cooperate with one another in the implementation of both fisheries management measures and EU environmental legislation, particularly at regional levels.[208] Similarly, Member States must coordinate their actions when adopting national fisheries conservation measures that apply in the coastal zone when such measures affect the fisheries interests of other Member States.[209] As previously pointed out, such measures must be adopted only after consulting the Commission and with other Member States where relevant, as well the ACs who represent stakeholder interests.[210]

Has the Paradigm Really Evolved?

The common pool character of marine living resources is long-since deeply embedded in the law and practice of the EU as it pertains to the management and regulation of fisheries. The aim of the revised CFP is to establish a legal framework that provides for environmental, economic, and social sustainability. The reform process revealed a tension between public ownership of a great natural asset and the vesting of industry with private property rights. At this point, one year after the adoption of the reform measures, the question may be

205 Council 8753/11 of 10.04.2011.

206 Article 47, Regulation 1380/2013.

207 Regulation (EU) No. 182/2011 of the European Parliament and of the Council of 16 February 2011 laying down the rules and general principles concerning mechanisms for control by Member States of the Commission's exercise of implementing powers OJ L 55, 28.2.2011.

208 Article 6(3), Regulation 1380/2013.

209 Article 6(4), Regulation 1380/2013.

210 *Ibid.*

legitimately posed: does the new legal paradigm provide appropriate mechanisms and procedures for the advancement of stakeholder interests?

We can deduce elements of the answer to this question from the analysis undertaken in this chapter. As seen, the term "stakeholder" is not a term-of-art in EU fisheries law with specific legal meaning. Accordingly, it can have several meanings in different management and regulatory contexts under the CFP. Similarly, stakeholder "involvement" in the CFP is very open-textured from both policy and legal perspectives and does not appear to tally exactly with the various schemes foreseen under the rubric of Union environmental law. For example, under the CFP, it consists predominantly of the various structures and processes that are in place for keeping the fishing industry and other stakeholders informed of policy decisions, as well as the content and shape of draft legislation at EU and national levels. The principal actors in this process are the EU institutions and national governments that operate through the good offices of the various statutory bodies concerned with fisheries management. In some Member States, however, there is a degree of "pluralistic and deliberative participation" by the fishing industry, particularly when they are vested with a direct role in quota allocation and utilization by means of a system of property rights or user entitlements. This may be contrasted with public participation in environmental decision-making under the Åarhus Convention, which does not extend to rule setting regarding resource allocation and utilization. Indeed, one authority has pointed out that the whole "process of monitoring the application of EU environmental law remains an almost entirely closed business between the EU Commission and the EU Member States," and therefore outside the purview of the European citizen as responsible environmental actors.[211]

The rationale for greater stakeholder involvement in the CFP is unequivocal and acknowledged by all concerned as a pre-requisite for the future success of the policy. As seen above, the principal stakeholder consultative bodies are the ACs, with a more localized role played in management by some industry led bodies including the POs. The Council supported the codification of stakeholder involvement in the policy process as one of the principles of good governance for the CFP.[212] At the behest of the majority of Member States, however, they did not codify the concepts of self- and co-management of

211 L. Krämer, "EU Enforcement of Environmental Laws: From Great Principles to Daily Practice—Improving Citizen Involvement," December 2013. Available at: http://www.clientearth.org/aarhus-centre/publications/eu-enforcement-of-environmental-laws-from-great-principles-to-daily-practice-improving-citizen-involvement-2418.

212 Article 3(f), Regulation 1380/2013.

fisheries in line with the initial views on rights-based management expressed by the Commission in the 2009 *Green Paper*.[213]

Following on from this conclusion, and in light of the predominantly positive experience of ITQs in many jurisdictions worldwide,[214] the greatest disappointment is that the reformed policy does not introduce a fully-fledged system of private property rights for the CFP. There are inherent difficulties, legal and otherwise, stemming from the Treaties in designing such a system and putting it into operation in practice at a pan-European level, including uncertainties stemming from EU neutrality on the property systems adopted by the Member States,[215] the principle of relative stability under the CFP, as well as the guarantees that arise for coastal communities under the Hague Resolution.[216] There is a degree of progress in EU fisheries law in so far as each Member State has full discretion in establishing or maintaining a TFC system on a voluntary basis. Although TFCs are perceived by the Council as a tool for the reduction of excessive fishing capacity in the fleet, they can at the same time enhance the role of industry in the management of fisheries. Furthermore, within the fishing opportunities assigned to them, Member States may also provide incentives to fishing vessels deploying selective fishing gear or using fishing techniques that aim to reduce their environmental impact.[217] The utility of TFCs in strengthening governance and enhancing the role of stakeholders ought to become increasingly evident as the new policy is progressively implemented by the Member States.

From the point of view of stakeholder empowerment, the second aspect of the reform that is notable is that the role of the Council and the Parliament in standard setting is largely undiminished. Both institutions control the overall scheme of EU decision and law-making processes as they apply to the management of fisheries. This outcome is no surprise and is somewhat inevitable as a result of the allocation of Union competence in relation to fisheries under the Treaties.[218] On the positive side, however, it should also be pointed out that as the policy moves further forward into the implementation phase, the role of Council and Parliament is firmly focused on objective setting with the experts

213 See *supra*.

214 See example from Maryland fishery, J. Holzer, "Property Rights and Choice: The Case of The Fishery," *American Journal of Agriculture Economics* 97(4) (2015): 1175–1191. First published online: 201.01.2015.

215 Article 345 TFEU. See *op. cit.* note.

216 See *supra*.

217 See *supra*.

218 Article 3(d), TFEU.

from the Member States having a greater say in relation to the adoption of implementation measures.

Perhaps an area, which promised so much in the reform process, but is unlikely to realize its full potential under the reformed CFP relates to the role of the ACs. At first sight, they appear to be cast in an ancillary and supporting role under the new governance structures, where they remain on the periphery of the decision-making bodies with no real executive functions under Annex III of the Basic Regulation. In particular, the Basic Regulation does not vest them with any innovative powers, apart from accentuating their consultative role in the formulation of new measures for the sustainable exploitation of marine biological resources.[219] Their role, however, should not be understated, as they have a legal mandate and a designated role in the formulation of multiannual plans for fisheries.[220]

The reformed CFP is partly focused on results based management, where the overarching framework is agreed by the Council and Parliament, and the Member States working in conjunction with industry to find the best technical solutions to particular fishery management decisions, such as quota allocation and utilization, and the prohibition on discards. Apart from fostering greater cooperation between industry and scientists, however, there is little evidence in the legal framework of efforts to widen industry access to scientific and technical expertise required for making fully informed fisheries management decisions. Neither does industry appear to be given responsibility for achieving specific management objectives or developing the experience and technical expertise to take on further management responsibilities. This finding does not accord with the positive role that industry has played in the management of specific fisheries such as the clam fishery in the Adriatic Sea.

One should not be overtly critical about the slow evolution of a rights-based management in EU fisheries law as there are some greenshoots evident in the new regime. Notably, the Basic Regulation foresees the delegation of a number of functions back to the Member States and through their good offices to others concerned with the management of fisheries such as POs.[221] If POs work together and with other stakeholders, appropriate rules can be developed using the procedure for joint-recommendations or the regionalization process as set out in Article 17 of the Basic Regulation. In this context, there is a clear link between regionalization and decentralization, and the future role of

219 Article 3(f), Regulation 1380/2013.

220 Recital 24, Preamble, Article 44, Regulation 1380/2013.

221 See discussion *supra*.

industry in meeting the obligations on catch-landings, discard plans,[222] as well as the targets set down in the relevant multiannual plan for specific fisheries. Looking forward, one other fundamental issue that needs to be resolved by national legislation, joint-recommendation, or by the regionalization process is reform of the rules governing representation and participation in the activities of POs, particularly in relation to local and industrial fisheries, where interests differ significantly. Furthermore, national legislation or regional measure will have to address matters, such as the requirements for industry to design and implement detailed management plans, as well as appropriate schemes to monitor the achievement of specific management targets, such as the utilization of quotas for fishing efforts, particularly where TFCs are used for this purpose. Significantly, the Basic Regulation provides a framework for granting multiannual Union financial assistance "tailored to the specific features of the fishing industry in individual Member States" that will be crucial to the further empowerment of POs. Finally, it should be noted that the Basic Regulation has a mechanism for linking the financial cost of enforcement with compliance. Viewed all together, these measures ought to increase the scope for industry and stakeholder participation in the management of fisheries under the CFP over the coming decades.

Winners and Losers under UNCLOS

In light of the theme of the 2014 LOSI conference, one may ask, who are the winners and losers in the most recent reform of the CFP in the context of UNCLOS? There is little doubt that the rule of law itself is a winner, as the Convention has almost universal acceptance at a global level and has had a profound impact on the shape and substance of EU law.[223] The EU and the Member States are also winners in many respects as the Convention has provided stability within the international legal system that facilitates the negotiation and conclusion

222 Recital 67, Preamble, Regulation 1380/2013.

223 The total number of parties to the Convention is 167, or 166 state parties plus the European Union. There remain 31 non-party states, 17 of which are land-locked. As of that date, the 14 non-party coastal states are: the United States (the only permanent member of the Security Council that is not a party to the treaty), Iran, Libya, North Korea, Syria, Venezuela, Colombia, Peru, El Salvador, Turkey, Cambodia, the United Arab Emirates, Eritrea and Israel.

of multilateral, regional and bilateral fisheries agreements.[224] The framework advanced by the Convention has also allowed the EU to demonstrate "imaginative leadership" on law of the sea matters, which was first called for by Professor Treves at the 1985 LOSI conference.[225]

Notably, over the intervening years, the EU has shown a good deal of leadership and a strong commitment to the international legal process. Today, all 28 Member States and the EU in its own right are party to the UNCLOS and the Fish Stocks Agreement, and the EU has accepted the FAO Compliance Agreement,[226] all of which have primacy over EU secondary legislation within the European legal order, including over the scheme established by the new Basic Regulation.[227] In the 1980s, the Convention promised so much by setting out the law governing the rights and duties of coastal States over natural resources in the EEZ and in relation to the continental shelf.[228] One of the original objectives of UNCLOS was to protect fisheries from the activities of distant-water fishing fleets under the open-access regime that then prevailed under international law.[229] In order to do this, the Convention vests coastal States with sovereign rights for exploring, exploiting, conserving and managing living resources in the EEZ.[230] In tandem, the Convention places an express obligation on coastal States to conserve marine living resources in the EEZ and to ensure that they are not endangered by over-exploitation.[231] Robin Churchill pointed out in 1985 that these obligations also arise for the EEC (the forerunner of the EU), which functions as a supranational coastal State for fisheries management purposes under the CFP.[232]

The aforementioned fundamental conservation obligations have not changed with the passage of time. The legal regime, however, continues to

---

224 European Commission Communication on External Dimension of the Common Fisheries Policy, COM(2011)0424, 13.07.2011.

225 Tullio Treves, "The EEC, the UN and the Law of the Sea," *Law of the Sea Institute Proceedings* (1987) at 518–526, and at 525 and 580.

226 *Op. cit.* note 5.

227 On the primacy of international law over secondary legislation see, Case C-61/94 *Commission v Germany* [1996] ECR I-3989, paragraph 52, and Case C-311/04 *Algemene Scheeps Agentuur Dordrecht* [2006] ECR I-609, paragraph 25.

228 Parts 5 and 6, UNCLOS.

229 M. Nordquist S. Nandan, S. Rosenne, N. Grandy, eds., *United Nations Convention on the Law of the Sea 1982: A Commentary*, Vol. II, 491–796, especially at 491–510.

230 Article 56, UNCLOS.

231 Articles 61, 62, 63, 64, 69 and 70, UNCLOS.

232 R. Churchill, "The EEC's Contribution to State Practice in the Field of Fisheries," *Law of the Sea Institute Proceedings* (1987) at 557–568.

evolve since the entry into force of the UNCLOS and further conservation obligations arise for the EU and its constituent Member States in relation to straddling and highly migratory fish stocks under the Fish Stocks Agreement, and in relation to high seas fisheries resources under a range of multilateral and regional agreements.[233] As an international legal actor, the EU is committed to halting the decline of fish stocks in line with the obligations undertaken at the 2002 Johannesburg World Summit on Sustainable Development.[234] Similarly, it must meet the fisheries conservation targets set down in the Strategic Plan for Biodiversity 2011–2020 adopted by the Conference of the Parties to the Convention on Biological Diversity in 2011,[235] which includes the attainment of MSY by 2015.[236] Perhaps reflecting the cold reality of depleted fish stocks and the exigencies of the political process within its own supranational law-making institutions, the EU is committed under the reformed CFP to meeting the latter objective "where possible" and for all stocks falling within the scope of the CFP on "a progressive incremental basis" by 2020 at the latest.[237]

Despite the laudatory and prescriptive nature of these obligations, we can conclude that in some ways fishers and civil society more widely appear to be losers under the CFP, as the EU marine resource management regime has not to date delivered sustainable fisheries. Neither has the creation of the EEZ regime under the UNCLOS nor the extension of coastal State jurisdiction over fishery resources averted the continued crises in European fisheries. The World Bank and the FAO have highlighted the catastrophic economic consequences of such failure at a world level on the basis that it is costing the global community about $50 billion per year, or put another way close to $2 trillion over the last three decades.[238] The World Bank has also postulated that society could re-capture a substantial portion of this loss by improving the structures and procedures for the governance of marine fisheries through the establishment of systems of private property rights and by empowering civil society in relation to the substance and form of fishery management decisions.[239]

233 See I. Popescu, "Beyond the European Seas: The External Dimension of the Common Fisheries Policy, European Parliament" (Nov. 2015), available at http://www.europarl.europa.eu/RegData/etudes/IDAN/2015/571323/EPRS_IDA(2015)571323_EN.pdf.

234 Report of the World Summit on Sustainable Development. Johannesburg, South Africa, 26 August to 4 September 2002. (New York: United Nations, 2002).

235 See Decision X/2 in UNEP/CBD/COP/10/27 (20 January 2011).

236 EU Biodiversity Strategy to 2020, COM(2011)244.

237 Article 2(2), Regulation 1380/2013.

238 World Bank and FAO, *The Sunken Billions: The Economic Justification for Fisheries Reform* (Washington D.C.: World Bank, 2009), p. xvii.

239 *Ibid.*

Although the EU remains a major player in world fisheries, it has made limited progress on the pathway to reform and towards the implementation of a more sustainable governance model. Again there are some positive and negative features in the complex equation adopted by the EU to resolve many of the thorny problems encountered in the practical aspects of fisheries management. On the positive side, what catches the eye is that the EU is striving to improve governance by managing fish stocks as ecological units and through the application of new normative tools including the precautionary principle and ecosystem-based management. On the negative side, however, the political leaders in the Council have all too often deviated from scientific advice and thus undermined the scientific basis for fishery management.[240] Moreover, MSY has proven to be an unsophisticated management tool, which contributes to the over-subsidisation and over-capitalization of the fishing industry, economic inefficiency and the establishment of weak governance structures.[241]

In view of these failures, the move by the EU towards ecosystem-based management must therefore be welcomed, as it is tacit recognition that effective fishery governance entails the management of a socio-ecological system, where there are clear linkages between people and the environment.[242] Fisheries governance under the CFP is not only about managing a natural resource it is also about shaping the human, social and natural environment. Ecosystem-based management also provides a rationale for making the industry more responsible and will help address some of the inherent weaknesses in international law including UNCLOS, which accords rights and duties to States and affords little recognition to the subjective rights and duties of persons as subjects of the law of the sea.[243] Indeed, the EU's Court of Justice has held that UNCLOS does not establish rules intended to apply directly and immediately to individuals.[244] In particular, the Convention does not confer rights upon ship owners, because a ship's international legal status is dependent on the flag State and not on the fact that it belongs to certain natural or legal persons.[245] The EU may inadvertently be addressing this shortcoming by moving the focus

240 *Op. cit.* note 22.

241 E. Hey, "The Persistence of a Concept: Maximum Sustainable Yield," *International Journal of Marine and Coastal Law* 27 (2012): 763–771.

242 *Ibid.* at 769.

243 I. Papanicolopulu, "The Law of the Sea Convention: No Place for Persons?" (2012) 27(4) *International Journal of Marine and Coastal Law* 27 (2012): 867–874.

244 Case C-308/06 *The Queen, on the application of International Association of Independent Tanker Owners (Intertanko) and Others v Secretary of State for Transport* [2008] ECR I-04057.

245 Articles 59–61, *ibid.*

of the regulatory regime towards controlling the activities of the nationals of the Member States. The extension of the scope of the CFP to cover the activities of the nationals of the Member States is of course a central aspect of the battle to combat IUU fishing.[246]

This in itself is possibly one the most significant developments on the landscape of EU fisheries law in recent years. Moreover, the exercise of jurisdiction by States over their nationals in relation to fisheries conservation and management obligations is increasingly scrutinized in advisory and contentious proceedings before international courts and tribunals. For instance, New Zealand raised the issue of state responsibility over nationals, beneficial owners and operators of fishing vessels, in its submission to ITLOS in the Request for an Advisory Opinion Submitted by the Sub-Regional Fisheries Commission (SRFC).[247] In particular, New Zealand pointed to the conservation obligations that arise under the Fish Stocks Agreement,[248] the Convention for the Conservation of Marine Living Resources,[249] the Convention on the Conservation and Management of Highly Migratory Fish Stocks in the Western and Central Pacific Ocean (WCPO Convention),[250] and the FAO International Plan of Action to Prevent, Deter, and Eliminate Illegal, Unreported and Unregulated Fishing (IPOA-IUU).[251] New Zealand contested that all of these instruments set down specific obligations on States to control the activities of their nationals on the basis of the doctrine of concurrent State responsibility on the part of the State of nationality of those operating the vessel.[252]

The EU argued that the aforementioned argument was inadmissible for the purpose of the SRFC advisory Opinion.[253] Nonetheless, it is pertinent to our discussion to note that that the CFP applies to activities carried out by nationals of Member States, without prejudice to the primary responsibility of the flag State.[254] The EU framework rules on law enforcement are aimed specifically at the "the development of a culture of compliance and co-operation

246 R. Churchill, D. Owen *op. cit.* note 3 at 70–71.

247 ITLOS Case 21, Written Statement of New Zealand, First Round, 27 November 2013.

248 Article 7(1), Fish Stocks Agreement.

249 Article 10, CCAMLR Convention.

250 Article 23, WCPO Convention.

251 Para. 18, IPOA-IUU.

252 Paras. 32–35, ITLOS Case 21, Written Statement of New Zealand, First Round, 27 November 2013.

253 Para. 15, ITLOS Case 21, Second Written Statement by the European Commission on behalf of the European Union, 13 March 2014.

254 Article 1(2)(d) Regulation 1380/2013.

among all operators and fishers."[255] Furthermore, EU law requires all Member States to control access to waters and resources by vessels flying their flag and *by their nationals* (emphasis added).[256] In short, there appears to be a gradual and perceptible shift in emphasis in both EU and international fisheries law towards the assertion of jurisdiction on the basis of nationality and the imposition of rights and duties on individuals and companies on the basis that they are subjects (as opposed to objects) with legal personality.[257]

The European Court of Justice has aided this shift towards greater accountability by holding that authorization is required from the European Commission for European flagged vessels to fish in third country waters and the prohibition the use of bare-boat charters for this purpose, thus closing the possibility of EU nationals utilizing flags of convenience to circumvent the terms and conditions that apply under fisheries partnership agreements concluded by the EU with third countries.[258] Moreover, the Court has also upheld the regulatory powers of the EU institutions to adopt emergency conservation measures prohibiting the fishing of bluefin tuna in the Mediterranean and Atlantic.[259] At the same time, the Court rejected a claim for damages by fishers affected by the conservation measures on the grounds that the Commission could legitimately restrict fishing activities to avoid a serious threat to the conservation and recovery of marine resources.[260] These judgments of the Court are foursquare with the conservation obligations that arise under UNCLOS and international fishery law.

The progressive development of EU law is of course an on-going process and the outcome of the most recent reform efforts under the CFP appear to fall well short of the call by the World Bank for the "empowerment of fisher communities, the establishment of secure user and property rights, and the investment in collective action by a strengthened civil society."[261] We can nonetheless see some progress in moving the EU legal paradigm towards a model whereby individuals and companies can pursue their rights more effectively and where they will be called to account for failure to discharge their

255 Article 36(2)(d) Regulation 1380/2013.

256 Article 5(2), Regulation (EC) No. 1224/2009, OJ L 343, 22.12.2009.

257 Case 26/62, *Van Gend en Loos v. Nederlandse Administratie der Belastingen* (1963).

258 Case C-565/13, *Ahlström and Others*, 9 October 2014.

259 Case C-611/12 P and in Joined Cases C-12/13 P and C-13/13 P *Giordano v Commission and Buono and Others v Commission and Syndicat des thoniers méditerranéens and Others v. Commission*, 14 October 2014.

260 *Ibid.*

261 World Bank and FAO at p. xxii.

responsibilities. The success of this governance model is very much contingent upon further action on the part of the Member States and the fishing industry stepping-up to its responsibilities.[262] Ultimately, if all parties embrace change, the vesting of stakeholders with rights and responsibilities under the reformed CFP will be viewed as an important and long-overdue development in EU fisheries law. Some may see it as further evidence of the convergence of human rights-based regimes and the law of the sea more generally, which is also an emerging feature in the case law of international courts and tribunals. What is more, the EU will have demonstrated the kind of leadership in law of the sea matters that Professor Treves called for over three decades ago, and has shown itself as an effective forum for the negotiation of complex issues associated with the implementation of UNCLOS and its associated agreements.

262 Fisheries Jurisdiction Case, *United Kingdom of Great Britain and Northern Ireland v. Iceland*, ICJ Reports 1974, p. 31, para. 72.

CHAPTER 2

West Africa & the New European Common Fisheries Policy: Impacts & Implications

*Katherine Seto**

Introduction

In 2009, the European Commission announced their intention to begin legislative reform of the European Common Fisheries Policy (CFP).[1] The CFP is one of several frameworks in which European Union (EU) member states have transferred certain sovereign legislative and regulatory powers to common European institutions in order to promote regional management and cooperation.[2] The CFP creates this framework for EU member states' fishing fleets, markets, and all waters belonging to those states. On July 13, 2011, the Commission released the *Proposal for a Regulation of the European Parliament and of the Council on the Common Fisheries Policy*, and On 30 May 2013 the European Parliament and the European Council reached an agreement on the reform of the CFP. The consolidated text of the new Basic Regulation was published on 10 October 2013, and with the new policy formally agreed upon by the Council and Parliament, the reformed CFP has been in effect since 1 January 2014.[3]

Considering the global importance of the European Union member states as both fishing nations and fish importers, the CFP is a framework that has

* Ms. Seto is a Doctoral Student at UC Berkeley in the Environmental Science, Policy, and Management Department. Her research focuses on the relationship between civil conflict and marine resources, as well as marine resource management in post-conflict Sub-Saharan Africa.

1 European Commission, *Green Paper: Reform of the Common Fisheries Policy*, (Brussels, April 22, 2009).

2 Juan C. Surís-Regueiro, Manuel M. Varela-Lafuente, and M. Dolores Garza-Gil, "Evolution and Perspectives of the Fisheries Structural Policy in the European Union," *Ocean & Coastal Management* 54, no. 8 (August 2011): 593.

3 Markus Salomon, Till Markus, and Miriam Dross, "Masterstroke or Paper Tiger—the Reform of the EU'S Common Fisheries Policy," *Marine Policy* 47 (July 2014): 76; Surís-Regueiro, Varela-Lafuente, and Garza-Gil, "Evolution and Perspectives of the Fisheries Structural Policy in the European Union;" "Review of the Balance of Competences Between the United Kingdom and the European Union | Fisheries Report," August 26, 2014, 1–75.

 | DOI 10.1163/9789004311442_004

global, not regional, impacts.[4] According to the FAO State of World Fisheries and Aquaculture 2014, the fisheries region with the highest average annual production per person is Europe, at 24.2 tonnes, and the EU is by far the largest market for fish and fish products, with imports valuing more than US$50 billion in 2013, or slightly less than half of world imports.[5] These characteristics have particular significance for West African fisheries, which have been increasingly exploited by EU fleets since World War II.[6] West Africa's close proximity to Europe, its former colonial ties to several EU member states, and its limited domestic demand for high-value fish products have rendered the region an ideal locale for EU fishing operations, as Europe's dependence on fish imports continues to grow.[7]

With these considerations in mind, this chapter has a threefold objective. First, in order to understand the objectives of the CFP and its emergence as a global fisheries regime, this chapter discusses its origins, evolution, and role in relation to the UN Convention on the Law of the Sea (UNCLOS). Second, the chapter outlines the CFP's structure and influence, and gives closer examination to the implications of specific historical CFP rules and regulations for West African states. Third, the chapter reviews recent CFP reform, and discusses potential implications of these reforms on the future of West African fisheries.

An Inexhaustible Mine of Wealth

> *It has been calculated that if no accident prevented the hatching of the eggs and each egg reached maturity, it would take only three years to fill the sea so that you could walk across the Atlantic dryshod on the backs of cod.*
>
> —ALEXANDRE DUMAS, *Le Grande Dictionnaire de Cuisine*, 1873

•••

4 Ronán Long, "The Role of Regional Advisory Councils in the European Common Fisheries Policy: Legal Constraints and Future Options," *The International Journal of Marine and Coastal Law* 25, no. 3 (January 1, 2010): 299–302.

5 Food and Agriculture Organization of the United Nations, *State of the World Fisheries and Aquaculture 2014*, (Food & Agriculture Org, 2014), 8, 31; Long, "The Role of Regional Advisory Councils in the European Common Fisheries Policy: Legal Constraints and Future Options."

6 Jacqueline Alder and Ussif R. Sumaila, "Western Africa: a Fish Basket of Europe Past and Present," *The Journal of Environment & Development* 13, no. 2 (June 1, 2004): 156–78.

7 Food and Agriculture Organization of the United Nations, *State of the World Fisheries and Aquaculture 2014, 14*; Alder and Sumaila, "Western Africa: a Fish Basket of Europe Past and Present," 156–158.

Any tendency to over-fishing will meet with its natural check in the diminution of the supply; this check will always come into operation long before anything like permanent exhaustion has occurred.

—T.H. HUXLEY to the International Fisheries Exhibition in London, 1883

•••

In addition to a highly productive soil, the seas which surround us afford an inexhaustible mine of wealth—a harvest, ripe for gathering at every time of the year—without the labour of tillage, without the expense of seed or manure, without the payment of rent or taxes. Every acre of those seas is far more productive of wholesome, palatable, and nutricious food than the same quantity of the richest land; they are fields which, perpetually "white to harvest" [ripe], require only the labourer's willing hand to reap that never failing crop which the bounty of Providence has kindly bestowed.... That the mine we have to work upon is in reality inexhaustible, a transient inspection will be sufficient to satisfy the most sceptical inquirer.

—HENRY SCHULTES, British political commentator, 1813

∴

For most of history, marine fish stocks have been considered one of the most abundant—indeed inexhaustible—resources available to humankind. Throughout the nineteenth century and well into the twentieth, the dominant conception of these stocks was that of a God-sent blessing, a rich and consistent source of food and revenue that was so vast and productive that no conceivable act of man could affect its supply. While this image may have been relatively understandable in previous centuries, the inception of industrial fleets of steam-powered trawlers in the late nineteenth century created a new era of fishing capability. In the 1930s and 1940s, another leap in fishing efficiency occurred with the arrival of large factory ships, modernized gear and vessel technology, and sonar.[8] The consequences of these developments become clear in global marine fisheries trends, such as the expansion of area and depth fished, rise in global fish catch, and increasing amounts of

8 A.L. Fridman, "The Development of Specialized Ships, Nets, and Equipment," in *The Role of Food, Agriculture, Forestry and Fisheries in Human Nutrition*, V.R. Squires, ed., Vol. 2, (Paris: eolss.net, 2009) 214; Peter Makoto Miyake, "A Brief History of the Tuna Fisheries of the World," in *Management of Tuna Fishing Capacity*, ed. William H. Bayliff, Juan Ignaciio de Leiva Moreno, and Jacek Majkowski, 15 ed., (Rome: Food & Agriculture Org., 2005), 25.

capital invested in fishing fleets.[9] Regardless of the dramatic increases in fishing capacity, however, the notion of inexhaustibility persisted well into the 1980s, despite widespread evidence of decline in catch per unit effort (CPUE) and in some cases, stock depletion.[10]

Having its origin in the regional cooperation and economic recovery movements of the 1950s and 1960s, the CFP emerged in the context of the notion of fisheries inexhaustibility. As a policy and regulatory framework for European fish production and consumption, the CFP has also adapted over time to adjust to new scientific, economic, and social circumstances. While much literature has examined these developments and adaptations with regard to EU member states and internal European relations,[11] this paper seeks to gain a better understanding of the impacts of the European CFP on external, or third states.

The Emergence of the European Common Fisheries Policy

There are four separate, but interrelated policies in the European Common Fisheries Policy: the structural policy, the market policy, the conservation policy, and the external fisheries policy.[12] The creation of the CFP lasted approximately 15 years, and January 25, 1983 is generally regarded as the official birth of the CFP. In reality, this is the date on which the highly controversial conservation policy was agreed, the market and structural policies having been adopted in 1970. The fourth policy, dealing with external or international fisheries, does not have an official agreement or adoption date, as it developed gradually

9 Wilf Swartz *et al.*, "The Spatial Expansion and Ecological Footprint of Fisheries (1950 to Present)," ed. Stuart A. Sandin, *PLoS ONE* 5, no. 12 (December 2, 2010), 1; Food and Agriculture Organization of the United Nations, *State of the World Fisheries and Aquaculture 2014*; Ransom A. Myers and Boris Worm, "Rapid Worldwide Depletion of Predatory Fish Communities," *Nature* 423, no. 6937 (2003): 280–83.

10 Frédéric Le Manach, "Past, Present and Future of Publicly-Funded European Union's Fishing Access Agreements in Developing Countries," (University of British Columbia, 2014), 28; F. Berkes, "ECOLOGY: Globalization, Roving Bandits, and Marine Resources," *Science* 311, no. 5767 (March 17, 2006): 1557; Scheiber, "Ocean Governance and the Marine Fisheries Crisis," 119–121; Food and Agriculture Organization of the United Nations, *The State of the World Fisheries and Aquaculture 2010*, (Rome: Food & Agriculture Org., 2010).

11 Till Markus, *European Fisheries Law*, (Groningen: Europa Law Publishing, 2009); M. Holden and D. Garrod, *The Common Fisheries Policy*, (Oxford: Fishing News Books, n.d.); Robin Churchill and Daniel Owen, *The EC Common Fisheries Policy*, (Oxford University Press, 2010).

12 Markus, *European Fisheries Law*.

out of the growing trend of establishing Exclusive Economic Zones (EEZ's) around 1977.[13]

The Origins of the Common Fisheries Policy

The basis for the CFP emerges from Article 3 of the Treaty of Rome—the 1957 Treaty that established the European Economic Community (EEC)—which states that, "For the purposes set out in Article 2, the activities of the Community shall include ... (d) the adoption of a common policy in the sphere of agriculture ...,"[14] and in Article 38 of the Treaty, in which the first paragraph states:

> The common market shall extend to agriculture and trade in agricultural products. "Agricultural products" means the products of the soil of stock—farming and of fisheries and products of first-stage processing directly related to these products.[15]

Although the mandate for a CFP was established in 1957, there was little movement toward establishing a policy throughout the 1950s and well into the 1960s, likely attributable to several factors. In the 1950s, five of the six member states (Belgium, France, Italy, the Netherlands, and Germany) caught 90 percent of their catches outside of their national waters—generally considered to be three miles from shore at that time—and the sixth member state, Luxembourg, was a landlocked State. Therefore little was to be gained by an agreement on shared fishing rights among member states' waters.[16] By the end of the 1960s, however, the Common Customs Tariff had begun to indirectly affect the fisheries sectors of both France and Italy. These States requested regulations that could help modernize their fleets and enable them to better compete. In 1967, a long paper titled "Basic principles for a common fisheries policy," was developed by the EC. Even following the release of the paper, however, States other than France and Italy had little incentive to assist these fleets, essentially modernizing their competitors.[17]

The application by Denmark, Norway, Ireland, and the United Kingdom to join the European Community was the pivotal event that led to the establishment

---

13 Holden and Garrod, *The Common Fisheries Policy*.

14 European Economic Community, *Treaty Establishing the European Economic Community*, (Rome, 1957).

15 *Ibid.*

16 Holden and Garrod, *The Common Fisheries Policy*; Churchill and Owen, *The EC Common Fisheries Policy*, p. 29.

17 *Ibid.*, pp. 4–5.

of the CFP.[18] All four were strong fishing nations, representing waters where the existing member states had highly active fisheries.[19] Therefore, around 1970, the incentives of EEC member states suddenly changed from maintaining the *status quo* to establishing a fisheries policy that new member states would have to accept wholesale as the *acquis communitaire*, newly-admitted states having no say in formulating preexisting policies.[20] The basis of the CFP was established in two regulations, barely passing in the final hour in 1970: Council Regulation (EEC) No. 2142/70 on the common market in fishery products and Council Regulation (EEC) No. 2141/70 establishing a common structural policy for the fishing industry.[21]

Following the adoption of these regulations, the European Court of Justice (ECJ) played a central role in the "communitarization" of the EC powers and the establishment of these powers in relation to fisheries.[22] First with the 1971 *ERTA* case, the ECJ affirmed the external competence of the Commission and referred to its exclusivity, indicating its precedence over the jurisdiction of Member States.[23] In the 1976 case of *Officier Van Justitie v. Kramer*,[24] the ECJ established the EEC's authority to enter into international agreements specifically relating to the conservation of the biological resources of the sea.[25] The 1981 judgment in *European Commission v. UK* further solidified these powers in ruling that the EEC had exclusive competence to adopt fisheries conservation measures in Member States' waters.[26]

---

18 Nienke van der Burgt, *The Contribution of International Fisheries Law to Human Development*, (Martinus Nijhoff Publishers, 2012), 232.

19 Markus, *European Fisheries Law*; Holden and Garrod, *The Common Fisheries Policy*; "Review of the Balance of Competences Between the United Kingdom and the European Union | Fisheries Report," August 26, 2014, 13.

20 Churchill and Owen, *The EC Common Fisheries Policy*, p. 5.

21 Holden and Garrod, *The Common Fisheries Policy*; Churchill and Owen, *The EC Common Fisheries Policy*, p. 5; Council of the European Communities, *Regulation (EEC) No. 2142/70 of the Council, Official Journal of the European Communities*, 1970; Council of the European Communities, *Regulation (EEC) No. 2141/70 of the Council, Official Journal of the European Communities*, 1970.

22 Van der Burgt, *The Contribution of International Fisheries Law to Human Development, 232.*

23 *Ibid.*, European Court of Justice, *Judgment of the Court of 31 March 1971. Commission Of the European Communities V Council Of the European Communities. European Agreement on Road Transport*, 264–84 (1971).

24 European Court of Justice, *Judgment of the Court of 14 July 1976. Cornelis Kramer and Others. Biological Resources of the Sea. Joined Cases 3, 4 and 6–76*, 1280–1314 (1976).

25 van der Burgt, *The Contribution of International Fisheries Law to Human Development*, 229.

26 European Court of Justice, Judgment of the Court of 5 May 1981. Commission of the European Communities v United Kingdom of Great Britain and Northern Ireland. Sea fisheries—Conservation measures, 1047–80 (1981).

The Structural Policy: Subsidies, Fleets, and Moving Targets

The primary objectives of the 1970 regulations, which formulated the structural and market policies and founded the CFP, were to eliminate the European deficit in the supply of fish and to promote the notion of European "auto-sufficiency."[27] Mike Holden, member of the European Commission Directorate General for Fisheries (DGXIV) 1979–1990, states that, "the apparently self-evident way in which to achieve this objective was to encourage the building of more vessels to catch more fish."[28] While the negative environmental and resource consequences of this approach may seem self-evident now, a few factors kept these considerations out of European common fisheries policymaking. First, at the time of regulation in 1970, and indeed until 1978, there were no scientific fisheries experts in the staff ranks of the Commission,[29] Second, as the conservation policy of the CFP was established thirteen years later in 1983, there was no formal process of reconciling the market and structural policies to conservation considerations.[30] Third, widespread perceptions of fisheries resources in 1970 were based on the lessons of the 1950s and 1960s, which demonstrated that massive increases in global fishing effort result in tremendous increases in catch. Pauly *et al.* state that this rapid increase in catch encouraged "an entire generation of managers and politicians to believe that launching more boats would automatically lead to higher catches."[31] In reality, the increase in effort and catch masked an actual decline in catch per unit effort (CPUE), which emerged in much of the world beginning in the 1950s.[32] The obsolete belief in the inexhaustibility of global marine fisheries from the days of Huxley continued to form policy into the late twentieth century.

In the case of European common fisheries, the emphasis on increasing vessels and fishing effort was embodied best in Article 11 of Regulation 2908/83, which gives priority funding to:

---

27 Van der Burgt, *The Contribution of International Fisheries Law to Human Development*; Holden and Garrod, *The Common Fisheries Policy*.

28 Holden and Garrod, *The Common Fisheries Policy*.

29 *Ibid.*

30 Emma Witbooi, "The Infusion of Sustainability into Bilateral Fisheries Agreements with Developing Countries: the European Union Example," *Marine Policy* 32, no. 4 (July 2008): 670.

31 Daniel Pauly et al., "Towards Sustainability in World Fisheries," *Nature* 418, no. 6898 (August 8, 2002): 689.

32 Food and Agriculture Organization of the United Nations, *Review of the State of World Marine Fishery Resources*, (Food & Agriculture Org., 2005); Myers and Worm, "Rapid Worldwide Depletion of Predatory Fish Communities."

1. the "purchase or construction of new fishing vessels,"
2. the replacement of vessels more than 12 years old or which were "irreparably damaged, broken up, or permanently withdrawn from use as fishing vessels," and
3. the commissioning of vessels in areas where fishing was "traditionally an important economic activity."[33]

Understandably, the result of these policies was a massive increase in the size of the European fleet from an estimated gross registered tonnage (GRT) of 794,000 in 1970 to 1,303,000 GRT in 1983 and 1,618,519 GRT in 1987.[34] With increasing fleet size, improved technology, and rising catches (though not CPUE), European fleets continued to expand effort until the decline of fish stocks became widely evident in the early 1980s.[35] At this time, it could be said that the objectives of the CFP expanded—if they were not fundamentally altered—to include direct conservation considerations for target fish stocks.[36] In response to the decline perceived in the fish stocks, the conservation policy was established and the structural policy was rapidly altered, promoting a policy tool known as multi-annual guidance programs (MAGPs).[37]

In 1983, Regulation No. 2908/83 redefined MAGPs from their former role as minor coordinating tools to programs that would achieve "a satisfactory balance between the fishing capacity to be deployed by the production facilities covered by the programmes and the stocks which are expected to be available during the period of validity of the programme."[38] The MAGPs were meant to set fishing capacity targets for all countries in the EEC as a means of reducing effort in the common fishery. There were four rounds of MAGPs negotiated: MAGP I (1983–1986), MAGP II (1987–1991), MAGP III (1992–1996), and MAGP IV

33 Council of the European Union, *31983R2908: Council Regulation (EEC) No. 2908/83 of 4 October 1983 on a Common Measure for Restructuring, Modernizing and Developing the Fishing Industry and for Developing Aquaculture*, 1983.

34 Holden and Garrod, *The Common Fisheries Policy*.

35 Clare Coffey, *Sustainable Development and the EC Fisheries Sector*, 1999, 11.

36 Witbooi, "The Infusion of Sustainability into Bilateral Fisheries Agreements with Developing Countries: the European Union Example," 670.

37 Holden and Garrod, *The Common Fisheries Policy*; Surís-Regueiro, Varela-Lafuente, and Garza-Gil, "Evolution and Perspectives of the Fisheries Structural Policy in the European Union," 594.

38 Council of the European Union, *31983R2908: Council Regulation (EEC) No. 2908/83 of 4 October 1983 on a Common Measure for Restructuring, Modernizing and Developing the Fishing Industry and for Developing Aquaculture*.

(1996–2002).[39] All of the MAGPs, however, failed to achieve their stated goals, with two of nine member states compliant in MAGP I, five of eleven compliant in MAGP II, and an overall decrease in fleet capacity of 5–12 percent rather than the expected 17–40 percent for MAGPs III and IV.[40] Though fleets in the 1990s and 2000s were no longer rapidly expanding, the stated reductions in vessel numbers masked actual increases in capacity caused by improved vessel efficiency and technology.[41] Holden states that "even for those Member States which met their targets, it must be a matter of some doubt as to what extent active fishing capacity had been reduced."[42]

These trends are echoed in the Financial Instrument for Fisheries Guidance (FIFG), which was meant to consolidate all public subsidies associated with the structural policy and act as the financial instrument for the CFP.[43] The FIFG was created in 1993 and had two programs: FIFG I (1994–1999) and FIFG II (2000–2006). Like the MAGPs, these programs were also considered unsuccessful, since reductions in fleet size and engine power were lower than the gains achieved through technological improvements.[44] A 2010 evaluation by the Pew Environment Group stated that "FIFG support for the construction of around 3,000 vessels and the modernization of nearly 8,000 vessels, compared to the scrapping of 6,000 vessels (a large proportion of which were small inshore vessels from Greece and Spain), is expected to have resulted in a net increase in fishing capacity."[45]

The third instrument of the CFP structural policy is the European Fisheries Fund (EFF), initiated in 2007 and extending to 2013. While any assessment of the EFFs effects is preliminary, trends in the structural policy have not much

39 Holden and Garrod, *The Common Fisheries Policy*; Surís-Regueiro, Varela-Lafuente, and Garza-Gil, "Evolution and Perspectives of the Fisheries Structural Policy in the European Union," 594.

40 Holden and Garrod, *The Common Fisheries Policy*; Surís-Regueiro, Varela-Lafuente, and Garza-Gil, "Evolution and Perspectives of the Fisheries Structural Policy in the European Union," 594–595. R. Cappell, T. Huntington, and G. Macfadyen, *FIFG 2000–2006 Shadow Evaluation*, March 2010, 1–2.

41 Surís-Regueiro, Varela-Lafuente, and Garza-Gil, "Evolution and Perspectives of the Fisheries Structural Policy in the European Union," 593–594; Commission of the European Communities, *Proposal for a Regulation of the European Parliament and of the Council on the Common Fisheries Policy*, (Brussels, February 28, 2012).

42 Holden and Garrod, *The Common Fisheries Policy*.

43 Cappell, Huntington, and Macfadyen, *FIFG 2000–2006 Shadow Evaluation*, 2–3.

44 Sebastian Villasante, "Global Assessment of the European Union Fishing Fleet: an Update," *Marine Policy* 34, no. 3 (May 2010): 663–70.

45 Cappell, Huntington, and Macfadyen, *FIFG 2000–2006 Shadow Evaluation*, i.

deviated during this phase, and in 2009, the European's Commission's Green Paper on the Reform of the CFP stated that "excess capacity is still the CFP's fundamental problem."[46]

The stated purpose of the MAGPS, FIFGS, and EFF was to align European fleet capacity to the actual potential of fisheries resources.[47] Considering the fact that absolute reductions in European fishing capacity have proven politically untenable, however, policies have been redirected.[48] The notion is that when actual reductions in active fishing capacity prove too difficult, the vessels that have been effectively "weeded out" of European fisheries may be redirected towards 1) new stocks (exploratory fishing), 2) new occupations (redeployment), or 3) new waters belonging to other states (joint ventures, access agreements or export of vessels). The potential for profitable exploitation of new fish stocks is exceedingly low, however, considering the global expansion of area fished and the increased ability of industrial fleets to reach previously inaccessible stocks.[49] Considering the high degree of subsidies which remain in the fishing sector, the large number of vessels, and the barriers to entry into other industries, redeployment is also an unlikely long-term solution.[50] Joint ventures have proven the most popular solution for removing fishing capacity from European fisheries—a proposal which has tremendous implications for West African fisheries and which is discussed below.

The Market Policy: Standards and Prices

Similar to the structural policy, the market policy was created in 1970 and, true to its origins, parallels the market system defined for European agricultural products.[51] In general, the objectives of the policy are:

46 Surís-Regueiro, Varela-Lafuente, and Garza-Gil, "Evolution and Perspectives of the Fisheries Structural Policy in the European Union;" European Commission, *Green Paper: Reform of the Common Fisheries Policy*.

47 Gerd Winter, ed., *Towards Sustainable Fisheries Law*, (Gland: IUCN, 2009), 277–284; Cappell, Huntington, and Macfadyen, *FIFG 2000–2006 Shadow Evaluation*, 1.

48 Holden and Garrod, *The Common Fisheries Policy*; Cappell, Huntington, and Macfadyen, *FIFG 2000–2006 Shadow Evaluation*; Anthony Acheampong, *Coherence Between EU Fisheries Agreements and EU Development Cooperation*, (Maastricht, 1997), 5.

49 Holden and Garrod, *The Common Fisheries Policy*; Daniel Pauly et al., "Fishing Down Marine Food Webs," *Science* 279, no. 5352 (February 6, 1998): 861; Swartz et al., "The Spatial Expansion and Ecological Footprint of Fisheries (1950 to Present)," 1–6.

50 Ernst & Young, *Interim Evaluation of the European Fisheries Fund (2007–2013)*, February 2011, http://ec.europa.eu/fisheries/documentation/studies/index_en.htm.

51 Churchill and Owen, *The EC Common Fisheries Policy*, pp. 4–5.

1. To establish marketing standards through freshness and size specifications;
2. To stabilize prices, avoid surpluses, and support producer incomes through the fixing of guide prices; and
3. To consider consumer interests by managing marketing by producer organizations.[52]

For the most part, the CFP market policy has been successful and uncontroversial within the EU, maintained in large part by the increasing deficit in supplies of fish species, which keep producer prices high for European fleets.[53] While the market policy has proven beneficial to European producers and consumers, however, the implications for foreign fisheries such as those throughout West Africa have been mixed.

The Conservation Policy: Reconsidering the Resource

The CFP conservation policy was established in 1983 and held as its primary objective "the conservation of biological resources of the sea and their balanced exploitation on a lasting basis and in appropriate economic and social conditions." Conservation approaches suggested in the policy included: effort restrictions, imposition of a total allowable catch (TAC), fishing gear restrictions, fish size and weight limitations, and several others.[54] The most influential regulation was the TAC, which allowed the European Council to set annual limits on various fisheries in EU waters, and apportion quotas to member states.[55]

In the face of continued declines in European fish stocks, the conservation policy underwent a series of changes under the 2002 reform. Specifically, Regulation 2371/2002 stated the objective of: "[the] sustainable exploitation of living aquatic resources in the context of sustainable development, taking account of the environmental, economic, and social aspects in a balanced manner."[56] Although this objective does not seem to substantially deviate from the 1983 objective, a few important differences are noteworthy. First, the use of

52 Holden and Garrod, *The Common Fisheries Policy*.

53 *Ibid.*

54 Witbooi, "The Infusion of Sustainability into Bilateral Fisheries Agreements with Developing Countries: the European Union Example," 669–670. "Review of the Balance of Competences Between the United Kingdom and the European Union | Fisheries Report," 14–16.

55 A. Karagiannakos, "Total Allowable Catch (TAC) and Quota Management System in the European Union," *Marine Policy* 20, no. 3 (1996): 235–36.

56 Council of the European Communities, *Council Regulation (EC) 2371/2002*, 2002.

the term "sustainable exploitation" specifically refers to the notion of "exploitation of a stock in such a way that the future exploitation of the stock will not be prejudiced and that it does not have a negative impact on the marine ecosystems."[57] Second, Regulation 2371/2002 requires the European Council to promote the ecosystem-based approach to fisheries management, and apply the precautionary principle. Third, while retaining the controversial TAC system, it supplements the policy with a multi-annual management scheme aimed at fostering sustainable exploitation.[58]

In addition to Regulation 2371/2002, a Community Action Plan to integrate environmental protection requirements into the CFP was also defined in 2002, to complement the regulation. The purpose of the Action Plan was to "translate such ideas [on environmental integration] into an Action Plan comprising objectives and principles, means, targets and timetables to enable the Common Fisheries Policy to address environmental challenges efficiently."[59]

Although the conservation policy was the last policy developed under the CFP in 1983, it has increased in importance in the past three decades and is now one of the overarching themes in the policy as a whole. Criticisms throughout the 1990s, primarily from the non-govenrment organization (NGO) community, targeting what was seen as the contradictory approaches of the structural and conservation policies led to major reforms in the early 2000s.[60] These reforms were also the product of larger European trends of policy integration and environmentalism.[61] While these reforms laid out some specific conservation tools for fisheries—such as the precautionary principle, ecosystem-based management, and TAC quotas—the major objective of the 2002 reforms was to re-situate the policies of the CFP into a broader approach of sustainability.[62]

---

57 *Ibid.*

58 Witbooi, "The Infusion of Sustainability into Bilateral Fisheries Agreements with Developing Countries: the European Union Example," 671–672.

59 Commission of the European Communities, *Communication from the Commission on the Reform of the Common Fisheries Policy ("Roadmap")*, May 28, 2002.

60 Eloise Todd and Ella Ritchie, "Environmental Non-Governmental Organizations and the Common Fisheries Policy," *Aquatic Conservation: Marine and Freshwater Ecosystems* 10, no. 2 (2000): 141–49.

61 Council of the European Communities, *Council Regulation (EEC) No. 3760/92, Official Journal of the European Communities*, 1992; Nigel Haigh, "Introducing the Concept of Sustainable Development into the Treaties of the European Union," in *The Transition to Sustainability the Politics of Agenda 21*, (The transition to sustainability: the politics of Agenda, 1998), 64–75.

62 L. van Hoof and J. van Tatenhove, "EU Marine Policy on the Move: the Tension Between Fisheries and Maritime Policy," *Marine Policy* 33, no. 4 (July 2009): 726–32.

The External Relations Policy: Exporting Capacity and Evolving Agreements

The origin of the external policy lies in the extension of coastal states' fishing rights from the previously recognized three nautical miles (nm) to 200 nm EEZs under UNCLOS.[63] With this drastic change, significant expanses of ocean that had previously been exploited by foreign fleets came under the sovereignty of coastal states' EEZs, necessitating the negotiation of fishing agreements in order to maintain access to previously exploited areas.[64] Indeed, the EEC's 1977 decision that all Member States would declare a 200 nm EEZ coincided with the decision that authorized the EEC to negotiate third party agreements.[65] The external fisheries policy was designed to cover international fisheries relations between EEC member states and third states (or those external to the EEC), in order to *maximize employment* from the European fishing sector and *ensure supply of fish* to European markets, while still *reducing fishing pressure* in European waters.[66]

The external fisheries policy accomplished this in two primary ways. First, in brokering international fisheries agreements via the EEC, rather than individual member states, the policy prevented European competition for fisheries access in third countries. Second, it ensured that as many European vessels as possible continued to fish outside common European waters, thereby 1) maintaining European supplies of fish and reducing imports from non-EC member states, 2) appeasing member states while still reducing capacity in European fisheries, 3) ensuring supply of species that do not occur in European waters, and 4) maximizing employment through the distant water fishing fleet.[67]

Therefore, while some believe this policy to be largely independent of the other three, the external policy can be viewed as fundamentally connected to the structural, market, and conservation policies, in that it protects European

63 Acheampong, *Coherence Between EU Fisheries Agreements and EU Development Cooperation*, 4; Holden and Garrod, *The Common Fisheries Policy*; United Nations, *United Nations Convention on the Law of the Sea (UNCLOS)*, (Montego Bay, 1982).

64 Witbooi, "The Infusion of Sustainability into Bilateral Fisheries Agreements with Developing Countries: the European Union Example," 672–673; Frédéric Le Manach et al., "Who Gets What? Developing a More Equitable Framework for EU Fishing Agreements," *Marine Policy* 38 (March 2013): 257–66.

65 Marine Resources Assessment Group MRAG, Ltd., *Policy Brief 6: Fisheries and Access Agreements*, n.d., http://www.mrag.co.uk/Recent_Publications.htm; Holden and Garrod, *The Common Fisheries Policy*.

66 *Ibid.*, Acheampong, *Coherence Between EU Fisheries Agreements and EU Development Cooperation*, 4–7.

67 Holden and Garrod, *The Common Fisheries Policy*.

producers and market prices by reducing imports, and enables displacement of European fishing fleets to other coastal state EEZ's, rather than requiring an absolute reduction in capacity.[68] In terms of these two objectives, the external fisheries policy was largely successful through the 1980s and into the 1990s.[69] In 1997, the European Commission estimated that approximately 40,000 fishing and processing jobs and 1300 vessels in the European Community were supported by agreements under the external fisheries policy.[70]

Despite their success with regard to EU member states, however, the external policy came under fierce criticism in the 1990s for a variety of reasons, related to third states. Issues such as unsustainable exploitation, lack of monitoring, control, and surveillance (MCS), impairment of developing economies, food insecurity, and perceived inconsistencies with other EU common policies in environment and development generated significant criticism of the external fisheries policy.[71] Notably, these criticisms coincided with tremendous growth in EU investment in third country fishing grounds, with the budget for international fisheries agreements increasing from 6 million ECU in 1981 to more than 280 million ECU 1996.[72] NGOs such as The Coalition for Fair Fisheries Agreements emphasized the incoherence in the fact that the European Commission's Directorate General for Fisheries was charged with gaining access for European vessels in third states' waters, while at the same time the Directorate General for Development aimed to develop the artisanal fisheries of these same states.[73]

In response to these criticisms, the external policy was also revised in 2002 under the second CFP reform. The reforms of the external relations policy were geared toward two primary objectives. The first is that of improving world governance of fisheries through existing international frameworks and regional cooperation mechanisms; this objective was targeted through the EU's renewed commitments of participation in and support of Regional Fisheries Organizations (RFOs) as well as support for enforcing and strengthening

68 Markus, *European Fisheries Law.*

69 Holden and Garrod, *The Common Fisheries Policy.*

70 Acheampong, *Coherence Between EU Fisheries Agreements and EU Development Cooperation*, 5.

71 Marine Resources Assessment Group MRAG, Ltd., *Policy Brief 6: Fisheries and Access Agreements.*

72 Commission of the European Communities, *Communication From the Commission to the Council and the European Parliament*, October 30, 1996; Hans Martin Havstein, "Change and Continuity: 40 Years of Reforming the Common Fisheries Policy," 2013, 33.

73 Acheampong, *Coherence Between EU Fisheries Agreements and EU Development Cooperation*, 3.

UNCLOS.[74] The second objective is that of improving bilateral agreements between EU member states and developing countries, by designing them more as holistic resource partnerships.[75] This second goal was addressed through the 2002 reforms, and further through the adoption of the Council Conclusions in 2004, which outline procedural mechanisms for the 2002 reforms.[76] These policies transitioned bilateral arrangements with third states away from a policy of "pay, fish, and leave" to that of Fisheries Partnership Agreements (FPAs), based on principles of sustainability and support for third state development.[77]

The FPAs were meant to signal a fundamental change in the legal and political relations between EU member states and third states. Rather than act as a fee-for-service arrangement, FPA goals meant to address the social, economic, and environmental aspects of the third state fisheries, and were "rooted in the ideals of greater coherence between access agreements and internal [EU] policies such as development cooperation, environment, and trade."[78] In other words, from 2002 onward, the FPAs were meant to redirect fisheries agreements from being simple commercial exchanges, and guide them "towards developing a sound fisheries management policy for the third country."[79] While some provisions of these agreements have improved, assessments conducted by the Directorate-General for Maritime Affairs and Fisheries of the European Commission (DG MARE) indicate that the reforms largely fell short

74 Directorate-General for Fisheries and Maritime Affairs and Maritime Affairs, *The European Community External Fisheries Policy*, (Brussels, December 5, 2005).

75 *Ibid.*

76 Witbooi, "The Infusion of Sustainability into Bilateral Fisheries Agreements with Developing Countries: the European Union Example," 673–674; Council of the European Communities, *Council Conclusions 11485/1/04 on the Commission Communication on an Integrated Framework for Fisheries Partnership Agreements with Third Countries*, (Brussels, July 15, 2004).

77 Directorate-General for Fisheries and Maritime Affairs and Affairs, *The European Community External Fisheries Policy*; Council of the European Communities, *Council Conclusions 11485/1/04 on the Commission Communication on an Integrated Framework for Fisheries Partnership Agreements with Third Countries*.

78 Council of the European Communities, *Council Conclusions 11485/1/04 on the Commission Communication on an Integrated Framework for Fisheries Partnership Agreements with Third Countries*, (Brussels, July 15, 2004); Witbooi, "The Infusion of Sustainability into Bilateral Fisheries Agreements with Developing Countries: the European Union Example," 673.

79 *Ibid.*, 673; Council of the European Communities, *Council Conclusions 11485/1/04 on the Commission Communication on an Integrated Framework for Fisheries Partnership Agreements with Third Countries*.

of "sound fisheries management policy."[80] What remains notable, however, is that the CFP reforms of the early 2000s expanded the external policy beyond the UNCLOS framework and situated it within the common European legal and regulatory regime. This was a significant move with regard to third party relations—including those with West African states—and some of the implications will be discussed below.

Europe and the West African Fisheries Legacy

The European Union has a unique relationship with West Africa, both historically and geographically, and many EU member states have fished the waters of the Central Eastern Atlantic for centuries. Yet twentieth century developments in de-colonization, industrial fishing, and the declaration of EEZs have redefined the terms on which these waters can, and should, be accessed for fisheries exploitation.[81] Following the colonial era, a series of treaties established the primary economic framework between EU and African-Caribbean-Pacific (ACP) states, fisheries and trade relations comprising two of the many sectors governed by this framework. First, the two Yaoundé Conventions (1963 and 1969), followed by the four Lomé Conventions (1975, 1981, 1985, 1989) specified relations regarding trade and development between EU and ACP countries.[82] Since 2000, the Cotonou Agreement has provided the political and legal frame for EU and ACP states on development and trade, with significant direct and indirect consequences for fisheries exploitation, production, and export.[83] A detailed analysis of the EU-ACP agreements is beyond the scope of this paper, though their implications and discussed throughout. Currently, the EU maintains access to fish stocks along most of the West African coast through some multilateral agreements, but primarily by means of bilateral fisheries

80 Le Manach, "Past, Present and Future of Publicly-Funded European Union's Fishing Access Agreements in Developing Countries."

81 Churchill and Owen, *The EC Common Fisheries Policy*, 342.

82 Le Manach, "Past, Present and Future of Publicly-Funded European Union's Fishing Access Agreements in Developing Countries," 95; van der Burgt, *The Contribution of International Fisheries Law to Human Development*, 259.

83 Le Manach, "Past, Present and Future of Publicly-Funded European Union's Fishing Access Agreements in Developing Countries, 96"; van der Burgt, *The Contribution of International Fisheries Law to Human Development*, 259.

agreements with individual West African states.[84] While these stocks supply fish for growing European markets, and employ European vessels and fishers, they also play an essential role in the economies, diets, livelihoods, and cultures of West African citizens. Many of these countries' populations are also dependent on unusually high proportions of dietary fish protein, such as Senegal (47 percent), Gambia (62 percent), Ghana (63 percent), and Sierra Leone (75 percent), as compared to a global average of 15 percent.[85] Considering the importance of these Central Eastern Atlantic fish stocks to both European and West African interests, the following section examines some of the consequences that the European CFP has historically had on West African states.

The CFP: Beyond Europe

The original goals of the European CFP structural policy were to eliminate the deficit in fish supply and promote European "auto-sufficiency." Following the decline of European fish stocks, and the consequent reforms, however, the primary goal of the structural policy since the early 1980s has been one of reducing capacity in the European fleet. One of the most popular solutions to this problem—and one that has had a tremendous impact on West African fish stocks—is that of access agreements and join ventures. Joint ventures may take a variety of different forms, however they generally indicate "an association of two or more parties, whether private or governmental, in order to undertake a commercial project in the fisheries sector, and to share in the risks and profits of that project".[86] These arrangments may involve setting up a separate joint company, or a simple contractual relationship. Historically, EU-West African joint ventures generally involve the African party gaining access to capital goods (i.e. vessels and fishing gear), existing export avenues and infrastructure, a potential source of employment, and a share in profits, while the European party gains access to shore-based resources, displaces effort from EU waters, and enables cheaper access to fish stocks by circumventing the fees associated with foreign access agreements.[87]

---

84 Witbooi, "The Infusion of Sustainability into Bilateral Fisheries Agreements with Developing Countries: the European Union Example," 669–670; van der Burgt, *The Contribution of International Fisheries Law to Human Development*, 10–12.

85 World Health Organization & Food and Agriculture Organization of the United Nations, *Diet, Nutrition and the Prevention of Chronic Diseases*, (Geneva, 2002).

86 Food and Agriculture Organization of the United Nations. Fishery Committee for the Eastern Central Atlantic, *Joint Ventures in Fishery Development in the CECAF Area*, 1974.

87 Churchill and Owen, *The EC Common Fisheries Policy*, 343–345.

While this arrangement may appear mutually beneficial, in reality, the consequences of joint ventures for developing countries are mixed. First, while joint ventures involve West African companies, the fish that are caught are usually bound for the export market—therefore while catches are recorded as domestically caught, they are oftentimes neither landed nor consumed for domestic benefit.[88] Second, since the proliferation of joint ventures led to increasing fish exports, the West African region transitioned from a net exporter of fish in the 1960s to a net importer in the 1990s, thereby reducing domestic food security and increasing reliance on market goods.[89] Third, while joint ventures occur between West African and European parties, the majority of agreements are of a private nature, and profits accrue to private companies.[90] Therefore, under joint ventures, the fish resource is exploited for the benefit of foreign and domestic private parties, with very low license fees (relative to the actual value of the catch) paid to the coastal state government, and an overall loss to coastal communities.[91] Furthermore, joint ventures enable EU-West African fisheries relations to be redirected away from the more regulated agreements brokered under the EU CFP external policy, and into the much more opaque realm of private bilateral agreements. In 2013, fewer than half of the EU's 700-vessel external fleet were involved in EU-negotiated FPAs, while over 400 vessels were involved in joint ventures, many operating in the same West African waters where FPAs had in fact already been negotiated.[92]

The general criticism of the structural policy with regard to West African states is simply that by displacing the European fleet from European waters to West African waters, the structural policy is simply shifting the declines seen in European stocks to the stocks of the Central Eastern Atlantic.[93] While succeeding in reducing the number of vessels actively fishing in EU waters, the policy

88 Alder and Sumaila, "Western Africa: a Fish Basket of Europe Past and Present," 156–157, 169–172.

89 *Ibid.*, 164.

90 Le Manach et al., "Who Gets What? Developing a More Equitable Framework for EU Fishing Agreements," 6–7.

91 Alder and Sumaila, "Western Africa: a Fish Basket of Europe Past and Present," 170–171; Cappell, Huntington, and Macfadyen, *FIFG 2000–2006 Shadow Evaluation*, 32–50; Le Manach et al., "Who Gets What? Developing a More Equitable Framework for EU Fishing Agreements."

92 Agritrade, *EU Common Fisheries Policy and Fisheries Partnership Agreements*, October 2013, http://agritrade.cta.int/, 2.

93 Churchill and Owen, *The EC Common Fisheries Policy*; Pauly et al., "Towards Sustainability in World Fisheries," 689–692; Cappell, Huntington, and Macfadyen, *FIFG 2000–2006 Shadow Evaluation*, 32–50.

simply moves that overcapacity to West African waters, while maintaining the benefit of the resource to Europe. Indeed, the argument is made stronger by the dearth of scientific information and monitoring, control, and surveillance within West African fisheries, as compared to the far more studied and regulated European fisheries. While the structural policy has consistently prioritized capacity reductions for the last thirty years, little to no actual reduction has taken place, and this overcapacity perpetuates the incentive for Community subsidies to distant water fishing operations.[94]

In response to these criticisms, the 2002 CFP reform tightened—and in 2004 formally terminated—the FIFG mechanism enabling joint ventures between EU companies and third state companies.[95] Since 2002, it has no longer been possible to set up temporary joint ventures, i.e. transfers of vessels for pilot joint ventures, aimed at relieving excess capacity in European waters. Since then, FIFG support of joint ventures has continued to decline.[96]

The Market Policy and West Africa

The main objectives of the CFP market policy are to protect consumer interests by establishing standards and managing producer marketing, and to protect producer interests by fixing guide prices and avoiding surpluses. While these policies have been largely uncontroversial to European member states, they have proven challenging to third states such as those in West Africa. The first significant challenge that West African states face with regard to the CFP market policy is the high sanitation, safety, and traceability rules required for access to European markets.[97] Ponte *et al.* (2007) summarize this challenge well:

94 Witbooi, "The Infusion of Sustainability into Bilateral Fisheries Agreements with Developing Countries: the European Union Example," 670–673; David Symes, "The European Community's Common Fisheries Policy," *Ocean & Coastal Management* 35, no. 2 (January 1, 1997): 137–55; Le Manach et al., "Who Gets What? Developing a More Equitable Framework for EU Fishing Agreements"; Churchill and Owen, *The EC Common Fisheries Policy*, 348.

95 Roman Grynberg, "WTO Fisheries Subsidies Negotiations: Implications for Fisheries Access Arrangements and Sustainable Management," *Marine Policy* 27, no. 6 (November 2003): 499–511; Stefano Ponte, Jesper Raakjaer, and Liam Campling, "Swimming Upstream: Market Access for African Fish Exports in the Context of WTO and EU Negotiations and Regulation," *Development Policy Review* 25, no. 1 (January 1, 2007): 113–38.

96 *Ibid.*

97 Churchill and Owen, *The EC Common Fisheries Policy*.

> The EU has institutionalized particularly challenging regulations in this respect. The basic framework for fisheries products was laid out in EC Directive 91/493 of 1991... It requires Member States and third countries to put in place systems of inspection and control to ensure the safety of fisheries products, including the implementation of Good Hygiene Practices (GHPs) and Hazard Analysis and Critical Control Point (HACCP) systems. Many EU fisheries-specific regulations have now been integrated within what is known as the new EU "hygiene package" of regulations... a bewildering array of rules and demands on the regulatory agencies and exporters in African countries, even without considering private standards on quality, packaging and processing, and the impact of eco-labeling on market access. Given the highly technical nature of compliance and its related costs, *this is probably the area that will pose the most serious challenges for African exporters and their industries and governments* [emphasis added].[98]

Although many health and sanitation standards are reasonable protections for food imports, some EU requirements have been criticized as protectionist and discriminatory to third states. Yet even when standards are not discriminatory, African states still face the administrative, legal, financial, and bureaucratic challenges of creating "competent authorities" with the capacity to ensure compliance with the EU's strict "hygiene package" and import requirements.[99] Considering these obstacles, it is unlikely that many West African states would be able to achieve these standards in the short term without substantial outside assistance.[100]

The second challenge, related to the issue of import standards, is that of processing revenue.[101] A 1999 IFREMER/CEMARE study estimated that EU vessels gained approximately €3 profit for every €1 spent on bilateral fishing agreements with African, Caribbean, and Pacific (ACP) countries simply through the processing and marketing of fish caught in those waters.[102] Since fish caught in the West African EEZs cannot meet EU market standards through local pro-

98 Ponte, Raakjaer, and Campling, "Swimming Upstream: Market Access for African Fish Exports in the Context of WTO and EU Negotiations and Regulation," 131–132.

99 *Ibid.*, 131–132.

100 *Ibid.*, 120–127–128.

101 Churchill and Owen, *The EC Common Fisheries Policy*.

102 Le Manach, "Past, Present and Future of Publicly-Funded European Union's Fishing Access Agreements in Developing Countries"; IFREMER, *Evaluation of the Fisheries Agreements Concluded by the European Community. Summary Report*, 1999.

cessing facilities, fish are shipped to the EU and all revenue from processing and marketing is obtained by European entities.[103] The unabashedly mercantilist slant of this policy is outlined in the European Community publication *European Distant Water Fleet: Some Principles and Some Data*, which claims that "agreements represent 40,650 jobs, over 83 percent of which depend upon 'southern agreements."[104] This second challenge reveals the difficulty that West African states face, not only in accessing revenue through the export of fish to the European market, but also in the intermediate revenue from processing those exported fish.

A third, and less direct, challenge lies in the subsidies that European companies receive, creating an added competitive disadvantage for third states. Millions of dollars in subsidies for fuel, processing, and marketing fish enable tremendous advantages for EU companies, and render the barriers to entry in foreign markets far too strong for competition from developing West African states.[105] Furthermore, these subsidies also have been criticized for enabling unfair advantage within West African waters, by allowing highly efficient EU vessels to outcompete local fishers and decreasing fish availability through high discard rates.[106]

The Conservation Policy and West Africa

In its initial 1983 form, the CFP conservation policy applied conservation measures such as restriction of effort, fishing gear, and fish size to European vessels fishing European waters, and left the conservation of stocks beyond the EU to the domain of UNCLOS.[107] The continued decline of global stocks and rising pressure from environmental groups, however, led to reforms in the early 2000s, which have profound significance for West African fisheries. Particularly,

103 IFREMER, *Evaluation of the Fisheries Agreements Concluded by the European Community. Summary Report*, 1999; Le Manach et al., "Who Gets What? Developing a More Equitable Framework for EU Fishing Agreements."

104 Directorate-General for Fisheries and Maritime Affairs, *European Distant Water Fleet: Some Principles and Some Data*, April 11, 2001.

105 Le Manach, "Past, Present and Future of Publicly-Funded European Union's Fishing Access Agreements in Developing Countries;" Antonius Gagern and Jeroen van den Bergh, "A Critical Review of Fishing Agreements with Tropical Developing Countries," *Marine Policy* 38 (March 2013): 375–86; D. Gascuel et al., "Decline of Demersal Resources in North-West Africa: an Analysis of Mauritanian Trawl-Survey Data Over the Past 25 Years," *African Journal of Marine Science* 29, no. 3 (December 2007): 331–45; Greenpeace, "The Price of Plunder," *Ocean Inquirer*, February 2012.

106 Churchill and Owen, *The EC Common Fisheries Policy*, 349.

107 United Nations, *United Nations Convention on the Law of the Sea (UNCLOS)*.

the Community Action Plan described above "obliges the Community to demonstrate the same commitment to environmental integration in its external activities as internally and requires the Community to adopt a strategy that contributes toward global sustainable fishing."[108] In other words, the regulations of the CFP Conservation Policy have "now broadened from focusing solely on EU waters to cover waters outside the EU in which EU vessels operate."[109]

While the importance of this extension of the CFP conservation policy to third states' waters cannot be underestimated, the implementation of this idea has been questionably successful. To date, European distant water fishing fleets have continued to receive substantial criticism for their environmentally detrimental activities, and West African waters have been particularly singled out for unsustainable practices from high discard rates to destructive fishing methods.[110]

The External Policy and West Africa

The external policy is the policy that has had the most obvious impact on West African states, as it has governed fisheries access agreements since the 1970s. The external policy emerged in order to respond to the growing trend of EEZ declarations, and aimed to maintain European fishing access to coastal waters which would be enclosed in third state EEZs. European demand for fish products, as well as the European fleet's capacity to exploit fish, far outstrip the reproductive capacity of fish stocks within European waters; therefore fishing access agreements were seen as the natural solution to occupy European vessels and supply continued fish to European markets.[111]

While the notion behind access agreements is one of mutual benefit, the combined advantages of EU subsidies, EU consolidated bargaining power, and stronger financial positions in negotiating agreements allowed the European fleet a tremendously asymmetrical position in these agreements, which—with very few exceptions—favored European interests far and above those

108 Commission of the European Communities, *Communication from the Commission Setting Out a Community Action Plan to Integrate Environmental Protection Requirements into the Common Fisheries Policy*, (Brussels, May 28, 2002).

109 Ponte, Raakjaer, and Campling, "Swimming Upstream: Market Access for African Fish Exports in the Context of WTO and EU Negotiations and Regulation," 127.

110 Vlad M. Kaczynski and David L. Fluharty, "European Policies in West Africa: Who Benefits From Fisheries Agreements?" *Marine Policy* 26, no. 2 (2002): 75–93; Ponte, Raakjaer, and Campling, "Swimming Upstream: Market Access for African Fish Exports in the Context of WTO and EU Negotiations and Regulation," 123, 126–128.

111 Kaczynski and Fluharty, "European Policies in West Africa: Who Benefits From Fisheries Agreements?" 76.

of their West African 'partners.'[112] Also, the notion of fisheries access agreements emerges from UNCLOS, which indicates that states that do not have the ability to exploit their fisheries resources to the point of their established TAC are obligated to allocate those fishing rights to other states.[113] Considering the fact that the EU provides approximately €120 million per year to West African countries via fishing agreements, West African states face a strong temptation to declare a TAC regardless of considerable gaps in scientific information regarding habitat, fish stocks, fishing pressure, etc.[114] Furthermore, since sustainability concerns were largely absent in fisheries agreements until the 2002 reform, and West African states generally lack adequate MCS and enforcement capabilities, fishing through access agreements allowed a significantly greater margin of environmental degradation than was possible in European waters.[115]

In general, the fee-for-access fishing agreements between the EU and developing states—especially those in West Africa—were fiercely criticized as exploitative, environmentally unsustainable, and contrary to development objectives.[116] In response to these criticisms, the European Commission insituted a number of policies in the 2002 reforms, including: sustainability impact assessments, stakeholder engagement, coherence between agreements and

112 ADE-PWC-EPU, *Evaluation of the Relationship Between Country Programmes and Fisheries Agreements*, Final Report to the European Commission, November 21, 2002; Council of the European Communities Committee on Fisheries, *Report on International Fisheries Agreements* (*Committee on Fisheries Report*), 1997; Ponte, Raakjaer, and Campling, "Swimming Upstream: Market Access for African Fish Exports in the Context of WTO and EU Negotiations and Regulation," 119–120; Witbooi, "The Infusion of Sustainability into Bilateral Fisheries Agreements with Developing Countries: the European Union Example," 172–174; Alder and Sumaila, "Western Africa: a Fish Basket of Europe Past and Present."; Winter, *Towards Sustainable Fisheries Law*.

113 United Nations, *United Nations Convention on the Law of the Sea* (*UNCLOS*).

114 Witbooi, "The Infusion of Sustainability into Bilateral Fisheries Agreements with Developing Countries: the European Union Example," 170–171. Ponte, Raakjaer, and Campling, "Swimming Upstream: Market Access for African Fish Exports in the Context of WTO and EU Negotiations and Regulation," 129; D.J. Agnew et al., *West Africa Regional Fisheries Project, Estimation of the Cost of Illegal Fishing in West Africa: Final Report*, 3rd ed., May 2010.

115 Witbooi, "The Infusion of Sustainability into Bilateral Fisheries Agreements with Developing Countries: the European Union Example," 670–671; IFREMER, *Evaluation of the Fisheries Agreements Concluded by the European Community. Summary Report*; Le Manach, "Past, Present and Future of Publicly-Funded European Union's Fishing Access Agreements in Developing Countries," 73–87.

116 Acheampong, *Coherence Between EU Fisheries Agreements and EU Development Cooperation*, 3; Witbooi, "The Infusion of Sustainability into Bilateral Fisheries Agreements with Developing Countries: the European Union Example," 170–171.

European policies, coherence between agreements and development policies in partner states, and an overall reform toward holistic "fisheries partnership agreements."[117] Although some scholars suggest that these reforms are, in fact, being progressively implemented,[118] others emphasize the fundamentally contradictory nature of agreements which are designed on the one hand to protect the interests of European fishing fleets, and other the other hand to promote sustainable fisheries in West African waters.[119]

The New CFP and West Africa: Changing Tides or Doldrums?

The European Common Fisheries Policy, discussed in this paper, has led to a host of consequences for the governments and populations of coastal West African states. The combined CFP policies have prevented development of domestic West African fleets, degraded habitat, unsustainably fished stocks, led to conflict between domestic and foreign fishers, handicapped third state marketing and processing entities, and provided meager compensation in comparison to the value of the product.[120] Alder and Sumaila summarize the consequences well, stating that, "the benefits of independence, EEZs, access fees, and management of coastal resources that most western African countries anticipated have not been realized ... In some cases, the negative impact on marine resources has been high compared to the benefits."[121]

Reforms in recent years, especially in 2002, have seen improvements in the CFP language regarding third states, conservation of stocks, destructive fishing methods, and regressive subsidies. Yet despite these changes, many NGOs,

117 Commission of the European Communities, *Communication From the Commission on an Integrated Framework for Fisheries Partnership Agreements with Third Countries*, (Brussels, December 23, 2002); Acheampong, *Coherence Between EU Fisheries Agreements and EU Development Cooperation*, 2–3.

118 Witbooi, "The Infusion of Sustainability into Bilateral Fisheries Agreements with Developing Countries: the European Union Example."

119 Ponte, Raakjaer, and Campling, "Swimming Upstream: Market Access for African Fish Exports in the Context of WTO and EU Negotiations and Regulation, 129;" van der Burgt, *The Contribution of International Fisheries Law to Human Development*, 250–255.

120 Witbooi, "The Infusion of Sustainability into Bilateral Fisheries Agreements with Developing Countries: the European Union Example," 672–675; Ponte, Raakjaer, and Campling, "Swimming Upstream: Market Access for African Fish Exports in the Context of WTO and EU Negotiations and Regulation;" Acheampong, *Coherence Between EU Fisheries Agreements and EU Development Cooperation*; Le Manach et al., "Who Gets What? Developing a More Equitable Framework for EU Fishing Agreements," 3–4.

121 Alder and Sumaila, "Western Africa: a Fish Basket of Europe Past and Present," 172.

journalists, and civil society members have expressed skepticism about the potential of the reforms to alter the current state of fisheries or North-South relations. Furthermore, scholars have emphasized the paradoxical aspects of the CFP, which seems to strive for European protections and profits on one hand, and development and conservation measures for West African waters on the other.[122] In essence, the internal incoherence in the CFP fundamentally emerges from the conflicting interests between short-term EU commercial concerns and long-term global concerns for the sustainability of fisheries.[123]

The New European Common Fisheries Policy

In 2009, the European Commission again announced the intention to conduct a legislative reform of the CFP, and The Green Paper on the Reform of the Common Fisheries Policy was published in April 2009, setting a time frame to address the reform and initiating a public consultation process.[124] The Commission's analysis, based on the Green Paper, "concluded that despite progress since the 2002 reform, the objectives to achieve sustainable fisheries in all its dimensions (environmental, economic and social) have not been met and the Green Paper identified a series of structural shortcomings of the current CFP...The European Parliament and the Council of Ministers supported this conclusion."[125] In July 2011, the Proposal for a Regulation of the European Parliament and of the Council on the Common Fisheries Policy was released, citing several significant problems with the current CFP, including: high discard levels, fleet overcapacity, overfishing, low compliance, poor integration of policies, poor data, poor profitability and market adaptability

122 Acheampong, *Coherence Between EU Fisheries Agreements and EU Development Cooperation*; Agritrade, "Fisheries Council Adopts Conclusions on the External Dimension of the CFP," April 23, 2012, http://agritrade.cta.int/; Long, "The Role of Regional Advisory Councils in the European Common Fisheries Policy: Legal Constraints and Future Options," 299–302.

123 Acheampong, *Coherence Between EU Fisheries Agreements and EU Development Cooperation*; Camilla Adelle, Julia Hertin, and Andrew Jordan, "Sustainable Development 'Outside' the European Union: What Role for Impact Assessment?" *European Environment* 16, no. 2 (March 1, 2006): 57–72; N. Sporrong, C. Coffey, and K. Bevins, *Fisheries Agreements with Third Countries—Is the EU Moving Towards Sustainable Development?*, 2002, 1.

124 European Commission, *Green Paper*, (Brussels, April 22, 2009); Surís-Regueiro, Varela-Lafuente, and Garza-Gil, "Evolution and Perspectives of the Fisheries Structural Policy in the European Union," 593–594.

125 Commission of the European Communities, *Proposal for a Regulation of the European Parliament and of the Council on the Common Fisheries Policy.*

in fleets, and overall unsustainability.[126] The July Proposal outlined a new basic regulation, a reformed market policy, and a communication on the external policy; in December, a proposal was added for a new structural policy, thereby establishing a process of reform for the majority of the components of the CFP.[127] The new market policy and basic regulation were agreed upon in May 2013, and came into effect from January 1, 2014. The new structural policy was agreed upon in October of 2013, and was published in May 2014.[128]

Major components of the EU CFP reform include:

1. *Implementation of Maximum Sustainable Yield* (*MSY*)—Article 2 of the new Basic Regulation established MSY as the new management target for setting TAC levels. This replaces the former approach of using minimum stock biomass and the precautionary principle to set "limits of reference" that may be exceeded for additional payment, and is specifically designed to increase the long term sustainability of the fisheries policy.[129] The timeline for achieving the MSY exploitation rate is 2015 where possible, and "on a progressive, incremental basis at the latest by 2020 for all stocks."[130]
2. *Revision of multi-annual management plans*—Article 9 of the new Basic Regulation revises the existing system of multi-annual management plans to emphasize their importance over annual plans, stress conservation meansures and management by MSY, and include quantifiable targets and clear timelines.[131]

---

126 *Ibid.*

127 Salomon, Markus, and Dross, "Masterstroke or Paper Tiger—the Reform of the EU'S Common Fisheries Policy" 77; European Commission, *Communication From the Commission to the Council Concerning a Consultation on Fishing Opportunities for 2014*, (Brussels, May 30, 2013); European Commission, "Proposal for a Regulation of the European Parliament and of the Council on the Common Fisheries Policy."

128 "Review of the Balance of Competences Between the United Kingdom and the European Union | Fisheries Report," 15–16.

129 Salomon, Markus, and Dross, "Masterstroke or Paper Tiger—the Reform of the EU'S Common Fisheries Policy," 77–78; Frédéric Le Manach et al., "European Union's Public Fishing Access Agreements in Developing Countries," ed. David Hyrenbach, *PLoS ONE* 8, no. 11 (November 27, 2013): 2.

130 Council of the European Communities, *Regulation (EU) No. 12007/13*, 2013.

131 Commission of the European Communities, *Proposal for a Regulation of the European Parliament and of the Council on the Common Fisheries Policy*; Salomon, Markus, and Dross, "Masterstroke or Paper Tiger—the Reform of the EU'S Common Fisheries Policy," 77–78.

3. *The discard ban*—In Article 15, the Basic regulation introduces a gradual implementation of an obligation to land "all catches of species which are subject to catch limits" and in the Mediterranean, "all catches of species which are subject to minimum sizes." Various listed species of pelagic, demersal, and other fish are required to come under the obligation between from 2014–2016, with some exceptions up to January 1, 2019.[132]
4. *Regionalization*—Article 18 creates a framework for regional cooperation around conservation measures and multiannual plans, and enables the creation of joint recommendations by member states to support implementation of the EC's acts, initiate legislation in the EC, and specify EU laws in national terms.[133]
5. *Capacity limitation*—Articles 21 and 22 require publicly available annual reports on the balance between fishing capacity and fishing opportunities. They further set an overall capacity limitation for the European fleet in term of a vessel's tonnage in GT (Gross Tonnage) and its power in kW (Kilowatt), and indicate the potential to establish a systems of transferable fishing concessions, usually in the form of individual transferable quotas (ITQs).[134]
6. *Sustainable Fisheries Partnership Agreements and Surplus Catch*—Article 31 stipulates that SFPAs will replace FPAs as the mechanism of fishing agreements with third states. These SFPAs are meant to emphasize democratic principles and human rights, and stress sustainability, as EU fishing vessels will only catch surplus catch, as described in UNCLOS Article 62.[135]
7. *Exclusivity in partnership agreements*—Article 31 of the Basic Regulation contains an exclusivity clause (5) whereby EU fishing vessels are forbidden to operate in the waters of a partner country unless authorized under the partnership agreement.[136]

132 National Assembly for Wales, *EU Policy Update (EU2013.5)*, October 2013; Council of the European Communities, *Regulation (EU) No. 12007/13*, 2013; Salomon, Markus, and Dross, "Masterstroke or Paper Tiger—the Reform of the EU'S Common Fisheries Policy," 78–79.

133 Commission of the European Communities, *Proposal for a Regulation of the European Parliament and of the Council on the Common Fisheries Policy*; National Assembly for Wales, *EU Policy Update (EU2013.5)*.

134 Commission of the European Communities, *Proposal for a Regulation of the European Parliament and of the Council on the Common Fisheries Policy*; Salomon, Markus, and Dross, "Masterstroke or Paper Tiger—the Reform of the EU'S Common Fisheries Policy," 81.

135 Council of the European Communities, *Regulation (EU) No. 12007/13*.

136 *Ibid.*

8. *New funding mechanism*—Regulation (EU) No. 508/2014 published in May 2014 established the European Maritime and Fisheries Fund (EMFF) the funding mechanism that will replace the current European Fisheries Fund (EFF). According to the EC, "compared to the European Fisheries Fund (EFF), the EMFF brings about a fundamental change of approach to public funding to the fisheries sector through a focus on collective actions and on the viability of coastal areas rather than fleet subsidies benefitting mostly vessel owners."[137]

Impacts and Implications for West Africa

Throughout the reform process, West African states have paid close attention to the European CFP negotiations. Although they have no formal vote within the EU, these States have lobbied ally states to advocate their positions.[138] In the build up to the reform, Greenpeace organized a campaign to bring West African fishermen to key locations in Europe to communicate the ways in which industrial European vessels impact their lives and communities.[139] While the consequences of the CFP reform will likely emerge in both predictable and unanticipated ways, it is worthwhile to conduct a preliminary analysis of the key reforms through the lens of West Africa.

With the establishment of MSY as the management target under the CFP, the EU has determined that a biologically relevant metric will determine the allowable catch under European juridiction. A similar reform outlined in Article 31(4) determined that Union fishing vessels shall only fish surplus stocks—where MSY is not fully exploited—in third country waters. While these reforms may seem like an encouraging shift toward conservation objectives, they can also be understood as a simple codification of existing international law into European terms.[140] In the Preamble, the Basic Reform states that:

> [existing] international instruments predominantly lay down conservation obligations, including obligations to take conservation and

137 European Commission, *The Social Dimension of the CFP Reform*, 2012; Till Markus, "Towards Sustainable Fisheries Subsidies: Entering a New Round of Reform Under the Common Fisheries Policy," *Marine Policy* 34, no. 6 (November 2010): 1117–24.

138 John Vidal and Fiona Harvey, "UK Steps in to Help West Africa in Fight to Overturn EU Fishing Abuses," *The Guardian*, March 18, 2012.

139 National Assembly for Wales, *EU Policy Update (EU2013.5)*.

140 Churchill and Owen, *The EC Common Fisheries Policy*; United Nations, *United Nations Convention on the Law of the Sea (UNCLOS)*.

> management measures designed to maintain or restore marine resources at levels which can produce the maximum sustainable yield both within sea areas under national jurisdiction and on the high seas, and to cooperate with other States to that end ... The CFP should, therefore, contribute to the Union's implementation of its international obligations under those international instruments.

While this codification may in fact result in very specific management and policy changes for European waters, the chances for those reforms to heavily impact West African fisheries remain slim for a number of reasons. First, the scientific information needed to set MSY, and therefore surplus, is rarely available for most West African stocks, especially on an annual basis.[141] Second, where such information is available, MSY is only pertinent in fisheries that set a TAC and have the ability to enforce it through limitations on entry or effort. Furthermore, while CFP jurisdiction extends to Union fishing vessels outside Union waters and nationals of EU member states, MSY as a management tool is only effective if applied to the fishery as a whole, not individual vessels comprising a fraction of the fishing effort. Although Article 31 states that surplus catch is dependent upon the "relevant information exchanged between the Union and the third country about the total fishing effort on the affected stocks by all fleets," the significant amount of illegal industrial fishing and unreported artisanal fishing conducted in West Africa serves to undermine any meaningful estimates of total fishing effort.[142] Furthermore, due to the asymmetrical negotiating power between the EU and West African states, where surpluses are calculated, they may play a reduced or controversial role in negotiation of agreements. The current negotiations between the EU and Mauritania were suspended on October 10, 2014 due to the parties' inability to resolve quote issues on stocks determined to be overexploited.[143] These same constraints are

---

141 Le Manach et al., "Who Gets What? Developing a More Equitable Framework for EU Fishing Agreements," 6–7.

142 IFREMER, *Evaluation of the Fisheries Agreements Concluded by the European Community. Summary Report*; Dyhia Belhabib et al., "Fisheries Catch Misreporting and Its Implications: the Case of Senegal," *Fisheries Research* 151 (March 2014): 1–11; Food and Agriculture Organization of the United Nations, *Report of the FAO Regional Workshop in the Elaboration of National Plans of Action to Prevent, Deter and Eliminate Illegal, Unreported and Unregulated Fishing—West African Subregion*, FAO Fisheries Report, (Rome: Food & Agriculture Org., 2006); David J. Agnew et al., "Estimating the Worldwide Extent of Illegal Fishing," ed. Stuart A. Sandin, *PLoS ONE* 4, no. 2 (2009).

143 Salomon, Markus, and Dross, "Masterstroke or Paper Tiger—the Reform of the EU'S Common Fisheries Policy," 82; Le Manach et al., "European Union's Public Fishing Access

also critical to the multi-annual management plan reforms, and similar challenges apply for the discard ban.

Article 15, outlining the discard ban, or landing obligation, states that the landing obligation will apply to "all catches of species *which are subject to catch limits*... caught during fishing activities in Union waters or by Union fishing vessels outside Union waters in waters *not subject to third countries' sovereignty or jurisdiction*" [emphasis added]. In other words, like the MSY and multi-annual plan reforms, the obligation to land discards would only apply to species that are already subject to determined catch limits, a management approach that relies heavily on scientific information, monitoring, and enforcement. Since only a small fraction of species caught as bycatch have determined legal catch limits, it can be understood that the discard ban would not apply to the tremendously high number of remaining fish, crustaceans, sharks, turtles, birds, and marine mammals.[144] It would also not apply to EU vessels fishing in third countries' EEZs, such as those in West Africa, though Article 31(3) does indicate that "the Union shall endeavor to include in Sustainable fisheries partnership agreements appropriate provisions on obligations to land fish and fishery products."[145]

Two other reforms, fleet capacity limitation and Sustainable Fisheries Partnership Agreements, also initially appear to reduce the historically negative impacts that European fleets have had in West Africa. As of the implementation of the reforms in January 2014, SFPAs replaced FPAs as the access agreement mechanism with third countries, with the objective to "establish a legal, environmental, economic and social governance framework for fishing activities carried out by Union fishing vessels in third country waters.[146] The language included in the Article is more aspirational than definitive or binding, however, calling for the Union to "endeavor to ensure... mutual benefit", "endeavor to include... obligations to land fish,"[147] and "include a clause concerning respect for democratic principles and human rights."[148] In the context of low capacity, monitoring, and enforcement that usually exists

Agreements in Developing Countries," 1; Commission of the European Communities, "Fisheries Partnership Agreement: EU-Mauritania Negotiations Suspended," *Ec.Europa. Eu*, October 10, 2014.

144 Churchill and Owen, *The EC Common Fisheries Policy*.

145 Council of the European Communities, *Regulation (EU) No. 12007/13*.

146 Commission of the European Communities, *Proposal for a Regulation of the European Parliament and of the Council on the Common Fisheries Policy*. Article 31(1).

147 *Ibid.*, Article 31(2–3).

148 *Ibid.*, Article 31(6).

in West African fisheries, soft language rarely begets concrete reform. The capacity limitation reform is similarly constructed, with Articles 21 and 22 calling for publicly available annual reports on the balance between fishing capacity and fishing opportunities and an overall limitation for the European fleet in terms of both gross tonnage and power.[149] Yet since the fishing capacity ceiling set out in Annex II sets the limit at the current levels of effort in the European fleet, no actual reduction in capacity is required. Furthermore, Article 22 also states that, "separate assessments shall be drawn up for fleets operating in the outermost regions and for vessels operating exclusively outside Union waters," indicating that the assessment of balance between capacity and fishing opportunities will be determined separately for third state waters such as those in West Africa.[150]

These policies, despite constituting the major components of the EU CFP reform, appear to hold little promise for altering relations with third states such as those on the Western coast of Africa. Regionalization, one of the reforms to garner the most excitement following the 2009 Green Paper, is also unlikely to impact West African fisheries, since it is primarily concerned with establishing a framework for regional cooperation *between EU member states*, and does not bestow an increase of powers on regions outside the European Union.

Two of the changes in the reform, however, merit closer examination for their potential implications in West Africa. Article 31 contains an exclusivity clause, which states that, "Union fishing vessels shall not operate in the waters of the third country with which a Sustainable fisheries partnership agreement is in force unless they are in possession of a fishing authorisation which has been issued in accordance with that agreement."[151] Prior to the current reforms, fewer than half of the European Union's external fleet had gained fishing access through an EU-negotiated Fisheries Partnership Agreement, while the majority had fished under privately negotiated joint ventures.[152] The new reform addresses this issue by prohibiting EU vessels from operating in the waters of a partnership country unless under the terms of the FPA and by complicating the terms under which vessels may reflag and return to the EU fishing fleet registry.[153] The terms of this exclusivity clause are currently under

149 *Ibid.*

150 *Ibid.*

151 *Ibid.*

152 Worldwide Fund for Nature (WWF), *WWF Response to the 2009 Green Paper*, ed. Helen Mclachlan, November 2009.

153 Council of the European Communities, *Regulation (EU) No. 12007/13*. Article 31(9).

examination in a European Court of Justice case questioning the agreement between Morocco and the European Union, and will likely have a major bearing on interpretation of Article 31 in the future.[154]

The second reform that merits consideration is the implementation of the European Maritime and Fisheries Fund (EMFF) that will replace the European Fisheries Fund (EFF) as the funding mechanism of the CFP. According to the EC, "compared to the European Fisheries Fund (EFF), the EMFF brings about a fundamental change of approach to public funding to the fisheries sector through a focus on collective actions and on the viability of coastal areas rather than fleet subsidies benefitting mostly vessel owners."[155] While much of the terminology of the EMFF remains loyal to its predecessors, the reforms determined that EMFF funds would not be allowed to subsidize the construction of new vessels. While this may present a strong positive outcome for West Africa, which has historically absorbed so much of the European fleet's overcapacity, the reforms fall short of prohibiting "capacity-enhancing" subsidies such as those that would modernize fishing gear and vessel technology.[156]

Conclusion

Despite the momentum that surrounded the EU CFP reform process,[157] a preliminary assessment of the major reforms promises little in way of altered relations between Europe and West Africa. While the EMFF reforms and the exclusivity clause indicate potential shifts in the way fisheries relations are negotiated in the region, the overall emphasis of reforms appears unequivocally domestic. Though some scholars suggest that the lack of emphasis on the

---

154 Case C-565/13: Request for a preliminary ruling from the Hovrätten för västra Sverige (Sweden) lodged on 4 November 2013—Kammaråklagaren v Ove Ahlström, Lennart Kjellberg, Fiskeri A.B. Ganthi and Fiskeri A.B. Nordic, 1–2 (n.d.).

155 Till Markus, "Towards Sustainable Fisheries Subsidies: Entering a New Round of Reform Under the Common Fisheries Policy," *Marine Policy* 34, no. 6 (November 2010): 1117–24; European Commission, *The Social Dimension of the CFP Reform*.

156 Agritrade, "EU Institutions Agree on European Fisheries Aid 2014–2020," March 19, 2014, http://agritrade.cta.int/; Salomon, Markus, and Dross, "Masterstroke or Paper Tiger—the Reform of the EU'S Common Fisheries Policy" 81–83; Greenpeace, "Parliament Fights Overfishing by Voting to Reform Subsidies," October 23, 2013, http://www.greenpeace.org/eu-unit/en/News/2013/Parliament-fights-overfishing-by-voting-to-reform-subsidies-/.

157 Le Manach et al., "European Union's Public Fishing Access Agreements in Developing Countries," 7.

external policy within the 2013 reforms was intentional,[158] early indications from leadership—including a statement by the European Commissioner for Maritime Affairs and Fisheries that the external dimension was "an intrinsic part of the reform"[159]—suggested the possibility of significant change.

Recent reforms seem to indicate little in the way of lasting consequences for West African waters, however many scholars have redirected their hopes for a change in relations away from the EU and toward West African domestic and regional policies.[160] Some scholars have suggested that the strongest potential for improvement lies in regionally-coordinated negotiation of agreements, improved scientific data, development of domestic fishing fleets, linking of access fees to resource rents, and insistence on the part of West African states that EU fleets land catches within the region, rather than within the EU or via transshipment at sea.[161] Furthermore, Salomon suggests that despite its shortcomings, the recent reform may have normative power, raising standards in European waters and in turn "strengthen[ing] the rule of law in international fisheries management."[162] The ultimate consequences of the EU CFP reforms will not be decided in the next year or even the next five years. Despite the recent reforms' limited attention to external relations, the European Community has demonstrated a growing understanding of the impacts of its policies throughout the world, and the trajectory of relations with third states is positive, if the speed is less than hoped for. While the EU remains a major force in global fisheries, the policies of other regions, such as West Africa, are of critical and increasing importance, and global cooperation will be necessary to achieve the strides needed to improve the state of global fisheries.

---

158 Salomon, Markus, and Dross, "Masterstroke or Paper Tiger—the Reform of the EU'S Common Fisheries Policy," 82.

159 Agritrade, "The External Dimension of the New CFP Is an 'Intrinsic Part of the Reform'," January 19, 2012, http://agritrade.cta.int/.

160 Surís-Regueiro, Varela-Lafuente, and Garza-Gil, "Evolution and Perspectives of the Fisheries Structural Policy in the European Union," 596–600; Le Manach et al., "Who Gets What? Developing a More Equitable Framework for EU Fishing Agreements," 6–9.

161 Acheampong, *Coherence Between EU Fisheries Agreements and EU Development Cooperation*, 15–21; Ponte, Raakjaer, and Campling, "Swimming Upstream: Market Access for African Fish Exports in the Context of WTO and EU Negotiations and Regulation," 133–134; Le Manach et al., "Who Gets What? Developing a More Equitable Framework for EU Fishing Agreements" 6–9; Le Manach, "Past, Present and Future of Publicly-Funded European Union's Fishing Access Agreements in Developing Countries."

162 Salomon, Markus, and Dross, "Masterstroke or Paper Tiger—the Reform of the EU's Common Fisheries Policy," 83.

CHAPTER 3

International Law at the Convergence of Two Epochs: Sea-Level Rise and the Law of the Sea for the Anthropocene

*Davor Vidas**

Introduction

Scientific evidence increasingly indicates that the Earth is undergoing a shift from the conditions of the most recent officially accepted geological time interval, the Holocene, to a new planetary state.[1] The Holocene is the latest, and formally still current, geological epoch. It comprises the past 11,700 years,[2] which have been marked by an exceptionally long period of relative environmental stability—including, for the past 6000 to 7000 years, generally stable sea levels. As such, the conditions of the Holocene were a key factor facilitating the development of human civilization and, ultimately, the emergence of the territorial division of the political world as we know it today.

---

* Research Professor in International Law and Director of the Law of the Sea Programme at the Fridtjof Nansen Institute, Norway. He is the Chair of the ILA Committee on International Law and Sea Level Rise, and a member of the Anthropocene Working Group of the International Commission on Stratigraphy. As of 2016, Professor Vidas is the Editor-in-Chief of a new journal monograph series, Brill Research Perspectives in the Law of the Sea.

This chapter is based on the presentation by the author at the LOSI Madrid Conference on 19 September 2014, which drew on the article by the author, on "Sea-Level Rise and International Law: At the Convergence of Two Epochs," *Climate Law*, 4 (2014): 70–84. For purposes of this chapter, findings contained in that article were supplemented and up-dated with the findings resulting from the author's more recent research, especially as contained in a review article: Davor Vidas *et al.*, "International Law for the Anthropocene?," *Anthropocene*, 9 (2015): 1–13.

1 See especially the recent article by the members of the Anthropocene Working Group: Colin Waters *et al.*, "The Anthropocene is functionally and stratigraphically distinct from the Holocene," *Science*, 351 (2016): 137.

2 Mike Walker *et al.*, "Formal Definition and Dating of the GSSP (Global Stratotype Section and Point) for the Base of the Holocene Using the Greenland NGRIP Ice Core, and Selected Auxiliary Records," *Journal of Quaternary Science*, 24 (2009): 3–17.

 | DOI 10.1163/9789004311442_005

Scientists increasingly argue, however, that because of the global environmental effects of economic development and increased human population, the Earth system has already left the Holocene and has entered a new epoch: the "Anthropocene."[3] Since its popularization in the early 2000s by Paul Crutzen (Nobel Prize laureate for his work on the atmospheric ozone),[4] the Anthropocene concept spread rapidly and became widely accepted as an unofficial scientific term referring to the human imprint on the Earth system, which is already regarded as so profound as to have reached geological magnitude. Since 2009, this concept has been under formal scrutiny within the discipline of geology: the Anthropocene Working Group (AWG) was then formed within the International Commission on Stratigraphy in order to examine the stratigraphic basis and the scientific justification for, and the utility of, possible formalization of the Anthropocene as the most recent geological time unit.[5] The preliminary findings and interim recommendations of the AWG are expected in the second half of 2018.

What is fundamentally new in the Anthropocene concept is its focus on the role of humans in the destabilization of the Earth system,[6] and not just the human impact on the environment, as in various earlier approaches.[7] Unlike the Holocene, which has been characterized, especially in its later stages, by the longest overall stability in environmental conditions on the Earth since the appearance of *Homo sapiens*, the Anthropocene is seen as characterized by change, uncertainty and, probably, considerable instability in

---

3 On the onset of the Anthropocene, see: Jan Zalasiewicz *et al.*, "When did the Anthropocene Begin? A Mid-Twentieth Century Boundary Level is Stratigraphically Optimal," *Quaternary International*, 383 (2015): 196–203.

4 See especially: Paul Crutzen and Eugene Stoermer, "The Anthropocene," *Global Change Newsletter*, 41 (2000): 17–18; and Paul Crutzen, "Geology of Mankind," *Nature*, 415 (2002): 23.

5 On the International Commission on Stratigraphy, see at: www.stratigraphy.org. On the Anthropocene Working Group, see at: http://quaternary.stratigraphy.org/workinggroups/anthropocene/.

6 On the magnitude of that destabilization, see also a recent article: Andrey Ganopolski *et al.*, "Critical insolation-CO2 Relation for Diagnosing Past and Future Glacial Inception," *Nature*, 529 (2016): 200–203—reporting a modelling study confirming findings that the Earth would already be entering a glacial interval (an "ice age") absent human impact on the CO2 concentration in the atmosphere from the 18th century on, and concluding that the onset of a glacial interval is, due to that impact, postponed by at least 100,000 years.

7 Clive Hamilton and Jacques Grinevald, "Was the Anthropocene Anticipated?," *The Anthropocene Review*, 2 (2015): 59–72.

the future behavior of the Earth system.[8] This perspective may have increasingly high relevance and, over time, may generate important consequences for the organization of international relations as currently regulated by international law.

This outcome is so because in the fundamentals of today's international law, stability operates not only at the abstract level of the political world that we have created, but also at the level of the (natural) conditions of the Earth system. The familiar level of understanding of "stability" in international law concerns the first aspect only: it relates to the conscious objective of working towards legally guaranteed stability in international relations, which themselves are vulnerable to frequent political change. The other level of stability is, however, implied: it is based on human experience of the generally stable environmental conditions of the late Holocene. Changes in that underlying element of stability contain the potential for new types of tension in relations between states.

One such core aspect relates to climate change and the resultant sea-level rise, as projected already for this century. This perspective may question some key aspects of international law, i.e., those that rely on the general stability of geographic conditions. Coastal geography, due to its perceived stability, serves as the objective circumstance on the basis of which the rights of states to maritime zones are determined, and maritime delimitation disputes resolved. A defined territory is indeed a constituent criterion of statehood under international law. In the not-too-distant future, important questions may arise about the sustainability of those aspects of international law, and changes may likely require a profound re-examination of various currently accepted paradigms of international law.

This chapter briefly addresses, first, the relevance of the Anthropocene perspective for current international law in general; and second, the Anthropocene manifestation of sea-level rise and its possible impact on the architecture of the law of the sea as it developed over the past centuries and as we know it today.

8 Jan Zalasiewicz, Paul Crutzen and Will Steffen, "The Anthropocene," in Felix Gradstein *et al.* (eds.), *The Geologic Time Scale 2012*, Vol. 2 (Amsterdam: Elsevier, 2012), 1033–1040; and Mark Williams *et al.*, "The Anthropocene Biosphere," *The Anthropocene Review*, 2 (2015): 196–219.

International Law at the Convergence of Two Epochs

Spheres of Tension in International Law of the Holocene

Modern international law, especially as it has developed since the mid-20th century, has been marked by two main spheres of constant tension. First, the tension between sovereign equality of states as the founding postulate of international law on the one hand, juxtaposed, on the other hand, with the political, military, strategic, economic and other differences between states. This tension between the sovereign equality of states and their geopolitical differences often has been presented as function of the "horizontal" nature of international law, and in counterpoint with the "vertical" reality of international politics due to the many differences among states.[9]

The second sphere of tension lies between the legal guarantees of territorial integrity of sovereign states on the one hand, and claims, nurtured by a myriad of factors from economic to ethnic, in the political division of the world, on the other. This sphere of tension stems from the fact that the world is territorially divided into many parts, each under sovereignty of a distinct state—currently some 200—delimited by interstate boundaries (albeit incomplete delimitations in some cases). The adjoining maritime areas of states with their own coasts are divided as well. What remains beyond is the high seas (which is one part of a water column under the sovereignty or sovereign rights of no state) and one part of the *seabed*, also lying beyond the limits of national jurisdiction (the international seabed area).[10] Competing claims to territories, land and maritime, have abounded through history, often leading to wars or to territorial changes based on political power. Therefore, facilitating the peaceful settlement of territorial disputes based on the rules of international law is among the highest historic achievements of human civilization.

In both the foregoing spheres of tension, the objective of international law has been to facilitate political stability in inter-state relations through international cooperation, and to maintain international peace and security.[11]

---

9 See, e.g., Richard Falk, *(Re)Imagining Humane Global Governance* (London: Routledge, 2014), at 87.

10 The Antarctic, defined as the area south of 60° S latitude, has a unique status under the 1959 Antarctic Treaty, which is the legal basis of the Antarctic Treaty System applying to this area; see Olav Schram Stokke and Davor Vidas (eds.), *Governing the Antarctic: The Effectiveness and the Legitimacy of the Antarctic Treaty System* (Cambridge University Press, 1996).

11 See UN Charter, Article 1(1) and (3).

The end of the Holocene and the onset of the Anthropocene will likely introduce a third, fundamentally different and profoundly new type of tension.

The Anthropocene Tension: The Earth System under Human-Induced Change

Showing relative stability throughout recent human history, the underlying condition of the Earth system has been taken as a given—and upon that premise our political structures were created in the course of the past centuries. The relationship of international law with the observed geographical features, and indeed with the overall geological dimension of the Earth system, has generally been confined to an implicit assumption about the undetermined, long-term horizon of current conditions—as an objective circumstance surrounding us since time immemorial. We are accustomed to understanding any change in the Earth system as a matter of geological time, with relevant events stretching through many millennia, or millions (if not billions) of years—as opposed to politically relevant time and the related pace of change, whether embodied in national election horizons, pre-war and post-war periods, newly emerging or dissolving world alliances, the adoption and entry into force of international treaties, creation of international organizations, or other milestones on that scale.

Many aspects of international law are based on such understanding of the stability of the Earth conditions.[12] Indeed, the definition of current international law is, in many respects, that of a system of rules resting on foundations that evolved under the circumstances of the late Holocene, assumed to be everlasting. International law takes the conditions of the Holocene for granted, and on that premise a huge edifice of international law has been constructed over the past several centuries. The conditions of the Anthropocene will bring a fundamental shift of the context in which international law operates—a shift in which the challenges are increasingly recognized as the consequences of natural, not only political, change. The change introduced in that underlying element of stability—and that is what the transition from Holocene to the Anthropocene conditions involves—contains the potential for an unprecedented type of tension in relations between states. This dynamic can spill

12 Exceptions found in international law to the overall conditions of stability are also based on experience: the changing courses of rivers (and the related rules on interstate boundaries in such situations); or the changing coastal geography in some large river deltas (and the related rules in the law of the sea).

over into and aggravate existing tensions in the regulation of inter-state relations under international law, between the territorial integrity of states and territorial claims—coupled with the fact of immense geopolitical differences, on the one hand, and sovereign equality of states as the founding postulate of international law, on the other.

International Law at the End of the Holocene

Various aspects of international law rely on an implicit pretext of the familiar conditions of stability. For example, a defined territory—indeed, *having* a territory at all—is a basic criterion of statehood under international law (as codified in Article 1 of the 1933 Montevideo Convention on Rights and Duties of States). The perspective of sea-level rise is but one of many symptoms in the outlook of an Anthropocene world; and challenges to the continuity of statehood of some low-lying small island states are bound to emerge. Moreover, the conceptual architecture of the law of the sea relies on the general stability of coastal geography and the resultant baselines, which serve as the key objective circumstance for determining the extent of various maritime zones of states, and for resolving maritime delimitation disputes. In the not-too-distant future, important questions may therefore arise, requiring re-examination of some presently accepted paradigms of international law.

With the challenges for international law increasingly recognized as the consequences of natural, and not only political change, new legal axioms will have to evolve. If international law is to be able to meet the new challenges of changing circumstances and achieve its overarching objectives of facilitating international cooperation and maintaining international peace and stability, humankind may have to organize society in the Anthropocene epoch differently from what we have known so far. This transformation will have to embrace some key principles of the international law architecture—with geography firmly and deeply embedded in its core.

In that broad framework of *our* needs and purposes looming on an Anthropocene horizon, the primacy given in international law to the assertion of sovereign rights over territory may have to be reconsidered, while the emphasis on population and human rights may have to gain in prominence and find expression in new forms of international law subjectivity. As international law rests on thick sediments of political power and accumulated vested interests, the challenges in changing the current course are deep-reaching, and involve many difficult questions. Whether—or, better, when—these issues will be raised in practice will depend on the changing severity of challenges over time.

International Law for the Anthropocene? An Outlook

With the possible formalization a new geological epoch and, indeed, the onset of the Anthropocene conditions, international law may become the subject of particular scrutiny for two reasons. First, certain segments of international law—the law of the sea in particular, but more generally the rules on the acquisition of territory—involve historical and ideological causal links with the development towards Anthropocene conditions.[13] Second, core aspects of international law rely on the stability of Holocene conditions, which have been implied as permanent—logically enough, given our experience thus far—but are now subject to reevaluation as being transitory, rather than permanent.

The Anthropocene contains in two ways the potential for raising profound questions as to the validity of currently accepted international law. The first way is a shorter-term perspective: the formalization of the Anthropocene as a new geological time-unit in the history of the Earth, ratified through due scientific process in stratigraphy, may significantly contribute to awareness-raising, prompting an increased focus on the implications for international law. The second aspect of implications for international law is directly related to the consequences of the changing conditions in the Anthropocene. Here the perspective is a longer-term one, even if some of the changing conditions, such as sea-level rise—although still uncertain as to the pace and magnitude—may become serious already in the course of the current century.

The complex and difficult issues that international law will need to face in responding to the phenomenon of sea-level rise in the context of the law of the sea—a matter on which the rest of this chapter focuses more closely—illustrate the dilemmas involved for international law at the end of the Holocene and onset of the Anthropocene conditions.

The International Law of the Sea and Sea-Level Rise

An Outline of the Development Leading to the Present Legal Framework

The development of the Law of the Sea as a branch of international law spans several centuries, ever since the overseas discoveries by Spain and Portugal in the late 15th and early 16th centuries, and the emergence of the new maritime

13 Davor Vidas, "The Anthropocene and the International Law of the Sea," *Philosophical Transactions of the Royal Society–A*, 369 (2011): 909–925.

powers of England and the Netherlands in the second half of the 16th and early 17th centuries. After the Second World War, the United States grew into a political and economic force decisively impacting law of the sea developments, prompted largely by technological advances. In the 1960s and 1970s, the demands of many developing states, including newly created ones, in turn led to the Third UN Conference on the Law of the Sea (UNCLOS III), from 1973–1982.

Much of this centuries-long development ultimately resulted in the single law of the sea framework: the 1982 UN Convention on the Law of the Sea (UNCLOS, or LOS Convention).[14] This international Convention came into force in 1994; it contains both codification of customary norms and contribution to the progressive development of international law.[15] The Convention regulates how seas and oceans are to be divided among coastal states (currently, some 150 states) in various maritime zones, including the territorial sea, the exclusive economic zone and the continental shelf. Coastal state sovereignty extends to the outer limit of the territorial sea; in EEZs and the continental shelf, certain sovereign rights and exclusive jurisdiction of the coastal state apply. Beyond those zones is the high seas, where all states (including land-locked ones)—and all ships flying their flags—enjoy freedom of the seas. Further, the seabed beyond the limits of coastal state jurisdiction is in certain respects, along with its resources, an international area regulated as the common heritage of mankind.

The LOS Convention is often referred to as the *Constitution* of the oceans: a legal framework that governs all major issues of the entire ocean space. Despite some initial difficulties in attracting broader support, the Convention is as of August 2, 2016 binding on 168 parties (167 states and the EU).[16] On the one hand, there is a broad political consensus that the LOS Convention contains the legal framework the integrity of which needs to be maintained. On the other hand, the pace of change is tremendous: population trends, advances

14 UN doc. A/CONF.62/122; text in *United Nations Treaty Series*, 1833, at 3; text reprinted in *International Legal Materials*, 21 (1982): 1261; available at <www.un.org/Depts/los>. The Convention was opened for signature on 10 December 1982, and entered into force on 16 November 1994.

15 In the past three decades, many rules contained in the LOS Convention grew into customary law, too; see further especially: Tullio Treves, "The Development of the Law of the Sea Since the Adoption of the UN Convention on the Law of the Sea: Achievements and Challenges for the Future," in Davor Vidas (ed.), *Law, Technology and Science for Oceans in Globalisation* (Boston/Leiden: Brill/Martinus Nijhoff, 2010), 41–58.

16 Beyond the EU, several parties to the LOS Convention are non-members of the UN: Cook Islands, Holy See, Niue, and the State of Palestine.

in technology, scientific achievements—all this with resultant uses of, and impacts on, the seas. The law of the sea, centred on the LOS Convention, is to some extent flexible and adjustable as a framework for future regulation, but is also deeply rooted in earlier developments.

The law of the sea has gradually developed as an exponent of two main driving forces, and the consequences of both are now captured in the LOS Convention. One driving force has been that of territorial appropriation of the seas. In some periods, territorial claims were reduced to a relatively narrow belt of the sea near the coast. In the post-Second World War period, however, the territorial driving force returned in the form of claims for segments of sovereignty—sovereign rights and exclusive jurisdiction—to be extended beyond the outer limits of the territorial sea.[17] These claims related to fisheries resources in vast areas of the water column, as well as to the mineral (primarily oil) resources of the continental shelf.[18] In consequence, there are now maritime zones in which, although found beyond the coastal state's (land and maritime) territory,[19] sovereign rights and exclusive jurisdiction of the coastal state apply: the EEZ and the continental shelf.

The other driving force in the development of the law of the sea was that of economic profit by functional as opposed to territorial access; and of securing strategic gains of naval powers in distant sea areas. Both were promoted by the concept of the "freedom of the seas," as initially conceived in Hugo Grotius' *Mare Liberum* of 1609.[20] This concept came to serve as an ideological platform for securing unimpeded international trade for emerging (and, thereafter, established) maritime powers, with the dual goals of maximizing

17 See especially Bernard Oxman, "The Territorial Temptation: A Siren Song at Sea," *American Journal of International Law*, 100 (2006): 830–851. The enormous complexity of the long-term historical background of offshore claims is analyzed in the classic treatise by Stefan A. Riesenfeld, *The Protection of Coastal Fisheries under International Law* (Washington: Carnegie Endowment, 1942).

18 The claims were contained in two unilateral acts issued by the United States in 1945: the US Presidential Proclamation No. 2667 and No. 2668.

19 In contrast to the EEZ and the continental shelf, the territorial sea is a part of state's territory.

20 Hugo Grotius, *Mare Liberum, sive de jure, quod Batavis competit ad Indicana commercia, dissertatio* (Leiden: Elzevir, 1609) (first published anonymously in April 1609, with the identity of the author disclosed later). For English translations see, respectively: David Armitage (ed.), *The Free Sea*, translated by Richard Hakluyt (the Younger), (Indianapolis: Liberty Fund, 2004); Robert Fenstra (ed.), *Hugo Grotius Mare Liberum 1609–2009* (Leiden/Boston: Brill, 2009); and James Brown Scott (ed.), *The Freedom of the Seas*, translated by Ralph van Deman Magofinn (New York: Oxford University Press, 1916).

profits for their economies and increasing their strategic dominance over new territories.[21]

The ideology launched in the early 17th century, expounded in *Mare Liberum*—arguing for the freedom of navigation in order to secure participation in international trade and, ultimately, accumulation of profit from overseas ventures—facilitated the development of the forces that were to lead to the Industrial Revolution and eventually, in the course of the 20th century, to the levels of development that have resulted in ever-greater human impacts on the Earth system.

The end of the Second World War in 1945 was also the time when the last "tectonic change" in the law of the sea began. Out of that war came various new technologies—many of which represented new applications for fossil fuels—and a commitment by governments in some industrially advanced countries to subsidize research and development. Only months after the war ended, the United States initiated articulation of the continental shelf as a concept of international law, in which geological information has played a crucial role. We still can observe the final outcomes: the current process of determination of the outer limits of the continental shelf beyond 200 nautical miles, facilitated by the work of the Commission on the Limits of the Continental Shelf.

The linkages between the development of the law of the sea and the onset of the Anthropocene epoch may be seen as two-fold.[22] First, there is a linkage of origin. The ideological foundations of the law of the sea, especially as found in *Mare Liberum* of the early 17th century, involve "deep-time" origins for the later processes, which have ultimately brought about the Anthropocene. Second, there may be a renewed linkage in interaction. Geological information has prompted key developments since the introduction of the continental shelf as a political and then international law concept in the mid-20th century; in the early 21st century geology—now in connection with the Anthropocene concept—may again provide an impetus for new law of the sea trends.

This process is now directing our attention to the changing ocean component of the Earth system in the Anthropocene. Consequences for the law of the sea may be seen in two main directions. First, there already are serious consequences for the biological and chemical conditions of the oceans across

21 For an in-depth study see: Martine Julia van Ittersum, *Profit and Principle: Hugo Grotius, Natural Rights Theories and the Rise of Dutch Power in the East Indies (1595–1615)* (Boston/Leiden: Brill, 2006).

22 Vidas, "The Anthropocene and the International Law of the Sea," 910.

different legal maritime zones.[23] Second, there are consequences of physical nature on the horizon, impacting on how the limits of various maritime zones will be determined in the foreseeable future.

Sea-Level Rise: Projections for the 21st Century

A rise in sea levels is one of the most certain outcomes of a warmer world. The complex role of the oceans in the climate system is now well acknowledged: the seas are a crucial heat buffer and absorb a significant share of the CO_2 released into the atmosphere. The seemingly insignificant increase in sea temperature in the past century already has led to global warming of the upper ocean layer. Water expands as it heats, contributing to sea-level rise. Globally, average sea level has risen by around 20 cm over the past century, with acceleration in the past few decades. A recent study of historical sea-level trends confirmed that the rise in sea levels over the past century is unmatched by any period under study in the past 6000 years.[24] Moreover, the acceleration, as observed in the last decades, is projected to continue over this century (and to carry on increasing).[25] The major impact, in the perspective of global warming, concerns the destabilization and gradual melting of Greenland and West Antarctic ice sheets, with studies confirming early signs already.[26]

However, a number of recent scientific reports differ as to the actual figures for sea-level rise, even for the present century. The results of process-based modelling as used for projections by the Intergovernmental Panel on Climate Change (IPCC) and those of semi-empirical model projections differ significantly, with the latter being almost twice the size of the former. The span of difference is illustrated by a review of the recent scientific literature on sea-level rise, which indicated that, in a world that has warmed by 4°C by the year 2100,

---

23 For further discussion of that aspect, which is not in the focus of this chapter, see: Davor Vidas, "Responsibility for the Seas," in Vidas (ed.), *Law, Technology and Science for Oceans in Globalisation*, 3–40.

24 Kurt Lambeck *et al.*, "Sea Level and Global Ice Volumes from the Last Glacial Maximum to the Holocene," PNAS, 111 (2014): 15296–15303.

25 Jonathan D. Woodruff *et al.*, "Coastal Flooding by Tropical Cyclones and Sea-Level Rise," *Nature*, 504 (2013): 44–52.

26 See reports: Intergovernmental Panel on Climate Change, *Climate Change 2014. Synthesis Report: Summary for Policymakers*, available at: https://www.ipcc.ch/pdf/assessment-report/ar5/syr/AR5_SYR_FINAL_SPM.pdf; and the World Bank, *Turn Down the Heat. Confronting the New Climate Normal* (Washington, DC: World Bank, 2014). See also studies: Carling Hay *et al.*, "Probabilistic Reanalysis of Twentieth-Century Sea-Level Rise," *Nature*, 517 (2015): 481–484; and Craig Rye *et al.*, "Rapid Sea-Level Rise along the Antarctic Margins in Response to Increased Glacial Discharge," *Nature Geoscience*, 7 (2014): 732–735.

the rise in global sea levels is estimated to become between 0.5 and 2 metres.[27] Will global warming eventually reach 2°C or perhaps 4°C above pre-industrial levels by the end of this century? Notwithstanding political declarations, there is no certainty today, and so estimates of sea-level rise differ significantly.

On the basis of currently available scientific knowledge, it can be concluded that a significant increase in sea levels can be expected in the present century. The issue, however, is not only about the amount of sea-level rise, but very much also about the *pace* of the phenomenon. Today, satellite technology has made remote sensing possible and thus enabled comparisons of sea-level rise with any recent reference year since 1993. Moreover, geological information derived from the Earth's past may also prove highly relevant. The key point here is that while sea-level rise is often seen as a phenomenon that progresses slowly, gradually, and linearly over a very long period, there is geological evidence to the contrary: some sea-level rises of the past have proceeded with significant "jumps" over a relatively short time-span of centennial—perhaps even decadal—scale,[28] followed by longer periods of still-stand or slower rise. The timing of future "jumps" is very difficult to predict.

While a measure of scientific uncertainty persists, increasing importance is attributed in more recent years to sea-level rise as an expected consequence of climate change. This finding is clear already from a simple comparison of the two latest IPCC Assessment Reports—issued in 2007 and in 2013. This regards both the extent to which these reports focused on the issue of sea-level rise, and the projections of sea-level rise presented. The most recent IPCC estimates indicate sea-level rise of up to 98 cm by 2100—a considerable increase over the previous, 2007, projections of up to 59 cm.[29] As explained by the IPCC in its latest report, the discrepancy is due primarily to new and improved modelling of land-ice contributions.[30]

---

27 Robert J. Nicholls *et al.*, "Sea-level Rise and Its Possible Impacts Given a 'Beyond 4 C World' in the Twenty-First Century," *Philosophical Transactions of the Royal Society—A*, 369 (2011): 161–181. However, as the authors indicate, rises greater than 1 m are deemed less likely.

28 Paul Blanchon and John Shaw, "Reef Drowning during the Last Deglaciation: Evidence for Catastrophic Sea-level Rise and Ice-sheet Collapse," *Geology*, 23 (1995): 4–8; see also Michael J. O'Leary *et al.*, "Ice Sheet Collapse Following a Prolonged Period of Stable Sea Level during the Last Interglacial," *Nature Geoscience* 9 (2013): 796–800.

29 Compare figures in the IPCC's 4th Assessment Report (*Climate Change 2007: Synthesis Report—Summary for Policymakers*, at 8) with those in the IPCCs 5th Assessment Report (IPCC, *Climate Change 2013: The Physical Science Basis—Summary for Policymakers*, October 2013, at 23).

30 IPCC, *Climate Change 2013*, at 18.

During the deliberations in UNCLOS III, held in the course of 1970s and concluded in the early 1980s, the negotiators of the Convention, judging from the content of its provisions, could have hardly foreseen substantial changes in coastal geography caused by a major natural phenomenon such as sea-level rise.[31] However, scientific studies available today contain projections of a significant increase in sea levels already in the present century. While the exact amount and pace of sea-level rise are uncertain, the projected range indicates that challenges for the law of the sea are bound to emerge.

International Law and the Perspective of Sea-Level Rise: Central Issues[32]

Since core aspects of current international law rest on geographical conditions that are generally perceived as relatively stable, a significant change in those conditions—as a consequence of sea-level rise—could come to pose major challenges to the sustainability of important aspects in our present-day system of international law. The severity of the challenges would depend on the progression of sea-level rise. Initially, drastic consequences would affect only some low-lying states—in particular, several Pacific and Indian Ocean island-states.

Naturally, the first part of international law to be affected by sea-level rise is the law of the sea, which regulates the maritime entitlements of states and the delimitation of their maritime zones. With rising sea levels, the baselines from which the breadth of the territorial sea is measured will move landward, affecting the outer limits of various maritime zones. Ultimately, sea-level rise may call into question the entire structure of the maritime zones under today's law of the sea. However, issues related to sea-level rise extend far beyond the law of the sea and the determination of baselines and maritime zones of coastal states. Even the early cases of affected low-lying island states, although limited in number, will necessarily give rise to fundamental questions regarding aspects of state territory and the principles of statehood under international law, as well as myriad questions raised by the forced transboundary migration of their populations.

31 The "Bangladesh exception"—Article 7(2) of the LOS Convention, applying primarily to highly unstable coastlines caused by river deltas and the related impact on straight baselines—is of limited reach and is targeted to specific situations. *Argumentum a contrario*, it confirms the main rule as further described below.

32 This chapter is limited to discussion of overall international law issues related to sea-level rise, including central questions of the law of the sea. For a discussion of relevant coastal management aspects and some specific law of the sea issues, see further in the Chapter 18 by Kerrylee Rogers and Clive Schofield, in this volume.

These challenges go deeper than simply questioning the appropriateness of certain existing rules of international law. The nature of the challenges is such that they call into question some basic axioms of international law. In the following, only a cursory overview of prominent examples is provided. The first two examples relate to essential aspects of the law-of-the-sea architecture, while the third concerns the principal basis of the architecture of international law as we know it today: the state itself.

The Land Dominates the Sea—By the Intermediary of the Coastal Front

The core axiom of the law of the sea, explaining the philosophy behind the maritime entitlements of the coastal states, is well captured in this observation by a leading expert:

> From the moment States were recognised as having rights over areas of sea... these rights have been based on two principles which have acquired an almost idiomatic force: the land dominates the sea and it dominates it by the intermediary of the coastal front.[33]

Accordingly, the basis for all maritime coastal zones as today codified in the LOS Convention is just one line—a line that is, in each case, determined by reliance on coastal geography. That line is the *baseline*. It either directly follows the coast, in which case it is called the "normal baseline;" or it depends on the specific configuration of the coastline and other coastal features (a chain of islands, fjords, and the like), in which case it is called the "straight" baseline, and everything landward of it is considered the internal waters of a coastal state. From the baselines (whether normal or straight), different maritime zones are measured. This objective criterion, which relies on a given coastal geography, serves not only as the basis for the various maritime zones of a coastal state, it is also central to the delimitation of maritime boundaries between states—to which we return below. The purpose of those rules of international law of the sea is to maintain certainty. The effect of the consequences of sea-level rise will be to introduce uncertainty, and increasingly exacerbate it.

---

33 Prosper Weil, "Geographic Considerations in Maritime Delimitation," in Jonathan I. Charney and Lewis M. Alexander (eds.), *International Maritime Boundaries*, Vol. 1 (Dordrecht: Martinus Nijhoff, 1993), at 115, drawing on: Prosper Weil, *The Law of Maritime Delimitation—Reflections* (Cambridge University Press, 1989), at 50.

In 2012, the Committee on Baselines of the International Law Association concluded:

> the normal baseline is ambulatory, moving seaward to reflect changes to the coast caused by accretion, land rise, and the construction of human-made structures ... and also landward to reflect changes caused by erosion and sea level rise. Under extreme circumstances, the latter category of change could result in total territorial loss and the consequent total loss of baselines and of the maritime zones measured from those baselines. The existing law of the normal baseline does not offer an adequate solution to this potentially serious problem.[34]

Attention has been drawn to this dilemma and the resultant issues in the international law literature since the late 1980s and early 1990s.[35] The LOS Convention does not provide a readily available solution: its negotiators apparently did not foresee substantial changes in coastal geography caused by a major natural phenomenon such as sea-level rise.[36] There is a further aspect of the LOS Convention to be kept in mind; as the UN Secretary-General appropriately remarked:

---

34 ILA, 75th Conference (Sofia, 2012), *Final Report of the International Committee on Baselines under the International Law of the Sea*, at 31. Text available at the ILA website, at <www.ila-hq.org/en/committees/index.cfm/cid/1028>.

35 See especially David D. Caron, "When Law Makes Climate Change Worse: Rethinking the Law of Baselines in Light of a Rising Sea Level," *Ecology Law Quarterly*, 17 (1990): 621–653; David Freestone, "International Law and Sea Level Rise," in Robin Churchill and David Freestone (eds.), *International Law and Global Climate Change* (Dordrecht: Martinus Nijhoff, 1991), 109–125; and Alfred H.A. Soons, "The Effects of a Rising Sea Level on Maritime Limits and Boundaries," *Netherlands International Law Review*, 37 (1990): 207–232. See also Eric Bird and Victor Prescott, "Rising Global Sea Levels and National Maritime Claims," *Marine Policy Reports*, 1 (1989): 177–196. An updated overview of the international law literature on this issue is provided in ILA, *Final Report of the International Committee on Baselines*, especially at 29–31.

36 As already noted, the "Bangladesh exception"—contained in Article 7(2) of the LOS Convention, and applying primarily to highly unstable coastlines caused by river deltas and the related impact on straight baselines—is of a limited reach, targeted to specific situations. On the negotiating history of Article 7(2) see Satya N. Nandan and Shabtai Rosenne (eds.), *United Nations Convention on the Law of the Sea, 1982: A Commentary*, vol. 2 (Dordrecht: Martinus Nijhoff, 1993), 97–101, stating that "the provisions in paragraph 2 were drafted with the specific case of the Ganges/Brahmaputra River delta in mind" (at 101).

> UNCLOS was not negotiated to correct geographical circumstances. To compensate partially for the latter, the Convention provides adequate remedies for situations where States are at a disadvantage.[37]

However, no remedies for the consequences of sea-level rise can be found in the Convention: it was tailored to the geographical circumstances of its own time, not the ones yet to come. In the 21st century, we may be facing the circumstances of that new, different time.

Objective and Predictable Criteria in the Determination of Maritime Boundaries[38]

Less than half of the world's maritime boundaries have been delimited to date. Unlike terrestrial delimitation, maritime delimitation involves, in addition to the settlement of the boundaries of sovereignty (internal waters and the territorial sea), also the settlement of the boundaries of states' other maritime zones, those conferring sovereign rights and jurisdiction, including continental shelves, EEZs, and some other zones of functional jurisdiction. As a consequence of this multiplicity of zones, the maritime political map of the world has remained profoundly incomplete—in fact, the number of maritime delimitation disputes is increasing as new issues emerge.

The role of international law is of the utmost importance in (maritime) boundary delimitation, as it enables the determination of boundaries based on the rule of law instead of recourse to force or political action. Reliance on international law thus emerges as a key factor in facilitating long-term stability in relations between neighboring states, as well as in entire regions. The basic consideration in maritime delimitation under international law is the importance accorded to neutral, objective, legal criteria, to enable predictability, together with an appreciation of the specific circumstances of each case, to achieve an equitable solution. By offering predictability and balance, the principles and rules of international law, consolidated through an

37 *Oceans and the Law of the Sea: Report of the Secretary-General*, UN Doc. A/59/62 of 4 March 2004, para. 41.

38 For a more complete treatment see Davor Vidas, "Consolidation or Deviation? On Trends and Challenges in the Settlement of Maritime Delimitation Disputes by International Courts and Tribunals," in Nerina Boschiero, Tullio Scovazzi, Cesare Pitea and Chiara Ragni (eds.), *International Courts and the Development of International Law: Essays in Honour of Tullio Treves* (The Hague: TMC Asser Press/Springer, 2013), 323–332, on which this section partly draws.

increasingly consistent judicial and arbitral practice, have significant potential as a factor for stability.

In a series of judgments over the past twenty to twenty-five years, the International Court of Justice has consolidated the importance of predictable, objectively determined criteria for delimitation, as opposed to subjective views. Reliance on the actual geographical features in each case has been repeatedly confirmed by the ICJ. For example, in its 2002 judgment in the case of the land and maritime boundary between Cameroon and Nigeria, the Court stated:

> The geographical configuration of the maritime areas that the Court is called upon to delimit is a given. It is not an element open to modification by the Court but a fact on the basis of which the Court must effect the delimitation.[39]

The 2006 Award in the Barbados/Trinidad and Tobago arbitration[40] contributed to this overall trend by providing a highly authoritative and clear restatement of the development of the law governing the delimitation of maritime boundaries. The Arbitral Tribunal observed:

> [t]he search for an approach that would accommodate both the need for predictability and stability within the rule of law and the need for flexibility in the outcome that could meet the requirements of equity resulted in the identification of a variety of criteria and methods of delimitation.[41]

The Tribunal then stated that with very few exceptions, "the quest for neutral criteria of a geographical character prevailed in the end over area-specific criteria such as geomorphological aspects or resource-specific criteria such as the distribution of fish stocks."[42] In that respect, geographical circumstances have been repeatedly accorded a predominant role by the ICJ and arbitral tribunals.

---

39 *Land and Maritime Boundary between Cameroon and Nigeria*, ICJ Judgment of 10 October 2002, para. 295.

40 *Award of the Arbitral Tribunal in the Matter of an Arbitration between Barbados and the Republic of the Trinidad and Tobago*, The Hague, 11 April 2006; text available at <www.pca-cpa.org/upload/files/Final%20Award.pdf>; reprinted in *International Legal Materials*, 45 (2006) 789–869.

41 *Ibid.*, paras. 221 and 232.

42 *Ibid.*, para. 228.

This trend has led to an increasingly consistent international juridical practice, in which other factors, if justified by specific circumstances, may also play a role, albeit a lesser one.

A possible consequence of projected sea-level rise—reversing the hitherto predictable and relatively stable characteristics attributed to coastal geography, so as to become a less stable factor and, ultimately, practically meaningless—will be the potential of major destabilization in inter-state relations. The fundamental purpose of the law of maritime delimitation is to facilitate expectations of certainty, or predictability, hence of stability. The consequences of climate change and sea-level rise could seriously undermine this objective.

Territorial Foundation of (the State in) International Law

Our present-day international legal world (still "Westphalian," as some term it) rests heavily on geography[43]—with state-based territory as a central feature of international law. This rule is clearly seen from Article 1 of the 1933 Montevideo Convention on Rights and Duties of States,[44] which contains the best-known formulation of the effectiveness criteria for statehood:[45] "(a) a permanent population; (b) a defined territory; (c) government; and (d) capacity to enter into relations with the other States." In addition to these essential definitional criteria of statehood, various other principles and rules of international law are also territorially based. Indeed, as succinctly observed by Daniel Bethlehem, geography stands at the very core of current international law, in which it is deeply embedded[46]—from specific rules, to the most fundamental principles.

While the theory of the creation of a state has been thoroughly developed in international law doctrine,[47] questions surrounding the termination of a state—often termed "state extinction"—are as a rule dealt with briefly and associated with the criteria on which the existence of a state can be

---

43 See Daniel Bethlehem, "The End of Geography: The Changing Nature of the International System and the Challenge to International Law," *The European Journal of International Law*, 25 (2014): 9–24.

44 *League of Nations Treaty Series*, 165, at 19.

45 See James Crawford, "State," in Rüdiger Wolfrum (ed.), *Max Planck Encyclopedia of Public International Law* (Oxford University Press, 2013), on-line edition, at para. 13; entry updated as of January 2011, available at <http://opil.ouplaw.com>.

46 Bethlehem, "The End of Geography," at 14.

47 See especially James Crawford, *The Creation of States in International Law*, 2nd ed. (Oxford University Press, 2006).

determined.[48] The logic behind this approach is that the "extinction of States must be determined by reference to [the] same criteria… on which the existence of a state can be determined."[49] However, Ziemele observes that state practice and doctrinal development on state extinction show that this logic can provide only a "starting point for further examination in the search for a modern concept of the notion."[50]

A conceptual change might, as has been indicated above, come about for the first time as the consequence of a natural phenomenon, such as sea-level rise. Ziemele refers to a basic rule applicable in determining state "extinction" in the following:

> Changes affecting the basic criteria of statehood such as territory, population, and government, separately or together, do not automatically affect the existence of a State *unless territory and/or population disappears*.[51]

In relation to the prospects of sea-level rise, Rosemary Rayfuse made a pioneering contribution to the literature of international law in a series of research articles;[52] and a comprehensive study on the subject has more recently come from Derek Wong.[53] Differences in their views can be seen as representing two streams of opinion in the literature on this issue.[54] Rayfuse observes that, with

48 For instance, in Crawford's 700-plus-page volume (ibid.), one of the shortest sections is on "The Extinction of States," to which altogether 18 pages are devoted.

49 See Ineta Ziemele, "States, Extinction of," in Rüdiger Wolfrum (ed.), *Max Planck Encyclopedia of Public International Law* (Oxford University Press, 2013), on-line edition, at para. 1; entry up-dated as of May 2007, available at <http://opil.ouplaw.com>.

50 *Ibid.*

51 *Ibid.*, at para. 3 (emphasis added).

52 See especially Rosemary Rayfuse, "W(h)ither Tuvalu? International Law and Disappearing States," *University of New South Wales Faculty of Law Research Series*, Paper 9, 2009 (based on a paper presented at the International Symposium on Islands and Oceans, Tokyo, January 2009); Rayfuse, "International Law and Disappearing States: Maritime Zones and the Criteria for Statehood," *Environmental Policy and Law*, 41 (2011): 281–287; and Rayfuse, "Sea Level Rise and Maritime Zones: Preserving the Maritime Entitlements of 'Disappearing' States," in Michael B. Gerrard and Gregory E. Wannier (eds.), *Threatened Island Nations: Legal Implications of Rising Seas and a Changing Climate* (Cambridge University Press, 2013), at 187–191.

53 Derek Wong, "Sovereignty Sunk? The Position of 'Sinking States' at International Law," *Melbourne Journal of International Law*, 14 (2013): 346–391.

54 Other notable contributions to the literature include: Maxine Burkett, "The Nation Ex-Situ: On Climate Change, Deterritorialized Nationhood, and the Post-Climate Era," *Climate Law*, 2 (2011): 345–374; Susin Park, "Climate Change and the Risk of Statelessness: The Situation

full inundation, "the criterion of territory will no longer be met and the claim to statehood will fail."[55] Further, in connection with the observation that the territory will become uninhabitable long before its total physical disappearance, she observes that in that case, too, "the criteria for statehood will cease to be met from the time of evacuation and the State will cease to exist."[56] Wong, however, considers this analysis problematic, since it assumes that failure to satisfy the requirements for the creation of states automatically results in the state's extinction.[57] He stresses that there is "a strong presumption in favor of the continued existence of a state."[58]

While this discussion may be based on the precedents of the past, on the use of analogies, and on differences in the interpretation of today's legal rules—the phenomenon of sea-level rise is something looming on the horizon of the future. It may be the change of context, rather than the change or interpretation of rules themselves, that will initially be controlling and ought to be examined first.

Core Dilemmas in Seeking a Response to Sea-Level Rise under the Law of the Sea

Seeking to find a solution to the perspective of sea-level rise facing the law of the sea, most law of the sea experts have so far proposed the development of a new rule of international law that would have the effect of either freezing the baselines, or permanently fixing the (outer) boundaries of maritime zones at today's status—by fixing them on a chart, or some other display.[59] In other words, proposals for dealing with sea-level rise aim at preserving a static legal

---

of Low-lying Island States," *UNHCR Legal and Protection Policy Series*, May 2011; and Jenny Grote Stoutenburg, "When Do States Disappear? Thresholds of Effective Statehood and the Continued Recognition of 'Deterritorialized' Island States," in Gerrard and Wannier (eds.), *Threatened Island Nations*, at 57–87. A most comprehensive discussion of that matter is, however, provided in: Alejandra Torres Camprubí, *Statehood under Water: Challenges of Sea-Level Rise to the Continuity of Pacific Island States* (Boston/Leiden: Brill/Martinus Nijhoff, 2016).

55 Rayfuse, "International Law and Disappearing States," at 284.

56 *Ibid.*

57 Wong, "Sovereignty Sunk?," at 360.

58 *Ibid.*, at 360 and 362.

59 See, e.g., Caron, "When Law Makes Climate Change Worse" (1990 and 2009); Soons, "The Effects of a Rising Sea Level"; José Luís Jesus, "Rocks, New-born Islands, Sea Level Rise and Maritime Space," in Jochen Abr. Frowein *et al.* (eds.), *Verhandeln für den Frieden. Negotiating for Peace: Liber Amicorum Tono Eitel* (Berlin: Springer, 2003), 579–603; Moritaka Hayashi, "Sea-Level Rise and the Law of the Sea: Future Options," in Davor Vidas (ed.), *The World Ocean in Globalisation* (Boston/Leiden: Brill/Martinus Nijhoff, 2011), 187–206.

situation in the face of an increasingly dynamic process of natural change. This quandary illustrates a core dilemma that is set to emerge: On the one hand, the natural processes of convergence between the Holocene and Anthropocene conditions will require a response or transformation, including the development of new legal axioms, in accordance with the needs and purposes in the new situation—rather than responding by analogy or precedent based on the earlier situation, no longer valid. On the other hand, however, the international legal order will always be in search of stability and, ultimately, solutions to facilitate peace and prevent conflict, therefore requiring gradual changes. In the case of the law of the sea and the methods for determining baselines and their domino effect on other maritime zones, a gradual solution between the "freezing" of current legal status and permanent fluctuation of the limits may be found in a new coordinating mechanism devised specifically with the effects of sea-level rise on baselines or, possibly, maritime zones in mind. The process of establishing a coordinating mechanism of that sort, as well as its composition and competencies, remains to be clarified.

The Law of the Sea at the End of the Holocene

The foundations of today's law of the sea are basically the product of centuries of often-antagonistic struggles among and between dominant human forces. These forces have produced impressive technological capabilities and made possible the modern way of life in industrialized societies, but they also seem to have contributed substantially towards threatening the stable conditions of the Holocene—and bringing about our entry into the Anthropocene.

The law of the sea is today a well-developed legal system, addressing human uses of the seas and oceans, organized through various maritime zones where the rights and duties of states (and the many stakeholders involved) are carefully balanced. The law of the sea system aims at facilitating stable international relations regarding the maritime spaces and their uses, as well as promoting the peaceful settlement of disputes that may arise.

Evolving over centuries and resulting in the current legal framework that emerged in the second half of the 20th century, the contemporary law of the sea is aimed primarily at addressing the changing political and economic circumstances—not changes in the overall natural conditions as well. This is a legal system tailored to the circumstances of the Holocene, implicitly held to be permanently valid.

With the profoundly different circumstances on the Anthropocene horizon, responding to the challenges for the law of the sea may increasingly involve more than merely amending or adjusting the rules of individual treaties, or adding new ones. The very foundations of this legal system and its current parameters may need re-evaluation. Certain cornerstone concepts of the law of the sea architecture, like the rules pertaining to baselines and the consequent determination of the maritime zones, may have to undergo thorough scrutiny and be supplemented by new rules—and with new coordinating mechanisms to enable their implementation.

Concluding Remarks: International Law Entering a New Epoch

The perspective of sea-level rise is not just one more new legal case to be solved. It is not a problem that can be dealt with merely by adjusting certain rules or adopting amendments to this or that international treaty. It is not a political controversy *per se*, either. It is a major natural phenomenon, unprecedented since the emergence of modern human civilization and the development of its organizing forms as currently regulated through a territorially based and state-centred system of international law. The context for that change in natural conditions is provided by the onset of the new epoch in the geological history of the planet, the Anthropocene, and by the gradual end of the familiar conditions of stability as prevailing in the Holocene.[60]

As we leave the relative stability of the Holocene and enter the more uncertain and less stable epoch of the Anthropocene, international law will need to adapt to new parameters provided by the factual circumstances. Under the new conditions likely to prevail in the Anthropocene of tomorrow, including the perspective of sea-level rise, the loss of state territory could be attributed to changes in the factual situation of natural conditions, and not only in consequence of political changes as so far. Here we are not discussing primarily questions of law but questions of fact—in the face of which law, if it proves unsuited to the factual situation, may become irrelevant or even absurd.

Many proposals on how to address the issues of sea-level rise aim at preserving a static legal situation in the face of an increasingly dynamic process of natural change. While proposals of that sort may be useful as temporary or transitory measures, the on-going processes of change in natural conditions will ultimately require a legal response or transformation in accordance with the needs in the newly created situation.

60 Waters *et al.*, "The Anthropocene is Functionally and Stratigraphically Distinct from the Holocene," 137.

The lawyer's instinct, or training, is to invoke precedents and search for analogies, in an attempt to prove or maintain a certain legal situation. However, the phenomenon discussed here is unprecedented—it has never been experienced since the invention of international law and the territorially-centred state. There are no suitable analogies available. Drawing on precedents of, or analogies with, the situation of the Holocene in order to solve quite a different situation in an Anthropocene context cannot yield any meaningful solution—at least not so in a longer-term perspective.

The international legal order will always be in search of stability and, ultimately, solutions to facilitate peace and prevent conflict. However, with a fundamental change of the context in which international law operates—as with the change of the conditions of the Holocene to those of the Anthropocene—new legal axioms will have to evolve. In the face of fundamentally changing facts, no forcing of presumptions, no invention of unsuitable analogies, and no artificial fixing of a "permanent" legal situation can in a longer run produce the ultimate objectives of international law: stability and peace.

It is international law, not the facts that will have to change. That transformation must embrace the fundamental principles of this legal system—in the core of which geography is firmly and deeply embedded.[61] If international law is to be adequate to meet the new challenges of changing circumstances and achieve the overall objectives of facilitating stability and peace, humankind will have to organize society differently, in line with its actual needs and purposes. The Anthropocene horizon, as illustrated here on the canvas of sea-level rise prospects, introduces a general dilemma that international law may need to face in the foreseeable future. To put it briefly here: *If the Earth is entering the Anthropocene, will international law stay in the Holocene*?

61 See Bethlehem, "The End of Geography," at 14.

CHAPTER 4

The Legal Protection of Underwater Cultural Heritage: Concerns and Proposals

*Mariano J. Aznar**

Introduction

On 1 November 2001, the UNESCO Convention for the Protection of Underwater Cultural Heritage (UNESCO Convention) was adopted.[1] As its preamble states, one of its aims is "to codify and progressively develop rules relating to the protection and preservation of underwater cultural heritage in conformity with international law and practice," including particularly the United Nations Convention on the Law of the Sea of 10 December 1982 (UNCLOS).[2] During the negotiating process and when voting its adoption, however, several States voiced some concerns,[3] particularly some maritime powers.[4]

---

* Professor of Public International Law, University Jaume I, Spain. Email: maznar@uji.es. The author would like to thank Sarah Dromgoole and Ole Varmer for their most valuable comments on an earlier draft of this article. The views expressed are however solely those of the author, as are any errors.

1 2562 UNTS 1, entered into force 2 January 2009 ("UNESCO Convention" or "the Convention" hereinafter). As for July 2016, the UNESCO Convention has 55 States parties (see the list *infra* n. 31). For the more recent analysis of the Convention and its legal implications see S. Dromgoole, *Underwater Cultural Heritage and International Law* (Cambridge: Cambridge University Press, 2013). For an analysis of the drafting of the Convention see R. Garabello, *La Convenzione UNESCO sulla protezione del patrimonio culturale subacqueo* (Milan: Giuffrè Editore, 2004).

2 Only two UNCLOS articles—149 and 303—deal with the "archaeological and historical objects" found at sea. For the text of both articles, see *infra* nn. 28 and 29.

3 See the position of most of these States as expressed in the final voting of the Convention in UNESCO, *Records of the General Conference, Proceedings of the 31st Session*, 2 November 2001, vol. 2, at 561–570 (available electronically at <http://unesdoc.unesco.org/images/0012/001289/128966m.pdf>, accessed 21 October 2014). Some abstaining States changed their initial position and decided to become a party (Paraguay and France) or have initiated the domestic process to become a party (the Netherlands and Germany).

4 See S. Dromgoole, "Reflections on the Position of the Major Maritime Powers with Respect to the UNESCO Convention on the Protection of the Underwater Cultural Heritage 2001," *Marine Policy* 38 (2013): 116–123; and O. Varmer, J. Gray and D. Alberg, "United States: Responses

 | DOI 10.1163/9789004311442_006

These concerns—which impeded (and still impede) the ratification of the Convention by some of these States—mostly relate two issues:[5] (1) how the UNESCO Convention affects the legal status of sunken State vessels (particularly warships); and (2) the impact that this Convention may have upon the jurisdictional equilibrium created by UNCLOS.

The concerns imply an assessment of the relationship between two successive treaties relating to a similar subject matter: the protection of underwater cultural heritage. Whereas article 303(4) of UNCLOS establishes that "[t]his article is without prejudice to other international agreements and rules of international law regarding the protection of objects of an archaeological and historical nature," article 3 of the UNESCO Convention states that:

> Nothing in this Convention shall prejudice the rights, jurisdiction and duties of States under international law, including the United Nations Convention on the Law of the Sea. This Convention shall be interpreted and applied in the context of and in a manner consistent with international law, including the United Nations Convention on the Law of the Sea.

Given that this last article seems to specify that the UNESCO Convention be subject to (or that it is not to be applied in a manner incompatible with) international law and UNCLOS, the provisions of the latter instrument prevail.[6] The problem is that customary law of the sea as codified by UNCLOS does not completely answer—at least explicitly—the two concerns posed by those reluctant States.

---

to the 2001 UNESCO Convention on the Protection of the Underwater Cultural Heritage," *Journal of Maritime Archaeology* 5 (2010): 129–141. See also the complete appraisal of the US laws and policies, a gap analysis in the context of the Convention and recommendations on how to fill those gaps in O. Varmer, "Closing the Gaps in Protection of Underwater Cultural Heritage on the Outer Continental Shelf," *Stanford Environmental Law Journal* 33 (2014): 251–286.

5 The UK remained also concerned about the Convention applying to all underwater traces of human existence—the so-called 'blanket protection' concern—which will not be however addressed in this paper. On this concern, see Dromgoole, *Underwater…, supra* n. 1, 93–94 and P.J. O'Keefe, *Shipwrecked Heritage. A Commentary on the UNESCO Convention on underwater cultural heritage* (London: Institute of Art and Law, 2002) at 27.

6 Art. 30(2), 1969 Vienna Convention on the Law of Treaties (1155 UNTS 331). Being "preserved" by Art. 303(4) UNCLOS, the "hierarchy" between the UNESCO Convention and UNCLOS is not affected by Art. 311 of the later.

This chapter revisits these two concerns, and elucidates the issues raised by those States, and addresses them in order to clarify the legal impact of the Convention and its consistency with international law including UNCLOS. This may, consequently, help to facilitate or ease the future ratification or acceptance of those reluctant States. The chapter also addresses possible implied solutions given by UNCLOS to those States neither able nor willing to become parties to the UNESCO Convention but desiring to adequately protect the underwater cultural heritage located in the maritime zones under their sovereignty or jurisdiction.[7]

The Main Concerns about the UNESCO Convention

The two main legal concerns about the UNESCO Convention were advanced during the negotiation process. Some States still persist and avoid accession to the Convention, although others have accepted its main principles and the rules of its Annex as best archaeological practice.[8]

The Treatment of Sunken State Vessels

The concern around the treatment of sunken State vessels arises from article 7(3) of the UNESCO Convention. This proviso reads as follows:

7 This paper will not, therefore, deal with the underwater cultural heritage located beyond the limits of national jurisdiction.

8 The majority of States have accepted the main principles of the Convention including consideration of *in situ* preservation as an initial policy preference and the rules of its Annex as an international standard or best practice for when it is determined that intrusive research and recovery is necessary or appropriate. Among the "reluctant" States, the UK has applauded the general principles of the text and, particularly, subscribed to the Norms annexed to the Convention. Actually, the British Government "has adopted the annex of the Convention as best practice for archaeology" (*Hansard Written Answers*, 24 January 2005, Col. 46W). See the last application of this policy in the Guidance "Protection and Management of Historic Military Wrecks outside UK Territorial Waters," April 2014, of the UK Department for Culture, Media & Sport and the UK Ministry of Defence (available electronically at <https://www.gov.uk/government/publications/protection-and-management-of-historic-military-wrecks-outside-uk-territorial-waters>, accessed 25 October 2014).

The United States, for its part, during the Second Meeting of the States Parties to the Convention, had the occasion to "re-affirm its support of the overall goal of this UNESCO Convention to protect underwater cultural heritage. The United States fully supports the Annex of Rules concerning activities directed at underwater cultural heritage." UNESCO Doc. UCH/09/2.MSP/220/4 REV., Annex, at 12.

> Within their archipelagic waters and territorial sea, in the exercise of their sovereignty and in recognition of general practice among States, States Parties, with a view to cooperating on the best methods of protecting State vessels and aircraft, *should* inform the flag State Party to this Convention and, if applicable, other States with a verifiable link, especially a cultural, historical or archaeological link, with respect to the discovery of such identifiable State vessels and Aircraft. [emphasis added]

The conditional tense finally included—"should" instead of "shall"—was the result of two opposing views during the negotiating process. The latter term was defended by historical and current maritime powers, which understood that a coastal State *must* respect the ownership and sovereign immunity of a foreign sunken State craft and communicate the discovery in its territorial sea (or archipelagic waters) of a sunken State vessel to its flag State.[9] Against this position, a significant number of coastal States maintained that nothing in international law obliges a State to inform another State of any circumstance that occurs within its territory unless otherwise agreed to or imposed by an international law rule.

To some extent, the conflicting relationship between sovereignty and property was behind the disagreement. To some States—most particularly the Latin American States—the location of the wreck was tantamount to inuring title to the coastal State. For example, Latin American and Caribbean countries asserted that underwater cultural heritage "is the property of the State in which it is found and through this it is the heritage of the Humanity."[10] This position was firmly opposed by those States defending the principle under which, according to international law, sunken State vessels—such as

---

9 Unless the name of vessel or flag State is readily apparent, this process would not appear to be trigged until evidence of the flag State is identified after what may be a lengthy and possibly complicated archaeological research process.

10 Santo Domingo Declaration, 16 June 1998, which resulted from the First Meeting of the Technical Commission on Underwater Cultural Heritage of the Forum of Ministers of Culture and Officials Responsible for Cultural Policy of Latin America and the Caribbean, reprinted in UNESCO, *Background Materials on the Protection of the Underwater Cultural Heritage* (vol. II, Paris, 2000), at 341–2.

However, a line of change could be identified from 2008 onwards, when the Meeting of Ministries of Culture of the *Conferencia Iberoamericana* adopted the Declaration of San Salvador (22–23 May 2008), in which it was decided to adopt a common position regarding the protection and conservation of underwater cultural heritage in the region and to call for a commission of experts to follow up on the question (para. 12). See the Declaration (in Spanish) at <http://www.oei.es/xicic.htm>, accessed 2 October 2014.

warships and vessels on government service—regardless of location or of the time elapsed, remain the property of the State owning them at the time of their sinking unless it explicitly and formally relinquishes ownership.[11]

The concern of some States, particularly the United States, is that the UNESCO Convention does not explicitly address or recognize "the regime under which the flag State must consent before its vessels can be the subject of recovery."[12] This issue includes concerns about respect of ownership and the sovereign immunity of sunken State vessels located in the internal waters and territorial waters of a coastal State.[13] Article 7(3) of the UNESCO Convention, however, does not automatically mean that the Convention does not respect the flag State consent regime or that ownership or title lies with the coastal State. On the contrary: the flag State consent regime and sovereign immunity are recognized throughout the Convention and the continuing title of the flag State is expressly provided for in article 2(8), under which the Convention does not modify "the rules of international law and State practice pertaining to sovereign immunities, nor any State's rights with respect to its State vessels and aircraft."

---

11 See this position as defended by the US, the UK, Spain, Japan, Germany, France or Russia at *Federal Register*, Vol. 69, No. 24, Thursday, February 5, 2004, Notices, at 5647–8. During the negotiation of the UNESCO Convention, a proposal submitted by the UK and Russia (and endorsed by the United States) tried to substitute "should inform" to "shall consult" and to add the following sentence: "such State vessels and aircraft shall not be recovered without the collaboration of the flag State, unless the vessel and aircraft have been expressly abandoned in accordance with the laws of that State." The proposal was rejected (see the proposal in UNESCO Doc. 31 C/COM.IV/DR.5, of 26 October 2001).

12 Statement of Robert C. Blumberg, Chair of U.S. Observer Delegation to the 31st UNESCO General Conference, to Commission IV of the General Conference, Regarding the U.S. Views on the UNESCO Convention on the Protection of Underwater Cultural Heritage, October 29, 2001, Paris, France (available electronically at <http://www.gc.noaa.gov/documents/gcil_heritage2_blumberg.pdf>, accessed 8 November 2014). It is worth noting that, on one hand, the US raises concerns about a requirement for reporting underwater cultural heritage discovered in the exclusive economic zone or on the continental shelf as possibly upsetting balance under UNCLOS and, on the other hand, it also raises concerns that if sunken State vessels are discovered within the territorial sea there is no requirement that it be reported to the flag State.

13 Actually, the UNESCO Convention's negotiators early accepted that the text would not address the question of title or property. See Report of the Meeting of Governmental Experts on the Draft Convention for the protection of the Underwater Cultural Heritage, UNESCO Doc. CLT-98/CONF.202/7, para. 13.

Consistent with UNCLOS, article 7(1) of the UNESCO Convention states that "States Parties, in the exercise of their sovereignty, have the exclusive right to regulate and authorize activities directed at underwater cultural heritage in their internal waters, archipelagic waters and territorial sea." But this provision relates only to the regulation and authorization of activities directed at the wrecks *within its territory* and does not automatically convey title to them to the coastal State,[14] nor does it change the interests of the flag State in its underwater cultural heritage under UNCLOS or otherwise.

An exception to the logical rule stated in article 7(1) of UNESCO Convention—the coastal State's exclusive right to regulate and authorize activities directed at underwater cultural heritage in their internal waters, archipelagic waters and territorial sea[15]—is the immunity of sunken warships, an exception allowed by articles 2(8) and 3 of the same Convention and confirmed by State practice and *opinio juris*.[16]

The logic of this approach could be further confirmed when dealing in particular with wrecks located in the exclusive economic zone (EEZ), the continental shelf or the Area. In the case of the EEZ and the continental shelf, article 10(7) simply states that "no activity directed at State vessels and aircraft shall be conducted without the agreement of the flag State." Without

14 For example, under the agreement regarding CSS *Alabama*, France recognized and respected the continued ownership, sovereign immunity and interest of the United States in its Civil War wreck while the US recognized the jurisdiction of France to regulate all activities directed at this wreck in the territorial sea of France (the text of the 1989 Agreement is available electronically at <http://www.history.navy.mil/branches/org12-9c.htm>, accessed 4 October 2014). This was also the position sustained, for example, by Spain around the legal title of the HMS *Sussex*, a British warship sunk in Spanish waters in 1694: when some problems arose with its recovery in 2007 by a private US company acting under the authorisation of the British Ministry of Defence, Spain did not asserted title over the wreck but it simply required to apply Spanish archaeological legislation to operations directed at an archaeological site located in Spanish waters. On these and other cases it may be seen M.J. Aznar, "Treasure hunters, sunken State vessels and the 2001 UNESCO Convention on the Protection of Underwater Cultural Heritage," *The International Journal of Marine and Coastal Law* 25 (2010): 209–236.

15 States have "exclusive competence" in regard to their own territory (*Isla de Palmas* [1928] 2 UNRIAA 829, at 838), including the maritime spaces within those territories (*Military and Paramilitary Activities in and against Nicaragua* (*Nicaragua v. United States of America*) (Merits) [1986] ICJ Rep 14, at 111 (para. 212)).

16 See further the Resolution on "The Legal Regime of Wrecks of Warships and Other State-owned Ships in International Law," adopted by the Institut de droit international on 29 August 2015, Tallinn Session (available at <http://www.justitiaetpace.org/idiE/resolutionsE/2015_Tallinn_09_en.pdf>).

flag State consent, it is admitted, however, that coastal States might otherwise adopt: (i) urgent measures to protect the wreck (including measures against the unauthorized salvage of the wreck)[17] and (ii) sovereign measures relating to the safety of navigation, which could damage the wreck. With regard to the Area, under article 12(7), "[n]o State Party shall undertake or authorize activities directed at State vessels and aircraft in the Area without the consent of the flag State." The recognition of the flag State's interest is therefore not affected, but reinforced. Furthermore, nothing in the Convention—which must be interpreted contextually—implies that the legal status of sunken State vessels changes by the mere fact that they are located in the internal waters, archipelagic waters or territorial seas of a coastal State instead of on the continental shelf or in the Area.[18]

The principles of sovereignty and immunity thus contextually irradiate throughout the Convention; these tenets of international law are (and must be) well balanced with the protection of sunken State vessels when considered

17 Art. 10(4) of the Convention states that "[w]ithout prejudice to the duty of all States Parties to protect underwater cultural heritage by way of all practicable measures taken in accordance with international law to prevent immediate danger to the underwater cultural heritage, including looting, the Coordinating State may take all practicable measures, and/or issue any necessary authorizations in conformity with this Convention and, if necessary prior to consultations, to prevent any immediate danger to the underwater cultural heritage, whether arising from human activities or any other cause, including looting. In taking such measures assistance may be requested from other States Parties." During the drafting of the Operational Guidelines of the UNESCO Convention, immediate danger was defined as the existence of conditions which can reasonably be expected to cause damage, destruction or looting, to a specific underwater cultural heritage within a short delay of time and which can be eliminated by taking safeguarding measures. This definition was not finally included in the Guidelines (archive on file with the author, who acted as member of the Spanish Delegation in the Working Group which drafted these Guidelines, adopted by Resolution 6/MSP 4 of the Conference of States Parties to the Convention, UNESCO doc. CLT /CEH/CHP/2013/OG/H/1, August 2013, and available electronically at <http://unesdoc.unesco.org/images/0022/002226/222670E.pdf>, accessed 22 October 2014).

18 As said by Craig Forrest, "[t]he actual regime for State owned vessels in the Convention reflects the resulting compromise between flag States and coastal States. Rather than focusing on the rights of the flag State, it was agreed that a balance would be drawn between the right of the flag State and the Coastal State." C. Forrest, "A New International Regime for the Protection of Underwater Cultural Heritage," *International and Comparative Law Quarterly* 51 (2002): 511–554, at 528–29.

as underwater cultural heritage under the definition provided for in article 1 of the Convention.

The Jurisdictional Equilibrium under the Law of the Sea Convention

When Article 303 UNCLOS was under negotiation during the Third UN Conference on the Law of the Sea (UNCLOS III), a proposal to extend coastal sovereign rights over archaeological objects upon the continental shelf and the EEZ attracted firm opposition from three maritime powers—the United States, the United Kingdom and The Netherlands. Their opposition is founded in concern over creeping jurisdiction:

> [the proposal] granted the coastal state rights over its continental shelf which were unrelated to the latter's natural resources and thus might pave the way for other exceptions, favoring creeping jurisdiction and, ultimately, lead to a regime of full coastal state sovereignty over the continental shelf.[19]

Since the adoption of UNCLOS, an increasing number of coastal States have declared domestically their jurisdiction to regulate the protection of the underwater cultural heritage in their EEZ and/or on their continental shelf.[20] This exercise of authority, however, is not necessarily inconsistent with UNCLOS.[21] To some extent, this consideration influenced the drafting of articles 9 and 10 of the UNESCO Convention.

---

19 Quoted in L. Caflisch, "Submarine Antiquities and the International Law of the Sea," *Netherlands Yearbook of International Law* 13 (1982): 3–32, at 17.

20 See, for example, the legislations of Argentina, Australia, Bangladesh, Barbados, Bermuda, Brazil, Canada, Cape Verde, China, Colombia, Cuba, Dominican Republic, DPR Korea, Greece, Guyana, Iceland, Indonesia, Iran, Ireland, Jamaica, Madagascar, Malaysia, Morocco, the Netherland, Pakistan, Philippines, Portugal, SFR Yugoslavia, Spain, Tanzania, United States, Vanuatu or Vietnam. See the practice cited in M.J. Aznar, "The Contiguous Zone as an Archaeological Maritime Zone," *The International Journal of Marine and Coastal Law* 29 (2014): 1–51.

21 For example, while the US *National Marine Sanctuary Act* (2000) asserts authority to protect certain natural and cultural heritage resources in special areas within the US exclusive economic zone, it expressly provides that any enforcement against a foreign flagged vessel or national must be consistent with international law, including consent of flag State (16 U.S.C. § 1435(a)). See Varmer, Gray and Alberg, *United States…*, *supra* n. 3, at 137–140.

The reluctance of most maritime powers arose when these articles raised concerns as to whether they would result in an enlargement of coastal State rights over activities in their EEZ and continental shelf.[22] The structure of the arguments is set forth as follows:

(a) Under paragraph 1(b) of article 9, in the case of discovery or activity directed at underwater cultural heritage in these zones, States Parties shall require the national or the master of the vessel to report such discovery or activity to them and to the coastal State or, alternatively, a State Party shall require the national or master of the vessel to report such discovery or activity to it and shall ensure the rapid and effective transmission of such reports to all other States Parties. For some States, this proviso would grant the coastal State the right to impose a reporting obligation on non-nationals and on masters of vessels flying a foreign flag to report such activities and discoveries. This outcome, however, could simply be seen as mere application of the principle upon which a State is recognized by international law to impose reporting obligations on its nationals and on vessels flying its flag wherever they may. There is nothing in the provision indicating that this is some extension of coastal State jurisdiction over underwater cultural heritage. Even if a coastal State did implement laws under this Convention to require foreign vessels and nationals to report finds of underwater cultural heritage within its EEZ and on its continental shelf, the enforcement could only be against foreign flagged vessels and nationals of states that have consented to requirement under the Convention or other international law. Such law of a coastal State would not apply to foreign flagged vessels and nationals of a State that are not parties to the Convention or consented to the requirement under some other law.

(b) A second set of jurisdictional concerns arises also from the role of the "coordinating State". Under the UNESCO Convention, the coordinating State—normally the coastal State unless decided otherwise[23]—will

---

22 For some States, this alleged enlargement was not solved with the "constructive ambiguities" deliberately included in some articles of the Convention. For Russia, for example, "[t]he ambiguity and vagueness of some articles of the Convention are, in our view, bound to give rise to conflict situations." UNESCO, *Records of the General Conference*…, *supra* n. 3, at 562.

23 Art. 10(3)(b) establishes that "[w]here there is a discovery of underwater cultural heritage or it is intended that activity shall be directed at underwater cultural heritage in a State Party's exclusive economic zone or on its continental shelf, that State Party shall […]

be the focal point in consultations with "all other States Parties which have declared an interest under Article 9, paragraph 5, on how best to protect the underwater cultural heritage."[24] The coordinating State—and this is the most controversial point—may further "take all practicable measures, and/or issue any necessary authorizations in conformity with this Convention and, if necessary prior to consultations, to prevent any immediate danger to the underwater cultural heritage, whether arising from human activities or any other cause, including looting."[25] It also shall implement measures of protection which have been agreed by the consulting States, shall issue all necessary authorizations for such agreed measures and may conduct any necessary preliminary research on the underwater cultural heritage and shall issue all necessary

---

coordinate such consultations as 'Coordinating State,' unless it expressly declares that it does not wish to do so, in which case the States Parties which have declared an interest under Article 9, paragraph 5, shall appoint a Coordinating State." However, if the coastal State does not want to take the lead in coordinating the various parties interests, perhaps because of the discovery of the flag State of underwater cultural heritage, there is a process for agreeing upon another State to be a coordinating State. So while the coastal State's right to regulate activities directed at underwater cultural heritage in its exclusive economic zone or on its continental shelf is firmly grounded on rights under UNCLOS, the respect of flag State jurisdiction, for example, and the possibility that such flag State may end up being coordinating State under certain circumstances further supports the view that the provisions in Article 10 does not upset the balance of flag State jurisdiction and coastal State jurisdiction under UNCLOS.

24 See Article 10(3)(a). Under Art. 9(5), "[a]ny State Party may declare to the State Party in whose exclusive economic zone or on whose continental shelf the underwater cultural heritage is located its interest in being consulted on how to ensure the effective protection of that underwater cultural heritage. Such declaration shall be based on a verifiable link, especially a cultural, historical or archaeological link, to the underwater cultural heritage concerned." The UNESCO Convention Operational Guidelines (see *supra* n. 17) includes a Form (Form 3) available for States parties to communicate their interest under Art. 9(5). In that Form, in order to demonstrate its condition of being an "interested States", a State must have to accompanying its declaration by (a) the results of scientific expertise, (b) historic documentation or (c) any other adequate documentation.

25 See article 10(4). Russia, for example, considered that this proviso may "be interpreted as broadening, beyond the limits laid down by the 1982 United Nations Convention on the Law of the Sea, the rights and jurisdiction of a coastal State over adjacent maritime areas. It can be regarded as offering a coastal State the right, unilaterally and at its own discretion, to take any measures, not excluding the use of force, on the pretext of protecting the underwater cultural heritage." UNESCO, *Records of the General Conference…*, *supra* n. 3, at 562.

> authorizations.[26] Again, these provisions are consistent with UNCLOS duties to protect underwater cultural heritage and cooperate for that purpose, and do not upset the delicate jurisdictional balance between flag State jurisdiction and coastal State jurisdiction under UNCLOS.
>
> Moreover, as it is expressly said in paragraph 6 of said Article 10, when adopting all these decisions, the coordinating State "shall act on behalf of the States Parties as a whole and not in its own interest" and "[a]ny such action shall not in itself constitute a basis for the assertion of any preferential or jurisdictional rights not provided for in international law, including the United Nations Convention on the Law of the Sea."

In summary, for the particular purpose of the protection of underwater cultural heritage in the exclusive economic zone or on the continental shelf, the UNESCO Convention is consistent with provisions in UNCLOS and seems to elaborate—perhaps using some "constructive ambiguities"—a regime for these marine zones that tries to translate the general principle of protection embodied in Article 303(1) UNCLOS and Article 2(3) of the UNESCO Convention, a protection that *shall* be performed either individually (paragraph 4) and through cooperation with other States (paragraph 2). If this obligation must be interpreted and applied in full conformity with UNCLOS (Article 3 of the UNESCO Convention) and no act or activity undertaken on the basis of this Convention may constitute grounds for claiming, contending or disputing any claim to national sovereignty or jurisdiction[27], then the UNESCO Convention may be consistent with the balance of jurisdiction under UNCLOS or be labelled as a "neutral" agreement with regard jurisdictional matters.

UNCLOS Explicit and Implied Alternatives

Only two UNCLOS Articles—149 and 303—address archaeological and historical objects found at sea. While Article 149 addresses these objects when found in the Area,[28] Article 303—located in the 'General provisions' part of

26 See article 10(5).

27 Article 2(11), UNESCO Convention.

28 Art. 149 says: 'All objects of an archaeological and historical nature found in the Area shall be preserved or disposed of for the benefit of mankind as a whole, particular regard being paid to the preferential rights of the State or country of origin, or the State of cultural origin, or the State of historical and archaeological origin.'

UNCLOS—applies to all maritime zones.[29] The legal regime thus established has been characterized, however, as "contradictory and counterproductive."[30]

The vague responsibilities to protect cultural objects in the Area, the identification and extent of the so-called preferential rights and rights of identifiable owners, the legal fiction and the presumption governing the regime of those objects in the contiguous zone or the application of the law of salvage and other rules of admiralty law to the underwater cultural heritage are among the conceivable critiques to the regime. The UNESCO Convention seems to be a plausible solution for a growing number of States that already adhere to its text.[31] Some States, however, still believe the UNESCO Convention does not adequately address these issues and other gaps present in the protection of underwater cultural heritage. Among the most important issues are how

29 Art. 303 wording is as follows: "1. States have the duty to protect objects of an archaeological and historical nature found at sea and shall cooperate for this purpose. 2. In order to control traffic in such objects, the coastal State may, in applying Article 33, presume that their removal from the seabed in the zone referred to in that article without its approval would result in an infringement within its territory or territorial sea of the laws and regulations referred to in that Article. 3. Nothing in this article affects the rights of identifiable owners, the law of salvage or other rules of admiralty, or laws and practices with respect to cultural exchanges. 4. This article is without prejudice to other international agreements and rules of international law regarding the protection of objects of an archaeological and historical nature."

30 T. Scovazzi, "A Contradictory and Counterproductive Regime," in *The Protection of the Underwater Cultural Heritage, Before and After the 2001 UNESCO Convention*, R. Garabello and T. Scovazzi, eds. (Leiden: Martinus Nijhoff, 2003): 3–18. See further B.H. Oxman, "Marine Archaeology and the International Law of the Sea," *Columbia Journal of Law and the Arts* 12 (1987–1988): 353–372; C. Lund, "Protection of the Under-Water Cultural Heritage," in *Il regime giuridico internazionale del Mare Mediterraneo*, U. Leanza, ed. (Milan: Giuffrè Editore, 1987): 351–354; or A. Strati, "Protection of the Underwater Cultural Heritage," in *Unresolved Issues and the New Challenges to the Law of the Sea*, Anastasia Strati *et al.*, eds. (The Hague: Martinus Nijhoff, 2006): 21–62.

31 As of July 2016, the Convention has 55 States parties: Albania, Algeria, Antigua and Barbuda, Argentina, Bahrain, Barbados, Belgium, Benin, Bosnia and Herzegovina, Bulgaria, Cambodia, Croatia, Cuba, Democratic Republic of the Congo, Ecuador, France, Gabon, Ghana, Grenada, Guatemala, Guinea Bissau, Guyana, Haiti, Honduras, Hungary, Iran (Islamic Republic of), Italy, Jamaica, Jordan, Lebanon, Libya, Lithuania, Madagascar, Mexico, Montenegro, Morocco, Namibia, Nigeria, Palestine, Panama, Paraguay, Portugal, Romania, Saint Kitts and Nevis, Saint Lucia, Saint Vincent and the Grenadines, Saudi Arabia, Slovakia, Slovenia, South Africa, Spain, Togo, Trinidad and Tobago, Tunisia and Ukraine. As far as this author knows, further ratifications are probably in the pipeline in Australia, Germany, the Netherlands, some countries in the Arab Gulf and Uruguay.

the legal regime applies beyond the outer limit of the territorial sea, and particularly in the EEZ and the continental shelf.

The Explicit and Evolving Solutions

Within its internal waters, archipelagic waters and territorial sea, the coastal State exercises full and complete sovereignty subject to UNCLOS and other rules of international law.[32] This rule means that the regulation of underwater cultural heritage in these maritime zones is governed by the coastal State's national regulation. The application of national laws to foreign flagged vessels and nationals must be consistent with international law—including UNCLOS. The Law of the Sea Convention includes a general duty on the part of coastal States to protect (and to cooperate in the protection of) underwater cultural heritage.[33] States thus apply their domestic cultural and heritage legislations and foreign States vessels, and nationals conducting activities in these zones are obliged to respect and comply with them.[34]

Some problems arise when dealing with the regime to be applied in the contiguous zone. Article 303(2) UNCLOS, referring to Article 33 of that instrument, establishes a complex regime based on a presumption and a legal fiction upon which the removal of underwater cultural heritage from the seabed of the contiguous zone without the approval of the coastal State would result in an infringement within its territory or territorial sea of its customs, fiscal, immigration or sanitary laws and regulations. This provision

32 See *supra* n. 15. Under Art. 7(1) of the UNESCO Convention "States Parties, in the exercise of their sovereignty, have the exclusive right to regulate and authorize activities directed at underwater cultural heritage in their internal waters, archipelagic waters and territorial sea."

33 Art. 303(1) UNCLOS, but also those obligation derived from other cultural agreements—like the 1972 UNESCO Convention concerning the Protection of the World Cultural and Natural Heritage (1037 UNTS 151) or the 1992 European Convention on the Protection of the Archaeological Heritage (revised) (*European Treaty Series* No. 143)—which obliged State parties to protect the cultural heritage located within their territories, including those maritime zones. See Michail Risvas, "The Duty to Cooperate and the Protection of Underwater Cultural Heritage," *Cambridge Journal of International and Comparative Law* 2 (2013): 562–590.

34 In a recent case before the International Tribunal for the Law of the Sea (ITLOS), the rights of the coastal State to enforce its domestic legislation in its internal waters and territorial sea, including that protecting the underwater cultural heritage, have been recognized and reinforced. See *The M/V "Louisa" Case* (*Saint Vincent and the Grenadines v. Kingdom of Spain*), Judgement of 28 May 2013, ITLOS Reports [2013], available electronically at <http://www.itlos.org/fileadmin/itlos/documents/cases/case_no_18_merits/judgment/C18_Judgment_28_05_13-orig.pdf>, accessed 23 October 2014.

has been viewed as somewhat of a stretch in that it creates grey legal zones and does not clarify the rights and duties of both coastal and other States in their activities directed at underwater cultural heritage in the adjacent contiguous zone.

State practice—both multilateral and unilateral—offers a general, quite uniform and constant evidence of the acceptance of a coastal State's right to legislate on and to enforce the protection of underwater cultural heritage located in its contiguous zone.[35] Consequently, it could be argued that current international law includes an archaeological maritime zone that extends to the outer limit of the State's declared contiguous zone. Article 8 of the UNESCO Convention seems to confirm this rule, as it states:

> Without prejudice to and in addition to Articles 9 and 10, and in accordance with Article 303, paragraph 2, of the United Nations Convention on the Law of the Sea, States Parties may regulate and authorize activities directed at underwater cultural heritage within their contiguous zone. In so doing, they shall require that the Rules be applied.

The UNESCO provision may be evidence of the general acceptance of the UNCLOS regime for protecting underwater cultural heritage in the contiguous zone to include the authority of the coastal State to regulate and authorize activities directed at that heritage.

Some Implied Solutions

There is still the challenge to adequately protect underwater cultural heritage beyond the outer limit of the contiguous zone. Given the limited (and vague) regime established by UNCLOS in articles 149 and 303, however, some of the complaints about the vagueness and limited guidance of Article 303 ignore how the treaty is a framework convention and that detail about rights and responsibilities in article 303 may be found in other provisions of the Convention. The Convention therefore may be helpful for coastal States desiring to protect the underwater cultural heritage located in their EEZ and on their continental shelf. Amid other possibilities, these solutions might be (1) to use the right of the coastal State to protect the marine environment in these two zones, (2) to explore the possibilities offered by the regime of marine scientific research to protect underwater cultural heritage and (3) to refer protection to article 59 UNCLOS as a case of non-attribution of rights and jurisdiction in the EEZ.

---

35 Aznar, *The Contiguous Zone, supra* n. 20.

1. Underwater Cultural Heritage and the Protection of Marine Environment

UNCLOS recognizes coastal States sovereign rights and jurisdiction over activities such as the exploration and exploitation of the natural resources of their EEZ and continental shelf.[36] In the case of the EEZ, article 56(1)(a) establishes that the coastal State has not only "sovereign rights for the purpose of exploring and exploiting" the natural resources (whether living or non-living) but the sovereign rights for the purpose of "conserving and managing" these natural resources. Although this last proviso has been labelled "to be a question of drafting rather than of substance,"[37] it seems to recognize broad coastal States' rights not only to the exploration and exploitation of the living resources, but also to conserve and manage these resources.

In the case of the continental shelf, the rights of the coastal State are recognized to be over activities directed at "the mineral and other non-living resources of the seabed and subsoil together with living organisms belonging to sedentary species, that is to say, organisms which, at the harvestable stage, either are immobile on or under the seabed or are unable to move except in constant physical contact with the seabed or the subsoil."[38]

Underwater cultural heritage has not been included among "natural resources." As the International Law Commission (ILC) affirmed, "[i]t is clearly understood that the rights of the coastal State do not cover objects such as wrecked ships and their cargoes (including bullion) lying in the seabed or covered by sand in the subsoil."[39] In most cases, however, the seabed, its subsoil and the sand embedding or covering archaeological objects are actually natural resources. Leaving aside sedentary species such as oysters, clams or abalones—protected by the continental shelf regime—the vast majority of non-sedentary fishing species live around reefs and submerged protrusions, and in or near man-made submerged structures. Some of them, using the words of a U.S. admiralty court, are "the remains of abandoned, two hundred year old shipwrecks, which have lain undisturbed for centuries under an undetermined amount of sand, [and that] reasonably can be characterized as natural resources for the purposes of the Federal Act."[40] Thus, activities directed

36 Arts. 56 and 77, UNCLOS, respectively.

37 R.R. Churchill and A.V. Lowe, *The Law of the Sea* (3rd ed., Manchester University Press: Manchester, 1999) at 166.

38 Article 77(4), UNCLOS.

39 *Yearbook of the International Law Commission*, vol. II, 1956, at 289.

40 *Subaqueous Exploration and Archaeology, Ltd. And Atlantic Ship Historical Society, Inc., v. The Unidentified, Wrecked and Abandoned Vessel et al.*, 577 F.Sup. 597, 613 (D. Md. 1983), aff'd, 765 F.2d 139 (4th Cir. 1985). The Federal Court refers to the 1953 Submerged Lands Act (67 Stat. 29, 43 U.S.C. Sec. 1301–1315).

at underwater cultural heritage may affect natural resources and, *vice versa*, exploitation of natural resources may affect underwater cultural heritage.

Under the UNESCO Convention, the definition and scope of underwater cultural heritage includes not only the archaeological objects (human remains, sites, structures, buildings, wrecks and their cargo, artifacts, etc.), but also their "archaeological and *natural* context" as well.[41] [emphasis added] Exactly as it occurs in land archaeology, where the removal of soil is necessary, underwater archaeological activities habitually produce an impact on the fragile marine environment that should be minimized.[42] Despite these efforts, activities directed to the underwater cultural heritage normally imply some destruction of the marine environment.

Faced with adverse activities directed to the underwater cultural heritage located in the EEZ or on their continental shelf, coastal States may set forth environmental standards, both international and domestic, applicable to their EEZ zone and continental shelf.[43] Complementing UNCLOS, some international rules have been adopted in a regional scale (the Mediterranean Sea[44] or the Caribbean Sea,[45] for example) using "special protected area"

---

41 Art. 1(1)(a) on the definition of underwater cultural heritage (emphasis added).

42 Hence, the annexed Rules to the UNESCO Convention also includes environmental protocols when dealing with an archaeological project: any project shall include and environmental policy (Rule 10(l)) to ensure that the seabed and marine life are not unduly disturbed (Rule 29); and any work shall include an assessment that evaluates the significance and vulnerability of the underwater cultural heritage and the surrounding natural environment (Rule 14), which shall include a background study of the environmental characteristics of the site (Rule 15). Between States parties to this Convention, hence, environmental standards may be used to reject any activity directed to the underwater cultural heritage located in the exclusive economic zone or on their continental shelf.

43 UNCLOS remains silent as to the measures a coastal State is entitled to take in case its sovereign rights in the exclusive economic zone and the continental are violated. Nonetheless, the ILC also considered the exercise of enforcement measures by the coastal State with a view to protecting its natural resources to fall within the scope of the "sovereign rights" notion (*Yearbook of the International Law Commission*, vol. II, 1956, at 297).

44 See, for example, the 1995 Protocol concerning Specially Protected Areas and Biological Diversity in the Mediterranean (2102 UNTS 203) or the 2008 Protocol on Integrated Coastal Marine Zone Management in the Mediterranean (text available at <http://195.97.36.231/dbases/webdocs/BCP/ProtocolICZM08_eng.pdf>, accessed 20 October 2014). These protocols were adopted in the general framework of the Barcelona Convention 1995 Convention for the Protection of the Marine Environment and the Coastal Region of the Mediterranean (the "Barcelona Convention," text available at <http://195.97.36.231/dbases/webdocs/BCP/bc95_Eng_p.pdf>, accessed 20 October 2014).

45 See, for example, the 1990 Protocol Concerning Specially Protected Areas and Wildlife (text available at http://www.cep.unep.org/ accessed 20 October 2014), adopted within

approach that, under agreement, may be established beyond 24 nm. In some other cases, it is the domestic legislation that goes beyond that limit and may be applied up to 200 nm.[46]

2. Underwater Cultural Heritage and Marine Scientific Research

UNCLOS contains no definition of marine scientific research. There is, however, general agreement that marine archaeology was not considered as marine scientific research during the negotiations for the treaty.[47] This decision was mainly due to a twofold argument: the limited recognition of marine archaeology as an independent scientific discipline in the 1980s, and the fact that marine scientific research was mainly confined to the natural environment and resources. It has been plausibly argued, however, that contemporary marine archaeology is a full scientific discipline and, that marine scientific research actually implies an indiscriminate collecting of data that may be used either for environmental or archaeological purposes.[48] Modern underwater technology—particularly the use of remote sensing equipment and different unmanned submersible vehicles—permits the gathering of massive information for the bottom of the seas which maybe directly transformed in a tradeable commodity.

As a matter of distinction, commentators generally agree to distinguish between pure or 'fundamental' research, which is conducted exclusively for peaceful purposes and in order to increase scientific knowledge for the benefit of all mankind, and 'applied' research, which is of direct significance for the exploration and exploitation of natural resources. Therefore, while pure research does not imply direct economic revenue, applied research is primarily focused on future and foreseeable commercial activities. The information gathered by modern underwater technology, when directed at an archaeological target (for example, a shipwreck), is likely not characterized as pure marine scientific research.

---

the general framework of the 1983 Convention for the Protection and Development of the Marine Environment in the Wider Caribbean Region (the "Cartagena Convention," text available at <htts://www.cep.unep.org>, accessed 20 October 2014).

46 This is done, as we have seen, in the US *National Marine Sanctuaries Act* (2000), under which an area of "marine environment" may be designed and regulated that includes "the exclusive economic zone, consistent with international law" (16 U.S.C. 1431, 1432, para. 3).

47 See, among others, Alfred Soons, *Marine Scientific Research and the Law of the Sea* (The Hague: TMC Asser Instituut, 1982): 275.

48 See S. Dromgoole, "Revisiting the Relationship between Marine Scientific Research and the Underwater Cultural Heritage," *The International Journal of Marine and Coastal Law* 25 (2010): 33–61.

Article 246 UNCLOS recognizes coastal State discretion to regulate both types of research in its EEZ and on its continental shelf. In both cases, coastal States, "in the exercise of their jurisdiction, have the right to regulate, authorize and conduct marine scientific research in their [EEZ] and on their continental shelf in accordance with the relevant provisions of [UNCLOS]";[49] and that research "shall be conducted with the consent of the coastal State." Whereas coastal States "shall, in normal circumstance, grant their consent" for pure research,[50] in cases of applied research, coastal States "may... in their discretion withhold their consent."[51]

Activities directed at underwater cultural heritage in the EEZ or on the continental shelf may then be subject to the coastal State's discretion, depending on the nature and purpose of that activity. If, for example, the latter is a research project supervised by a scientific institution, in cooperation with States and following archaeological protocols internationally recognized by the scientific community,[52] the coastal State should grant its consent in normal circumstances. This type of project is carried out exclusively for peaceful purposes, and in order to increase scientific knowledge for the benefit of all mankind.

Commercial activities, however, directed at underwater cultural heritage, such as work performed by treasure hunters, is not pure research notwithstanding the recent efforts of some companies to introduce themselves as archaeological endeavours. Recent cases in Spain and the United Kingdom, as well as other well known cases on the west coast of Africa, the Indian Ocean and in the Caribbean, show how these treasure hunters do not respect archaeological protocols.[53] The purpose of these individuals and companies is not to increase scientific knowledge for the benefit of all mankind, but to

---

49 Paragraph 1.

50 Paragraph 3.

51 Paragraph 5.

52 For example those listed in the Charter on the Protection and Management of Underwater Cultural Heritage ("Charter of Sofia") adopted by the International Council on Monuments and Sites (ICOMOS) in its 11th General Assembly in Sofia, October 1996 and exported *mutatis mutandi* to the Annex of the 2001 UNESCO Convention. See the Charter at <http://www.international.icomos.org/charters/underwater_e.pdf>, accessed 13 October 2014.

53 A good example could be the attitude of one of these treasure hunter companies (Odyssey Marine Exploration, Inc.) when excavating in 2007 the wreck of the Spanish Royal Navy frigate Nuestra Señora de las Mercedes about 1,000 meters deep on the current Portuguese continental shelf, without the sovereign owner's permission and without any archaeological care. As a US District Judge acknowledged, that action "irreparably"

increase their economic benefits. Therefore, coastal States, in order to protect underwater cultural heritage located in their EEZ or on their continental shelf, have the discretion to grant or withhold any permit to those actors. Furthermore, the regime of marine scientific research grants research rights only to States, not private actors. Private companies such as treasure hunters are not inherently entitled to be granted such permits.

3. Protection of the Underwater Cultural Heritage as a Non-Attributed Regime

A final question to be addressed is the consequence derived from the definition of the protection of the underwater cultural heritage as a non-attributed regime under UNCLOS.[54] Article 59 regulates the basis for the resolution of conflicts regarding the attribution of rights and jurisdiction in the EEZ in cases where the Convention does not attribute these rights or jurisdiction to the coastal State or to other States within the zone. When a conflict arises between the interests of the coastal State and any other State(s), "the conflict should be resolved on the basis of equity and in the light of all the relevant circumstances, taking into account the respective importance of the interests involved to the parties as well as to the international community as a whole."[55]

Three variables apply to resolve these possible conflicts: (a) equity and relevant circumstances,[56] (b) interests of the parties involved, and (c) interest of the international community as a whole.

(a) In order to find an equitable solution to the possible conflict, some relevant circumstances may be taken into account: the purpose (commercial or scientific) of the activity directed to the underwater cultural heritage, the flag State, in case of an activity directed to a shipwreck or aircraft and its possible consideration as public property thus subject to the immunity principle,[57] the cultural and historic link of different States to the archaeological site, or the exact location of the site and its possible inter-

---

disturbed the site. See *Odyssey Marine Exploration, Inc. v. Unidentified Shipwrecked Vessel*, 657 F.3d 1159 (11th Cir. 2011), *cert. denied*, 132 S. Ct. 2379 (2012).

54 See Churchill and Lowe, *supra* n. 37, at 152, note 22.

55 Art. 59, UNCLOS.

56 "Equity" and "relevant circumstances" may be perfectly melted in a single variable since equitable principles are generally applied in the law of the sea taking into account all the relevant circumstance. See as a jurisprudential point of departure in the *North Sea Continental Shelf* cases (ICJ Reports [1969], at 53, para. 101).

57 The 2001 UNESCO Convention establishes that "no activity directed at State vessels and aircraft shall be conducted without the agreement of the flag State and the collaboration of the Coordinating State." (Art. 10(7)).

ference with other legitimate interests (international navigation, security, protection of marine environment, other uses of the EEZ or the continental shelf, etc.)[58] Whether some of these variables may solve the conflict, the solution must be generally found through cooperation. This is the UNESCO Convention's option proposed in its Articles 9 and 10 under a basic scheme of identification of interest, combined with information sharing, consultations, cooperative protection, and management of sites.

(b) With regard the interests of the parties involved, it has been suggested that "[g]iven the functional nature of the exclusive economic zone, where economic interests are the principal concern, this formula would normally favour the coastal State. Where conflicts arise on issues not involving the exploration for and exploitation of resources, the formula would tend to favour the interests of other States or of the international community as a whole."[59] Some types of activities directed to underwater cultural heritage may be characterized as commercial activities and therefore subject to the jurisdiction of the coastal State. Furthermore, these activities are normally characterized by the use of artificial installations and structures, the operation of drilling (or similar purpose) artifacts, etc. In these cases, UNCLOS attributes the exclusive rights to regulate these activities exclusively to the coastal State.[60]

Even when these activities have a scientific, non-economic motive, however, they may be qualified as marine scientific research and, as such, fall also under the coastal State's jurisdiction in its exclusive economic zone or on its continental shelf.

(c) Perhaps the basic criteria to be used when protecting underwater cultural heritage in the EEZ is the interest of the international community as a whole. Underwater cultural heritage, as any cultural heritage, must be generally protected. States generally "have the duty to protect objects of an archaeological and historical nature found at sea and shall

58 For the later circumstance, the UNESCO Convention provides that "[a] State Party in whose exclusive economic zone or on whose continental shelf underwater cultural heritage is located has the right to prohibit or authorize any activity directed at such heritage to prevent interference with its sovereign rights or jurisdiction as provided for by international law including the United Nations Convention on the Law of the Sea." (Art. 10(4)).

59 Myron Nordquist, ed., *United Nations Convention on the Law of the Sea: A Commentary* (Vol. 2, Dordrecht/Boston/London: Martinus Nijhoff, 1993): 569.

60 Arts. 60, 80 and 81 UNCLOS.

cooperate for this purpose,"[61] and particularly, "shall, individually or jointly as appropriate, take all appropriate measures in conformity with this Convention and with international law that are necessary to protect underwater cultural heritage, using for this purpose the best practicable means at their disposal and in accordance with their capabilities."[62] If coastal States have a protective and a proactive domestic legislation on cultural heritage (an increasing trend otherwise), nothing impedes a fair application of its rules to the underwater cultural heritage located in its EEZ or on its continental shelf, without impairing legitimate rights of third States under current law of the sea, including UNCLOS.[63] If coastal States are parties to the UNESCO Convention, they have a particularly enhanced regime that governs activities directed to underwater cultural heritage located in the waters under their jurisdiction beyond 24 nm.

Some Tentative Conclusions

The 2001 UNESCO Convention is an international agreement that is consistent with UNCLOS, which the negotiating parties used as their framework to build upon some of the more vague provisions regarding underwater cultural heritage. Both treaties, however, are by-products of their respective epochs. UNCLOS was drafted and adopted when maritime powers were concerned about free navigation, and coastal States were concerned about control of exploitation of natural resources off their coast. The treaty was negotiated when underwater technology was not as advanced or relatively available for deep water exploitation, and underwater cultural heritage was an exotic subject of interest. The UNESCO Convention, on the other hand, arrived when seas and oceans already were spaces where integral policies are to be developed, when technology is unveiling information unimaginable a few decades ago, and where, thanks to evolving scientific methods, underwater remains are indispensable to understanding our common history. For example, the

61 Art. 303(1), UNCLOS.

62 Art. 2(4), UNESCO Convention.

63 And third States, "[i]n exercising their rights and performing their duties under [UNCLOS] in the exclusive economic zone [of another State,] shall have due regard to the rights and duties of the coastal State and shall comply with the laws and regulations adopted by the coastal State in accordance with the provisions of this Convention and other rules of international law in so far as they are not incompatible with [Part V UNCLOS]." (Art. 58(3), UNCLOS).

discovery and scientific management of the *Titanic*'s remains and recognition that deep water technology had developed to exploit underwater cultural heritage set the stage for development of the UNESCO Convention and its protection of that heritage.

The Convention has attracted some criticism and created certain legal concerns from different angles. Criticism comes from private actors, such as treasure hunters and commercial individuals, that believe the Convention edges out their commercial activities.[64] Even longstanding salvage disputes about well-known wrecks are beginning to be resolved under archaeological and public law criteria, rather than by strictly private commercial interests.[65]

Notwithstanding the relevance of these criticisms, sunken State vessels and the jurisdictional concerns are indeed more important. This chapter suggests that these concerns may be dispelled with an evolving and contextual interpretation of the UNESCO Convention as its "constructive ambiguities" are being implemented consistent with international law including UNCLOS. This interpretation must include the overarching principle: the duty of all States to protect underwater cultural heritage, enshrined in Article 303(1) UNCLOS.

Several other principles and rules would act as building blocks of the protective regime proposed by the UNESCO Convention, interpreted and applied in the context of and in a manner consistent with international law, particularly UNCLOS: territorial sovereignty, immunity, scientific precaution, and cooperation. Whether States are convinced that the fine jurisdictional balance

64 For a recent opposing view, see among others J.B. Segarra, "Above Us the Waves: Defending the Expansive Jurisdictional Reach of American Admiralty Courts in Determining the Recovery Rights to Ancient or Historic Wrecks," *Journal of Maritime Law and Commerce* 43 (2012): 349–391. See, previously, D.J. Bederman, "Rethinking the Legal Status of Sunken Warships," *Ocean Development and International Law* 31 (2000): 97–125.

It must be reminded that under Art. 4 of the Convention, "[a]ny activity relating to underwater cultural heritage to which this Convention applies *shall not be subject* to the law of salvage or law of finds, unless it: (a) is authorized by the competent authorities, *and* (b) is in full conformity with this Convention, *and* (c) ensures that any recovery of the underwater cultural heritage achieves its maximum protection" (emphasis added, calling attention to the negative tense of this article's drafting and the cumulative criteria adopted by the Convention trying to exclude, almost completely, the application of salvage law to underwater cultural heritage).

65 The case of the RMS *Titanic* is paramount. See the US NOAA, Guidelines for Research, Exploration, and Salvage of R.M.S. Titanic (NOAA Guidelines), 66 *Fed. Reg.* 18906 (12 April 2001); and the recent interim decision of the salvage award at *RMS Titanic, Inc. v. The Wrecked and Abandoned Vessel*, F. Supp. 2d, 2010 WL3239112 (E.D.Va., 2010). See M.J. Aznar and O. Varmer. "The *Titanic* as Underwater Cultural Heritage: Challenges to Its Legal International Protection," *Ocean Development and International Law* 44 (2013): 96–112.

achieved by UNCLOS is not compromised by the new regime created by the UNESCO Convention, especially for the EEZ and the continental shelf, then coastal States may protect in close cooperation with other interested States the underwater cultural heritage located in these marine zones. A particular role must be recognized for flag States over their sunken State vessels, irrespective of the location of their wreckage.[66] For these cases, collaboration among States is the best solution, as evidenced by several bilateral agreements already concluded.[67] As the UNESCO Convention is nested in UNCLOS, future underwater cultural heritage agreements (regional or bilateral) may be rooted in the 2001 Convention. Article 6(1) encourages States parties "to enter into bilateral, regional or other multilateral agreements or develop existing agreements, for the preservation of underwater cultural heritage." These new agreements shall be in full conformity with the Convention, and may not dilute its universal character. Furthermore, they should ensure better protection of underwater cultural heritage than that adopted in the Convention.

Cooperation within the legal framework created by UNCLOS and the UNESCO Convention, or framing new, more adequate agreements is thus the cornerstone for the protection of underwater cultural heritage.[68] UNCLOS, as the codified expression of the international law of the sea, may accom-

66 This has been accepted by US courts sitting in admiralty in two cases involving Royal Spanish Navy frigates: the *Juno* and *La Galga*, wrecked in 1750 and 1802 in the US contiguous zone (*Sea Hunt, Inc. v. Unidentified Shipwrecked Vessel or Vessels*, 221 F.3d 634 (4th Cir. 2000), *cert. denied*, 148 L. Ed. 2d 956, 121 S.Ct. 1079 (2001)) and the *Nuestra Señora de las Mercedes*, wrecked in 1804 in the Portuguese continental shelf (see *supra* n. 53).

67 For example the Exchange of Notes constituting an Agreement between the Government of the United Kingdom of Great Britain and Northern Ireland and the Government of Italy regarding the Salvage of HMS *Spartan* (1952), the Agreement between the Netherlands and Australia concerning old Dutch shipwrecks (1972), the Exchange of Notes between South Africa and the United Kingdom Concerning the Regulation of the Terms of Settlement of the Salvaging Wreck of the HMS *Birkenhead* (1989), the Agreement between the Government of the United States of America and the Government of the French Republic Concerning the Wreck of the CSS *Alabama* (1989), the Memorandum of Understanding Between the Government of Great Britain and Canada Pertaining to the Shipwrecks HMS *Erebus* and HMS *Terror* (1997), or the Agreement between the French Republic and the Government of the United States of America concerning the Wreck of *La Belle* (2003).

68 These agreements may be either of a soft law nature: for example, Spain has recently signed two Memorandums of Understanding with the US (2013) and with Mexico (2014) for the cooperation in the protection of the underwater cultural heritage located in their respective waters (archives on file with the author).

modate—explicitly and implicitly—the general interest of States protecting underwater cultural heritage. Some of its clauses—particularly those referring to the protection of marine environment and those regulating marine scientific research—may be constructively interpreted in order to protect underwater cultural heritage in the exclusive economic zone or on the continental shelf. Additionally, Article 59 may help to resolve conflicts of interest when protecting that heritage.

A more evolved regime may be found in the UNESCO Convention—which, as any other treaty, is to be applied only among States parties—or may be created when drafting new agreements protecting particular archaeological sites or regulating that protection for a more ample marine area (including a regional sea).[69] As a by-product of these international agreements, States have the duty to implement domestically their international obligations so created, thus establishing a multi-layered array of rules protecting underwater cultural heritage.

69 For example, all riparians States to the Adriatic Sea are States parties to the UNESCO Convention. A particular regime for that Sea could be envisaged among these States, establishing a rather more tailored regime if so decided.

CHAPTER 5

Protection and Management Policy of Underwater Cultural Heritage in Korea

Seoung Wook Park and Chang Soo Choe***

Introduction

The UNESCO Convention on the Protection of Underwater Cultural Heritage (hereafter, the UNESCO Convention) was adopted in 2001.[1] Although a powerful group of maritime States abstained from the vote, or voted against adoption, there are some indications that some northern European maritime states, such as France, now have changed their position on the treaty.[2] The UNESCO Convention has been gradually approved by an increasing number of states and it has entered into force. States Parties have been in the process of implementing the agreement within their domestic legal system, and this chapter examines national legislation and the issues within the Republic of Korea.

As a peninsula state surrounded by water on three sides, Korea has 2,413 km of coastline and plentiful streams and rivers. For more than five thousand years Koreans have interacted with countries on other continents, as well as its neighbors in northeast Asia. Due to this historical and geographical context, Korea has an extensive underwater cultural heritage (UCH).[3] This legacy contains histories of the international trade and cultural exchange between Korea and the rest of the world.[4]

* Division Leader, Policy Research Section, Ocean Policy Institute, Korea Institute of Ocean Science and Technology (KIOST).

** Professor of Public Administration at the Cyber Hankuk University of Foreign Studies (CUFS) in Seoul, Korea, and adjunct professor at Central Officials Training Institute of Korea (COTI).

1 Convention on the Protection of the Underwater Cultural Heritage, Paris, Nov. 2, 2001, entered into force Jan. 9, 2009, available at http://portal.unesco.org.

2 Sarah Dromgoole, "Reflections on the position of the major maritime powers with respect to the UNESCO Convention on the Protection of the Underwater Cultural Heritage 2001," *Marine Policy* 38 (2013): 117.

3 Seok Yong Lee, "Legal Status and Protection of the Underwater Cultural Heritage and Sunken Warships in International Law," *Anam Law Review* 34 (2011): 989–990. (*In Korean*).

4 For example, the *Shinan Underwater Cultural Heritage* that was found in the Yellow Sea of Korea in 1975 included 28 tons of coins, more than 20,000 Chinese and Korean celadon, and household items including wooden shoes from Japan.

 | DOI 10.1163/9789004311442_007

This chapter explores the policy and legal framework for UCH protection in Korea. This study unfolds in three main parts. Part I provides a brief account of the development of administrative and legal system for UCH protection in Korea; Part II highlights contributions and limitations of the UNESCO Convention to UCH within the Korean legal framework; and, the third Part discusses issues and challenges in the Korean legal framework for UCH protection in relation to the UNESCO Convention.

Development of the Administrative and Legal System in Korea

It is no exaggeration to say that the modern history of Korea's underwater cultural heritage began in 1975 with a fisherman's discovery of celadon ware off the coast of *Shinan*, an island off the southwestern coast of Korea. With the excavation of the *Shinan Underwater Cultural Heritage*, the Korean government decided to develop the capacity for underwater archaeology. The *Mokpo Conservation Center* was opened in 1981 for the conservation treatment of large shipwrecks of the past, while the excavation of the *Shinan Underwater Cultural Heritage* was in progress. The Center was established under the Cultural Heritage Management Bureau in the Ministry of Culture and Tourism. As more underwater cultural assets were discovered in hundreds of locations, the Korean government began to devote more sustained attention to the national history preserved in its vast underwater cultural heritage.

At the early stage of the development in institutional arrangements for UCH protection, Korea focused on the development of administrative mechanisms rather than establishing a systematic legal framework. In 1990 the name of the *Mokpo Conservation Center* was changed to the *Mokpo Maritime Heritage Conservation Center*, and in 1994 it was reorganized it into the *National Maritime Museum*. Later, the functions of the institutions were extended to include inspection and research of UCH, and the Museum was elevated to the status of the *National Institute of Maritime Cultural Heritage* (NIMCH). Meanwhile, the Cultural Heritage Management Bureau was separated from the Ministry of Culture and Tourism and became the *Cultural Heritage Administration* (CHA) in 1995. As an umbrella organization of the NIMCH, the CHA is responsible for managing and protecting all manner of national cultural assets, including UCH. Under the supervision of the CHA, the NIMCH performs its functions to develop national UCH excavations, conservations, and research, and to cooperate with other concerned parties. The NIMCH is located in Mokpo City in southwestern Korea, and it has a branch office, *West Sea Cultural Heritage Preservation Center*, in Taean.

In 2010 the Korean government established a legal system for the protection of UCH. The National Assembly of Korea comprehensively revised existing legislation to adopt the Cultural Heritage Protection Act (*Korea CHP Act*). The Protection and Inspection of Buried Cultural Heritage Act (*Korea BCHP Act*) was enacted in the same year to maintain and accede to the original form of national culture by preserving buried cultural heritage and to efficiently protect, inspect and manage it. As there is no independent legislation for UCH, it is treated as a part of buried cultural heritage in Korea. By stipulating in Art. 2 that "the term buried cultural heritage" means "tangible cultural heritage buried or distributed underground or underwater, the *Korea BCHP Act* classifies buried cultural heritage into two categories; underground cultural heritage and underwater cultural heritage. Furthermore, the *Korea BCHP Act* emphasizes the protection of UCH rather than excavation if there is a risk of damage to UCH. Article 4 and 5(1) read that "an area in which buried cultural heritage is recognized to exist, as prescribed by Presidential Decree shall be protected so that the original form is not to be damaged...." There is a duty to protect UCH as well: "a person who intends to plan and implement a development plan, such as the State, local government, etc., shall not allow buried cultural heritage to be damaged."

Based on the provisions of the *Korea CHP Act*,[5] the CHA is required to establish a comprehensive Master Plan for Cultural Heritage every five years, which includes a comprehensive master plan for UCH in order to systematically develop, manage, and conserve national underwater cultural assets. As a part of the Cultural Heritage Master Plan of the period 2007–2011, over the past five years the NIMCH conducted an exploration and investigation survey along the coastline of the southwestern region in Korea.

In sum, Korea has been carrying out UCH conservation for over 30 years and accumulating experience in practice and in theoretic research. The content, subject and the methods of protection of Korean UCH have been changed in

---

5 Article 6(1) of the CHP Act: "The Administrator of the Cultural Heritage Administration shall establish a comprehensive master plan addressing the following matters every five years, following consultation with the competent Mayor/Do Governor for the preservation, management and utilization of cultural heritage 1. Basic direction-setting and objectives for the preservation of cultural heritage; 2. Analysis and evaluation of the former master plan for cultural heritage; 3. Matters concerning the repair, maintenance and restoration of cultural heritage; 4. Matters concerning a historic and cultural environment of cultural heritage; 5. Matters concerning the safety management of cultural heritage; 6. Matters concerning the informatization of cultural heritage data; 7. Matters concerning raising financial resources used to preserve cultural heritage; 8. Other necessary matters for the preservation, management and utilization of cultural heritage."

many aspects, but the national legislation and policy support are still not sufficient to effectively enhance the state's capacity for protecting UCH. Several reasons exist that reduce the effectiveness of the legislation. First, national policy direction as a whole focuses on economic development rather than conserving social and cultural heritage. Second, there is a lack of concern on the part of policy-makers in both executive and legislative branches of national government. Third, there is insufficient academic and practical infrastructure.

Contributions and Issues of the Convention

The UNESCO Convention is the most complete international regime for the UCH protection, but presently it is not a powerful and popular international instrument.[6] As of 2015, most of the major maritime powers have not ratified the UNESCO Convention.[7] As the offshoot of the *United Nations Convention on the Law of the Sea 1982* (UNCLOS),[8] however, the UNESCO Convention includes both valuable and problematic elements for international cooperation among States Parties with regard to protection of UCH in practice.

Contribution to UCH Protection

As the principal international agreement on the protection of UCH, the UNESCO Convention provides some valuable provisions.

First, the Convention articulates an international consensus to protect UCH as stipulated in its Article 2(1), (2), and (3).[9] Eighty-seven Member states agreed to adopt the UNESCO Convention, while four States (Norway, the Russian Federation, Turkey, and Venezuela) voted against the instrument and

---

6 Dromgoole, "Reflections," 23–27.

7 Four States voted against the Convention (Norway, Russia, Turkey, and Venezuela), and fifteen State abstained (Brazil, Colombia, Czech Republic, France, Germany, Greece, Guinea-Bissau, Iceland, Israel, Netherlands, Paraguay, Sweden, Switzerland, United Kingdom, and Uruguay), but France later ratified the treaty. The United States declined to accede to the UCH Convention.

8 United Nations Convention on the Law of the Sea, Montego Bay Dec. 10, 1982, entered into force Nov. 10, 1994, UN Doc. A/CONF.62/122, 21 ILM 1621–1354 (1982), 1833 UNTS 397. See also Dromgoole, *supra*, note 2.

9 Article 2(1), (2) and (3) of the UNESCO Convention: "This Convention aims to ensure and strengthen the protection of underwater cultural heritage." "State Parties shall cooperate in the protection of underwater cultural heritage." And "States Parties shall preserve underwater cultural heritage for the benefit of humanity in conformity with the provision of this Convention.

fifteen abstained from the vote. Although some major maritime powers were against the treaty and have not ratified it, for the most part even these States have shown support for the basic principles and objectives of the UNESCO Convention.

Second, consistent with the principle of sovereign immunity in international law, the UNESCO Convention establishes a basic principle[10] on the rights of the coastal States and the sovereignty of State vessels,[11] followed by provisions elaborating how to apply this principle to different marine areas. In international waters, archipelagic waters and territorial sea, States Parties have the exclusive right to regulate and authorize activities directed at the UCH, but the coastal State "should inform" the flag State Party on these measures.[12] In the Exclusive Economic Zone (hereafter EEZ) or on the Continental Shelf (hereafter CS) activities directed at State vessels shall be conducted with the agreement of the flag State.[13] In the Area, the consent of flag State is needed for all the activities, which demonstrates respect the principle of sovereign immunity.[14]

Third, the UNESCO Convention acknowledges that the UCH is of public interest and has unique value for humanity. Hence, it should not be treated as an economic or natural resource in the seabed. Commercial exploitation of UCH for trade is fundamentally incompatible with the UNESCO Convention. Forty-eight States Parties to the Convention have respected the principle of non-commercial exploitation, and the principle is a feature of recent bilateral

---

10 Article 2(8) of the UNESCO Convention: "Consistent with State practice and international law, including the United Nations Convention on the Law of the Sea, nothing in this Convention shall be interpreted as modifying the rules of international law and State practice pertaining to sovereign immunities, not any State's rights with respect to its State vessels and aircraft."

11 Article 1(8) of the UNESCO Convention: "State vessels and aircraft" means warships and other vessels or aircraft that were owned or operated at the time of sinking only for government non-commercial purposes, that are identified as such and that meet the definition of underwater cultural heritage."

12 Article 7(1) of the UNESCO Convention: "State Parties, in the exercise of their sovereignty, have the exclusive right to regulate and authorize activities directed at underwater cultural heritage in their internal waters, archipelagic waters and territorial sea."

13 Article 10(7) of the UNESCO Convention: "Subject to the provisions of paragraph 2 and 4 of this Article, no activity directed at State vessels and aircraft shall be conducted without the agreement of the flag State and the collaboration of the Coordinating State."

14 Article 12(7): "No State Party shall undertake or authorize activities directed at State vessels and aircraft in the Area without the consent of the flag State."

and multilateral agreements concerning the UCH protection, which also seek to bar commercial exploitation.[15] Along with the principle of sovereign immunity, the principle of non-commercial exploitation is commonly recognized by the legal systems of Member States.

Fourth, the UNESCO Convention includes the provisions of the "Coordinating State." The UCH Coordination State scheme is regulated by Articles 9 to 12 of the UNESCO Convention. In the EEZ or on the CS, the "Coordinating State" is the State that locates the UCH, "unless it expressively declares that it does not wish to do so." In that case, the "Coordinating State" is the one that has a cultural, historical, or archaeological link to the concerned UCH. On the basis of this principle the Director-General of the UNESCO shall appoint a State Party to be a "Coordinating State" in the Area. The right of a "Coordinating State" is to implement measures of protection of UCH, to coordinate cooperation and consultation among States Parties and to communicate their decisions while acting on behalf of all interested States Parties.

The fifth valuable contribution of the UNESCO Convention is establishment of a report and notification system. The Convention requires any national of a State Party or any vessel flying the flag of a State Party, who discovers or intends to engage in activities directed at UCH located in the Area (i.e., beyond the outer limits of national jurisdiction) to report such activities to the State Party.[16] States Parties have different reporting and notification obligations in different maritime areas, according to the coastal state's competence and authority. In the territorial sea or the archipelagic waters, the coastal States have an obligation to inform the flag State or other States with an effective link to the UCH, with a view to protecting State vessels and aircraft.[17] In the EEZ, or on the CS

15 The 2004 international agreement among United Kingdom, France and Canada on *R.M.S. Titanic* is a good example.

16 Article 11(1) of the UNESCO Convention: "State Parties have a responsibility to protect underwater cultural heritage in the Are in conformity with this Convention and Article 149 of the United Nations Convention on the Law of the Sea. Accordingly, when a national, or a vessel flying the flag of a State Party, discovers or intends to engage in activities directed at underwater cultural heritage located in the Area, that State Party shall require its national or the master of the vessel, to report such discovery or activity to it.

17 Article 7(3) of the UNESCO Convention: "Within their archipelagic waters and territorial sea, in the exercise of their sovereignty and in recognition of general practice among States, States Parties, with a view to cooperating on the best of protecting State vessels and aircraft, should inform the flag State Party to this Convention and, if applicable, other States with a verifiable link, especially a cultural, historical, or archaeological link, with respect to the discovery of such identifiable State vessels and aircraft.

or in the Area, the State Party has an obligation to report UCH discoveries and activities to the other State concerned, or to all other State Parties.[18]

Limitations to UCH Protection

In contrast to these valuable contributions to the state of the law of UCH protection, the UNESCO Convention also has some limitations that impede effective protection of UCH.

The most important limitation is that the UNESCO Convention does not regulate the ownership of UCH. Throughout the text, the issue of UCH ownership is never addressed. It was decided early in the negotiations that the UNESCO Convention would not deal with matters of ownership. Accordingly, property title issues remain outside its scope.[19]

Yet, the UNESCO Convention does explain its relationship with the law of salvage and the law of finds. Article 4 of the UNESCO Convention stipulates:

> Any activities relating to UCH to which this Convention applies shall not be subject to the law of salvage or law of finds, unless it: (a) is authorized by the competent authorities, and (b) is in full conformity of this Convention, and (c) ensures that any recovery of the UCH achieves its maximum protection.

Second, there is some argument on the rights of the coastal State for UCH protection in the EEZ and on the CS. The UNESCO Convention provides that all States, Parties have a responsibility to protect UCH, and that a coastal State has the right to prohibit or authorize activities concerning UCH within its contiguous zone in order to prevent interference with its sovereign rights or jurisdiction.[20]

---

18 Article 9.1(b) of the UNESCO Convention: "Reporting and notification in the exclusive economic zone and on the continental shelf. (i) State Parties shall .... report such discovery or activity to that other State Party; (ii) alternatively, a State Party shall ... report such discovery or activity to all other State Parties." Article 11.3 of the UNESCO Convention: "Reporting and notification in the Area ... The Director-General shall promptly make available to all States Parties as such information supplied by States Parties."

19 Keun-Chen Fu, "Current Legal Developments China: A Chinese Perspective on the UNESCO Convention on the Protection of the Underwater Cultural Heritage," *The International Journal of Marine and Coastal Law* 18 (2003): 113.

20 Article 8 of the UNESCO Convention states: "Without prejudice to and in addition to Articles 9 and 10, and in accordance with Article 303, paragraph 2, of the United Nations Convention on the Law of the Sea, States Parties may regulate and authorize activities directed at underwater cultural heritage within their contiguous zone. In so doing, they

This provision can be problematic in terms of rights of coastal States and flag States, however.[21] Sarah Dromgoole, for example, insists that Article 9 and 10 of the UNESCO Convention are problematic because there are a number of constructive ambiguities in the wording of the text that allow for different interpretations. Any shift in the rights and jurisdiction of coastal States in the EEZ and on the CS from matters related purely to natural resources to matters related to UCH may precipitate open other functional competence of the coastal State to jurisdictional controversy. The general intention of Article 10 of the UNESCO Convention is to afford the coastal State frequent opportunity to play the role of Coordinating State on behalf of the States Parties as a whole and not in its own interest. The Coordinating State also will implement measures agreed with other interested States Parties, and where there is an immediate danger, the Coordinating State may take measures without prior consultation. Furthermore, there is the possibility that the Coordinating State could exercise a veto over activities to prevent them from taking place.

In sum, provisions concerned with the roles and rights of coastal States may infringe on the delicate balance between the rights and jurisdiction of coastal States and flag States in the EEZ and CS. While provisions of the UNESCO Convention contribute to international cooperation for UCH protection, some may generate new conflicts over the division of rights among States Parties.

Issues and Challenges in Korea's UCH Protection

The current legal and administrative system for UCH in Korea is not sufficient to cope with recent developments in international society. Furthermore, Korean regulations are not synchronized with the UNESCO Convention. This section addresses issues and challenges in Korea's approach to UCH.

---

shall require that the Rules be applied." Article 9(1) of the UNESCO Convention states: "All State Parties have a responsibility to protect underwater cultural heritage in the exclusive economic zone and continental shelf in conformity with this Convention." Article 10(1) & (2) of the UNESCO Convention: "No authorization shall be granted for an activity directed at underwater cultural heritage located in the exclusive economic zone or on the continental shelf except in conformity with the provision of this Article." "A State Party in whose exclusive economic zone or on whose continental shelf underwater cultural heritage is located has the right to prohibit or authorize any activity directed at such heritage to prevent interference with its sovereign rights or jurisdiction as provided for by international law including the United Nations Convention on the Law of the Sea.

21 Dromgoole, "Reflections," passim.

The first issue of concern is the definition of UCH in Korea. The negotiations considered whether to include a "significance criterion" in the definition of UCH in the UNESCO Convention. The conclusion of the debate was a compromise: UCH designation requires that material has been underwater for at least one hundred years, and that it has a cultural, historical, or archaeological character.[22] The *Korea BCHP Act* also states that UCH is "tangible cultural heritage contained in the structure, natural cases and fossils formed and deposited in underwater including seas, lakes and rivers and outstanding geological values as prescribed by Presidential Decree."

There is, however, a difference in age limit between the provision in the *Korea BCHP Act* and the one in the UNESCO Convention. While the *Korea BCHP Act* does not include any provision for a specific age limit of UCH, the UNESCO Convention set 100 years for it. Underwater cultural heritage less than 100 years old, might not be protected by the UNESCO Convention, if it were to be applied to supersede the *Korea BCHP Act*.

The second issue is concerned with ownership and the Coordinating State scheme of the UNESCO Convention regarding treasures in the EEZ and the CS. As shown in Figure 5.1, in Northeast Asia there are overlapping EEZ claims between Korea and China in the Yellow Sea and between Korea and Japan in the East Sea. In these areas of overlapping EEZs it is uncertain which State will be a coordinating State and which state will have ownership.

The position of the government of Korea is that the median line principle should be used to delimit EEZ claims. Korea's *Exclusive Economic Zone Act* stipulates in Article 5 that Korea shall not exercise its jurisdiction outside the median line of overlapping claimed area unless there is an agreement on the maritime boundary delimitation between the State Parties.

China's position on this matter is unclear.[23] The Regulation of the People's Republic of China on Protection and Administration of Underwater Cultural Relics (*China UCH Regulation*) enacted in 1989 and The Law of the People's Republic of China on the Protection of Cultural Relics (*China CR Law*) regulate ownership of UCH in China. Article 5 of the *China CR Law* states that all cultural relics remaining underground or in the internal waters or territorial

22 Article 1(1)(a) of the UNESCO Convention: "Underwater cultural heritage means all traces of human existence having a cultural, historical or archaeological character which have been partially or totally under water, periodically or continuously, for at least 100 years ..."

23 Hee-Cheol Yang. "Research on the Factors to Consider in the Yellow Sea Maritime Boundary Delimitation between Korea and China, and Countermeasures to It," *Korean Journal of International Law Studies* 57 (2012): 107–145. (*In Korean*).

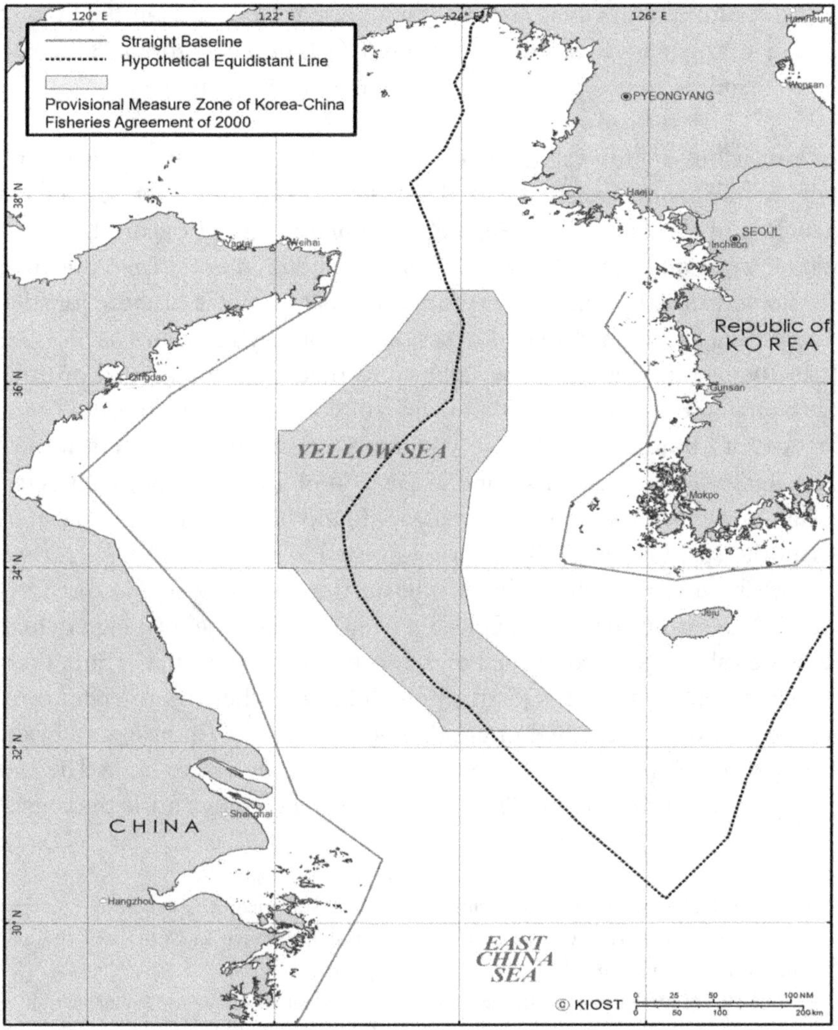

FIGURE 5.1 *Sea Borders in Yellow Sea.*

seas within boundaries of the China are owned by the State. Article 1 of *China UCH Regulation* states that the ownership of underwater cultural relics shall reside in China and that China shall exercise jurisdiction over them in two circumstances:

(1) all the cultural relics of Chinese origin, or of unidentified origin, or foreign origin that remain in the Chinese internal waters and territorial waters.

(2) cultural relics that are of Chinese origin or of unidentified origin that remain in sea areas outside the Chinese territorial waters but under Chinese jurisdiction according to the Chinese law.

These provisions are too vague to be applied with a sufficient degree of predictability, and are inconsistent with the UNESCO Convention. By Application of Article 2 of the Law of the People's Republic of China on the Exclusive Economic Zone and the Continental Shelf (*China EEZ and CS Law*) the areas of the sea "outside the Chinese territorial waters but under Chinese jurisdiction according to the Chinese law" describe the country's EEZ and CS.[24]

These two situations and provisions concerned can be analyzed in three aspects. First, there is no doubt that ownership of this UCH belongs to China if the UCH of Chinese origin is discovered in Chinese internal waters, territorial sea, and EEZ or on its CS. Second, if the UCH of foreign origin, such as foreign state-owned shipwrecks or aircraft, are found in Chinese internal waters or territorial sea, Beijing shall exercise jurisdiction over them. In such case, however, Chinese authorities should inform the flag State of the discovery of the UCH according to the principle of territoriality and the sovereign right of state-owned shipwrecks confirmed in the UNESCO Convention.[25] Third, there is a question of whether it is justified for Chinese authorities to claim ownership of UCH of unidentified origin that is found in the Chinese EEZ or CS. Neither UNCLOS[26] nor the UNESCO Convention addresses this issue. The Law of the Sea Convention and the *China EEZ and CS Law*[27] only entitle the coastal

24 Lina Liu & Shuguang Liu, "From Difference to Convergence: A Comparative Analysis of the Legal Protection of the Underwater Cultural Heritage in China and the 2001 Convention," *Comparative Maritime Law* 52 (2013): 162.

25 Article 7(1): "State Parties, in the exercise of their sovereignty, have the exclusive right to regulate and authorize activities directed at underwater cultural heritage in their internal waters, archipelagic waters and territorial sea.

26 Article 56(1) of UNCLOS: "In the exclusive economic zone, the coastal State has: (a) sovereign rights for the purpose of exploring, conserving and managing the natural resources," and Article 77(1) of UNCLOS: The coastal State exercises over the continental shelf sovereign rights for the purpose of exploring it and exploiting its natural resource."

27 Article 3 of the China EEZ and CS Law: "The People's Republic of China exercises sovereign rights for the purpose of exploring, exploiting, conserving and managing natural resources in water areas above the seabed, in seabed and subsoil of the exclusive economic zone as well as for the purpose of other economic activities of exploitation and exploration such as utilization of seawater, sea current and wind power to produce energy. In Article 6, it is stated that "The competent authorities of the People's Republic of China shall have the power in its exclusive economic zone to conserve and manage the trans-boundary

State to exercise sovereign rights over natural resources. The unidentified UCH, for which the establishment of a verifiable link with a certain State through current technology has failed, is definitely a part of the cultural heritage of humanity rather than natural resources of the coastal State.

With the geographical condition and differences in legal provisions of both countries described above, historical experiences among States in Northeast Asia may affect the issues of ownership and Coordinating State Status of unidentified UCH. In particular, three states in the region, China, Korea, and Japan, have a history of conflict as well as deep cultural and economic exchange. These geographical conditions and historical experiences have been a major source of dispute among these countries. Recently, Korea and China, for example, have revisited territorial disputes with Japan.

There is another issue related to the notification procedure. The *Korea BCHP Act* does not regulate UCH located on the CS, seabed, or outside the territorial sea. Given that the UNESCO Convention stipulates that States Parties have different reporting and notification obligations in the distinct maritime zones, Korea's legal framework for UCH protection is incomplete. In addition, there are several issues that are not provided in the *Korea BCHP Act* with regard to UCH protection compared to the UNESCO Convention, such as designating UCH protection areas, standardization of recovery criteria and procedure, and the creation of a preventive system against illicit recovery, export, and entry of UCH.

Concluding Remarks

Since the excavation of *Shinan Underwater Cultural Heritage* started in 1975, Korea has developed administrative and legal frameworks for UCH. While Korea has focused on the establishment of an administrative mechanism for UCH protection in the earlier period, it was not until 2010 that it developed legal arrangements to respond to the changing international and domestic environments concerned with the protection of UCH. There are, however, many issues remain to be addressed in order to cope with international development in UCH protection. First, Korea should begin cooperation with neighboring states to prevent conflicts over UCH. This will be a difficult task because it is subject to disputes over maritime boundary delimitation and regional politics.

---

population, highly migrating fishes, marine mammals, anadromous spawning population originating in rivers of the People's Republic of China, and downstream spawning fishes spreading most of their life cycles in the water areas of the People's Republic of China.

Korea, like China, Russia, and Japan has not ratified the UNESCO Convention yet. This is because the contents of the UNESCO Convention may affect the pending issues of maritime jurisdiction and sunken ships in Northeast Asia. These issues are volatile in domestic politics as well as in international context in Northeast Asian states. None of them wants to be bound by the UNESCO Convention in order to remain free to maximize their respective national interests. In this situation, Korea is unlikely to move in for ratification before neighboring states do it.

Nevertheless, it is necessary to supplement the *KOREA BCHP ACT* to increase protection of UCH.

Given that the situation, there are two options for Korea. The first option is to add more provisions on UCH in the *Korea BCHP Act* that better correspond to the provisions of the UNESCO Convention. For example, a new provision on reporting and notification can be inserted into the *Korea BCHP Act* in a way that stipulates separate reporting notification procedures for UCH that is located in the EEZ, on the CS, and on the seabed or outside the territorial sea. The other option is to stipulate key issues in the *Korea BCHP Act* at a minimum level of explicitness and at the same time to state that remaining issues, which are not critical, will be consistent with the provisions both of the UNESCO Convention and the UNCLOS. The most important thing is to raise awareness of public and policy-makers, as any improvement in legal and administrative arrangements for better protection of UCH requires both.

PART 2

Maritime Security

CHAPTER 6

The Laws of Civil Disobedience in the Maritime Domain

*James Kraska**

Introduction

Protest and civil disobedience on the high seas occur within a legal space that lies between two bodies of international law—international human rights norms governing the right of free expression, petition, and political participation—and the principles of good order and safety in the global maritime commons, embodied in the international law of the sea. Disruptive speech by highly motivated elements of civil society, such as environmental protesters, poses a particularly challenging set of issues and choices over free speech and public safety, and human rights versus national sovereignty and sovereign maritime rights. This lacuna exposes a key difference between law that regulates protest on land and protest at sea. Efforts to address civil disobedience on land unfold within clear boundaries of sovereignty and jurisdiction. Police authorities have a dual role to protect and sometimes separate protesters from other citizens in order to preserve public safety and order, while they simultaneously protect the right of protesters to free expression. In contrast, civil disobedience at sea occurs under conditions and geographic areas of shared authority, overlapping jurisdiction, and functional governance divided between flag states and coastal states. If a ship is entering or leaving port, then port authorities may also come into play.

If the game of chess serves as a metaphor for legal issues surrounding protest on land; then protest at sea is akin to three-dimensional chess.[1]

The amorphous conditions on the oceans, with sometimes opaque legal doctrine and authority and the practical difficulty of maintaining maritime order, emboldens protesters and activists to use tactics that they would not or could not effect on land. "Green" activists are driven to "direct action" to push

* Howard S. Levie Professor of International Law and Research Director, Stockton Center for the Study of International Law, U.S. Naval War College.

1 Three dimensional (3D) chess was a feature of the television series, Star Trek. *See* http://www.chessvariants.org/3d.dir/startrek.html.

 | DOI 10.1163/9789004311442_008

the envelope of safe and acceptable behavior in order to promote social causes that they deem ineffectively addressed by society or government. Flag states, for example, routinely fail to enforce on their ships internationally accepted standards of vessel source pollution and marine environmental protection, seafarer labor standards, and marine animal and habitat conservation. Coastal States, which are charged with the exercise of stewardship, exceed maximum sustainable yields on fishing and ignore land-based sources of marine pollution. On the other hand, seafarers regard these protesters as little better than pirates or brigands who recklessly flaunt maritime mores and timeless rules of the road to endanger lives and property. Sympathy with their message makes authorities feckless and timid in the face of direct action, opening the door to incidents at sea. Frustration runs high on both ends and can lead to violence.

Coastal States may exercise sovereignty over the water only to a distance of 12 nautical miles from a lawfully drawn baseline, which generally runs along the shoreline at the low water mark.[2] Most of the oceans lie beyond the territorial sovereignty or jurisdiction of any state, so national law has a tenuous reach beyond the territorial sea. States may apply their laws on board ships that fly their flag, and they also may use jurisdictional devices, such as safety and security zones, constructive presence, or passive personality or nationality, to extend their laws to a limited range of activities and functional purposes beyond the territorial sea. These efforts, however, are idiosyncratic and derive from foreign relations law in national courts and coastal state implementation of the international law of the sea.

In cases of maritime civil disobedience, the rules concerning maritime safety at sea also must accommodate international human rights law. This chapter explores whether these two areas of law can be reconciled to form a coherent framework that preserves the maritime order of the law of the sea, while also promoting the free expression and liberal values that undergird human rights law.

Whereas international human rights law is relatively new and focused on ensuring human dignity and notions of individual and social justice, the international law of the sea emerged from the Enlightenment, with a lineage that dates back to the ancient world.[3] The law of the sea is focused on securing

2 Art. 3, United Nations Convention on the Law of the Sea, done at Montego Bay, Jamaica, 10 December 1982, entered into force 10 November 1994, UN Doc. A/CONF.62/122, 21 I.L.M. 1621–1354 (1982), 1833 U.N.T.S. 397 (UNCLOS).

3 Percy Thomas Fenn, Jr., "Justinian and the Freedom of the Sea," *American Journal of International Law* 19 (Oct. 1925): 716.

a public order of the oceans in order to maintain stable maritime political boundaries and facilitate seaborne trade. A cornerstone of international law and a fundamental construction of world order, the law of the sea applies throughout 70 percent of the surface of the earth, the surface, water column and seabed of the ocean, and the airspace above it. Although human rights law is more recent than the law of the sea, it also has deep roots in ancient Jewish and Christian theology, the Greek legacy of democracy, Roman law of Citizenship, and it is even more ingrained in the collective political conscience.[4] By preserving the right of people to petition for change or reform, the rights of protest and free expression are central to participatory democracy, social progress, and political stability.[5]

In the land territory of a state, the line separating lawful protest and unlawful violence or threats to civil order is not always clear. In such cases, however, police authorities have jurisdiction over both the protesters and the targets of their dissent. The principle of state sovereignty means that states have wide latitude to shape laws concerning free expression that best promote cultural values and varying levels of security. National laws and courts balance competing interests among police and protesters, and between activists and third party targets of protest. International law, however, has a different role within the community of nations. In monist systems, public international law is reflected directly in national laws; in dualist systems, international law is implemented through specific legislation. In either case, the role of the state is more circumspect and less absolute in development and enforcement of international law than national law. Furthermore, whereas national laws typically bind individual citizens and groups, international law generally binds states as subjects of the law.

Consequently, the legal environment at sea is more complex than on land. Rather than having to consider only the competing laws and divergent public policies within a single polity, incidents of maritime protest typically fall within the jurisdiction of several states. Protestors at sea may be from one or more states, and they may not share nationality, either with the people they are protesting or law enforcement authorities on scene. Vessels that operate together as a coordinated protest may be registered in different countries, with

4 *See generally* Dinah L. Shelton, "An Introduction to the History of International Human Rights Law: Lectures given at the International Institute of Human Rights, Strasbourg, France," *George Washington University Legal Studies Research Paper No. 346* (July 2003).

5 Alexander Tsesis, "Free Speech Constitutionalism," *University of Illinois Law Review* 2015 (2015): 1015, 1034.

separate sovereigns responsible for regulation of their behavior.[6] Adversaries may come from widely contradictory legal and political systems that reflect conflicting values and opposing cultural norms. It is daunting to navigate among these competing authorities and epistemic communities.

The debate has circulated around the text of a 2008 draft Code of Conduct offered by the Government of Japan for consideration by the member States of the International Maritime Organization (IMO). The Code followed resolutions adopted by the Japanese Diet that condemned "dangerous actions at sea," including collisions of ships by non-State activists.[7] The Code of Conduct is the only instrument that explicitly balances the rights of free expression at sea with the imperative for navigational safety. Because the rules within the draft Code reflect existing obligations, however, it was not adopted at IMO. Instead, in 2010 the Maritime Safety Committee adopted a resolution that stipulated the importance of compliance with long-standing norms of seamanship, flag State jurisdiction and responsibilities, and enumerated instruments, such as the Collision Regulations and the Convention on the Suppression of Unlawful Acts against the Safety of Maritime Navigation.

Human Rights Law

National laws, tempered by international norms, set forth the boundaries and limits of civil disobedience, and balance the need for public order and national security with the individual right of free expression, free assembly, and meaningful participation in the political process. The right of protest is considered one of the most fundamental human rights. The imperative to protect political conscience and expression while preserving public order poses one of the most vexing problems in jurisprudence.

The Heritage of Civil Disobedience

The heritage of civil disobedience runs deep in Western culture. In ancient Greece, Sophocles' character Antigone was torn between her moral code and the law.[8] She had a religious duty to bury the body of her brother, while the

6 See, e.g., James Kraska, "Rule Selection in the Case of Israel's Naval Blockade of Gaza: Law of Naval Warfare or Law of the Sea?," *Yearbook of International Humanitarian Law* 13 (2010): 367, 370.

7 IMO Doc. NAV 54/10/1, Code of Conduct During Demonstrations/Campaigns Against Ships on the High Seas, Apr. 25, 2008, para. 8.

8 Sophocles, Antigone (J.E. Thomas, Trans.) (Clayton: Prestwick, 2005).

state commanded that his body remain unburied and cast on an open plain outside the city walls, picked at by wild dogs and vultures. Creon's law required Antigone to yield her conscience and abandon what she knew to be rationale and moral. Tragically, she chose to obey her conscience and was punished. Socrates considered disobedience of the law as a threat to the fabric of society, risking the erosion of the very foundations of social order. On the other hand, Henry David Thoreau wrote a classic dictum on the obligation to protest oppression: "under a government which imprisons any unjustly, the true place for a just man is also a prison."[9]

Human rights lingered throughout the Enlightenment and the Renaissance periods, but it was not until the atrocities of the early-twentieth century that they became pillars of international law and politics. The right to protest was recognized by the community of nations after World War II, nearly one century after the strategic upheaval of the Revolutions of 1848, which in any event, despite their frightening disruption of Continental stability, lacked coherence or much success.[10] The United Nations system, however, incorporated human rights into its seminal documents, including the Charter of the United Nations. Civil disobedience and protest has evolved from the stigma that it is purely anarchic and destructive to social fabric of society toward a more nuanced view that disobedience can sometimes serve a positive social and political function. Furthermore, there is the perception that a healthy legal system can accommodate some measure of departure from rules without branding the acts as unlawful.[11] In the contemporary era, the power and impact of movements of civil disobedience has been nothing short of breathtaking.

In recent years the process of transformatory socio-political change brought about through civil protest only has accelerated. The "Autumn of Nations" movement in Eastern Europe in 1989 felled the Soviet Union and formed a new identity for the Eastern bloc states and the Russian Federation.[12] The people's revolution in Yugoslavia in 2000,[13] the Rose Revolution in Georgia

9 Henry David Thoreau, *Civil Disobedience*, in WALDEN AND OTHER WRITINGS 646 (New York: Random House, Brooks Atkinson, ed. 1937) (2000).

10 Priscilla Smith Robertson, *Revolutions of 1848* (Princeton: Princeton University Press 1968): 419.

11 Mortimer R. Kadish and Sanford H. Kadish, *Discretion to Disobey* 69–102 (1973).

12 Serge Schmemann, "Clamor in the East: The Border is Open," *N. Y. Times*, Nov. 11, 1989, at p. 1.

13 Roger Cohen, "The World: Stare Down; Serbs Shed Their Fear and Make a Revolution," *N. Y. Times*, Oct. 8, 2000, § 4, p. 1.

in 2003,[14] Ukraine's Orange Revolution in 2004,[15] and the Tulip Revolution in Kyrgyzstan in 2005[16] generated massive street protests that overthrew established autocracies.

In the wake of the implosion of the world economy in 2008, Spanish *indignados* ("the indignant"),[17] Greek anti-austerity protests, and Montreal's *casseroles* have redrawn the political map. The Arab Spring, which was ignited in Tunisia on December 18, 2010 and led to the downfall of the government, spread to Benghazi, Libya, in February 2011, and cast that country into civil war.[18] The consequences of the movement transformed the Maghreb and still reverberate from Egypt to southern Yemen and Syria.

Originating in the Western world, "Occupy" protests and encampments sprang up in 70 countries. In the United States, active encampments appeared in Washington, D.C. and Oakland, California, and as far away as Hong Kong. One of the principle tenets of the Occupy movement was to purposefully revive the idea of the commons—areas open to everyone to exchange ideas and represent different perspectives.[19] In this respect, the Occupy experience leverages an essential aspect of shared space, and presages the issues that arise from protest in the ocean commons. Just as the Occupy encampments are part of the statement of the movement, so too are vessels at sea, which serve as a staging areas for displays of free expression and civil disobedience. In both cases, participants champion fundamental human rights. Similarly, the presence of a vessel at sea, often with flags and placards to convey political messages, serve as a statement of protest. Moreover, just as their land-based counterparts mix speech and employ direct action to further their agenda, maritime protesters pursue a variety of environmental and political causes on the water.

---

14 Peter Baker, "Opposition Leader Sweeps Georgia's Presidential Vote," *Washington Post*, Jan. 5, 2005, at p. A11.

15 Clifford J. Levy, "Orange Revolution Hero Claims Victory in Ukraine," *International Herald Tribune*, Oct. 2, 2007, at p. 5.

16 Richard Spencer, "Revolution Takes Hold in Kyrgyzstan," *Daily Telegraph* (London), Mar. 22, 2005, at p. 14.

17 Alan Clendenning, "Spain Beset by Protests of Economic Policy," *Washington Post*, May 13, 2012, at p. A16.

18 Anthony Shadid, After Arab Revolts, Reigns of Uncertainty, *N. Y. Times*, Aug. 25, 2011, at p. 1.

19 *Suppressing Protest: Human Rights Violations in the U. S. Response to Occupy Wall Street* 10 (The Global Justice Clinic (NYU School of Law) and the Walter Leitner International Human Rights Clinic (Fordham Law School, 2012).

Protest as a Human Right

International human rights law broadly construes the concept of freedom of speech and freedom of expression. The civil right to free speech and assembly is reflected in the International Covenant on Civil and Political Rights (ICCPR) and the Universal Declaration on Human Rights (UNDHR). All persons also enjoy the right of peaceful assembly. The right to assembly may not be restricted, except by rules "imposed in conformity with the law and which are necessary ... in the interests of national security or public safety. ..."[20] The right to freedom of thought and conscience is the cornerstone of these rights.[21] Freedom of peaceful assembly, petition, and the right to engage in political expression are considered as fundamental to human rights, and are protected in international law. These rights are to be broadly interpreted:

> Right to freedom of expression ... shall include freedom to seek, receive and impart information, and ideas of all kinds, regardless of frontiers, either orally, in writing or in print, in the form of art, or through any other media of his choice[22]

Three regional human rights treaties, the European Convention on Human Rights,[23] the African Charter on Human and Peoples' Rights,[24] and the American Convention on Human Rights[25] contain similar language on the human right of free expression. Such human rights derive from the "inherent dignity and equal and inalienable rights of all humans,"[26] and may be considered the foundation of freedom, justice, and peace in the world."[27] The right of free expression is constrained, however, by special duties or responsibilities.

---

20 Art. 21, International Covenant on Civil and Political Rights, done at New York, 16 December 1966, entered into force New York, entered into force 23 March 1976, 999 U.N.T.S. 171 and 1057 U.N.T.S. 407, 6 I.L.M. 368 (1967) (ICCPR). *See also* Articles 19 and 20, 1948 Universal Declaration of Human Rights, GA Res. 217A (III), UN Doc. A/810 at 71 (1948) (UNDHR).

21 *Ibid.*, para. 18(1).

22 *Ibid.*, para. 19(2).

23 Convention for the Protection of Human Rights and Fundamental Freedoms, Nov. 4, 1950, E.T.S. 5; 213 U.N.T.S. 221, entered into force Sept. 3, 1953, Article 10(1).

24 African Charter on Human and Peoples' Rights, June 27, 1981, 1520 U.N.T.S. 217, 245; 21 I.L.M. 58, 59 (1982), Article 9.

25 American Convention on Human Rights, Nov. 21, 1969, O.A.S. T.S. No. 36; 1144 U.N.T.S. 143; S. Treaty Doc. No. 95–21, 9 I.L.M. 99 (1969), Article 13.

26 *Ibid.*, Pream. and UN Doc. GA Res. 2200A (XXI), Dec. 16, 1966.

27 Pream. ICCPR and UN Doc. GA Res. 2200A (XXI), Dec. 16, 1966.

States may adopt and enforce laws to limit the right in order to protect national security or maintain public order.[28] National laws differ, but each state is obligated to adopt rules that give effect to the ICCPR.[29] States may derogate from their commitments under the ICCPR only in cases of public emergency which threatened the life of the Nation. . . ."[30]

In a 1998 case, for example, the European Court of Human Rights (ECHR) contended with two protesters, Rush and Steel, who deliberately walked into the line of fire of grouse hunters to demonstrate opposition to bird hunting. The Court was willing to treat physical behavior on the part of the protesters as a form of expression protected under Article 10(1) of the European Convention on Human Rights.[31] Where the conduct threatened public order, however, the Court afforded authorities considerable latitude to restrict or terminate the exercise of that type of symbolic expression. The ECHR held:

> The Court recalls that the . . . [men] were arrested while protesting against a grouse shoot and the extension of a motorway respectively. It is true that these protests took the form of physically impeding the activities of which the applicants disapproved, but the Court considers nonetheless that they constituted expressions of opinion within the meaning of Article 10 [of the European Convention on Human Rights]. The measures taken against the applicants were, therefore, interferences with their right to freedom of expression.[32]

The Court also found that the arrests and detention of the activists were "prescribed by law" in order to prevent disorder and protect the rights of others. The judges were divided, however, on the issue of whether the government's actions were "necessary in a democratic society."[33] In a 5–4 majority, the Court

28 ICCPR, para. 19(3)(b).

29 *Ibid.*, para. 2(2).

30 *Ibid.*, para. 4(1). No derogation from fundamental human rights is permitted, however. These rights are reflected in articles 6, 7, 8 (paragraphs 1 and 2), 11, 15, 16 and 18 of the ICCPR. *Ibid.*, para. 4(2).

31 European Convention for the Protection of Human Rights and Fundamental Freedoms, as amended by Protocols Nos. 11 and 14, entered into force 3 September 1953, ETS 5, 213 U.N.T.S. 222.

32 *Case of Steel and Others v. United Kingdom*, 67/1997/851/1058, Sept. 23, 1998, para. 92. *See also, Chorherr v. Austria* (1993) 17 EHRR 358, A.266–B and Alastair Mowbray, *Cases and Materials on the European Convention on Human Rights* 706 (Oxford: Oxford University Press 2d. ed. 2007).

33 *Case of Steel and Others v. United Kingdom*, 67/1997/851/1058, Sept. 23, 1998, para. 101.

determined that Ms. Steel's detention was not disproportionate, since her behavior created a "danger of serious physical injury to herself and others" and risked "culminating in disorder and violence."[34] In dissent, Judges Vlaticos and Makarczyk found that detention for 44 hours followed by a sentence of 28 days in prison in response to Steel's "extreme manner" of jumping up and down in front of a grouse hunter to prevent him from killing a "feathered friend" was "manifestly extreme" and a violation of the ECHR.

The Court found in favor of Ms. Needham, Mr. Polden and Mr. Cole. Police conduct was not lawful, as the Court was not "satisfied that the police had grounds to reasonably apprehend" the applicants.[35] It was unclear whether their part in the protests would cause a breach of the peace, and therefore whether police interference with their right of free expression was "disproportionate to the aims of preventing disorder and protecting the rights of others...."[36]

The right to petition may be protected, but the means of doing so are not unlimited. In the United States, the Supreme Court has recognized reasonable time, place, and manner restrictions on free speech.[37] European courts have adopted a similar approach. In *Grigoriades v. Greece* (1997), the ECHR addressed the limits of freedom of expression for an officer in the Greek armed forces.[38] The applicant was convicted in a military proceeding of "insulting the army," but the European Court ruled that the sentence of three months imposed on him constituted an interference with his freedom of expression under Article 10(1) of the European Convention on Human Rights. Although the ECHR agreed with the government that the interference in the applicant's free expression was "prescribed by law" and in pursuit of a legitimate governmental aim, which was good order and discipline in the armed forces, it held in favor of the applicant that his prosecution and punishment was not justified as "necessary in a democratic society."[39] The applicant, however, was exonerated because his intemperate remarks were contained in a letter sent to his commanding officer that was critical of the army as an institution. It was not insulting to any person in particular, and most importantly, it was not made public.[40]

---

34 *Ibid.*, para. 105.

35 *Ibid.*, para. 110.

36 *Ibid.*, para. 110.

37 *Ward v. Rock against Racism*, 491 U.S. 781, 791 (1989).

38 *Grigoriades v. Greece* (1997) 27 EHRR 464, 1977-VII p. 2575, http://www.bailii.org/eu/cases/ECHR/1997/93.html.

39 *Case of Steel and Others v. United Kingdom*, *supra*, note 33, paras. 32–42.

40 *Ibid.*, paras. 44–48.

Separating lawful protest with unlawful crime is often difficult, as it requires lawmakers, state police authorities, and judicial systems to draw lines that maximize individual conscience and autonomy while protecting the interests of the greater society. Courts in Europe and North America have the longest and most durable history of human rights jurisprudence, and they set the standard for judicial review, although their records are less than perfect. More generally, there is a vast gulf that separates the idealized standards of human rights and the reality of either incomplete compliance or abject failure to comply in many parts of the world. Furthermore, while international human rights law is imperfectly applied by states, overlapping and sometimes competing national jurisdictions at sea produces even greater challenges than on land.

On the other hand, the mores and norms reinforced through a millennia of seafarer practice may provide a basis for fashioning an international approach. Concepts such as a duty to assist mariners in distress, obligations to avoid collision at sea, and the authority of the master of a vessel and the flag state, provide a framework for analysis.

Maritime Protest

Protests at sea have coalesced around concern for the marine environment, including protection of marine mammals. Since the 1990s, environmental protesters have become especially sure-footed in their campaigns to improve the health and quality of the marine environment. Greenpeace and other non-governmental organizations have shaped international negotiations, prepared statutory law and treaty text for governments, and influenced the political narrative and outcomes nationally and globally. Two issues have particular salience: the debate over whaling, which is a product of a split in the International Whaling Commission (IWC), and plans for oil and gas development in the Arctic Ocean, which are made by possible, ironically enough, by climate change that reduces ice cover and provides greater access for offshore development. For several decades, the IWC has been evenly divided between conservationist states, such as Japan and Norway that view whales as a resource to be sustainably harvested, and protectionist perspectives, which see marine mammals as a class of animals under threat and in need of being saved. The IWC is deadlocked by the "normative gulf" between the two approaches.[41]

---

41 *See* Harry N. Scheiber, "Historical Memory, Cultural Claims, and Environmental Ethics in the Jurisprudence of Whaling Regulation," *Ocean and Coastal Management* 38 (1998): 5–40 and Gary Nagtzaam, "The Temptation of the Impossible: Normative Contestation and Deadlock at the International Whaling Commission," *Appalachian Natural Resources Law Journal* 8 (2013–2014): 109, 139–146.

In response to the lack of momentum to protect cetaceans, non-governmental environmental groups have initiated more ardent or confrontational methods. Australian activists from Sea Shepherd Conservation Society, for example, pursue Japanese whalers on the high seas. Similarly, Greenpeace protesters harass offshore oil developments in the Arctic Ocean as a last-ditch effort to preserve a pristine environment from damage by industrialization.

Greenpeace has a long history of maritime activism. In July 1999, for example, the group towed a raft carrying a large inflatable elephant into the path of the *Pacific Teal*, a British ship carrying plutonium nuclear fuel bound for Japan.[42] The vessel was driven back to the English port of Barrow. Less than two years later, the self-declared "Nuclear-Free Tasman Flotilla" was unsuccessful in locating and obstructing MOX fuel transport ships by using a string of Australian yachts held fast.[43] Eighteen months later, however, MOX cargo passed again through the Tasman Sea and Australian Greenpeace activists swarmed the ship with Zodiac inflatable small boats.

More recently, Greenpeace has been active in efforts to preserve the Arctic marine environment. On February 24, 2012, seven Greenpeace activists (including actress and activist Lucy Lawless) boarded the Shell Oil drilling ship Noble Discovery while the ship lay in Port Taranki, New Plymouth, New Zealand. The ship was slated to get underway for the Chukchi Sea off the coast of Alaska. The protesters hung banners on the ship that declared, "Save the Arctic" and "Stop Shell." In September 2013, the Greenpeace Motor Yacht (M/Y) *Arctic Sunrise* entered Russia's exclusive economic zone (EEZ) to protest offshore oil development in the Prirazlomnoya oil field, which is located in the Pechora Sea south of the island of Novaya Zemlya. *Arctic Sunrise* is a Netherlands-flagged 949-gross ton icebreaker.[44] The Prirazlomnoy oil field holds an estimated 610 million to 1.4 billion barrels of oil, and Greenpeace sought "direct action" to stop the drilling program. Thirty environmental activists and crew from 18 foreign states were on board the ship.

The protesters also targeted Gazprom's Offshore Ice-resistant Fixed Platform (OIRFP) "Prirazlomnoya." The Prirazlomnoya OIRFP is the first Arctic-class oil drilling platform in the world, which lies stationary approximately 32 miles

42 "UK 'Regret' over Japan Nuclear Row," *BBC News*, July 10, 1999, http://news.bbc.co.uk/2/hi/uk_news/707983.stm.

43 Larry Schwartz, "The Hunt for Red Hot Cargo," *Sunday Age* (Melbourne), January 28, 2001, p. 4.

44 The ship is owned by Stichting Phoenix, Dorpsstraat 3, 1151 AC Broek in Waterland, Amsterdam, The Netherlands, which is owned by Stichting Greenpeace Council, Amsterdam, under IMO number 7382902.

North of Nenets Autonomous Okrug in water depths of 19–20 meters.[45] On 16 September 2013, the Russian Coast Guard vessel *Ladoga* warned the *Arctic Sunrise* over bridge-to-bridge radio against infringement of Russian rules adopted in accordance with UNCLOS to ensure safety of shipping in the vicinity of the Prirazlomnaya.

The next day, *Arctic Sunrise* set a course toward the Prirazlomnaya rig and conducted maneuvers "on the boundary area adjacent to the platform."[46] The *Ladoga* warned that the protesters were not permitted to enter the area within a radius of 3 nautical miles (nm) around the OIRFP where there is a danger to shipping.[47] The 3-nm zone is measured from the center point of the Prirazlomnoya platform. The associated Notice to Mariners states:

> Vessels should remain outside this safety zone. However, if there is a need for passage through the rig safety zone, permission must be obtained from the operator of the drilling platform before entering this safety zone. There will be a frequent shuttle tanker service between the platform and Murmansk, about 750 miles [East/Northeast]. The Prirazlomnaya Offshore Platform is also equipped with [Automatic Identification System].[48]

The *Ladoga* also warned the ship away from a declared inner 500-meter safety zone that also surrounds the Prirazlomnaya rig. On 18 September, the *Arctic Sunrise* launched rigid hull inflatable speedboats near the perimeter of the 3-nm zone, and the craft headed toward the Prirazlomnaya platform. Three of the boats were carrying a large fiberglass "safety pod" designed to be suspended from the oil rig and provide shelter for the activists. Several Greenpeace protesters successfully boarded the Prirazlomnaya rig from these small boats. Subsequently, activists who had scaled the rig were arrested by officers of the Russian Coast Guard. The Russian coast guard later claimed

---

45 Nat'l Geospatial Agency, *Publication 183*, Sailing Directions (Enroute), North Coast of Russia (Springfield: National Geospatial-Intelligence Agency 2014), at p. 60 (Pub. 183).

46 Annex 7, Submission of dispute to arbitration in case of the Arctic Sunrise, The Kingdom of the Netherlands v. The Russian Federation, International Tribunal for the Law of the Sea Case No. 22, 4 October 2013 (ITLOS Case No. 22 Submission).

47 Hydrographic Office, *Notice to Mariners No. 51* (London 2011) states: "[v]essels should not enter the safety zone of the marine ice-stable platform without permission of the operator of the platform." *See also* Annex 2 to Annex 8, ITLOS Case No. 22 Submission, ¶ 8.

48 Pub. 183, *supra*, note 45, at p. 60.

Greenpeace maneuvers were in contravention of the International Regulations for Preventing Collisions at Sea, 1972. Russia claimed the master of the *Arctic Sunrise* was unresponsive to Russian coast guard inquiries, the ship and its associated speedboats endangered shipping and the safety of the OIRFP, and put at risk Russian divers in the water.[49]

On 19 September, Russian authorities boarded by helicopter and arrested the vessel *Arctic Sunrise* as it navigated within the Russian EEZ.[50] The Russian coast guard fired 11 warning shots at the *Arctic Sunrise*.[51] The ship was 5 nm from the Prirazlomnaya rig, and the crew of nearly thirty individuals was detained. The vessel and crew were transferred to the port of Murmansk, and detained by Russian authorities.[52] The protesters were charged with maritime piracy, and the criminal complaint later was commuted to charges of "hooliganism."[53] The Netherlands brought suit at the International Tribunal for the Law of the Sea (ITLOS), for provisional measures, including release of the ship and its crew, under article 290 of UNCLOS. Russia declined to participate in the proceedings. On November 22, 2013, ITLOS ordered by a vote of 19 to 2 release of the ship and detained persons upon posting of a bond of €3.6 million. The case gradually became moot as the crew members were released after about 100 days in captivity, and the vessel was unexpectedly

49 Annex 7, ITLOS Case No. 22 Submission.

50 *Ibid.*

51 Greenpeace International, Statement of facts concerning the Boarding and Detention of the MY *Arctic Sunrise* and the Judicial Proceedings against all 30 persons on board, at p. 47, paras. 22 and 25, *reprinted at* Annex 8, ITLOS Case No. 22 Submission.

52 Tracy McVeigh, "The Arctic 30: Sailors, Journalists and Scientists Held in Russian Jail," *The Observer* (England), October 13, 2013, p. 30.

53 Article 227(3), ("piracy committed by an organized group"), The Criminal Code of The Russian Federation No. 63-Fz of June 13, 1996 (with Amendments and Additions of May 27, June 25, 1998, February 9, 15, March 18, July 9, 1999, March 9, 20, June 19, August 7, November 17, December 29, 2001, March 4, 14, May 7, June 25, July 24, 25, October 31, 2002, March 11, April 8, July 4, 7, December 8, 2003, July 21, 26, December 28, 2004, July 21, December 19, 2005, January 5, July 27, December 4, 30, 2006, April 9, May 10, July 24, November 4, December 1, 6, 2007, February 14, April 8, May 13, July 22, November 25, December 22, 25, 30, 2008, February 13, April 28, June 3, 29, July 24, 27, 29, October 30, November 3, 9, December 17, 27, 29, 2009, February 21, March 29, April 5, 7, May 6, 19, June 17, July 1, 22, 27, October 4, November 29, December 9, 23, 28, 29, 2010, March 7, April 6, May 4, July 11, 20, 21, November 7, 21, December 6, 7, 2011, February 29, March 1, 2012), a by the State Duma on May 24, 1996 and adopted by the Federation Council on June 5, 1996.

released from the port of Murmansk the following summer.[54] One of the activists concluded, "They didn't lock us up for what we did. They locked us up for what we stood for."[55] But the case underscores the contemporary collision between human rights and maritime safety and good order at sea. Maritime protests raise questions about the protection of the human right of free expression and the limit of civil disobedience within the regimes of the international law of the sea.[56]

The Law of the Sea

The structure of the international law of the sea is designed to accommodate the interests of the world community in safe and environmentally sound navigation and sustainable resource exploitation and management. The rules create a stability of expectations that builds confidence and prevents conflict from interactions at sea between ships and aircraft of other nations, facilitates international commerce and global prosperity, and provides a framework for claiming and using ocean resources. Maritime law and order is derived from customary international law and treaty law; and much of state practice is codified in the United Nations Convention on the Law of the Sea (UNCLOS).[57]

Flag State and Coastal State Jurisdiction

Generally, flag states rather than coastal states, have jurisdiction over persons on board ships that fly their flag. The principle of exclusive flag state jurisdiction is a fundamental concept of customary international law. This principle means that a vessel exercising high seas freedoms is generally subject only to the jurisdiction of the flag state. Vessels are not subject to boarding, search, seizure or arrest by any nation other than their flag State, unless a

54 Updates from the Arctic Sunrise Activists, Internet Website of Greenpeace International, http://www.greenpeace.org/international/en/news/features/From-peaceful-action-to-dramatic-seizure-a-timeline-of-events-since-the-Arctic-Sunrise-took-action-September-18-CET/#a14.

55 *Ibid.*

56 The term "regime" is used to denote "principles, norms, rules and decision-making procedures around which actor expectations converge in a given issue-area." Stephen D. Krasner, "Structural Causes and Regime Consequences: Regimes as Intervening Variables," *International Organization* 36 (Spring 1982): 185.

57 J. Ashley Roach, "Today's Customary International Law of the Sea," *Ocean Development and International Law* 45 (2014): 239–259.

jurisdictional exception applies.[58] The exceptions to flag state consent for bordering are found in UNCLOS and the Charter of the United Nations, and include stateless vessels, slave trafficking, maritime piracy, illegal broadcast into the coastal State, the inherent right of self-defense, and, sometimes controversially, the consent of the master of the vessel. If a flag state consents to the exercise of a coastal state's jurisdiction or waives objection to enforcement, for example, such a vessel may be subject to the jurisdiction of a coastal state. Exclusive flag state jurisdiction is especially strong in the EEZ, where foreign flagged ship and aircraft exercise high seas freedom of navigation and freedom of overflight, and other internationally lawful uses of the sea associated with the operation of ships and aircraft.[59] In such areas, coastal states exercise sovereign rights to explore, exploit, conserve, and manage the living resources to a distance of 200 nm from lawfully drawn baselines, which normally are coterminous with the low water mark.[60]

Coastal states may conduct maritime law enforcement operations in their EEZ to protect their sovereign rights and jurisdiction over natural resources and related matters. Disputes, such as the *Arctic Sunrise* incident, often arise when a coastal state asserts law enforcement jurisdiction over persons on board foreign-flagged ships engaged in maritime protest in the coastal state's EEZ. The scenario creates a clash not only between human rights and law of the sea, but also within the law of the sea between coastal state rights and jurisdiction and exclusive flag state jurisdiction.

Article 56(1) sets forth the general rights, duties, and jurisdiction of the coastal State within its EEZ, which include:

> (a) sovereign rights for the purpose of exploring and exploiting, conserving and managing the natural resources, whether living or non-living, of the waters superjacent to the seabed and of the seabed and its subsoil, and with regard to other activities for the economic exploitation and exploration of the zone, such as the production of energy from the water, currents and winds;
> (b) jurisdiction as provided for in the relevant provisions of this Convention with regard to:
> ....

---

58 Clarence Alexand § er, *Law of Arrest in Criminal Proceedings* vol. 2 (1949): 1833, 1844–1845, § 734. *See also* Yoshifumi Tanaka, *The International Law of the Sea* (Cambridge: Cambridge University Press, 2012): 152–155.

59 Article 87, UNCLOS.

60 Article 73, UNCLOS.

(iii) the protection and preservation of the marine environment;
(c) other rights and duties provided for in this Convention.

Article 58(3) is an important corollary to article 56, and it provides for the rights and duties of other states in the EEZ:

> In exercising their rights and performing their duties under this Convention in the exclusive economic zone, States shall have due regard to the rights and duties of the coastal State and shall comply with the laws and regulations adopted by the coastal State in accordance with the provisions of this Convention and other rules of international law in so far as they are not incompatible with this Part.

In accordance with article 73, coastal states are entitled to enforce compliance with laws and regulations "in the exercise of its sovereign rights to explore, exploit, conserve and manage the living resources" in the EEZ. Coastal state enforcement measures generally relate to those that address protection of living resources, such as fishing regulations. Domestic enforcement must be consistent with the law of sea provisions, particularly those relating the navigational freedoms.

Coastal states and flag states must have "due regard" for the rights of the other in the exercise of their own respective rights in the EEZ. The term "due regard" is a staple of maritime and aerospace governance, as the activities of each state should respect the legitimate activities of every other user of the global commons. The term does not connote a substantive or new obligation on the part of the international community (or the coastal state), but rather is properly understood as a procedural duty to observe existing rights that pertain to other states.[61]

"Due regard" consists of two elements: an awareness of and consideration for other state's interests and a weighting of those interests or sources of authority.[62] The precursor to "due regard" is the concept of "reasonable regard,"

61 An example of the concept of "due regard" may be drawn from administrative law. In the United States, a person who appears before an administrative law judge is entitled to procedure that is not arbitrary and capricious. The legal standard, "arbitrary and capricious" does not have content independent of the administrative issue being considered by the judge.

62 George K. Walker, "Defining Terms in the 1982 Law of the Sea Convention IV: The Last Round of Definitions Proposed by the International Law Association (American Branch) Law of the Sea Committee," *California Western International Law Journal* 36 (Fall 2005): 133, 174.

which is in Article 2 of the 1958 High Seas Convention.[63] The Convention states high seas freedoms "shall be exercised by all States with reasonable regard to the interests of other States in their exercise of the freedom of the high seas."[64] The ICJ has determined that the provision of "reasonable regard" is declaratory of customary international law.[65]

The *Virginia Commentary* explains the requirement for "due regard" in Article 58(3) as:

> [A] qualification of the rights of States in exercising freedoms of the high seas. The standard "due regard" requires all States, in exercising their high seas freedoms, to be aware of and consider the interests of other States in using the high seas. As the [International Law Commission] stated in its Commentary in 1956, 'States are bound to refrain from any acts that might adversely affect the use of the high seas by nationals of other States.' The construction in paragraph 2 recognized that all States have the right to exercise high seas freedoms, and balances consideration for the rights and interests of all States in this regard.[66]

Article 56(2), in turn, requires the coastal state to have due regard for the rights and duties of other states operating in the EEZ. This provision mirrors other sections of the treaty that stipulate corresponding rights and duties in the various maritime zones, including article 24(1) (territorial sea); article 42(2) and 44 (straits used for international navigation); article 51 (archipelagic waters), article 58 (EEZ); article 78(2) (continental shelf) and article 87(2) (high seas).[67]

---

63 Art. 2, Convention on the High Seas, done at Geneva 29 April 1958, 450 U.N.T.S. 6465.

64 The 1958 Convention on the High Seas uses the term "due regard" in article 26, which requires the coastal state to pay "due regard" to cables and pipelines already positioned on the sea bed. Similarly, the term is included in article 24 of the 1958 Geneva Convention on the Territorial Sea and the Contiguous Zone, done at Geneva, 29 April 1958, entered into force 10 September 1964, 15 U.S.T. 1606, 516 U.N.T.S. 205 and article 3 of the 1958 Convention on the Continental Shelf, done at Geneva, 29 April 1958, entered into force, 10 June 1964, 15 U.S.T. 471, 499 U.N.T.S. 311.

65 Fisheries Jurisdiction Case (*U.K. v. Iceland*), 1974 I.C.J. 3, 22, 29.

66 III *United Nations Convention on the Law of the Sea: A Commentary* (Boston: Brill, Satya N. Nandan, C.B.E. & Shabtai Rosenne, eds., 1995): ¶ 87.9(1) at 86.

67 II *United Nations Convention on the Law of the Sea: A Commentary* (Boston: Brill, Satya N. Nandan, C.B.E. & Shabtai Rosenne, eds., 1993): ¶ 56(11)(f) at 544–45.

During the drafting of the Convention, the term "reasonable regard" evolved first to "due consideration" before settling at "due regard."[68] The U.S. Navy views the "reasonable regard" and "due regard" standards as essentially identical, and "require any nation to be cognizant of the interests of others in using a high seas area, and to abstain from nonessential, exclusive uses which substantially interfere with the exercise of other nation's high seas freedom."[69] Professor George K. Walker suggests that "due regard" requires all states, in exercising their high seas freedoms, to be aware of and consider the interests of other states in using the high seas, and to refrain from activities that interfere with the exercise by other sates of the freedom of the high seas."[70] This "due regard" analysis indicates that when maritime protest in the EEZ does not interfere with resource-related rights of the coastal state, then it would be an activity protected under article 58 and 87 as part of freedom of navigation and other internationally lawful uses of the sea. In such case, the coastal state must give due regard to those navigational interests. Only if the coastal states' resource rights are affected, such as interference with its maritime activities or infrastructure, such as aids to navigation, fishing, aquaculture or offshore energy, may the coastal state regulate the maritime protest. In such cases, one of the staple responses of the coastal state is to deal with protesters as "pirates." In the case of the Arctic Sunrise, Russia initially charged the 30 activists with the crime of maritime piracy.

The Crime of Piracy

Piracy is any illegal act of violence, detention or depredation committed for private ends by crew or passengers of a private ship or aircraft against another ship, persons or crew, and committed outside of a State's territorial waters, constitutes the crime of piracy. The definition emerged from consolidation of custom and state practice, and it first appeared in a multilateral treaty in the 1958 Convention on the High Seas.[71] A slightly modified definition of piracy was included in UNCLOS. The United States is a State Party to the 1958 Convention, but remains one of the few states that have not joined the 1982 treaty. Still,

68 III *United Nations Convention on the Law of the Sea: A Commentary, supra,* note 66, ¶ 87.8 at 80. *See also,* R.R. Churchill & A.V. Lowe, *The Law of the Sea* (Huntington, NY: Juris Publishing, Inc. 3d ed. 1999): 206.

69 Department of the Navy, *Commander's Handbook on the Law of Naval Operations* (2007): ¶ 2.4.3.

70 Walker, "Defining Terms in the 1982 Law of the Sea Convention," *supra,* note 62, at 133, 174.

71 See Harvard Research in International Law, Draft Convention on Piracy, with Comments, American Journal of International Law Supplement 26 (1932): 749 and subsequent Convention on the High Seas, 29 April 1958, 450 U.N.T.S. 82.

the definition of piracy contained in both treaties is virtually identical and accepted by U.S. and foreign courts.[72]

Article 101 of UNCLOS defines piracy as consisting of the following acts:

> (a) any illegal acts of violence or detention, or any act of depredation, committed for private ends by the crew or the passengers of a private ship or private aircraft, and directed—
>
> i. on the high seas, against another ship or aircraft, or against persons or property on board such ship or aircraft;
> ii. against a ship, aircraft, persons or property in a place outside the jurisdiction of any State;
>
> (b) any act of voluntary participation in the operation of a ship or of an aircraft with knowledge of the facts making it a pirate ship or aircraft;
> (c) any act of inciting or intentionally facilitating an act described in subparagraph (a) or (b).

As UNCLOS has evolved into the constitution for the oceans, the terms of the treaty are enforceable as either conventional law or customary law in virtually every country. Over the past decade, the multinational experience in naval suppression of Somali piracy in the Western Indian Ocean has produced extensive national case law to support contemporary prosecution of maritime piracy. Maritime piracy cases involving classic crimes of violence, kidnapping, and murder, have been upheld in the courts in the United States, Europe, and Asia.[73] There is much less jurisprudence, however, on questions on the periphery of the standard definition of piracy. At what point does direct action by nongovernmental groups cross the line from free expression to unlawful piracy?

In the 1986 litigation *Castle John v. NV Mabeco*, Belgian courts explored the issue of whether environmental activism qualifies as piracy.[74] The courts determined that protests were pursued for "private ends," even though the activist may be motivated by public policy. In a case of first impression, in

72 United States v. Dire, 680 F.3d 446, 469 (4th Cir. 2012).

73 For the United Kingdom, *see Re Piracy Jure Gentium* [1934] AC 586 and The Merchant Shopping and Maritime Security Act 1997; for the United States, *see United States v. Dire*, 680 F.3d 446, 455 (4th Cir. 2012), cert. denied, 133 S. Ct. 982 (2013); for Republic of Korea, *see* Seokwoo Lee and Young Kil Park, "International Decision: *Republic of Korea v. Araye*, No. 2011 Do 12927, Supreme Court of Republic of Korea, December 22, 2011," *American Journal of International Law* 106 (2012): 630–636; and for Japan, *see* Jun Tsuruta, "The First Prosecution of Somali Pirates under the Japanese Piracy Act: The *Guanabara* Case," *Journal of East Asia and International Law*, 7(1) (Spring 2014): 243–248.

74 Cour de Cassation [Cass.] [Court of Cassation] *Castle John v. NV Mabeco*, Dec. 19, 1986, 77 I.L.R. 537 (Belg.).

the 1986 litigation *Castle John v. NV Mabeco*, Belgian courts explored the issue of whether environmental activism qualifies as piracy.[75] The court held that the conduct was pursued for "private ends, even though the activist may be motivated by concerns of public policy." Greenpeace boarded, occupied, and damaged two vessels in order to attract attention to the issue of vessel source discharge of noxious waste on the high seas.[76] The court held that Greenpeace activities constituted violent acts that qualified as maritime piracy under Article 15(1) of the Geneva Convention.[77] This case still reflects good law. The United States has been even more accepting of private litigants pursuing claims based on a violation of international law. The key to the litigation lies in the U.S. interpretation of the Alien Tort Statute, which is an old American law with new applicability.

The U.S. Ninth Circuit considered a civil case against the Sea Shepherd Conservation Society (SSCS) for its harassment of Japanese whaling vessels. The Japanese whaling consortium Institute of Cetacean Research sued SSCS under the Alien Tort Statute for injunctive and declaratory relief. Sea Shepherd Conservation Society employed "direct action" interference and disruption as a form of protest against Japanese whaling activities, which were done purportedly for scientific purposes. The Alien Tort Statute provides a cause of action for "a tort ... committed in violation of the law of nations or a treaty of the United States."[78]

The Institute of Cetacean Research argued that the acts of SSCS Shepherd were tantamount to piracy in violation of international law of the sea. The district court construed "private ends" as those aimed at a pecuniary interest or "financial enrichment." The district court also provided a narrow interpretation of what constitutes "violence" at sea when it concluded that Sea Shepherd's conduct was not "violent" because it targeted ships and equipment rather than people. Thus, the district court rejected the request for preliminary injunction and dismissed the claim of piracy. Institute of Cetacean Research appealed the district court decision, and the case was heard in the Ninth circuit on October 9, 2012. The Ninth circuit also asserted jurisdiction over the piracy claims, since the district court's rationale for dismissing the preliminary injunction was "inextricably intertwined" with the claims of piracy.

The Ninth circuit rejected the narrow definition of piracy adopted by the district court, and instead viewed the term "private" as the antonym of "public." Private ends include "those pursued on personal, moral or philosophical

75 *Ibid.*, at 540.

76 *Ibid.*, at 537.

77 *Ibid.*, at 538.

78 Alien Tort Claims Act, 28 U.S.C. § 1350.

grounds," including Sea Shepherd's environmental goals.[79] The court was unwilling to consider conduct aimed at serving the public good as protected by "public ends." This broad reading of private conduct is consistent with earlier U.S. case law that regarded piracy as "an act of hostility … being committed by a vessel not commissioned and engaged in lawful warfare."[80] The Ninth circuit also gave "considerable weight" to the *Castle John* decision in Belgium.[81] Essentially any conduct at sea that is not sanctioned by the state constitutes private conduct and is not cloaked in public protection. This analysis is historically grounded in the distinction between mere pirates that were unlawful brigands, and privateers, which were different only in that they held letters of marque that authorized their pillaging.

The Ninth circuit also rejected the district court's narrow interpretation of "violence," noting that article 101 of UNCLOS proscribes violence against another ship, persons, or property. "Ramming ships, fouling propellers and hurling fiery and acid-filled projectiles easily qualify as violent activities, even if they could somehow only be directed only at inanimate objects."[82] The Ninth circuit concluded that SSCS were indeed pirates:

> You don't need a peg leg or an eye patch. When you ram ships; hurl glass containers of acid; drag metal-reinforced ropes in the water to damage propellers and rudders; launch smoke bombs and flares with hooks; and point high-powered lasers at other ship, you are, without a doubt, a pirate, no matter how high-minded you believe your purpose to be.[83]

Finally, the Ninth circuit also rejected the district court's reliance on an Australian default judgment against the Institute of Cetacean Research in a separate case brought by SSCS. The Australian court enjoined the Institute of Cetacean Research from conducting whaling in the coastal waters of Antarctica—waters over which Australia asserts a controversial claim of sovereignty. The U.S. circuit court, however, would not give comity to the Australian decision. "Whatever the status of Cetacean's whaling under Australian law,

79 The Institute of Cetacean Research v. Sea Shepherd Conservation Society, No. 12-35266 (9th Cir. 2013), p. 5.

80 *Harmony v. United States*, 43 U.S. (2 How.) 210, 232 (1844).

81 *Abbott v. Abbott*, 130 S. Ct. 1983, 1993 (2010).

82 Cetacean v. Sea Shepherd, *supra*, note 79, p. 6. The circuit court also cited Webster's New International Dictionary definition of plain meaning of the word "violence," which includes behavior not directed at another person, such as when a man violently pounds a table with his fist. *Ibid.*

83 *Ibid.*, 5.

it gives Sea Shepherd no license to engage in piracy. It is for Australia, not Sea Shepherd, to police Australia's court orders."[84] Australia's judgment was not recognized because comity applies only if the foreign court has competent jurisdiction over the case.[85] The Ninth circuit court noted that the United States has protested Australia's claim of sovereignty over the waters off the coast of Antarctica.[86] According comity to Australia's court decisions on the issue would implicitly recognize Australia's jurisdiction over Antarctic coastal waters, contrary to U.S. policy.[87] Doing so, the circuit court held, would invade the exclusive domain of the Executive branch to conduct the nation's foreign affairs.[88]

Safety Zones

Except for the crime of piracy, which is a crime of universal jurisdiction,[89] offenses committed at sea generally fall within the jurisdiction of the flag State of the ship on which or from which the acts were committed. There is a more narrowly prescribed class of offenses, however, that may also fall within the jurisdiction of the coastal State. For example, coastal States may prescribe and enforce certain offenses if the consequences of the crime extend to the coastal State or if the crime is "of a kind to disturb the peace" ashore, or the "good order of the territorial sea."[90] Likewise, in the contiguous zone, coastal States may enforce customs, fiscal, immigration, and sanitary laws and regulations.[91] Similarly, fisheries and other resource crimes, and even some environmental offenses, may be prosecuted by the coastal State in its exclusive economic zone.[92]

---

84 *Ibid.*, at 13.

85 *Asvesta v. Petroutsas*, 580 F.3d 1000, 1011 (9th Cir. 2009).

86 Note from U.S. Deputy Representative to the United Nations, to Secretary-General of the United Nations (Dec. 3, 2004); Note from Embassy of the United States, to Australian Department of Foreign Affairs and Trade (Mar. 31, 1995), cited in, Institute of Cetacean Research v. Sea Shepherd Conservation Society, No. 12-35266, D.C. No. 2:11-cv-02043-RAJ, slip op. at 13 (9th Cir. Ct. Feb. 25, 2013).

87 James Kraska, *Maritime Power and the Law of the Sea* (Oxford: Oxford University Press 2011): 292–294.

88 *United States v. Hooker*, 607 F.2d 286, 289 (9th Cir. 1979).

89 John Edward Oster, Speech of the Honorable John Marshall, *The Political and Economic Doctrines of John Marshall* (New York: The Neal Publishing Company, 1914): 225, 230.

90 Article 27(1)(a)–(b), UNCLOS, *supra*, note 2.

91 *Ibid.*, Article 33.

92 *Ibid.*, Article 56.

Coastal states also may enforce safety zones around artificial installations in the EEZ and on the continental shelf, such as the 2013 Greenpeace assault on Russia's Prirazlomnaya oil rig.[93] Exploitation of the resources of the continental shelf should be carried out with the least impact on freedom of navigation. The issue of coastal State authority with regard to offshore artificial islands, installations and structures was addressed within the context of governance of the continental shelf during negotiations at the First United Nations Conference on the Law of the Sea in 1958.

States understood that coastal States should have authority to establish safety zones at a "reasonable distance" from installations on the continental shelf, but there was an equally strong sense among states, including the United Kingdom, Sweden, and Spain an open-ended rule was too vague.[94] Working with industry, The Netherlands proposed a 50-meter safety zones.[95] The Netherlands believed that a 50-meter zone would ensure that coastal States could not impose unnecessarily large zones that encroach on freedom of the seas.[96] West Germany supported this narrower zone.[97] The zone would be measured from a radius from the center point of the installation, or a belt measured from the outer edge and surrounding the entire structure.[98] The U.K. proposed a 500-meter safety zone,[99] as did Yugoslavia.[100] The broader 500-meter zone was thought to implement a margin of safety to separate shipping from installations.[101] The International Law Commission also suggested a maximum 500-meter safety zone, and this text was adopted by the Fourth Committee as Article 71.[102]

Since aircraft may endanger installations even more than by ships, Yugoslavia proposed an air safety zone to a height of 1,000 meters surrounding the installation.[103] The idea of air safety zones, however, appeared to fall outside of the competence of a conference focused on the law of the sea. In the end, the 1958 Convention incorporated a 500-meter safety zone, as measured

---

93 *Ibid.*, Article 56 and 77.

94 UN Doc. A/CONF.13/42 (1958) at p. 3 (Sweden), p. 4 (U.K.), p. 7 (Spain), and p. 12 (Turkey).

95 UN Doc. A/CONF.13/42 (1958), at p. 81, 84–85 ¶ 7 and UN Doc. A/CONF. 13/C.4/L.22.

96 UN Doc. A/CONF.13/C.4/L.22 (1958) and A/CONF.13/42 (1958), p. 86, ¶ 17.

97 UN Doc. A/CONF.13/42 (1958), pp. 84–85, ¶ 7.

98 *Ibid.*, p. 86, ¶ 18.

99 UN Doc. A/CONF.13/C.4/L.28 (1958).

100 UN Doc. A/CONF.13/C.4/L.15 (1958) and UN Doc. A/CONF.13/C.4/L.16 and Add.l.

101 UN Doc. A/CONF.13/42 (1958), at p. 82, ¶ 10.

102 Text of the articles and final clauses adopted by the Fourth Committee, UN Doc. A/CONF.13/L.12, annex (1958).

103 UN Doc. A/CONF.13/42 (1958), p. 84, ¶ 5.

from each point of their outer edge.[104] This method of determining the outer extent of the safety zone was carried over into UNCLOS.[105]

The safety zone applies only to surface ship navigation, and it may not be used to interfere with vessel navigation, especially in sea lanes.[106] Importantly, the state was afforded authority not just to establish safety zones, but legal competence to take measures in those zones that are "necessary for their protection."[107] Ships of all nationalities must respect safety zones.[108]

Although historically the establishment and use of artificial islands, installations and structures was considered an activity on the continental shelf, the contemporary regulations reside in Part V (Exclusive Economic Zone). In the EEZ, coastal States have sovereign rights over the resources of the water column, and for other activities related to the economic exploitation of the zone, including production of energy from water, currents, and winds. Coastal State jurisdiction extends to the establishment of artificial islands, installations, and structures, marine scientific research, and protection of the marine environment, "as provided for in the relevant provisions" of the Convention. Article 60 of UNCLOS contains provisions for establishment and use of artificial islands, installations and structures in the EEZ. The rights of the coastal State are enumerated in Article 56, and include sovereign "rights and jurisdiction" for the living and non-living resources, and production of energy from water, currents and winds. The coastal State has exclusive right to "construct and to authorize and regulate the construction, operation and use" of such objects in the EEZ, but only for economic purposes.[109] The coastal State also has the exclusive right to construct, authorize, and use installations and structures that "may interfere with the exercise of the rights of the coastal State in the zone."[110]

---

104 Article 3, Convention on the Continental Shelf, Convention on the Continental Shelf, 29 April 1958, 499 U.N.T.S. 312.

105 UNCLOS, *supra*, note 2, Article 60(5).

106 International Law Commission, Articles concerning the Law of the Sea with commentaries 1956, Article 71 and associated Commentary, *Yearbook of the International Law Commission*, 1956, vol. 2. II (1956): 265, 299–300.

107 Article 3, Convention on the Continental Shelf, *supra*, note 64, Article 2.

108 *Ibid.*, Article 3.

109 The actual text states that in the EEZ, the "coastal State shall have the exclusive right to construct and to authorize and regulate the construction, operation, and use of... installations and structures for the purposes provided for in article 56 and other economic purposes." Article 60(1)(b), UNCLOS, *supra*, note 2.

110 *Ibid.*, Article 60(1)(c).

The coastal State has exclusive and comprehensive jurisdiction over such installations, and may establish a 500 meter safety zone around them.[111] Safety zones should be promulgated by Notices to Mariners to advise vessels of the location of offshore installations and structures, the breadth of the safety zone, and any fairways available.[112] The coastal State should require operators of offshore installations and structures to "prevent infringement" of them through effective light and sound signals, radar beacon transponders ("racons"), permanent visual look-out and radar watch, listening for and warning ships on VHF channel 16 and the establishment of vessel traffic services.[113] Actions by vessels that jeopardize safety, including infringement of safety zones, should be reported to the coastal State.[114]

In accordance with Article 60(6) of UNCLOS, foreign-flagged vessels are required to observe such safety zones, and comply with coastal State routing measures established "in the area."[115] The right to exercise high seas freedoms in the EEZ are tempered by the rule that all vessels shall comply with generally accepted international standards inside the safety zone.[116] Furthermore, ships navigating "in the vicinity" of offshore infrastructure should navigate with caution. These vessels have a duty to give due consideration to safe speed and observe appropriate passing distances, to take into account prevailing weather conditions, and "the presence of other vessels and dangers."[117] The obligation includes "early and substantial" action to avoid installations and structures, and to provide information to the coastal State to facilitate safety.[118] Thus, it is clear that coastal State jurisdiction over artificial installations emanates from enforcement of a 500-meter safety zone. All ships are required to respect safety zones, and comply with "generally accepted international standards," regarding navigation in the "vicinity" of the artificial island, installation, or structure.[119] The law of the sea is ambiguous as to the outer reaches of the obligation by other states, in this regard, or whether "vicinity" extends beyond the limits of the 500-meter zone itself.

---

111 *Ibid.*, Article 60(2).

112 IMO Doc. A.671(16), Safety zones and safety of navigation around offshore installations and structures, 19 October 1989 (IMO Assembly Resolution A 16/Res.671) 30 November 1989, Annex, ¶ 1.1.

113 *Ibid.*, Annex ¶ 1.3.

114 *Ibid.*, Annex ¶ 1.4.

115 *Ibid.*, Annex ¶ 2.3.

116 *Ibid.*, ¶ 1(d) and Article 60(6), UNCLOS, *supra*, note 2.

117 IMO Doc. A.671(16), Annex ¶ 2.1–2.3 and Article 60(2), UNCLOS, *supra*, note 2.

118 *Ibid.*, Annex ¶ 2.2.

119 UNCLOS, *supra*, note 2, Article 60(6).

Hot Pursuit

The 1958 Convention on the High Seas states that coastal States may conduct hot pursuit of a foreign ship when there is "good reason" to believe that the ship has violated the laws and regulations of the coastal State.[120] The term "good reason" is a standard of proof required for a valid reason in accordance with the coastal State's authorities in UNCLOS, and not to be mistaken for a substantive rationale. The ship may be pursued when either it or any other associated boats or craft working with it as a team, have violated the laws of the coastal State in an area under coastal State jurisdiction.

> Hot pursuit is not deemed to have begun unless the pursuing ship has satisfied itself by such practicable means as may be available that the ship pursued or one of its boats or other craft working as a team and using the ship pursued as a mother ship are within the limits of the territorial sea, or as the case may be within the contiguous zone....[121]

The right of hot pursuit was imported into UNCLOS, and now reflects customary international law.[122] A coastal State may, as a general proposition, pursue and seize a non-national vessel suspected of having committed an offense in the territorial sea, contiguous zone, EEZ, or continental shelf in cases in which the violator flees to the high seas to avoid arrest.[123] Hot pursuit involves the chase of a foreign vessel following a violation of the law committed by a vessel in an area under the maritime jurisdiction of the pursuing State. The right of hot pursuit emerged from common law jurisprudence.[124] The right of hot pursuit may be commenced within internal waters, the territorial sea, or the contiguous zone,[125] but it also applies to violations of the EEZ or on the continental shelf, including "safety zones around installations continental shelf installations."[126] In any event, hot pursuit may be engaged only for violation of the rights for the protection of which the zone was established. For example, hot pursuit may be undertaken beyond the EEZ for violation of

120 Article 23(1), Convention on the High Seas, *supra*, note 63.

121 *Ibid.*, Article 23(3).

122 Article 111, UNCLOS, *supra*, note 2.

123 *Ibid.*, Article 111(1)–(2).

124 *"The King. v. The North" American Journal of International Law* 2 (1908): 688, 699 and D.P. O'Connell, *Law of the Sea* (Oxford: Oxford University Press 1982): 1076.

125 Article 111(1), UNCLOS, *supra*, note 2.

126 *Ibid.*, Article 111(2).

coastal State fisheries laws, or some other lawful purposes, and not for violation of a coastal State law that is incompatible with UNCLOS.[127]

Foreign-flagged ships may be subject to hot pursuit beyond a coastal State's safety zone for conduct committed inside the zone. For example, in the 2001 case of the Togo-registered *South Tomi*, Australian and South African authorities chased and then apprehended the ship after a fourteen-day pursuit that covered 3,300 nm. Similarly, the Uruguayan-flagged fishing vessel *Viarsa I* was seized by South Africa and the United Kingdom following a twenty-day hot pursuit.

If a perpetrator enters into the safety zone of an offshore installation and then flees beyond the outer limits of the zone, they may be pursued, even if there is not positive proof that they committed an offense. Hot pursuit, for example, may be undertaken when coastal state authorities have "good reason to believe" the ship or its passengers or crew have violated its laws and regulations.[128] Hot pursuit may stretch onto the high seas, and in any area of the oceans except areas under the sovereignty of a coastal state, such as the territorial sea, archipelagic waters, or internal waters of another state.[129]

Constructive Presence

Under the doctrine of constructive presence, coastal States may assert jurisdiction over ships and persons who have committed acts outside the jurisdiction of the coastal State, but that but that generates effects or have a nexus to illegal acts conducted within it.[130] Enforcement jurisdiction over ships that commit crimes beyond coastal State jurisdiction, but that generate effects within that jurisdiction may be obtained through the doctrine of constructive presence. The doctrine of constructive presence was expressed in a landmark 1877 case off the coast of Australia. In 1877 the American ship *Catalpa* sent its boats to the shores of... Australia to assist in the escape of convicts. In their opinion to the Colonial Office on the incident the Law Officers advised that no sound distinction could be drawn between the *Catalpa* and its boats, and that, the

127 M/V Saiga (No. 2) Case, p. 1354, ¶ 146.

128 Article 111(1), UNCLOS, *supra*, note 2.

129 *Ibid.*, Article 111(3).

130 D.P. O'Connell, *The International Law of the Sea* vol. 2 (Oxford: Oxford University Press 1982): 1075–76, 1093 and Andrew Brian Gough, "Canadian Criminal Jurisdiction over the Seas: Current State and Present and Future Sources for Expansion," *Ocean Yearbook* 22 (2008): 479.

ship having therefore violated British territory, it was liable to be pursued and stopped on the high seas.[131]

The doctrine was again invoked again in the 1888 *Araunah* case that involved illegal fishing. Churchill and Lowe trace the subsequent development of the doctrine of constructive presence into the twentieth century:

> According to later refinements of the doctrine, illustrated, for example, by some of the diplomatic exchanges concerning the seizure of *the Henry L. Marshall* for violation of the American liquor laws in 1922 and by the Italian case of the *Sito* in 1957, it had no application where a foreign ship communicated with the shore not by means of its own boats but by means of boats sent out from the shore. In such cases the exclusive jurisdiction of the flag State over its ships on the high seas was said to remain intact.[132]

The law is unclear how far from the coastal State the doctrine of constructive jurisdiction extends, except insofar as it applies to conduct that generates effects inside the zone or area under coastal State jurisdiction. In the *Arctic Sunrise* case, for example, Greenpeace activists used the ship as a base of operations from which to storm the Prirazlomnaya OIRFP lying in the Russian EEZ. Russian authorities arrested the interlopers on the platform, and also pursued the ship well beyond the safety zone of the offshore installation. In such case coastal state law enforcement may invoke hot pursuit if it is satisfied that the pursued vessel, or one of its boats or associated craft, work as a team. In the case of the *R. v. Mills and Others*, for example, U.K. authorities arrested a drug trafficking mother ship, *Poseidon*, which had discharged a load of illegal drugs onto the *Delvan* on the high seas before *Delvan* entered into a British port.[133] The teamwork between the *Poseidon* and *Delvan* was sufficient to form a mother ship relationship that made both vessels subject to U.K. jurisdiction. In short, the coastal State must demonstrate a reasonable belief that the ship

131 A.V. Lowe, "The Development of the Concept of the Contiguous Zone," *British Yearbook of International Law* 52 (1981): 117.

132 Churchill & Lowe, *The Law of the Sea, supra*, note 68, at 133.

133 William C. Gilmore, "Hot Pursuit: The Case of R. v. Mills and Others," *International and Comparative Law Quarterly* 44 (1995): 949. *See also R. v. Sullivan and Soleyman*, Nova Scotia Supreme Court Appeal Division, No. 1, January 2, 1986, pp. 300–308; No. 2 N.S. Supreme Court, Trial Division, January 28, 1986, pp. 308–14; No. 3 February 6, 1986, pp. 315–19; No. 4, February 11, 1986, pp. 319–322; No. 5, March 24, 1986, pp. 322–26, in Nova Scotia Reports (second series), vol. 73, cited (1986), 73, N.S.R. (3d) 308. The original decision by Judge Glube is from 8 August 1985, C.J.T.D.

outside areas under its jurisdiction acts as a mother ship, and that the conduct of the ship outside of the zone has a nexus to the activity inside the safety zone. In the case of the Arctic Sunrise, once Russian authorities determined that the ship was a mother ship for the Greenpeace speedboats operating inside the safety zone, it could assert jurisdiction over it under theory of constructive presence.

Complementary Maritime Regimes

The international law of the sea and UNCLOS form the foundation of maritime law, and they are complemented by additional treaties and regimes that may apply in cases of maritime civil disobedience. This section addresses the application of the Collision Regulations, the Convention on the Suppression of Unlawful Acts against the Safety of Maritime Navigation, and Japan's proposed Code of Conduct during Demonstrations and Campaigns against Ships on the High Seas. These instruments provide additional normative and empirical context for civil disobedience in the maritime domain.

Collision Regulations

Perhaps the most basic obligation of seafarers is to operate in a manner that does not endanger other vessels. Maritime protesters may not depart from safe navigational practices. The Institute of Cetacean Research complained that ships operated by Sea Shepherd Conservation Society deliberately collided with Japanese whaling ships. The evidence suggests that on a number of occasions SCSS executed unsafe navigation against ICR vessels in violation of international collision regulations. The *International Regulations for Preventing Collisions at Sea*, 1972 (COLREGS) contain basic rules for navigational safety governing surface and subsurface vessels.[134] The COLREGS prohibit interference with the navigation of another vessel and contain obligatory and universal norms for navigation of ships to avoid collision.[135] The rules apply to all areas of the oceans, although coastal states may establish different rules inside the territorial sea, archipelagic waters, and internal waters. Rule 6 requires maintaining a safe speed as to avoid collision. Rule 7 and Rule 8 deal with avoiding collision. Rule 18 requires a vessel to keep out of the way of a vessel

---

134 Convention on International Regulations for Preventing Collisions at Sea, Oct. 20, 1972, entered into force July 15, 1977, 28 U.S.T. 3459, T.I.A.S. 8587, 1050 U.N.T.S. 17.

135 *Crowley Marine Services, Inc. v. Maritrans, Inc.*, 530 F.3d 1169, 1172–73 (9th Cir. 2008) (COLREGS are a "universal system of sea traffic rules").

restricted in its ability to maneuver. Rule 17(b) requires that a vessel take action to avoid collision, even if it has the right of way.[136]

COLREGS may serve as the basis for addressing some instances of maritime civil disobedience that go beyond free expression and create a navigational hazard through intentionally dangerous seamanship. It is important to note, however, that flag States generally are responsible for enforcement of the provisions in COLREGS. Coastal States may be able to adopt and enforce national laws within the territorial sea that reflect the rules in COLREGS, although this jurisdictional hook to nab offenders is fraught with political peril. To avoid international incidents, flag States should implement their obligations under COLREGS and enforce the provisions against vessels that fly their flag.

Convention on the Suppression of Unlawful Acts

The crime of piracy in UNCLOS has been supplemented by the Convention for the Suppression of Unlawful Acts against the Safety of Maritime Navigation (SUA). The first SUA Convention was negotiated in 1985 in response to the lack of effective coordination among states after the hijacking committed on board the *Achille Lauro*. During the assault, Palestinian terrorists seized control of the cruise ship Achille Lauro, and ultimately killed a disabled American passenger, Leon Klinghoffer, before fleeing to Egypt.[137]

At the time of the attack, many states did not have criminal legislation for extradition or prosecution for the offense of vessel hijacking. Article 3 of the treaty commits states to create criminal offenses for persons who "unlawfully and intentionally... perform an act of violence against a person on board a ship if that act is likely to endanger the safe navigation of that ship," or destroys cargo, or places devices on a ship that likewise might endanger the vessel.[138] These offenses require mental culpability on the part of the offender who must act "unlawfully" (not conforming to law or without authority) and "intentionally (deliberately or purposefully). Instead of contending with the thorny question of who constitutes a terrorist, the text focuses on proscribing specific conduct.

136 This doctrine requires vessels to take steps to avoid *in extremis* situations.

137 Lionel Barber, "Crisis Ends in Tragedy for New Yorker's Family; Pirates Surrender Ship; 1 American Killed," *Washington Post*, October 10, 1985 p. A1.

138 Art. 3(1), Convention for the Suppression of Unlawful Acts against the Safety of Maritime Navigation, 27 I.L.M. 668 (1988); 1678 U.N.T.S. 221.

The preamble to the Convention illuminates some of the international law authorities considered in drafting the Convention, such as the Charter of the United Nations, the Universal Declaration of Human Rights and the International Covenant on Civil and Political Rights. The treaty also acknowledges UN General Assembly resolution 40/61 of December 9, 1985, which "unequivocally condemns, as criminal, all acts, methods and practices of terrorism wherever and by whomever committed. ..." The same UNGA resolution invited the IMO to "study the problem of terrorism aboard or against ships with a view to making recommendations on appropriate measures." State parties are obligated to either criminally prosecute suspected offenders or extradite them to another country for prosecution.[139] Although the 1988 SUA Convention included a duty among states parties to cooperate in the prevention of the offenses set forth in article 3, the treaty was primarily an instrument to ensure the prosecution or extradition of suspected offenders.[140] The main purpose of the Convention is to ensure that appropriate action is taken against persons who commit unlawful acts against ships, including acts of violence against persons on board ships.

The master of a ship of a State party may deliver to the authorities of another state (the receiving state) any person who he has "reasonable grounds" to believe has committed an offense under the Convention.[141] While making such delivery, the master shall give notification of his intention to do so, "if possible before entering the territorial sea of the receiving state."[142] The receiving state is required to accept delivery, but there is a rather wide exception that applies. In cases in which the receiving state has "grounds" to believe the Convention does not apply, it may refuse accepting the suspect. The standard for developing "grounds" is undefined, and there is no qualification that the grounds must be reasonable or made in good faith, although general principles of law suggest they must be. A corresponding SUA treaty applies to artificial installations and fixed platforms on the continental shelf, so both ships and offshore infrastructure are covered by the 1988 regime.

---

139 *Ibid.*, Art. 6(4).

140 *See* Malvina Halberstam, "Terrorism on the High Seas: The Achille Lauro, Piracy and the IMO Convention on Maritime Safety," *American Journal of International Law* 82 (1988): 269, 292.

141 Art. 8, 1988 SUA Convention, *supra*, note 138.

142 *Ibid.*, Art. 8(1).

After the attacks of 9/11, member states of the International Maritime Organization amended the treaty through the Protocols to the Convention on the Suppression of Unlawful Acts against the Safety of Maritime Navigation.[143] The 2005 Protocols were adopted by a Diplomatic Conference of the IMO on 14 October 2005 and entered into force on 28 July 2010. The amended treaty is called the 2005 SUA Convention, and reflects the 2005 Protocol amendments to the 1988 Convention. The 1988 Convention and the 2005 Protocol to the Convention are read as a single, integrated treaty called the 2005 SUA Convention, which include the consolidated text of the Convention for the Suppression of Unlawful Acts against the Safety of Maritime Navigation and of the Protocol of 2005 to the Convention.[144]

Whereas the 1988 treaty helps states cooperate to take criminal action after such attacks, the 2005 protocol commits states to cooperate to prevent such attacks from occurring. The offenses in article 3 were expanded to include seizing or exercising control over a ship by force, destroying or damaging maritime navigational facilities or seriously interfering with their operation, communicating information which is known to be false that endangers the safe navigation of the ship, or injuring or killing any person in connection with the commission of any of the offenses.

Article 3*ter* prohibits the transportation of a person on board a ship knowing that the person has committed an act that constitutes an offence under the 2005 SUA Convention *or* any other offense that is a feature of any one of the nine counterterrorism treaties listed in the Annex, which was formed by Article 7 of the 2005 Protocol. The treaties listed in the Annex are:

(1) Convention for the Suppression of Unlawful Seizure of Aircraft, done at The Hague on 16 December 1970;
(2) Convention for the Suppression of Unlawful Acts against the Safety of Civil Aviation, done at Montreal on 23 September 1971;
(3) Convention on the Prevention and Punishment of Crimes against Internationally Protected Persons, including Diplomatic Agents, adopted by the General Assembly of the United Nations on 14 December 1973;
(4) International Convention against the Taking of Hostages, adopted by the General Assembly of the United Nations on 17 December 1979;

143 Adoption of the Final Act and Any Instruments, Recommendations, and Resolutions Resulting from the Work of the Conference: Protocol of 2005 to the Convention for the Suppression of Unlawful Acts Against the Safety of Maritime Navigation, IMO Doc. LEG/CONF.15/21 (Nov. 1, 2005).

144 *Ibid.*, Art. 15.

(5) Convention on the Physical Protection of Nuclear Material, done at Vienna on 26 October 1979;

(6) Protocol for the Suppression of Unlawful Acts of Violence at Airports Serving International Civil Aviation, supplementary to the Convention for the Suppression of Unlawful Acts against the Safety of Civil Aviation, done at Montreal on 24 February 1988;

(7) Protocol for the Suppression of Unlawful Acts against the Safety of Fixed Platforms Located on the Continental Shelf, done at Rome, 10 March 1988;

(8) International Convention for the Suppression of Terrorist Bombings, adopted by the General Assembly of the United Nations on 15 December 1997;

(9) International Convention for the Suppression of the Financing of Terrorism, adopted by the General Assembly of the United Nations on 9 December 1999.

The last part of amended Article 3 includes a bevy of inchoate crimes, including attempt, accomplice, criminal organization or leadership in a criminal enterprise, or making a contribution to the commission of one of the offenses. Some of the additional offenses in the 2005 Protocol contain multiple *mens rea* elements. The provision of article 3*quarter* (a)(ii), for example, requires that an offense is committed if the offender serves as an accomplice with the knowledge that the group in which he participates is violating an offense in one of the nine anti-terrorism conventions identified in Article 3*ter*. Thus, the layers of intent create a fairly high burden on the part of prosecuting governments, as it may be impossible to demonstrate specific knowledge.

Article 8*bis* of the 2005 Convention includes important safeguards that apply when a state party takes enforcement measures against a ship, including boarding. The safeguards include: not endangering the safety of life at sea; ensuring that all persons on board are treated in a manner which preserves human dignity and in keeping with human rights law; taking due account of safety and security of the ship and its cargo; ensuring that measures taken are environmentally sound; and taking reasonable efforts to avoid a ship being unduly detained or delayed. The use of force is to be avoided except when necessary to ensure the safety of officials and persons on board, or where the officials are obstructed to the execution of authorized actions. These important safeguards help to ensure the protection of human rights at sea.

The SUA Convention is an appropriate template for flag State, coastal State and port State implementation. As the number of States party expands, the agreement will have a greater potential effect by widening the pool of flag state registries, coastal State authorities and port State authorities that are subject to its terms. The 1988 treaty and 2005 protocol are conventional law agreements, so they bind states that specifically have subscribed to their terms; they are no enforceable as a matter of customary international law.

IMO Code of Conduct

The greatest legal challenge in addressing maritime civil disobedience is that the governing international law creates State responsibilities, and so they are subject to all of the vagaries of spotty or imperfect implementation. In some respects, this situation is because states, rather than individuals, are the subjects of international law. The rules bind states, and states are obligated to enforce them against their people and ships. Although coastal States and port States also may prescribe and enforce some internationally accepted civil or even criminal rules against vessels and persons temporarily in the territorial sea or in port, most offenses are best dealt with by the flag State in accordance with the doctrine of exclusive flag State jurisdiction.

In an attempt to clarify some of these issues, Japan proposed a set of guidelines at the International Maritime Organization (IMO). In response to the campaign of direct action demonstrations by the SSCS against Japanese whaling vessels, Japan has sought adoption at the IMO of an international code of conduct that would apply to demonstrations and campaigns against ships on the high seas.[145] The IMO is the United Nations specialized agency for maritime matters, and is regarded as the "competent international organization" for adoption of global international shipping rules and standards, especially concerning marine environmental protection.[146]

Japan seeks to broaden the coalition of states affected by maritime protest, so the draft Code of Conduct applies to protests against fishing operations, research activities, refueling and transfer of personnel, equipment, and cargo, such as nuclear material.[147] Japan wants to protect ship crews and ensure safety of navigation, although protection of whaling vessels may draw little sympathy from states that are inclined to protect rather than hunt marine mammals.

The 82nd session of the Maritime Safety Committee (MSC) agreed to add the Japanese concept as a high priority agenda item to the Flag State Implementation Sub-Committee work program, and the Navigation Sub-Committee was

145 IMO Doc. NAV 54/10/1, Code of Conduct During Demonstrations/Campaigns Against Ships on the High Seas, Apr. 25, 2008, Annex.

146 Article 59, Convention on the International Maritime Organization, done at Geneva, 6 March 1948, 1276 U.N.T.S. 468 (IMO Convention).

147 Code of Conduct During Demonstrations, *supra*, note 145, Annex, para. 3.6.

assigned as coordinator. The Navigation Sub-Committee envisioned a two-session work cycle to complete the item.[148] The 83rd session of the MSC approved the work item, and the 53rd session of the Navigation Sub-Committee agreed to begin discussions on the item during its 54th session.

During the 54th session of the Navigation Sub-Committee, Japan offered draft text of a proposed recommendatory (i.e., not mandatory) code of conduct to assure the safety of crew and maritime navigation during protests at sea by non-state activist groups. By offering the Code as a recommended instrument, Japan hoped it would gain wide traction because States would not be under legal obligation to implement it. The draft Code draws upon relevant principles of international law, including COLREGS and UNCLOS, and also recognizes the sanctity of the right of free expression laid out in the 1948 UDHR and the 1966 ICCPR. Consequently, the draft Code is the only instrument that incorporates and balances the equities of all of the actors involved in maritime protest.

The draft applies all vessels on the high seas,[149] and provides that "demonstrations [or] campaigns should be conducted in a manner consistent with the COLREGS and have due regard for the safety and preservation of human life and the environment at sea."[150] The Code of Conduct is intended to set forth principles for how demonstrations and campaigns at sea may be conducted.

Demonstration activities are restrained and regulated by the Code in five distinct areas: (1) requirement of non-violence, (2) assurance of safety of the crew and navigation of the vessels subject to protest, (3) maintenance of appropriate distance between protest ships and vessels subject to protest in order to avoid collision, (4) protection of radio communications against abuse or interference, and (5) compliance with other agreements, such as COLREGS and UNCLOS. Japan's foremost concern is to avoid violence. The Code calls for a prohibition on the use or threat of violence, either directly or indirectly, that place at risk human life, property or safe navigation.[151] Furthermore, demonstrations should not utilize "weapons, toxic or poisonous substances and other

---

148 IMO Doc. MSC 82/21/8.

149 A "vessel" includes "every description of water craft, including non-displacement craft, WIG craft and seaplanes, used or capable of being used as a means of transportation on the water." *Ibid.*, Annex, para. 2.6.

150 Code of Conduct During Demonstrations, *supra*, note 145, Annex, para. 2.3–4. The term "due regard" is present in UNCLOS articles 27(4), 39(3)(a), 56(2), 58(3), 60(3), 66(3)(a), 79(5), 87(2), 142(1), 148, 161(4), 162(2)(d), 163(2), 167(2), 234, and 267, as well as represented in Appendices II, article 2(1), IV, articles 5(1) and 5(2). The origins of the term predate the 1982 Convention.

151 Code of Conduct During Demonstrations, *supra*, note 145, Annex, para. 3.1.1.

hazardous materials."[152] Sea Shepherd, for example, uses "stink bombs" made of butyric acid—essentially putrid butter—to contaminate the decks of whaling ships and despoil whale blubber and flesh, in order to make it unmarketable.

Second, the Code also expresses concern for the safety of the crew and navigation of both the protest vessel and the subject of the protest. The "first priority" should be given to protecting life at sea.[153] Protesters should "take all measures possible to ensure their actions... do not affect the normal safe operation of vessels subject to protest."[154] Third, an "appropriate distance" should be maintained between the protesting ship and the ship subject to protest.[155] This separation should be "far enough for the vessels subject to protest to carry out safe navigation and their own operation without any disturbance...."[156] Demonstrations and campaigns should be suspended if the ship subject to protest issues a "warning and/or call for attention," related to the safety of the ship or its crew.[157] The provision would empower ships subject to protest with authority to stop it. Fourth, because radio communications are critical to all ships at sea for ensuring safety of navigation, they should not be used to convey campaign messages and slogans.[158] Finally, the other rules of international maritime law, and in particular COLREGS and UNCLOS, should be observed.[159]

The Code of Conduct amplifies the rules of COLREGS to ensure that protests will not interfere with the operation of vessels targeted by the protesters. Demonstrations that require other vessels to alter course or speed are specifically prohibited.[160]

The Code also stresses the responsibilities of flag states of protest vessels, and tries to draw them into the equation to enforce maritime law and dissuade protesters from interfering with navigation of other ships. Under the Code of Conduct, flag states should take "appropriate and sufficient measures" to ensure that demonstrations are peaceful and non-violent.[161] Flag states also should be prepared to respond to inquiries from the flag state of ships subject

152 *Ibid.*, Annex, para. 3.1.2.
153 *Ibid.*, Annex, para. 3.2.1.
154 *Ibid.*, Annex, para. 3.2.2.
155 *Ibid.*, Annex, para. 3.3.1.
156 *Ibid.*, Annex, para. 3.3.1.
157 *Ibid.*, Annex, para. 3.3.2.
158 *Ibid.*, Annex, para. 3.4.
159 *Ibid.*, Annex, para. 3.5.
160 *Ibid.*, Annex, para. 3.6.2.
161 *Ibid.*, Annex, para. 4.1.

to protest when their "orderly navigation" is being disrupted. This provision appears to elevate communications on the issue to a more formal level. The text of the Code of Conduct concerning communication between flag states is similar to provisions in SUA requiring states to coordinate with one another.

Greenpeace opposed adoption of the Code of Conduct, arguing that existing instruments already obligated all states to enforce COLREGS. Although the organization has distanced itself from protests that threaten navigation, it also was concerned that the draft Code might have a "chilling effect" on lawful free expression.[162] Recognizing that it is "indisputable" that all vessels at sea must comply with safe practices of navigation, Greenpeace, it would be "strange to require a different standard of safety from ships engaged in advocacy. . . ."[163]

Greenpeace also challenged the factual basis upon which the Code was proffered. Japan claimed that its factory whaling ship, Nisshin Maru, was deliberately rammed by the Greenpeace ship Arctic Sunrise, on 8 January 2006.[164] An investigation and review of film footage of the incident by Dutch flag state authorities, however, did not result in charges against members of the organization for a violation of law.[165] Furthermore, foreign nations may pursue claims against vessels under Article 94 of UNCLOS by filing a claim with the flag State—in this case, the Netherlands. Japan did not take such action.[166]

The IMO Sub-Committee on Safety of Navigation did not adopt the Code at NAV 54, and instead passed draft a resolution, which it sent to the IMO Flag State Implementation (FSI) subcommittee for comment.[167] Comments were considered at NAV 55 in the summer of 2009, and the draft resolution was forwarded to MSC for adoption.[168] Since 2009 was an IMO Assembly year, there was no fall meeting of MSC, so the next session for MSC to convene was MSC 87 in May 2010, when the resolution was adopted.[169]

---

162 IMO Doc. NAV 54/10, Code of Conduct During Demonstrations/Campaigns against Ships on the High seas; Reflections on a proposal for a code of conduct during demonstrations/campaigns against ships on the high seas, 26 March 2008, para. 4. For the right to free expression, *see* instruments in notes 29–32, *supra*.

163 *Ibid.*, para. 6.

164 IMO Doc 82/21/8, *supra*, note 148.

165 IMO Doc. NAV 54/10, *supra*, note 162, para. 8.

166 *Ibid.*

167 IMO Doc. NAV 54/WP.9, Code of Conduct During Demonstrations/Campaigns against Ships on High Seas, 3 July 2008.

168 IMO Doc. NAV 54/25, Report to the Maritime Safety Committee, 14 August, 2008, pp. 29–32.

169 IMO Res. MSC.303(87) 17 May 2010, *reprinted in* IMO Doc. MSC 87/26/Add.1, Annex 22.

IMO Resolution 303(87) calls upon governments to urge:

- "persons and entities under their jurisdiction" from actions that "intentionally imperil human life" or property during demonstrations, protests or confrontations on the high seas;[170]
- vessels that fly their flag to comply with IMO instruments concerning safety of navigation, security and safety of life at sea;[171]
- all vessels to comply with COLREG and the Convention on the Safety of Life at Sea to avoid collisions;[172]
- all vessels during maritime demonstrations or protests or confrontations to comply with the International Telecommunication Union Radio Regulations in their communications;[173]

Furthermore, governments were urged to establish jurisdiction over SUA Convention offenses, and to conduct inquiries into marine casualties and incidents.[174] The resolution delivered less to the Japanese delegation than a Code of Conduct, but nonetheless, as an IMO instrument, it reemphasized the normative standards for safe operations during maritime protests or confrontations.

Conclusion

The right to free expression is not absolute and may be impeded, but interference in the freedom of expression is not unlimited. Article 19(2)(3) of the ICCPR states that the exercise of the right to free expression can be modified or restricted, if the limitations are set forth in law and "are necessary" to respect the rights or reputations of others, for the protection of security or public order or to protect public health and morals. The European Court of Human Rights has ruled that even non-binding regulations imposed on a practicing lawyer that lack any sanction, including admonishment, may pose unlawful censure by influencing the types of factual and legal arguments he may express.[175] Since the Code of Conduct has not been adopted, States are subject to the

170 *Ibid.*, para. 3(1).

171 *Ibid.*, para. 3(2).

172 *Ibid.*, para. 3(3).

173 *Ibid.*, para. 3(4).

174 *Ibid.*, paras. 4–5.

175 *Steur v. the Netherlands*, 28 October 2003, Application No. 39657/98 (European Court of Human Rights), para. 29.

aforementioned binding legal instruments, including UNCLOS, COLREGS and SUA. There are ample authorities for flag States, and in narrowly prescribed circumstances, coastal States and port States, to take appropriate action to protect safety of navigation from direct action protest. On the other hand, the human rights instruments, some of which are legally binding and others that are not, form a bedrock set of norms that also must be observed. The right to free expression is among the most fundamental human rights, and the use of the police authority of the state to restrict demonstrations at sea should be measured and prudent to protect life and property, both of the protesters and the targets of their protest.

Perhaps the most interesting element of this analysis is that international human rights law is designed to form a contract between the individual and the state, whereas the law of the sea expresses a relationship between one state (e.g. a flag state) and another state (e.g. a coastal state). Consequently, individual seafarers operate in essence as the agent of the flag State. While coastal State action against a foreign-flagged vessel invokes a state-to-state relationship, it also implicates individual human rights of the seafarers on board the ship. There is no doubt that States are faced with difficult and uncomfortable events orchestrated by maritime civil disobedience, as that is one of the principal purposes of free expression. States must be careful to ensure that their legitimate police action to prevent violence or danger to ships, people, and property at sea does not become a broader effort to stifle free speech. Likewise, police authorities have an obligation to specifically and generally deter destructive protesters that commit acts of violence or endanger the lives of seafarers.

The Code of Conduct has not gained traction within the IMO, and it was not adopted by Member States. This outcome means that States and protesters have a small pool of legal authorities to guide them. In short, although the risk posed to navigation by "direct action" protest is in the eye of the beholder, the law inures in favor of safety of navigation in cases that threaten vessel safety. There are two important caveats. First, mere political speech, such as signs and speaking—even through amplification—is protected in international human rights law in a comprehensive fashion. This finding raises interesting follow-on questions about the interplay at sea between national laws of authoritarian states, such as Saudi Arabia or China, and the human right of free expression of ships near the coast. Any State may enforce laws against illegal broadcast at sea either within or even beyond the EEZ, so attempts to broadcast radio transmissions into a State are, at least formally, subject to coastal State authority.[176]

176 Article 110, UNCLOS, *supra*, note 2.

Protesters on board a foreign-flagged ship in the coastal state's EEZ have wide latitude to express their views. Second, coastal states may protect their offshore infrastructure and sovereign rights and jurisdiction over natural resources in the EEZ, even against foreign-flagged protesters who may place those interests at risk. This authority, however, which flows from article 56, UNCLOS, does not trump basic norms of international human rights.

Maritime law enforcement efforts even against violent or disruptive protesters, must observe principles of human rights. The minimum force necessary may be used as a last resort; human life and property shall be protected. Human rights norms apply throughout the entire course of interaction between coastal state authorities and protesters conducting activist maritime civil disobedience from initial encounter, through enforcement interaction, and subsequent detention and criminal trial proceedings. Consequently, coastal state rights and jurisdiction may be protected, but not at the expense of the duties owed to individuals in human rights law, regardless of nationality, to free expression. Violations of human dignity and non-discrimination, and departures from the norm against degrading or inhuman treatment, are always forbidden.[177]

---

177 UNDHR, Articles 1, 2, 5, 6, and 7.

CHAPTER 7

Anti-Piracy Responses in the Gulf of Guinea: Addressing the Legal Deficit

*Kamal-Deen Ali**

Introduction

Any assumption that piracy in the Gulf of Guinea was a transient occurrence soon to end now can be firmly discounted.[1] Following the decline in piracy off the coast of Somalia, in 2013 the Gulf of Guinea rose to second place (after Southeast Asia) in the global piracy table.[2] Since then, the West African maritime region has shown enduring attributes of piracy, demonstrated by the spread of incidents and accompanying violence in a series of attacks in 2014. Indeed, from 2010 reports of both the International Maritime Organization (IMO) and the International Maritime Bureau (IMB) have rated the coast of Nigeria as the most dangerous marine space in the world. The indications in 2015 suggest that there is no end in sight to the violence inflicted on seafarers in the region. The hijacking of the fishing vessel, *Lu Rong Yuang Yu 917*, off the coast of Ghana in February 2015 by suspected Nigerian pirates, for example, resulted in the death of one crew member of vessel, several more injured.[3]

* Captain in the Ghana Navy as well as the Director of Research at the Ghana Armed Forces Command and Staff College and the Navy's Legal Director. Dr. Ali is also the Executive Director of the Centre for Maritime Law and Security (CEMLAWS) Africa and an Adjunct Lecturer in Law and other departments at the University of Ghana.

1 The geographical scope of the Gulf of Guinea region may vary depending on the issue or interest at stake. The Gulf of Guinea region is defined here as the coastal states stretching from Senegal to Angola and covering the Economic Community of West African States (ECOWAS) and the Economic Community of Central States (ECCAS). States falling within the region are Cape Verde, Senegal, Gambia, Guinea Bissau, Guinea, Sierra Leone, Liberia, Cote d'Ivoire, Ghana, Togo, Benin, Nigeria, Cameroon, Equatorial Guinea, Sao Tome and Principe, Gabon, Congo Brazzaville, Democratic Republic of Congo and Angola.

2 IMO, *Reports on Acts of Piracy and Armed Robbery Against Ships: Annual Report 2013*, MSC.4/Circ. 208, 1 March 2013.

3 IMB, *Piracy and Armed Robbery against Ships: Report for the Period 1 January to 30 March 2015*. Ghana News Agency, "Pirates Hijacked Fishing Vessel," 6 February 2015, http://www.ghananewsagency.org/human-interest/pirates-hijack-fishing-vessel-85450.

 | DOI 10.1163/9789004311442_009

Other incidents of attacks also occurred in 2015 where vessels were successful hijacked and the life of the crew was in grave danger.[4]

Gulf of Guinea piracy remains a threat to the safety and security of the maritime global commons, even in the face of the adoption, with wide international support, of the Code of Conduct for the repression of piracy and illicit activities by States in the region in 2013.[5] One of the major gaps confronting current and future counter-piracy efforts in the region is the legal complexity within which Gulf of Guinea piracy is cast. This chapter analyzes the legal challenges in the region, and the corresponding responses required to address the challenges. In doing so, it sets the background and explores the trends and dynamics of maritime piracy in the region.

Trends of Gulf of Guinea Piracy and Prevailing Responses

While this chapter focuses on the legal complexity of piracy in the Gulf of Guinea, analyses of the trends and incidents of attacks in the region is essential to understanding the legal challenges. In fact, the peculiar piracy dynamics in the region produce the salient legal challenges.[6]

Incidents and Trends in the Gulf of Guinea

The global piracy statistics of 2013 show where Gulf of Guinea stands relative to worldwide trends. In 2013, there were few reported incidents off the Somali coast and no ship was successfully hijacked.[7] This low incidence of piracy is compared to over 237 incidents in Somalia in 2011.[8] Since 2008, Somali piracy

4 IMB, *Piracy and Armed Robbery against Ships: Report for the Period 1 January to 30 March 2015.*

5 Summit of Heads of State and Governments of ECOWAS and ECCAS, Yaoundé, Cameroon on 24–25 June 2013. *Ban welcomes anti-piracy strategy adopted by leaders from West, Central Africa,* UN News Centre, http://www.un.org/apps/news/story.asp?NewsID=45281, 29 December 2013. *At UN-backed meeting, African States seek to combat piracy in Gulf of Guinea,* UN News Centre, Available: http://www.un.org/apps/news/story.asp?NewsID=44440&Cr=piracy&Cr1=.

6 Ali Kamal-Deen, "The Anatomy of Gulf of Guinea Piracy," *Naval War College Review* 68, No. 1 (2015): 93–118.

7 IMO, *Reports on Acts of Piracy and Armed Robbery Against Ships: Annual Report 2013,* MSC.4/Circ. 208, 1 March 2013. See also Jon Huggins and Lisa Kane-Hartnett, *Somali Piracy—Are We at the End Game?,* Oceans Beyond Piracy Working Paper, http://oceansbeyondpiracy.org/publications/somali-piracy-are-we-end-game.

8 IMB, *Somali Pirate Clampdown Caused Drop in Global Piracy IMB Reveals,* 15 January 2014, *available at,* http://www.icc-ccs.org/news/904-somali-pirate-clampdown-caused-drop-in-global-piracy-imb-reveals.

has continuously decline. Within the same period, however, piracy in the Gulf of Guinea has steadily increased posing a significant threat to the safety and security of global sealines of communication (SLOC), although the incidence of piracy in the Gulf of Guinea is still lower than it was off the coast of Somalia at its peak.[9]

A good assessment of the piracy threat goes beyond the figures of reported incidents, and includes a more nuanced understanding of the dynamics and trends.[10] Gulf of Guinea piracy came into the limelight in 2007 following 60 reported incidents. Over 80 percent of the incidents resulted in an actual boarding and successful hijacking of a ship.[11] The rate of successful hijackings in the region remains at 75–90 percent of reported cases. This high rate of success by pirates distinguishes the Gulf of Guinea from other regions, where the number of reported incidents may be higher, but the number of successful hijackings are lower. Furthermore, the piracy threat has widened, as Gulf of Guinea pirates move farther out to sea, and spread horizontally throughout the maritime region.[12] Significantly, at the close of 2013 the region recorded more piracy attacks on the high seas than in previous years.[13] The 2013 hijacking of the Maltese-flagged vessel, *Cotton*, for example, was the first attack of its kind along the coast and port area of Gabon. Then in 2014, Angola and Ghana registered their first incidents of attacks, demonstrating that there is a deepening threat of piracy.[14]

Threats to Energy Security

Piracy in the Gulf of Guinea also poses a threat to global energy security, especially since the region is increasingly a source of energy supplies.[15] In the Gulf of Guinea, oil and gas typically are processed with Floating, Production,

---

9 Peter J. Pham, "Africa's Other Dangerous Waters: Piracy in the Gulf of Guinea," *World Defence Review* 3 (Dec. 2009). Scott Baldauf, "Pirates Take New Territory: West African Gulf of Guinea," *The Christian Science Monitor*, 15 Jan. 2010.

10 Ali Kamal-Deen, "The Anatomy of Gulf of Guinea Piracy," *Naval War College Review* 68, No. 1 (2015): 93–118.

11 IMB, *Piracy and Armed Robbery against Ships: Annual Report 2007*.

12 Kamal-Deen, "The Anatomy of Gulf of Guinea Piracy," *supra*, note 10.

13 IMO, *Reports on Acts of Piracy and Armed Robbery Against Ships—Annual Report 2013*, MSC.4/Circ. 208, 1 March 2013.

14 The *MT Kerala*, was hijacked off Angola in January 2014 and its oil cargo worth US $8 million was stolen. Then in the months of June and July 2014 the coast of Ghana registered three cases of hijacking, two of which were on oil tankers. IMO, *Reports of Piracy and Armed Robbery against Ships January 2014*, MSC.4/Circ. 206, 7 March 2014.

15 *See* Damian Ondo Mañe, *Emergence of the Gulf of Guinea in the Global Economy: Prospects and Challenges*, IMF Working Paper, WP/05/235, 2005. George F. Jedrzej and Manuel

Storage and Offloading units (FPSOs). These are either converted double hull oil tankers or structures specifically designed for the purpose.[16] The FPSO technology is widely regarded as one of the greatest breakthroughs of the twenty-first century that has made deep-sea oil and gas production viable.[17] The Gulf of Guinea is host to a growing number of FPSO platforms including *Agbami*,[18] *Dalia*,[19] *Nkrumah*[20] and *Serpentina*.[21] *FPSO Agbami*, located offshore Nigeria, is the largest in the world, with a storage capacity of over 2.1 million barrels.

The security of these multi-million dollar platforms is at risk because of insurgents and pirates.[22] For example, in 2008, Nigeria's Niger Delta insurgents, who for all practical purposes are engaged in piracy,[23] successfully attacked the *FPSO Bonga* over 120 km off the coast of Nigeria.[24] This incident came after earlier attacks. The *Bulford Dolphin*, a mobile drilling rig, was attacked in April 2007 by insurgents.[25] In May 2007 *Mystras* was also attacked, and three days

---

Paulo, "A New Scramble for African Oil? Historical, Political and Business Perspectives," *African Affairs* 106, No. 423 (April 2007): 229–251.

16 Neil Ford, *The Star of the Gulf of Guinea*, July 20 2005, http://www.gasandoil.com/news/2005/08/cna53395.

17 *Ibid.*

18 *FPSO Agbami*, off the coast of Nigeria's Niger Delta region has a processing capacity of 250,000 barrels of oil per day and a storage capacity of 2.1 million barrels of oil. The contract for building it was awarded in 2005 and it arrived in the oil field in 2007. See, http://www.offshore-technology.com/projects/agbami/.

19 Located in ultra-deep waters of Angola (about 80nm), Dalia is one of the world's largest deep water oil developments. The Dalia FPSO floating production, storage and offloading (FPSO) vessel used in the field is 300 meters long, 60 meters wide and 32 meters high. It weighs in at an equally impressive 416,000 metric tons, nearly 30,000 metric tons for the topsides alone. Its living quarters can accommodate up to 190 people. See, http://www.total.com/en/our-energies/oil/exploration-and-production/projects-and-achievements/dalia-940847.html (accessed 12 August 2011).

20 The FPSO *Kwame Nkrumah MV21* is installed in approximately 1,100 meters water depth on the Jubilee Field. It has been named after the first president of the country and arrived for operations in June 2010.

21 FPSO *Serpentina* was built in 1974. It was delivered to the Zifiro fields on 1 November 2005 under the order of Equatorial Guinea Inc., an ExxonMobil subsidiary ExxonMobil subsidiary.

22 On threats to offshore energy, see Mikhail Kashubsky, "Protecting Offshore Oil and Gas Installations: Security Threats and Countervailing Measures," *Journal of Energy Security* (August 2013), www.ensec.org/.

23 Kamal-Deen, "The Anatomy of Gulf of Guinea Piracy," *supra*, note 10.

24 "Bonga Deepwater Project, Niger Delta," *Offshore Technology*, www.offshore-technology.com/.

25 *Bulford Dolphin* (Singapore) was attacked on 1 April 2007, sixty-four kilometers off the Nigerian coast. The kidnapped worker was released four days later; IMB, *Reports on Acts*

later *Trident VIII* was targeted.[26] In addition to the physical damage and personal injuries inflicted by the insurgents, these incidents affected the operation of the platforms.[27] The *Bonga* incident had especially serious implications for Nigeria, the wider Gulf of Guinea, and beyond, as global oil prices rose after the attack, which reduced Nigeria's production.[28]

Security Council Resolutions and the Regional Response

As in the case of piracy off the coast of Somalia, the United Nations Security Council (UNSC) has adopted resolutions concerning piracy in the Gulf of Guinea. Resolution 2018, for example, was adopted in 2011. It expresses concerns about the rising insecurity in the region, and its impact on regional and global security. The resolution calls on regional States and the global community to take action to subdue the threat.[29] This resolution was followed by Resolution 2039 the following year. Resolution 2039 also highlights the threat of piracy in the region and reinforces the urgency of effective regional and global counter-piracy measures.[30] In between the adoption of these two resolutions, the UN Secretary General dispatched a team to the region to assess the situation.[31]

---

of Piracy and Armed Robbery against Ships: Annual Report 2007, January 2008), p. 66. The IMB report identifies the rig as *Bueford Dolphin*, but the author's research (which included a review of the comparative IMO annual report for 2007, MSC.4/Circ. 115 (10 April 2008) confirms that the correct name is *Bulford Dolphin*.

26 *Mystras*, an FPSO, was attacked on 3 May 2007, while *Trident VIII*, a mobile offshore drilling rig, was attacked 5 May. In both cases people were kidnapped. IMB, *Reports on Acts of Piracy and Armed Robbery against Ships: Annual Report 2007* (London: January 2008): 67. See also the listing of these incidents in IMO, *Piracy and Armed Robbery: Annual Report 2007*, MSC.4/Circ. 115, 10 (April, 2008): 11–12.

27 S. Bajpai and Jai Gupta, "Securing Oil and Gas Infrastructure," *Journal of Petroleum Science and Engineering* 55 (2007): 174–86.

28 For a detailed analysis of the *Bonga* incident and its effect on global energy and insurance industries, see Mikhail Kashubsky, "Offshore Energy Force Majeur: Nigeria's Local Problems with Global Consequences," *Maritime Studies* (May–June 2008): 20–26. See also Alexander Kwiatkowski and Julie Ziegler, "Shell Shuts Nigeria's Bonga Oil Field after Attack," *ODAC: The Oil Depletion Analysis Centre*, 19 June 2008, www.odac-info.org/; and Jeff Veil, "Nigeria: The Significance of the Bonga Offshore Oil Platform Attack," *Oil Drum*, 24 June 2008, www.theoildrum.com/.

29 United Nations Security Council (UNSC), *Resolution 2018—Piracy and Security in Africa*, (31 October 2011).

30 United Nations, *Resolution 2039—Peace Consolidation in West Africa*, (29 February 2012).

31 United Nations, *Report of the United Nations Assessment Mission on Piracy in the Gulf of Guinea*, UNSC S/2012/45, (19 January 2012).

In 2013, the states of the Gulf of Guinea adopted the Code of Conduct for the repression of piracy in June 2013 (Yaoundé Code of the Conduct) which covers the broader Gulf of Guinea region.[32] The Yaoundé framework is anchored on the maritime security architecture of the member States of the Economic Community of Central African States (ECCAS), adopted earlier in 2009.[33] The ECCAS framework divided the maritime space of the seven ECCAS coastal states into three zones, each with a Multinational Coordination Centre that are networked with a Regional Coordination Centre, known by its French acronym CRESMAC (Centre Régional de Sécurité Maritime de l'Afrique Centrale).[34] Under the umbrella of the Yaoundé Code of Conduct, member states of the Economic Community of West African States (ECOWAS) are mirroring the ECCAS arrangement. The states have also organized the ECOWAS region into three zones, starting with a pilot zone.[35] The ECOWAS and ECCAS regions are on track to become coordinated through the Inter-regional Coordination Centre (ICC) being established in Yaoundé, Cameroon.

Mapping the Legal Complexity in the Gulf of Guinea

The regional arrangements discussed above demonstrate a commitment by Gulf of Guinea states to address the menace of regional piracy. Realistically, however, the success and efficacy of both regional and global counter-piracy

---

32 Summit of Heads of State and Governments of ECOWAS and ECCAS, Yaoundé, Cameroon on 24–25 June 2013. *Ban welcomes anti-piracy strategy adopted by leaders from West, Central Africa,* UN News Centre Available: http://www.un.org/apps/news/story.asp?NewsID=45281, 29 December 2013. *At UN—backed meeting, African States seek to combat piracy in Gulf of Guinea,* UN News Centre, http://www.un.org/apps/news/story.asp?NewsID=44440&Cr=piracy&Cr1=, 29 December 2013.

33 The coastal States of ECCAS are Angola, Cameroon, Congo-Brazzaville, DR Congo, Gabon, Equatorial Guinea and Sao Tomé & Principe. The landlocked States include the Central Africa Republic, Burundi and Chad. Rwanda was originally a member State of ECCAS, but withdrew from the Community as a result of major disagreements with other member States, especially Angola. Indeed, the two States fought on opposite sides during the war in DR Congo. Whilst some media reports have indicated that Rwanda has officially rejoined ECCAS, the official web page of ECCAS does not list Rwanda as a member State. See "Economic Community of Central African States," http://www.au.int/en/recs/eccas.

34 Protocole Relatif a la Strategie de Securitisation des Interest Vitaux en Mer des Etats de la CEEAC du Golfe de Guinea (Yaoundé, Cameroon, 24 October 2009).

35 ECOWAS, Final Communique, Forty-third Ordinary Session of the ECOWAS Authority of Heads of State and Government, Abuja, Nigeria, 17–18 July 2013, paragraph 37 and 38.

responses must go beyond surveillance and information sharing. Unless these efforts are supported by a robust legal architecture, anti-piracy efforts will not be effective. Consequently, it is especially important to be aware of the peculiar and distinct legal complexity of by piracy in the Gulf of Guinea. The threat in West Africa can be contrasted with the Somali experience.

The Somali Experience

Piracy off the coast of Somalia, and the corresponding international response, led to significant developments in international and national to counter maritime piracy.[36] These legal developments were borne out the peculiarity of the situation in Somalia. First, the Somali state was subject to widespread violence and instability since 1991, and it has been without a functioning government. As a result of the collapse of the state, Somali authorities have been unable to effectively combat acts of piracy and armed robbery, and they certainly lack any supporting legal regime for doing so. The problem is compounded by the fact that the State of Somalia claims a 200 nm territorial sea of 200.[37] The territorial sea claim is inconsistent with the 12 nm maximum breadth stipulated by international law, and it added to the difficulty in implementing counter-piracy measures because it blurred the distinction between piracy and armed robbery at sea.[38]

To obviate these legal and practical difficulties, beginning in 2008, the UNSC adopted multiple resolutions authorizing anti-piracy measures in Somalia. The normative implications of these resolutions, especially Resolutions 1846 and 1851, have attracted considerable debate in international law.[39] Resolution 1846 granted authority to naval forces to enter Somali territorial waters and

---

36 On Somali piracy, the international responses and challenges, see Douglas Guilfoyle, ed., *Modern Piracy: Legal Challenges and Responses* (Cheltenham, UK; Northampton, MA, USA: Edward Elgar Publishing, 2013); Bibi van Ginkel and Frans-Paul van der Putten, *The International Response to Somali Piracy: Challenges and Opportunities* (Leiden; Boston: Brill/Nijhoff, 2010); and Martin Murphy, *Somalia, The New Barbary?: Piracy and Islam in the Horn of Africa* (London: Hurst Publishers, 2010).

37 Law No. 37 on the Territorial Sea and Ports, of 10 September 1972, available at http://www.un.org/Depts/los/LEGISLATIONANDTREATIES/PDFFILES/SOM_1972_Law.pdf, 15 July 2013.

38 Tullio Treves, "Piracy, Law of the Sea, and Use of Force: Developments off the Coast of Somalia," *European Journal of International Law* 20, No. 2 (2009): 399–414.

39 For analysis of different legal interpretations of these resolutions see Sam Bateman, "UNCLOS and its Limitations as the Foundation for a Regional Maritime Security Regime," *Institute of Defence and Strategic Studies Singapore*, Working Paper 111 (April 2006). See also Douglas Guilfoyle, "Somali Pirates as Agents of Change in International

use "all necessary means" to repress piracy while Resolution 1851 permits the pursuit of pirates into Somali land territory.[40] Although the resolutions do not stipulate specific measures, their scope has been interpreted to include the right to seize and dispose of boats, vessels, arms and related equipment being used by the pirates or robbers, as well as the right to arrest and prosecute such offenders.[41]

Somali piracy also raised questions about the adequacy of legislation and the comprehensiveness of the criminal justice system for combating the crime of piracy at national, bilateral, regional and multilateral levels.[42] This led to a UN commissioned report that evaluated the suitability of prosecuting pirates using a regional court or through an international court of the likeness of the International Criminal Court.[43] The overall recommendation, however, went in favour of using a national process. Thus, foreign naval forces patrolling the Indian Ocean had to establish mechanisms for the transfer of arrested Somali pirates to neighboring States for prosecution.[44] This automatically ignited legal issues and challenges within the domestic legal system of the prosecuting States.[45] The United Nations Office on Drugs and Crime (UNODC) subsequently developed a piracy prosecution model within a broader Maritime Crime Programme (MPC) for the Indian Ocean.[46]

In addition to the above multilateral legal developments, the US Appellate Court also made ground-breaking contribution to the law on piracy with the view that while the offence of facilitating piracy must be in aid of acts committed on the high seas, the associated offence of facilitating (or aiding and abetting) is not contingent on the person having physically participated in the

Law-making and Organization," *Cambridge Journal of International and Comparative Law* 1, No. 3 (2012): 81–106.

40 United Nations, Security Council Resolution 1846 (6026th Meeting, 2 December 2008). United Nations, Security Council Resolution 1851 (6046th Meeting, 16 December 2008).

41 See para. 10(b), UNSC Resolution 1846, 26 December 2008.

42 For a discussion of these issues, see Guilfoyle, ed., *Modern Piracy*, 35–61, 147–202.

43 "Report of the Secretary General on Modalities for the Establishment of Specialized Somali Anti-Piracy Courts," UNSC S/2011/360, 15 June 2011. See also "Report of the Special Adviser to the Secretary-General on Legal Issues Related to Piracy off the Coast of Somalia," Annexed to UNSC S/2011/30, Letter dated 25 January 2011.

44 See generally, Guilfoyle, ed., *Modern Piracy*.

45 Musili Wambua, "The Jurisdictional Challenges to the Prosecution of Piracy Cases in Kenya: Mixed Fortunes for a Perfect Model in the Global War against Piracy," *World Maritime University Journal of Maritime Affairs* 11 (2012): 95–113.

46 UNODC, "Maritime Crime Programme—Indian Ocean," http://www.unodc.org/unodc/en/piracy/indian-ocean-division.html.

crime on the high seas.[47] According to the Court, it is sufficient that the person has played a role in facilitating the piracy.

Emerging Legal Complexity in the Gulf of Guinea

Whereas the legal developments and experiences from Somalia have relevance for the Gulf of Guinea, piracy in the region presents distinct legal challenges. The complexities are evolving, hence their scope and interactions are far from settled. Nonetheless, current and future counter-piracy responses (national, regional and international) have to navigate these legal conundrums.

The successful prosecution of any crime depends, foremost, on the ability to establish jurisdiction over the suspects, followed by an assembling of the requisite evidence to prove the crime. Although Somali pirates operated with considerable skill and resolve, aided by sophisticated criminal networking, their *modus operandi* fits largely into the paradigm of traditional piracy.[48] In this context, a ship is hijacked, and held in a safe haven while payment of a ransom is negotiated. This profile makes both the establishment of jurisdiction and the relevant collection of evidence for prosecution uncomplicated, especially when the pirates are generally Somalis. The process of ransom negotiations, payment of the money, which may be transferred directly to the pirates or through middlemen and through banks, leave footprints that aids investigation and criminal prosecution of pirates.

The Gulf of Guinea presents a different scenario, with its hybrid of traditional and insurgent piracy. Unlike Somalia, pirates in West Africa frequently disable ship's equipment and take control of the ship. This model necessitates not just combating "crimes of piracy,"[49] but also those offenses provided under the 1988 Convention for the Suppression of Unlawful Acts at Sea against the Safety of Maritime Navigation (SUA) and its 2005 Protocols that are directed at safeguarding the safety and security of the ship.[50] This gives rise to *mix*

---

47 United States v. Ali, (DC Cir. 11 June 2013).

48 On the sophistication of Somali piracy, see Raymond Gilpin, "Counting the Costs of Somali Piracy," *United States Institute of Peace Working Paper*, June 2009; Lauren Ploch, Christopher M. Blanchard, Ronald O'Rourke, Chuck R. Mason and Rawle O. King, "Piracy off the Horn of Africa," *US Congressional Research Service*, 2009; and "Pirates Have Better Equipment," *Fairplay*, 2 November 2006.

49 The phrase "crimes of piracy" is used here because in addition to the normal offence of piracy, related crimes of assault, causing harm, or depending on the situation, even murder, may arise from a piracy incident.

50 The SUA Framework is analyzed more closely in the last section. The 1988 Convention for the Suppression of Unlawful Acts against the Safety of Maritime Navigation, 10 March 1988, Rome, Italy, *entry into force* 01 March 1992, 1678 U.N.T.S. 201, 27 I.L.M. 672 (1988)

jurisdiction in relation to the "heads" of crime and therefore multiple criminal investigation and prosecution requirements.

There are further complications arising from the mix jurisdiction. Piratical incidents in the Gulf of Guinea are both inward and outward. These characteristics make the distinction between piracy and armed robbery at sea problematic, and renders the jurisdictional interests complex. For example, there are instances where Nigerian pirates enter the territorial sea, and at times, port areas of other states (especially Benin and Togo) to hijack ships, which then are commandeered to the shores of the Niger Delta.[51] From the perspective of Benin or Togo this scenario is an example of an *inward-outward* case of piracy, but from Nigeria's perspective, it is an example of *outward-inward* piracy. Both States would have immediate and significant jurisdiction founded on *sovereignty*, and an interest in the protection of their national security and economic interest. This situation poses a dilemma of overlapping jurisdiction over the same offense. This outcome, however, may be positive because it allows, in theory, any of the states to investigate and prosecute. The reality, however, may be different. In practice, the evidence could be split, with some landing in the possession of Benin and some under the possession of Nigeria. The witnesses likely are dispersed, and more significantly, if any of the state refuses to cooperate or views the crime as exclusively falling within its sovereignty (armed robbery at sea), the jurisdictional fault lines could be even more problematic.

There are also indications that unlike their Somalia counterparts, Gulf of Guinea pirates, including those cases in the Niger Delta, may be of different nationalities. Therefore, additional jurisdiction may be founded (and should be founded) on the basis of nationality by third states, especially where suspects are within the territory of the state.

All this aside, the flag state and the state of nationality of any of the victims of the crime, and generally all states, would have jurisdiction on the basis of international law. These intricate overlays lead to a conundrum of jurisdictional interests and imperatives with grave implications for effective

---

(1988 SUA Convention) and 2005 Protocol for the Suppression of Unlawful Acts against the Safety of Maritime Navigation, 14 October 2005, IMO Doc. LEG/CONF. 15/21, Adoption of the Final Act and Any Instruments, Recommendations and Resolutions Resulting from the Work of the Conference, Protocol of 2005 to the Convention for the Suppression of Unlawful Acts against the Safety of Maritime Navigation, 1 November 2005 (2005 SUA Convention).

51 IMO, *Reports on Acts of Piracy and Armed Robbery against Ships: Acts Reported during June 2011*, MSC.4/Circ. 173, 11 July 2011, and *Acts Reported during July 2011*, MSC.4/Circ. 174, 8 August 2011.

investigation, evidence collection, procurement of witnesses and successful criminal prosecution.

Another key challenge in the Gulf of Guinea context relates to the articles, and proceeds of crime. Stealing petroleum cargo is the primary motivation of pirates in the Gulf of Guinea. The cargo is often transferred and sold at sea, while the hijacked ship released. The duration of the crime generally is short, compared to the kidnapping for ransom model that prevails along the Somali coast. More significantly, it is often difficult to trace both the article and proceeds of the time. This additional limiting factor impedes successful prosecution of piracy offenses in the Gulf of Guinea.

Addressing the Legal Challenges in the Gulf of Guinea

These trends and legal analysis indicate that the Gulf of Guinea exhibits an enduring profile of piracy that is complicated to combat through the law enforcement mechanism. It is not a case of "one size fits all," where the experiences and initiatives from Somalia could migrate to the Gulf of Guinea. What is required is an approach that uniquely addresses the peculiar legal challenges present in the region. First and foremost, an anti-strategy in the Gulf of Guinea must provide for comprehensive piracy legislation, but should also address additional legal challenges identified above.

Piracy Legislation

Piracy is a crime of universal jurisdiction, and this rule is codified in the United Nations Convention on the Law of the Sea (UNLOS).[52] Article 100 of UNCLOS encapsulates two interrelated obligations regarding piracy.[53] First, States are required to repress piracy at the national level, and also cooperate with other States in the suppression of piracy at the regional and international level. To give practical effect to the first obligation, the Gulf of Guinea States are obligated to enact and enforce laws that cover all aspects of the crime of piracy.[54]

52 See Articles 100–105, United Nations Convention on the Law of the Sea (UNCLOS), 10 December 1982, entered into force 16 November 1994, 1833 U.N.T.S. 397, *reprinted in* 21 *I.L.M.* 1261 (1982) (UNCLOS). The Convention has 167 ratifications as of January 2015.

53 Article 100 of UNCLOS provides: "All States shall cooperate to the fullest possible extent in the repression of piracy on the high seas or in any other place outside the jurisdiction of any State."

54 *See* Ch. 4, sec. 4.2.1.2 and 4.2.1.3. *See also* Eugene Kontorovich, "A Guantanamo on the Sea: The Difficulty of Prosecuting Pirates and Terrorists," *California Law Review* 98 (2010):

With respect to the second obligation, cooperative instruments and structures should be established which, at a minimum, will facilitate the sharing of information, and also possibly lead to joint anti-piracy patrols in the region.

A review of the national legislation database of the United Nations Division on the Law of the Sea (UNDOALOS), however, shows that Liberia and Togo are the only States in the region that have enacted piracy legislations in the Gulf of Guinea.[55] It is only since January 2013 that Nigeria is reported to have initiated a process for the enactment of a law to combat piracy and other maritime crimes.[56] It has also been noted that the definition of the crime of piracy in the national laws of Benin is out-dated and not consistent with the provisions of LOSC.[57] So is Ghana.[58]

In sum, there is a legislative deficit with respect to the crime of piracy in the Gulf of Guinea. Even if Gulf of Guinea States are able to conduct maritime law enforcement patrols off their coasts, the efficacy of these deployments is seriously undermined by the absence of domestic legislation to prosecute and punish activities that pose a threat to maritime security. This situation likely results in what is termed "catch and release," in which counter-piracy naval forces frequently released apprehended pirate suspects because of legal obstacles to effective criminal prosecution. "Catch and release" further entrenched insecurity in the initial few years of international counter-piracy patrols off the coast of Somalia.[59] The states in the Gulf of Guinea, therefore, must create an effective counter-piracy legal regime by first enacting offences for piracy with

---

243–76 and Diana Chang, "Piracy Laws and the Effective Prosecution of Pirates," *Boston College International and Comparative Law Review* 33 (2010): 273–188.

55 See Penal Code of Liberia, 2008, and Code de la Marine Marchande, Togo, 1971, United Nations Division of Ocean Affairs and Law of the Sea (UNDOALOS), Legislation and Treaties Data Base, http://www.un.org/Depts/los/LEGISLATIONANDTREATIES/index.htm, 1 January 2013.

56 See "NIMASA Seeks Legal Backing to Fight Piracy," *This Day*, 14 February 2013.

57 United Nations, Report of the United Nations Assessment Mission on Piracy in the Gulf of Guinea, Security Council S/2012/45, 19 January 2012, p. 5.

58 Citifm online, "Ghana's piracy laws must be reviewed—Maritime Expert," http://citifmonline.com/2015/01/20/ghanas-piracy-laws-must-be-reviewed-maritime-expert/#sthash.dy8bWiSk.dpuf.

59 For a discussion of "catch and release" and the difficulties associated with piracy prosecution in Somalia, *see* Eugene Kontorovich, "A Guantanamo on the Sea: The Difficulty of Prosecuting Pirates and Terrorists," *California Law Review* 98 (2010): 243–76. *See also* Elizabeth Andersen, Benjamin Brockman-Hawe and Patricia Goff, *Suppressing Maritime Piracy: Exploring the Options in International Law*, 27 July 2013, http://acuns.org/wp-content/uploads/2012/06/SuppressingMaritimePiracyExploringOptionsIntlLaw.pdf.

accompanying penalties, and second, establishing and maintaining effective criminal justice systems.

Ratifying and Incorporating the SUA Framework

The 1988 SUA Convention provides the basis for responding to insurgent-type attacks against ships in the Gulf of Guinea.[60] Such attacks have the tendency to falling outside the scope of UNCLOS, as they may be deemed to lack the element of a "private motive," which is one the essential ingredients of the crime of piracy. The 1988 SUA Convention has dispensed with the private motive requirement, and thus gives Gulf of Guinea states the opportunity to effectively respond to both classical maritime piracy and insurgent-type maritime attacks. State Parties to UNCLOS are also under an obligation to criminally prosecute offenders in their custody, even if the offense was not committed off their coast. This jurisdictional hook solves the piracy conundrum in the Gulf of Guinea, where Niger Delta pirates commit attacks off the coasts of other States, and then flee into Nigerian waters as a safe haven. Under the SUA framework, Nigeria is obligated to prosecute the pirates, or otherwise extradite them to any state that makes an extradition request. The 1988 SUA Protocol[61] also provides a means for securing offshore oil platforms, and by implication, energy security in the Gulf of Guinea.[62]

Despite the relevance of the SUA framework, the ratification and implementation of the instruments by Gulf of Guinea States has been unsatisfactory. Only Cote d'Ivoire is a party to all the SUA instruments, and just since 2012.[63]

Angola, Cameroon, Congo, Gabon and Nigeria all are major oil producing states with substantial offshore infrastructure, but none have ratified the 1988 SUA Fixed Platform Protocol, even though it was negotiated precisely to help states safeguard offshore oil platforms and installations.[64] Those states that have ratified the SUA instruments have generally failed to implement them

---

60 Generally on the SUA framework and the suppression of terrorism, see Helmut Tuerk, "Combating Terrorism—The Suppression of Unlawful Acts Against the Safety of Maritime Navigation," *University of Miami International and Comparitive Law Review* 15 (2008): 337–67.

61 1988 Protocol for the Suppression of Unlawful Acts against the Safety of Fixed Platforms Located on the Continental Shelf, 10 March 1988, entry into force 1 March 1992, 1678 U.N.T.S. 201, 27 I.L.M. 672 (1988) (SUA Fixed Platform Protocol, 1988).

62 For a detailed discussion of the Protocol's provisions relating to offshore installations and structures, see Stuart Kaye, "International Measures to Protect Oil Platforms, Pipelines and Submarine Cable from Attack," *Tulane Maritime Law Journal* 31, No. 2 (2007): 377–423.

63 IMO Status of Multilateral Convention and Instruments as at June 30, 2013.

64 *Ibid.*

in their domestic legal system.[65] For example, although Benin, Cote d'Ivoire, Ghana and Nigeria have all ratified the 1988 SUA Convention, not one has adopted implementing legislation.[66] Based on the information available to the United Nations Division of Ocean Affairs and Law of the Sea (UNDOALOS), Liberia is the only state in the region to have taken steps to implement the 1988 SUA instruments through amendment of its penal code in 2008.[67] Even this effort, however, falls short of a comprehensive domestic regime, as Liberia has failed to ratify the 2005 Protocols to the SUA Convention. The 2005 Protocols provide an enhanced regime for the protection of shipping and offshore installations, and therefore are crucial for Liberia, particularly if one considers its status as the second largest flag state in the world.[68]

The inadequate implementation of the SUA framework in the Gulf of Guinea has serious implications for regional maritime security. Gulf of Guinea States are incapable of taking effective enforcement measures at the national level, which exposes offshore platforms and the shipping industry to attack. Therefore, it is imperative for Gulf of Guinea States to ratify and implement the 2005 SUA instruments within their domestic legal and policy frameworks. This action should be followed by the development of regional responses, also made on the basis of the SUA instruments.

Investigating Piracy and Networks

Robust investigation is essential to combating piracy in the Gulf of Guinea. The difficulty of effective criminal investigation however, is especially acute in the region. Most of the significant cases of hijacking in the Gulf of Guinea span different national jurisdictions. The vessel is often hijacked in the coast of one State and either the ship or the article or proceeds of the crime, sent to

---

65 This fact was been confirmed in an address delivered at the UN by the Secretary General of the Maritime Organisation for West and Central Africa (MOWCA). *See* Magnus T. Addico, *Maritime Security Threats and Responses in West and Central Africa* (New York: UNICPOLOS, 23–27 June 2008).

66 For Nigeria, *see* "NIMASA Seeks Legal Backing to Fight Piracy," *This Day*, 14 February 2013. For Benin, see United Nations, *Report of the United Nations Assessment Mission on Piracy in the Gulf of Guinea*, Security Council S/2012/45, 19 January 2012. For Cote d'Ivoire and Ghana, *see* Interview with National Authorities, on file with Author.

67 Liberia, An Act to Amend the Penal Code of 1976 Adding the Offences of Armed Robbery, Terrorism and Hijacking as Capital Offences, 22 July 2008.

68 Rating of Flag States by the United Nations Commission on Trade and Development (UNCTAD) in 2012. *See* UNCTAD, *Review of Maritime Transport* (2012): 44.

another State. This model is particularly salient for the case oil cargo, which typically end up in another State within the region or beyond.[69]

Responding to such a crime profile requires thorough and well-coordinated investigation.[70] The skills and mechanisms available to institutions such as the International Criminal Police Organization (INTERPOL), however, can be of great value. Since 1985, INTERPOL is committed to the investigation of maritime piracy incidents.[71] This decision gave birth to Project BADA, which is INTERPOL's main vehicle for cooperation and investigation of piracy attacks, financing and the proceeds.[72] The IMO also emphasized the requirement for thorough investigations of piracy incidents, and it has provided guidelines for doing so.[73] Gulf of Guinea States and the global community should maximize these processes.

Prosecution and Extradition

The IMO Guidelines of May 2011 reiterate the need to capture, prosecute and sentence persons involved in piracy and armed robbery.[74] Where the State is not interested in prosecuting suspected pirates, appropriate structures should be put in place that permits or even obligates the extradition of suspects. In contrast to the generally permissive language of some treaties, the 1988 SUA Convention imposes strong obligations on States parties in whose territory a SUA offender is found to take the person into custody, irrespective of whether the State wants to prosecute the offender.[75] The State also must extradite the suspect to a requesting State Party on the basis of an existing extradition treaty, or by using the Convention as the basis for extradition.[76] Where the State fails

69 Patrick Dele Cole, a politician from the oil-rich Niger Delta region, said 90 percent of the stolen oil is being shipped illegally out of the country. *See* "Ghana Oil Watch, Stolen Nigeria oil goes to Balkans and Singapore," 23 October 2012, http://ghanaoilwatch.org/index.php/ghana-oil-and-gas-news/2754-stolen-nigeria-oil-goes-to-balkans-and-singapore. *See also* Christina Katsouris and Aaron Sayne, *Nigeria's Criminal Crude: International Options to Combat Export of Stolen Oil* (London: Chatham House, September 2013).

70 Henri Fouche and Jacques Meye, "Investigating Sea Piracy: Crime Scene Challenges," *World Maritime University Journal of Maritime Affairs* 11 (2012): 33–50.

71 54th INTERPOL General Assembly, AGM/54/RES/1,—Washington, D.C., 1–8 October 1985.

72 "INTERPOL to create task force against maritime piracy," INTERPOL, http://www.interpol.int/News-and-media/News/2010/PR004.

73 IMO Resolution A.1025 (26), MSC./Circ. 1404, 23 May 2011.

74 *Ibid.*

75 1988 SUA Convention, Article 7.

76 *Ibid.*, Articles 10 and 11.

to extradite the suspect, the State is "obliged without exception" to criminally prosecute the suspect.[77]

Implementing the Yaoundé Code of Conduct

In June 2013, the Gulf of Guinea States adopted the Code of Conduct concerning the repression of piracy and illicit maritime activity (Yaoundé Code of Conduct).[78] The Yaoundé Code of Conduct has become the lead framework for addressing the scourge of piracy in the region. The Parties undertakes to adopt appropriate national maritime security policies to safeguard maritime trade from all forms of unlawful acts, and to have appropriate national legislation, practices and procedures, to operation of port facilities and ships at all security levels.[79] The Code also stresses the requirement for arrest and prosecution of persons who have committed piracy or who are reasonably suspected of having done so.[80]

The Yaoundé Code of Conduct by itself, however, has not made a remarkable addition to the empirical framework of anti-piracy. Whereas the envisaged cooperative structures and information sharing mechanisms under the Code are important, the usefulness of the process depends on the four levels of analysis carried out in this paper: comprehensive piracy legislation; ratification of the SUA framework by Gulf of Guinea States and incorporation into their domestic legal framework; thorough investigation of incidents; and the exercise of effective jurisdiction and prosecution of persons involved in piracy.

Conclusion

Recent statistics show that piracy in the Gulf of Guinea is a major threat to peace and security. While Somalia qualifies as a piracy enterprise, the Gulf of Guinea can be appropriately classified as a piracy complex: Ships are hijacked,

---

77 *Ibid.*, Article 10.

78 Summit of Heads of State and Governments of ECOWAS and ECCAS, Yaoundé, Cameroon on 24–25 June 2013. *Ban Welcomes Anti-Piracy Strategy Adopted by leaders from West, Central Africa*, UN News Centre, 29 December 2013, http://www.un.org/apps/news/story.asp?NewsID=45281 and *At UN-Backed Meeting, African States Seek to Combat Piracy in Gulf of Guinea*, UN News Centre, 29 December 2013, http://www.un.org/apps/news/story.asp?NewsID=44440&Cr=piracy&Cr1=.

79 *See* Article 3, Yaoundé Code of Conduct 2013.

80 *Ibid.*, Article 5.

cargo stolen, and violence against crews is on the increase, including kidnappings for ransom. Unlike other regions, the activities of pirates in the Gulf of Guinea are seriously undermining global energy security. Following resolutions by the UNSC calling on Gulf of Guinea to address the growing threat, regional states adopted the Yaoundé Code of Conduct. The Code provides for sharing of information, and also envisages joint patrols and surveillance.

To effectively combat piracy, however, it is important to have robust mechanisms for criminal prosecution. States in the Gulf of Guinea may rely on some lessons learned from the international experience in Somalia. The analyses in this chapter has shown, however, that the Gulf of Guinea present a different set of legal challenges that must be understood and addressed on their own terms. The recommendations advanced in this chapter enhance the effectiveness of maritime law enforcement efforts against piracy in the Gulf of Guinea.

CHAPTER 8

Flag States' Liability for Wrongful Acts by Private Military and Security Companies on Board Ships

*Vasco Becker-Weinberg**

Introduction

With most of the world's trade depending on maritime transportation, ensuring maritime security is a difficult transnational undertaking. Presently, several important shipping routes, such as those in the South China Sea, the Indian Ocean and the Gulf of Guinea, are the setting for maritime crime, especially piracy and armed robbery at sea.[1]

"Piracy" is defined in article 101 of the United Nations Convention on the Law of the Sea (UNCLOS),[2] while the International Maritime Organization (IMO) identifies 'armed robbery against ships' as "any illegal act of violence or detention or any act of depredation, or threat thereof, other than an act of piracy, committed for private ends and directed against a ship or against persons or property on board such a ship, within a State's internal waters,

* Professor at the Faculty of Law of the Universidade NOVA de Lisboa and a qualified lawyer at the Portuguese Bar Association. Dr. Becker-Weinberg is the co-coordinator of the LLM program on "Law of the Sea and the Sea-Economy."

1 Although in the last decade there has been an resurgence in notoriety of these incidents, it should be pointed out that the IMO has been addressing the matter for many years, *e.g.* IMO Resolution A.545(13), adopted on 17 November 1983, *Measures to Prevent Acts of Piracy and Armed Robbery against Ships*; IMO Resolution A.686(17), adopted on 6 November 1991, *Prevention and Suppression of Acts of Piracy and Armed Robbery against Ships*; IMO Resolution A.738(18), adopted on 4 November 1993, *Measures to Prevent and Suppress Piracy and Armed Robbery against Ships*.

2 Adopted in Montego Bay on 10 December 1982 and entered into force on 16 November 1994, published at 1833 U.N.T.S. 3.—"Piracy consists of any of the following acts: (a) any illegal acts of violence or detention, or any act of depredation, committed for private ends by the crew or the passengers of a private ship or a private aircraft, and directed: (i) on the high seas, against another ship or aircraft, or against persons or property on board such ship or aircraft; (ii) against a ship, aircraft, persons or property in a place outside the jurisdiction of any State; (b) any act of voluntary participation in the operation of a ship or of an aircraft with knowledge of facts making it a pirate ship or aircraft; (c) any act of inciting or of intentionally facilitating an act described in subparagraph (a) or (b)."

 | DOI 10.1163/9789004311442_010

archipelagic waters and territorial sea;" or "any act of inciting or of intentionally facilitating an act described above."[3]

In recent decades the practice of States and international organizations has been in response to two events: the hijacking of the *Achille Lauro*[4] and the terrorist attacks of 9/11 in the United States.[5] These events have shaped the work of the IMO on maritime security.[6] These two events served as the genesis for the 1988 Convention for the Suppression of Unlawful Acts against the

---

3 Section 2.2 of the *Code of Practice for the Investigation of Crimes of Piracy and Armed Robbery against Ships* approved by IMO Resolution A.1205(26), adopted on 2 December 2009. Also *see* IMO Resolution A.1044(27), adopted on 30 November 2011, *Piracy and Armed Robbery against Ships in Waters Off the Coast of Somalia*; IMO MSC.1/Circ.1333, 26 June 2009, *Recommendations to Governments for Preventing and Suppressing Piracy and Armed Robbery against Ships.*

4 The *Achille Lauro* incident has in occasion been wrongfully defined by some States as piracy, even though it did not fulfil the relevant requirements. *See* José Luis Jesus, "Protection of Foreign Ships against Piracy and Terrorism at Sea: Legal Aspects," *International Journal of Marine and Coastal Law* 18 (2003): 388–389. Also *see* L.C. Green, "The *Santa Maria*: Rebels or Pirates?" *British Year Book of International Law* 37 (1961): 496–505; Larry A. McCullough, "International and Domestic Criminal Law Issues in the *Achille Lauro* Incident: A Functional Analysis," *Naval Law Review* 36 (1986): 53–108.

5 *See* Christopher C. Joyner, "Suppression of Terrorism on the High Seas: The 1988 IMO Convention on the Safety of Maritime Navigation," *Israel Yearbook on Human Rights* 19 (1989): 348; Natalie Klein, *Maritime Security and the Law of the Sea* (Oxford: Oxford University Press, 2011), 147–151; Stuart Kaye, "Maritime Security in the Post-9/11 World: A New Creeping Jurisdiction in the Law of the Sea?," in Clive Schofield, Seokwoo Lee and Moon-Sang Kwon, eds., *The Limits of Maritime Jurisdiction* (Leiden, Boston: Martinus Nijhoff Publishers, 2013), 327–348.

6 IMO resolution A.584(1X), *Measures to Prevent Unlawful Acts which Threaten the Safety of Ships and the Security of their Passengers and Crews*; MSC/Circ.443, *Measures to Prevent Unlawful Acts against Passengers and Crew on Board Ships*; IMO resolution A.924(22), of 20 November 2001, *Review of Measures and Procedures to Prevent Acts of Terrorism which Threaten the Security of Passengers and Crews and the Safety of Ships.* The aim was to revise international legal and technical measures to prevent and suppress terrorist acts against ships at sea and in port, as well as to improve security on board ships and on port. The legal framework for maritime security was also further developed with the amendments to the 1974 *Safety of Life at Sea Convention* and in particular of Chapter XI which introduced the *International Ship and Port Facility Security Code* (ISPS Code). The objectives of the ISPS Code included the establishment of an international legal framework to detect threats and take preventive measures against security incidents affecting ships or port facilities used in international trade (article 1.2.1). The ISPS Code does not allow the boarding of vessels at sea by non-flag States. *See* Malvina Halberstam, "Terrorism on the High Seas: The *Achille Lauro*, Piracy and the IMO Convention on Maritime Safety," *American Journal of International Law* 82 (1988): 291–292; Hartmut G. Hesse, "Maritime Security in a Multilateral Context: IMO Activities to Enhance Maritime Security," *International Journal of Marine and Coastal Law* 18–3 (2003): 327–340.

Safety of Maritime Navigation and Protocol[7] and the 2005 Protocol to the 1988 SUA Convention were adopted.[8] Despite these and even much earlier efforts,[9] States have not achieved a comprehensive legal framework for combating terrorism at sea, countering other forms of maritime crime, or establishing a framework that effectively ensures maritime security.[10]

Traditional enforcement mechanisms recognized under international law have not been able to provide a sufficient and wide-ranging response to the exercise of jurisdiction and law enforcement against criminality at sea, even though UNCLOS provides a main framework of reference. This is mostly due to the legal, economic and political challenges that are difficult to fully overcome. For example, although interception (interdiction) and boarding operations offer the clearest response available against proliferation of seaborne weapons of mass destruction, and although many flag States are unable to exercise jurisdiction effectively over ships flying their flag, flag States are reluctant to allow interception and boarding by personnel from foreign warships.[11] In peacetime,

---

7 Adopted in Rome on 10 March 1988 and entered into force on 1 March 1992, published at 1678 U.N.T.S. 221. The *1988 SUA Convention and the 1988 Protocol for the Suppression of Unlawful Acts against the Safety of Fixed Platforms Located on the Continental Shelf* (1988 SUA Convention) had the purpose of ensuring that appropriate action was taken against persons committing unlawful acts against ships and fixed platforms on the continental shelf, namely the seizure of ships by force, acts of violence against persons on board ships, and the placing of devices on board a ship to destroy or damage it. The 1988 SUA Convention followed the models of the *Convention for the Suppression of Unlawful Seizure of Aircraft*, adopted in The Hague on 16 December 1970 and entered into force on 14 October 1971, published at 860 U.N.T.S. 105, and the *Convention for the Suppression of Unlawful Acts against the Safety of Civil Aviation*, adopted in Montreal on 23 September 1971 and entered into force on 26 January 1973, published at 974 U.N.T.S. 178.

8 Adopted in London on 14 October 2005 and entered into force on 28 July 2010, IMO Doc. LEG/CONF.15/21 (2005 SUA Convention).

9 *Convention for the Prevention and Punishment of Terrorism*, on 16 November 1937, League of Nations Doc. C.546M.383 (1937), not in force. On the historical evolution of the concept of international terrorism, *see* Reuven Young, "Defining Terrorism: The Evolution of Terrorism as a Legal Concept in International Law and Its Influence on Definitions in Domestic Legislation," *Boston College International and Comparative Law Review* 29 (2006): 23–68. Also *see* United Nations, *A More Secure World: Our Shared Responsibility, Report of the Secretary-General's High-level Panel on Threats, Challenges and Change* (2004), 51–52, paras. 157–164.

10 Jesus, "Protection of Foreign Ships," 387.

11 Vasco Becker-Weinberg and Guglielmo Verdirame, "Proliferation of Weapons of Mass Destruction and Shipping Interdiction," in Marc Weller, eds., *Handbook on the Prohibition of the Use of Force* (Oxford: Oxford University Press, 2014), 1018–1019.

generally, maritime interdiction operations (MIO) are possible only with the consent of the respective flag States and within certain safeguards.[12] Indeed, one of the main concerns in the drafting of the 2005 Protocol to the 1988 SUA Convention was that the principle of freedom of navigation and the right of flag State jurisdiction should not be hampered.[13] Other factors that often contribute to a lack of capacity by States to conduct MIO is a lack of information-sharing or the effective implementation of international legal instruments and the adoption of measures under domestic law.

In recent years, the increasing professionalism of pirates and of other criminals, with access to abundant material and financial resources, has made criminality at sea extremely rewarding and more difficult to counter. In light of these factors, the shipping industry has undertaken the right and responsibility to exercise self-protection,[14] namely through a combination of preventive and protective measures,[15] including emplacement of privately contracted armed security personnel (PCASP) on board ships. The first line of defense is to make the ship a more difficult target. IMO recognizes that measures of avoidance, evasion, deterrence and delay remain the best practices of self-protection against piracy.[16] The European Parliament also underlined that the use of armed personnel could not "substitute the necessary comprehensive solution to the multifaceted threat from piracy," calling upon the European Commission

12 Article 8*bis*(10).

13 Robert C. Beckman, "The 1988 SUA Convention and 2005 SUA Protocol: Tools to Combat Piracy, Armed Robbery, and Maritime Terrorism," in Rupert Herbert-Bruns, Sam Bateman and Peter Lehr, eds., *Lloyd's MIU Handbook of Maritime Security* (Boca Raton, London, New York: CRC Press, 2009), 194–196.

14 Douglas Guilfoyle, "Piracy off Somalia and Counter-Piracy Efforts," in Douglas Guilfoyle, eds., *Modern Piracy: Legal Challenges and Responses* (Cheltenham, UK, Northampton, MA, USA: Edward Elgar, 2013), 49–58.

15 *See Best Management Practices for Protection against Somalia Based Piracy: Suggested Planning and Operational Practices for Ship Operators, and Masters of Ships Transiting the High Risk Area*, version 4 (Livingstone, Edinburgh: Witherby Publishing Group Ltd., 2011), Annex 2 to IMO MSC31/Circ.1339, 14 September 2011, *Piracy and Armed Robbery against Ships in Waters off the Coast of Somalia: Best Management Practices for Protection against Somalia Based Piracy*; IMO MSC.1/Circ.1334, 23 June 2009, *Piracy and Armed Robbery against Ships: Guidance to Shipowners and Ship Operators, Shipmasters and Crews on Preventing and Suppressing Acts of Piracy and Armed Robbery against Ships*; IMO MSC.1/Circ.1332, 16 June 2009, *Piracy and Armed Robbery against Ships in Waters off the Coast of Somalia* and Annex 1 (*Suggested Planning and Operational Practices for Owners, Operators, Managers and Masters of Ships Transiting the Gulf of Aden and off the Coast of Somalia*).

16 IMO Resolution MSC.324(89), adopted on 20 May 2011, *Implementation of Best Management Practice Guidance.*

and Council "to work towards shaping an EU approach to the use of certified armed personnel on board in order to ensure proper implementation of the IMO guidelines in this regard."[17]

The use of private and military security companies (PMSCs) and the rendering of their services to the shipping industry is acknowledged as an efficient method for addressing threats at sea.[18] The IMO defined PMSCs as "*private contractors employed to provide security personnel, both armed and unarmed, on board for protection against piracy*."[19]

Private maritime security services are responsible for the significant drop in successful piracy attacks in the Indian Ocean.[20] The hiring of these companies, however, raises important legal questions concerning the use of force at sea and, in particular, the legitimate powers of PCASP on board ships.[21] This was the case, for example, in the 2012 incident involving the Italian oil-tanker

---

17 *European Parliament Rresolution of 10 May 2012 on Maritime Piracy* (*2011/2962*(*RSP*)), 2013/C 261 E/06, Official Journal of the European Union, C 261 E/34, 10.9.2013, para. 29.

18 James Kraska emphasis the fact that there has not been a successful hijacking of any ship carrying on board armed security. *See* James Kraska, "International and Comparative Regulation of Private Maritime Security Companies Employed in Counter-Piracy," in Douglas Guilfoyle, ed., *Modern Piracy: Legal Challenges and Responses* (Cheltenham, UK, Northampton, MA, USA: Edward Elgar, 2013), 249. Carolin Liss states that "the lines between national and international security have become blurred and security challenges are today met by a range of diverse state and non-state actors," and that "[t]hese new non-state responses to security are a challenge to the traditional perception of the role of the state;" *see* Carolin Liss, "Losing Control? The Privatization of Anti-Piracy Services in Southeast Asia," *Australian Journal of International Affairs* 63–3 (2009): 391–392.

19 Section 2 of IMO MSC.1/Circ.1405/Rev.2, 25 May 2012, *Revised Interim Guidance to Shipowners, Ship Operators and Shipmasters on the Use of Privately Contracted Armed Security Personnel on Board Ships in the High Risk Area.*

20 United Nations, *Report of the Working Group on the Use of Mercenaries as a Means of Violating Human Rights and Impeding the Exercise of the Right of Peoples to Self-Determination,* UN Doc. A/HRC/24/45/Add.2, 1 July 2013, paras. 60–65. For a recent analysis of the effectiveness of PMSCs actions, *see* Achilles Skordas, "The Dark Side of Counter-Piracy Policies," in Panos Koutrakos and Achilles Skordas, eds., *The Law and Practice of Piracy at Sea: European and International Perspectives* (Oxford and Portland, Oregon: Hart Publishing, 2014), 320–322.

21 On the use of force in operations at sea, *see* Tulio Treves, "Piracy and the International Law of the Sea," in Douglas Guilfoyle, ed., *Modern Piracy: Legal Challenges and Responses* (Cheltenham, UK, Northampton, MA, USA: Edward Elgar, 2013), 142–145; Andrew Murdoch and Douglas Guilfoyle, "Capture and Disruption Operations: The Use of Force in Counter-Piracy off Somalia," in Douglas Guilfoyle, ed., *Modern Piracy: Legal Challenges and Responses* (Cheltenham, UK, Northampton, MA, USA: Edward Elgar, 2013), 150–153, 166–169; Alexander Proelss, "Piracy and the Use of Force," in Panos Koutrakos and

Enrica Lexie. Privately-contracted Italian navy marines on board the ship killed two Indian fishermen off the coast of Kerala after mistaking them for pirates.[22] Although the matter concerned the use of authorized military personnel on board ships, the IMO had already in 2009 referred to the need for flag States to provide clarity regarding their policy on the use of military or law enforcement officers on board ships flying their flag.[23]

The use of deadly force at sea is always a last resort. Most legal issues concerning the use of force at sea occur over disputes arising from the interdiction of foreign vessels.[24] Consequently, several guidelines and recommendations have been developed to standardize the use of force and the rendering of military and security services on board ships. These guidelines and recommendations were developed by States, the IMO and other organizations[25] and have received significant support from the PMSC community.[26] The new rules address a wide variety of issues, including on the appropriate use of force and the carriage of firearms and ammunition on board ships. Although there is presently some fragmentation of these soft-law instruments they strive to set a common dominator for the regulation of the activities of PMSCs and of armed personnel on board ships.

The efficiency of such recommendations and guidelines depends on flag States, as well as port and coastal States. Without these latter States adopting the necessary legislative and enforcement measures at the national level, it is impossible to ensure that PMSCs and armed personnel on board ships will

---

Achilles Skordas, eds., *The Law and Practice of Piracy at Sea: European and International Perspectives* (Oxford and Portland, Oregon: Hart Publishing, 2014), 53–66.

22 The incident sparked serious diplomatic controversy between Italy and India but also the European Union. The officers were arrested and charged with murder and are still pending trial. Italy claims to have exclusive jurisdiction. On the *Enrica Lexie* incident, *see* Noah Black, "Criminal Jurisdiction over Maritime Security in the Indian Ocean," *Cornell International Law Journal Online* 1: 77–82 [online: http://cornellilj.org/wp-content/uploads/2013/11/Black-Criminal-Jurisdiction-Martime-Security-final.pdf (accessed September 2014)].

23 IMO MSC.1/Circ.1333, 26 June 2009, para. 8. Also *see* IMO MSC.1/Circ.1334.

24 *See* Douglas Guilfoyle, *Shipping Interdiction and the Law of the Sea* (Cambridge: Cambridge University Press, 2009), 271–294.

25 *See* International Organization for Standardization, *Standards for Guidelines and the Certification of Private Maritime Security Companies*: ISO/PAS 28007:2012, *Ships and Marine Technology—Guidelines for Private Maritime Security Companies (PMSC) Providing Privately Contracted Armed Security Personnel (PCASP) on Board Ships (and Pro Forma Contract)*.

26 For example, *see* Security Association for the Maritime Industry [online: http://www.seasecurity.org (accessed on September 2014)].

comply with the rules applicable to the protection of human rights. States must lay the legal foundations for the effective exercise and enforcement of criminal jurisdiction at sea, in addition to coordinating the implementing of measures to avoid creating the conditions for abuses and violations of human rights to occur. States therefore should strive to achieve legal harmonization on the use of force by PMSCs and armed personnel, as well as the treatment under national law of human rights abuses or violations, namely through the enactment of legislation on liability of PMSCs and armed personnel on board. Meaningful cooperation between States' judicial and judiciary bodies on the enforcement of common standards is indispensable to avoiding a "permissive environment for wrongful acts by companies of all kinds without adequate sanctioning and reparation."[27]

Ultimately, States have a legal duty to ensure that PMSCs and PCASP are accountable and subject to civil liability and criminal prosecution for human rights abuses or violations resulting from actions or omissions, at least those committed on board ships flying their flag. States differ on whether they allow for a legal cause of action to be brought in their national courts for such conduct or omission when they take place on board foreign ships. This is a matter of extraterritorial jurisdiction.

Outsourcing of Military and Security Services

Non-State actors now are undertaking roles that traditionally belonged solely to States, such as law-making, administration and compliance, enforcement and adjudication.[28] These roles have also included the rendering of military and security services during peacetime, consisting of armed guarding and protection of persons, objects, convoys, buildings and other places, as well as the

27 United Nations, "Promotion and Protection of All Human Rights, Civil, Political, Economic, Social and Cultural Rights, Including the Right to Development. Business and Human Rights: Further Steps towards the Operationalization of the 'Protect, Respect and Remedy' Framework," *Report of the Special Representative of the Secretary-General on the Issue of Human Rights and Transnational Corporations and other Business Enterprises*, John Ruggie, UN Doc. A/HRC/8/5, 7 April 2008, para. 3. Also *see* International Law Association, Report of 74th Conference, The Hague Conference, 15–19 August 2010, *Non State Actors*, by Professor Malgosia Fitzmaurice and Dr. Cedric Ryngaert, 634.

28 In a footnote the UN Resolution (Security Council) n. 1540, 28 April 2004, refers to "Non-State actor" as "an individual or entity, not acting under the lawful authority of any State in conducting activities which come within the scope of this resolution."

maintenance and operation of weapons systems, prisoner detention facilities, and providing consultancy and training to local forces and security personnel.

Outsourcing of military and security services is not prohibited by international law and it is a convenient way for States to reduce the risk to their forces.[29] The United Nations, for example, uses PMSCs in a variety of activities and locations around the world, although it has underlined its concern for their operations and the actions of their personnel,[30] including with respect to the hiring of mercenaries.[31]

Most PMSCs are business-orientated multinational enterprises with links to more than one legal system. The relations they establish with States, or for that matter the shipping industry, constitute commercial transactions, which can adopt the form of services-contracts.[32] These contracts are not international agreements, and international law is inapplicable—except for the basic standards of human rights that always apply to individual and State conduct. PMSCs are also not subjects of international law since they do not have a separate or distinct international legal personality.[33]

Maritime Jurisdiction and the Use of Force at Sea

UNCLOS sets the limits of maritime jurisdiction, but does not address directly the issue of the use of force. This is evidenced, for example, in the case of

29 For a critical view with respect to employing PMSCs to undertake humanitarian intervention, *see* James Pattison, "Outsourcing the Responsibility to Protect: Humanitarian Intervention and Private Military and Security Companies," *International Theory* 2, No. 1 (2010): 1–31. Pattison underlines as major objections the financial interests of PMSCs, rather than humanitarian, and the undermining of democratic control and legal accountability of these companies (at 17).

30 United Nations, *Concept Note: Expert Panel Event on the Use of Private Military and Security Companies (PMSCs) by the United Nations*, 31 July 2013.

31 United Nations, *Report of the Working Group on the Use of Mercenaries as a Means of Violating Human Rights and Impeding the Exercise of the Right of Peoples to Self-Determination*, UN Doc. A/HRC/24/45/Add.1, 5 August 2013, paras. 15–18.

32 International Law Association, "Private International Law Aspect of Civil Litigation for Human Rights Violation," interim report by Mr. Edward Ho and Mr. Jacob van de Velden, The Hague Conference (2010), at 574. Also *see* OECD *Guidelines for Multinational Enterprises* (2000); International Labour Organization, *Principles Concerning Multinational Enterprises* (2006).

33 *Reparation for injuries suffered in the service of the United Nations, Advisory Opinion: I.C.J. Reports 1949*, 174, at 178–179.

piracy, where universal jurisdiction only occurs in the high seas and exclusive economic zone (EEZ). Likewise, in the case of acts committed in territorial waters, coastal States may be unwilling or unable to prosecute and punish offenders. There is no State obligation to pass legislation or to secure criminal jurisdiction.[34] Armed robbery, for example, does not give rise to universal jurisdiction, regardless of where its takes place.

Douglas Guilfoyle on addressing the issue of foreign law enforcement operations in the territorial sea, points out that: "[a]s a matter of general principle, once such [State's] officials have entered a foreign state's jurisdiction they are 'subject as a matter of legal substance to the laws of the receiving states' and should obey them even if they benefit from an immunity from local enforcement and adjudicative jurisdiction. (…) The question of what local law is applicable is complex and can only be answered by reference to local law."[35]

The sea is divided into areas that are subject to some measure of sovereignty or jurisdiction by coastal States, and areas that are beyond it and cannot be claimed by or allocated to any State. The first group comprises internal waters, the territorial sea, international straits, archipelagic waters, the contiguous zone, the EEZ, and the continental shelf.[36] The sovereignty and jurisdiction enjoyed by Sates in these areas varies. In some cases (*e.g.* the territorial sea, internal waters) the coastal State has rights of sovereignty; in others (*e.g.* the continental shelf, the EEZ and, where established, the contiguous zone) they may exercise only limited sovereign rights and jurisdiction. The second group, which is beyond the permanent jurisdictional reach of any State, comprises the high seas and the Area.[37] In these areas, flag States have certain rights. In all areas, including even internal waters, the flag States rights are significant, and often dispositive.

The rights of foreign ships also vary among legal areas. Most importantly, a foreign ship enjoys the right of innocent passage through the territorial sea and archipelagic waters without being intercepted on condition that passage is innocent,[38] *i.e.* that passage is not prejudicial to the peace, good order or

---

34 On shortcomings of the piracy regime under UNCLOS, *see* Jesus, "Protection of Foreign Ships," 380–381.

35 Guilfoyle, *Shipping Interdiction*, 283–284.

36 Articles 2, 8, 34, 46, 33, 55 and 76 of UNCLOS.

37 Articles 1(1) and 86 of UNCLOS.

38 Articles 17, 18, 24(1), 52(1) and 54 of UNCLOS. Similarly, all ships enjoy the right of transit passage in straits used for international navigation, *see* articles 38, 39 and 45 of UNCLOS. On the sovereign immunity of warships and government vessels and the exercise of coastal State jurisdiction, *see* articles 29 to 32 of UNCLOS.

security of the coastal State and that the ship does not engage in any threat or use of force against its sovereignty, territorial integrity or political independence, or in any other manner in violation of the principles of international law embodied in the United Nations Charter.[39] There is no general right of innocent passage in internal waters; there the coastal State may board a foreign ship, and enforce its laws against her and those on board.[40] The right of innocent passage, however, raises interesting questions about the legal competence of the coastal State to regulate PCASP on board foreign ships.[41]

There are exceptions to the key principles of freedom of navigation and the exclusive jurisdiction of the flag State—universal jurisdiction to seize ships engaged in piracy being the most ancient.[42] Although States must cooperate in the repression of piracy, there is no obligation but merely a right to intercept and board foreign ships on suspicion of being engaged in piracy.[43] There is no obligation for States to establish rules on criminal jurisdiction with respect to piracy, although all States have a duty to repress piracy on the high seas and "shall cooperate to the fullest extent possible."[44] In cases of slave trafficking and unlawful broadcasting at sea, States have jurisdiction to arrest the ship,[45] while in the event of illicit traffic in narcotic drugs and psychotropic substances the lawful interdiction of trafficking ships depends on the request

39 Article 19(1), (2)(a) of UNCLOS. On the definition of innocent passage, *see* R.R. Churchill and A.V. Lowe, *The Law of the Sea* (Manchester: Manchester University Press, 1999), 82–85.

40 An exception is provided in Article 8(2) of UNCLOS, as well as other areas subject to non-suspendable innocent passage.

41 See James Kraska, "Excessive Coastal State Jurisdiction: Shipboard Armed Security Personnel," in *Jurisdiction Over Ships: Post-UNCLOS Developments in the Law of the Sea*, Henrik Ringbom, ed., (Leiden, Boston: Brill/Nijhoff Publishers, 2015): 167, 176–181.

42 Article 92(2), and 58(1) and (3), 78(2) and 87(1)(a) and (2) of UNCLOS. *See* also *S.S. Lotus (France v. Turkey)*, 1927 P.C.I.J. (ser. A) No. 9, at 25 (Sept. 7), paras. 64–65.

43 Article 105 of UNCLOS. Early piracy cases include the *Marianna Flora* [24 U.S. (11 Wheat.) 1 (1826)]. The *Marianna Flora* was a ship of Portuguese registry that was approached by a U.S. ship. On approach, the *Marianna Flora* fired on the U.S. ship, mistaking it for a pirate ship. As a result, the *Marianna Flora* was seized and brought into port on the grounds of piratical aggression. In this case, the Court considered that the seizure of the *Marianna Flora* was a lawful exercise of the right to approach any vessels descried at sea, for the purpose of ascertaining their real characters.

44 Article 100 of UNCLOS.

45 Articles 99, 109(3) and 110 of UNCLOS. The slavery exception emerged in the XIX century and had been definitively accepted by the beginning of the XX century. It played an important role in the emergence of modern humanitarianism, *see* Guglielmo Verdirame, *The UN and Human Rights* (Cambridge: Cambridge University Press, 2011), 36–47.

of the respective flag State or the intended destination of the illicit cargo.[46] Warships have a general right of approach and visit in respect of ships without nationality.[47] In case of pollution or threat of pollution in the high seas, the principle is also that of flag State jurisdiction, but port and coastal States may adopt some measures on foreign ships in certain circumstances and subject to safeguards.[48]

States also may pursue a foreign ship suspected of having violated its laws and regulations.[49] The right of hot pursuit gives teeth to the coastal States' jurisdictional prerogatives, as non-abiding ships might otherwise find refuge upon reaching the high seas.[50] Pursuit, boarding and seizure may require the use of force when the vessel is not cooperative. The questions lays, however, in determining if the use of force in these cases should be characterised as use of force for the purposes of Article 2(4) of the United Nations Charter, or should it instead as law enforcement.[51]

---

46 Article 108(2) of UNCLOS and Article 17 of the *United Nations Convention against Illicit Traffic in Narcotic Drugs and Psychotropic Substances*, adopted in Vienna on 20 December 1988 and entered into force on 11 November 1990, published at 1582 U.N.T.S. 95. Also *see* Article 27(1)(d) of UNCLOS. *See* W.C. Gilmore, "Drug Trafficking by Sea: the 1988 United Nations Convention against Illicit Traffic in Narcotic Drugs and Psychotropic Substances," *Marine Policy* 15 (1991): 183–192.

47 Article 110(1)(d) of UNCLOS.

48 Articles, 19(2)(h), 21(1)(f), 25, 27(5), 42(1)(b) and (2), 56(1)(b)(iii) and (2), 87(2), 94(3)(a), 194(3)(b) and (4), 211, 218, 220 and 224 to 227 of UNCLOS.

49 Article 111(1) of UNCLOS.

50 R.C. Reuland, "The Customary Right of Hot Pursuit onto the High Seas: Annotations to Article 111 of the Law of the Sea Convention," *Virginia Journal of International Law* 33 (1992–1993): 589. *See* also C.H. Allen, "Doctrine of Hot Pursuit: A Functional Interpretation Adaptable to Emerging Maritime Law Enforcement Technologies and Practices," *Ocean Development and International Law* 20 (1989): 312–321; Klein, *Maritime Security*, 109–114. On the origin and development of the doctrine of hot pursuit, *see* C. John Colombos, *The International Law of the Sea* (New York: David McKay Company Inc., 1967), 168–175.

51 The ICJ dealt with these questions in the *Fisheries Jurisdiction case* (*Spain v. Canada*). Spain submitted that the boarding of a Spanish vessel on the high seas, while pursuant to powers conferred on the Canadian Navy by Canadian legislation, contravened international law. Canada maintained that the ICJ had no jurisdiction by virtue of a reservation in the Canadian declaration of acceptance of the Court's compulsory jurisdiction, which excluded "disputes arising out of or concerning conservation and management measures taken by Canada with respect to vessels fishing in the NAFO Regulatory Area, as defined in the Convention on Future Multilateral Co-operation in the Northwest Atlantic Fisheries, 1978, and the enforcement of such measures." Spain's position was instead that the use of force against its vessel had amounted to a violation of article 2(4) of the Charter and gave rise to a "*separate cause of action not caught by the reservation.*" The ICJ ruled in favour of Canada. It held that: "the use of force authorized by the Canadian legislation and

The use of force against foreign vessels is subject to the general criteria of necessity and reasonableness: if there are no other practicable means to board, search, seize and bring into port the suspected ship, the State may be entitled to use reasonable force in light of the refusal to comply with the arrest.[52] In the *Saiga* case, the International Tribunal for the Law of the Sea underlined the importance of a graduated response in trying to stop a suspected ship.[53] Other conditions include the arrest of the ship by a warship or government ship,[54] and the adoption of measures that do not exceed the legitimate use of force, such as the deliberate sinking of an unarmed ship,[55] or the endangering of human life.[56] The assessment of the necessity and reasonableness of the use of force will depend on the circumstances of each case, taking into account the alleged offence and the conduct of the ship.[57] The State will be liable for any use of unjustified force.[58]

Force may be used to stop and board a ship in the exercise of the right of approach and visit under Article 110 of UNCLOS, or to seize a pirate ship under

---

regulations falls within the ambit of what is commonly understood as enforcement of conservation and management measures. This is so notwithstanding that the reservation does not in terms mention the use of force. Boarding, inspection, arrest and minimum use of force for those purposes are all contained within the concept of enforcement of conservation and management measures according to a 'natural and reasonable' interpretation of this concept." *See Fisheries Jurisdiction* (*Spain v. Canada*), Jurisdiction of the Court, Judgment, I.C.J. Reports 1998, 432 (*Fisheries Jurisdiction*), paras. 14, 78, 84.

52 I.T.L.O.S., *The M/V "Saiga" (No. 2) Case, (Saint Vincent and the Grenadines v. Guinea)*, 1 July 1999, para. 155. *See* L. de La Fayette, "ITLOS and the Saga of the *Saiga*: Peaceful Settlement of a Law of the Sea Dispute," *International Journal of Marine and Coastal Law* 15–3 (2000): 385–387.

53 *The M/V "Saiga," ibid.*, para. 156.

54 Articles 107(1), 110(1) and (2) and 111(5) of UNCLOS. Also *see* Article 224 of UNCLOS. A definition of warships is provided in Article 29 of UNCLOS.

55 *S.S. "I'm Alone" (Canada v. United States), Joint Final Report of the Commissioners in the Case of the "I'm Alone," dated January 5, 1935, and filed with the Secretary of State at Washington and the Minister of External Affairs for Canada at Ottawa, January 9, 1935*, 3 *United Nations Reports of International Arbitral Awards* 1609, at 1617, 1618.

56 *The Red Crusader (Commission of Enquiry, Denmark-United Kingdom)* (1962) 35 ILR 485.

57 Reuland, "The Customary," 585.

58 Articles 110(3) and 111(8) of UNCLOS. *See* also article 304 of UNCLOS and articles 21(18), 22(1)(f) and 35 of the *Agreement for the Implementation of the Provisions of the United Nations Convention on the Law of the Sea of 10 December 1982 Relating to the Conservation and Management of Straddling Fish Stocks and Highly Migratory Fish Stocks*, adopted in New York on 4 December 1995 and entered into force on 11 December 2001, published at 2167 U.N.T.S. 3.

Article 105.[59] But there is a line to be drawn between the use of force pursuant to State competence recognized under the law of the sea, and the use of force that falls outside of these purposes and is therefore regulated by reference to the international law on the use of force proper (*jus ad bellum*).

There are two situations where the law on the use of force may be engaged: first, the manifest and forcible abuse of rights of visit and seizure provided for under the law of the sea and, second, the forcible exercise of these rights outside the legal basis contemplated under the law of the sea. In both of these cases, it is still important to distinguish, as the International Court of Justice (ICJ) held in *Nicaragua*, "the most grave forms of the use of force (those constituting an armed attack) from other less grave forms."[60] The former will trigger the right of self-defence in terms of Article 51 of the United Nations Charter, the latter will not.[61]

59 Tullio Treves, "Piracy, Law of the Sea, and Use of Force: Developments off the Coast of Somalia," *European Journal of International Law* 20 (2009): 413.

60 *Military and Paramilitary Activities in and against Nicaragua* (*Nicaragua v. United States of America*) *I.C.J. Reports 1986*, p. 14 at para. 191.

61 Applying the *Nicaragua* principle in the *Oil Platforms* case, the ICJ held that the various incidents complained of by the US, including the missile attack on an oil tanker, did not constitute an armed attack, *see Oil Platforms* (*Islamic Republic of Iran v. United States of America*) *I.C.J. Reports 2003*, p. 161 at para. 64. This finding was premised on the attribution to Iran of the specific attack on the oil tanker and of the other incidents in question, but in the previous paragraphs of the judgment the Court had held that the evidence did not support the contentions of the US regarding attribution. In a separate finding, the Court did not exclude that the mining of a single military vessel might amount to an armed attack but, in this respect too, did not find that the evidence on Iran's responsibility was inconclusive (*ibid.*, at paras. 65–72). A dissonant note is the award in *Guyana v. Suriname*, in which Surinamese patrol boats ordered an oil rig and its service vessels to leave an area of the sea disputed between Guyana and Suriname. The oil rig did not fly the Surinamese flag, but it operated on the basis of a concession granted by the Government of Suriname. The Tribunal accepted Guyana's argument that this situation was different from that in *Fisheries Jurisdiction*, and that the action undertaken by the Surinamese patrol boats 'seemed more akin to a threat of military action rather than a mere law enforcement activity, *see Guyana v. Suriname*, Award of the Arbitral Tribunal Constituted Pursuant to Article 287, and in accordance with Annex VII of UNCLOS (17 September 2007), para. 445. The *Guyana v. Suriname* decision has attracted criticism because of its failure to identify the legal basis for this conclusion clearly. Treves, "Piracy," 414. Moreover, even accepting the basis for distinguishing this case from *Fisheries Jurisdiction*, its findings would still sit uncomfortably with the ICJ's approach to the definition of armed attack in *Nicaragua* and *Oil Platforms*.

The exercise of States' rights and the performance of their duties under the law of the sea are subject to the prohibition on the use of force.[62] UNCLOS does not include a set of general principles and rules on maritime security which could be invoked to justify any general right of interdiction. Security concerns are mentioned in the context of the right of innocent passage through the territorial sea.[63] Elsewhere, UNCLOS provides that the high seas, the Area, the EEZ and the continental shelf are reserved for peaceful purposes, and that marine scientific research and the use of research installations shall be exclusively for peaceful purposes,[64] while military uses are not within the scope of the Convention.[65] The use of armed personnel and of weapons on board ships is also not regulated in the law of the sea, except through flag State obligations. These matters are subject to the law of the relevant flag State and port and coastal States may adopt legislation to that effect with respect to the access to ports and to the territorial sea.[66] Moreover, the law of the sea does not address private security at sea.

Although this framework is applicable to States and not to non-State actors, the legal constraints applicable to States regarding the use of force at sea may not be more lenient when force is exercised by PMSCs. Otherwise, States would be able to outsource their liability. Nonetheless, in this case, the applicable law is that of the flag, port or coastal State, depending on where the ship is located; to the extent that port and coastal State laws apply, they typically do so in conjunction with flag State law.

The outsourcing of military and security services is not always regulated, allowing for PMSCs to operate without any supervision by the competent national authorities. Yet, the fact that States do not adopt measures under

62 Article 301 of UNCLOS.

63 Articles 19(1) and 25(3) of UNCLOS.

64 Articles 58(2), 88 and 141 of UNCLOS deal, respectively, with the EEZ, the high seas and the Area. Articles 239, 240(a), 242(1), 246(3) and 258 of UNCLOS with marine scientific research. On the concept of "peaceful purposes" and its applicability to UNCLOS, *see* B.A. Boczek, "Peaceful purposes provisions of the United Nations Convention on the Law of the Sea," *Ocean Development and International Law* 20 (1989): 364, 368–381.

65 Donald R. Rothwell and Natalie Klein, "Maritime Security and the Law of the Sea," in Natalie Klein, Joanna Mossop and Donald R. Rothwell, eds., *Maritime Security: International Law and Policy Perspectives from Australia and New Zealand* (London, New York: Routledge, 2010), 28–29.

66 Nonetheless, PMSCs have used practices to avoid potential violation of the laws and regulations of port and coastal States, such as dumping their weapons at sea or using offshore platforms as deposit. *See* Anna Petrig, "The Use of Force and Firearms by Private Maritime Security Companies against Suspected Pirates," *International and Comparative Law Quarterly* 62 (2013): 686–687.

domestic law to regulate PMSCs and the services they provide, or enact legislation on the use of force at sea in compliance with international law to prevent human rights abuses by PCASP does not mean that there is a legal void or that PMSCs and their personnel may not be accountable. On the one hand, flag States may not exempt themselves from complying with international law and enforce the relevant rules on board ships flying their flags, and on the other hand, PMSCs and their personnel may not act on board ships as if there is no law.[67]

Rules of International Law Applicable to PMSCs

Many elements of the international business community have been increasingly supportive of the protection of human rights.[68] In the case of PMSCs, the international legal rules applicable to the outsourcing of military and security services emerged from these services being provided during or after armed conflict. Excluding the appraisal of matters dealing with the criminal responsibility of individuals for international crimes, the applicable rules and principles of international law in each one of these situations are those resulting from international humanitarian law (IHL) and international human rights law (IHRL). Although both are complementary, only IHL is applicable to non-State actors in conflict situations. As for the responsibility of armed personnel for violations of human rights, their status is determined by IHL on a case-by-case basis taking into consideration the nature and circumstances of the functions in which they are involved, their authority and control and not merely those engaged by a formal link with the respective company.

Whether the rules of IHRL apply to peacetime PMSCs and their armed personnel during peacetime is not a clear matter.[69] It is generally considered that IHRL only binds States to individuals in post-conflict situations. Non-State

67 Andrew Clapham, *Human Rights Obligations of Non-State Actors* (Oxford: Oxford University Press, 2006), 31.

68 International Organization of Employers, *Business and Human Rights: The Role of Business in Weak Governance Zones: Business Proposals for Effective Way of Addressing Dilemma Situations in Weak Governance Zones*, paper prepared in collaboration with the International Chamber of Commerce and the Business and Industry Advisory Committee, December 2006, at 4. Also *see* Chatham House, *Business and Human Rights: Closing the Gap—A Summary of the Chatham House International Law Discussion Group meeting held on 6 November 2008*, at 4.

69 Yael Ronen, "Entities that Can be States But Do Not Claim to Be," in Duncan French, ed., *Statehood and Self-Determination: Reconciling Tradition and Modernity in International Law* (Cambridge: Cambridge University Press, 2013), 23–24.

actors do not incur in direct human rights obligations enforceable under international law, except regarding peremptory norms or obligations included in human rights treaties in force in the jurisdiction of the relevant State. However, instruments on international human rights obligations that might be applicable to non-State actors would consist mainly of the Universal Declaration of Human Rights (UDHR). Although not a treaty, many of the UDHR's provisions reflect general principles of law.

The UDHR determines that none of its provisions "may be interpreted as implying to *any State, group or person* any right to engage in any activity or to perform any act aimed at the destruction of any right and freedoms set forth herein" (emphasis added).[70] Also, in its preamble the UDHR recognizes the need for the protection of human rights by the rule of law and the promotion of universal respect for and observance of human rights and fundamental freedoms, as well as setting the UDHR as "a common standard of achievement for all peoples and all nations, to the end that every individual and every organ of society (...) shall strive (...) by progressive measures, national and international, to secure their universal and effective recognition and observance, both among the peoples of Member States themselves and among the peoples of territories under their jurisdiction." Yet, despite certain aspects of the UDHR having entered into customary international law, the relevant practice for the formation of *opinio juris* is that of States, regardless of calls to include non-State actors in the formation of international customary law.[71]

The UDHR gives little support to the conclusion that PMSCs, as non-State actors, are subject to direct human rights obligations and provide protection against their conduct. In this case, however, human rights obligations would only apply indirectly through States, seeing that only the latter are under the obligation to protect individuals. Indeed, international legally binding instruments and jurisprudence have focused on the accountability of States for preventing human rights abuses by those within their jurisdiction and not on that of PMSCs or their corporate liability for human rights violations.[72]

PMSCs retain a significant degree of discretion and in only very few instances public international law has recognized the possibility for companies and individuals responding before an international tribunal.[73] For example, significant efforts have been made in the field of improving labour conditions and recognizing the rights of workers, particularly in the case of activities of

70 Article 30 of the UDHR.

71 Roozbeh B. Baker, "Customary International Law in the 21st Century: Old Challenges and New Debates," *European Journal of International Law* 21 (2010): 173.

72 Crawford, *Brownlie's*, 655–656.

73 For example, *see* articles 187(c), (d) and (e), and 291(2) of UNCLOS.

multinational enterprises operating in developing countries. These efforts envisaged encouraging enterprises to make a positive contribution to economic and social progress and to minimize and resolve the difficulties resulting from their operations.[74]

There have been several soft-law instruments that established general non-binding commitments, such as the Montreux Document[75] and the International Code of Conduct for Private Security Service Providers.[76] In fact, the most significant changes with regard to human rights and non-State actors can be seen in the more frequent and increasingly diverse use of codes of conduct and the use of extraterritorial regulation of non-State actors.[77] Yet, the Montreux Document expressly recognizes the shortcomings of international law and states that it does not affect existing international legal obligations.

---

74 *Tripartite Declaration of Principles Concerning Multinational Enterprises and Social Policy*, adopted by the Governing Body of the International Labour Office at its 204th Session, November 1977 and amended at its 279th Session, November 2000, and 295th Session, March 2006. These mostly taking into account the United Nations resolutions advocating the establishment of a New International Economic Order.

75 *Montreux Document on Pertinent International Legal Obligations and Good Practices for States Related Operations of Private Military and Security Companies During Armed Conflict*. The Montreux Declaration is a text containing rules and good practices relating to PMSCs operating in armed conflict and that is the result of an international process launched in 2006 by the Government of Switzerland and the International Committee of the Red Cross, intended to promote respect for international humanitarian law and human rights law. It was signed on 17 September 2008 by 17 States: Afghanistan, Angola, Australia, Austria, Canada, China, France, Germany, Iraq, Poland, Sierra Leone, South Africa, Sweden, Switzerland, the UK, Ukraine and the USA. *See Letter dated 2 October 2008 from the Permanent Representative of Switzerland to the United Nations addressed to the Secretary-General*, General Assembly, 63rd Session, Status of the Protocols Additional to the Geneva Conventions of 1949 and relating to the protections of victims of armed conflicts, UN Doc. A/63/467-S/2008/636, 6 October 2008.

76 The Code of Conduct sets forth a commonly-agreed set of principles for private security service providers and to establish a foundation to translate those principles into related standards as well as governance and oversight mechanisms. The Code of Conduct endorses the principles of the United Nations, *Guiding Principles on Business and Human Rights: Implementing the United Nations "Protect, Respect and Remedy" Framework*, United Nations Human Rights Office of the High Commissioner (2011) [online: http://www.icoc-psp.org/uploads/INTERNATIONAL_CODE_OF_CONDUCT_Final_without_Company_Names.pdf (accessed September 2014)].

77 August Reinisch, "The changing international legal framework for dealing with non-State actors," in P. Alston, eds., *Non-State Actors and Human Rights* (Oxford: Oxford University Press, 2005), 42–61.

Therefore, while soft law might be relevant at different levels, it does not alter the existing rules potentially applicable to PMSCs.[78]

There is no legal framework binding PMSCs to human rights protection despite general recognition that their activities should be properly regulated and subjected to public international law.[79] In this respect, a national legislation survey by the United Nations concluded that that there is a widely divergent level of regulation of PMSCs among States, and there is a need for an international legally binding instrument in order to ensure effective regulation.[80]

---

78 James Kraska concludes that "although a framework exists that enables industry and governments to make informed decisions, the disparate guidance for privately contracted armed security companies and their personnel has not been consolidates into a single sources. Kraska, "International," 247–248. *See* IMO MSC.1/Circ.1443, 25 May 2012, IMO MSC.1/Circ.1443, 25 May 2012, *Interim Guidance to Private Maritime Security Companies Providing Privately Contracted Armed Security Personnel on Board Ships in the High Risk Area*, para. 2.1.

79 Philippe Sands, *International Environmental Law: Emerging Trends and Implications for Transnational Corporations* (New York: United Nations Publ.: 1993), xv. The Chatham House discussion group considered the use of extraterritorial jurisdiction and universal jurisdiction for human rights breaches in the context of corporate structures was not simple and would be difficult to implement in practice due to political considerations, Chatham House, *Business*, 6. The United Nations has also recognized that companies are moving from good intentions to significant actions with respect to human rights, even though recognizing at the same time that there is still much to be done. In this respect, the *United Nations Global Compact* identifies the *principle* that businesses should support and respect the protection of internationally proclaimed human rights and make sure that they are not complicit in human rights abuses. *See* United Nations, *United Nations Global Compact, Global Corporate Sustainability Report 2013*. Also *see* United Nations, *Promotion of All Human Rights, Civil, Political, Economic, Social and Cultural Rights, Including the Right to Development; Business and Human Rights: Towards Operationalizing the "Protect, Respect and Remedy" Framework*, (Report of the Special Representative of the Secretary-General on the issue of human rights and transnational corporations and other business enterprises, UN Doc. A/HRC/11/13, 22 April 2009). On the efforts of the United Nations to regulate the conduct of transnational companies, *see* Elisa Morgera, *Corporate Accountability in International Environmental Law* (Oxford: Oxford University Press, 2009), 78–118.

80 United Nations, "Use of Mercenaries as a Means of Violating Human Rights and Impeding the Exercise of the Right of Peoples to Self-Determination," Note by the Secretary General, 68th Session, UN Doc. A/68/339, 20 August 2013, paras. 73–74: "The Working Group refers, by analogy, to the recent successful negotiations of States that led to the adoption of the Arms Trade Treaty as an example of States' commitment to regulate the international transfer of conventional weapons through a legally binding instrument. The Working Group notes that a possible convention on the regulation of the activities of private military and security companies could be a similar success story in that it would regulate

The Extraterritorial Dimension of the Duty to Protect Human Rights

PMSCs that operate in the State where they are registered or where they have the principal place of management are exclusively subject to the law of that State (the home state). Home states may, however, regulate extraterritorial activities of PMSCs operating abroad, including on board foreign ships where a State other than the home state exercises jurisdiction. They are not required to do so, nor are they prohibited, provided there is a recognized jurisdictional basis and connection with the home state to justify exterritoriality.

Extraterritoriality is not founded on a legally binding duty for home states to exercise control over PMSCs operating in other jurisdictions, either existing directly or through a general obligation for 'all' states to protect human rights. IHRL also does not regulate the activities of non-state actors that violate human rights in their extraterritorial activities. For example, when the *United Nations Committee on the Elimination of Racial Discrimination* encourages the United States to take appropriate legislative or administrative measures to prevent acts by U.S. transnational corporations that have a negative impact on the enjoyment of rights of indigenous peoples abroad and also to explore ways to hold these corporations accountable, it does so without going any further than making recommendations.[81]

The U.S. *Aliens Tort Claims Act of 1789* allows action to be brought for torts (civil wrongs) against transnational corporations for violations of international human rights committed abroad, thus creating an important tool to increase corporate accountability, even though it is only applicable to those internationally agreed upon and recognized rights. Over the years, several actions have been put forward after the 'rediscovery' of the Act with the *Filátriga v. Peña-Irala* case,[82] despite some cases having been unsuccessful at obtain-

---

the international transfer of certain services that pose threats to the enjoyment of human rights that are at least as serious as those posed by the international transfer of conventional weapons. Similarly, to export control arrangements, an international convention to control the private military and security service export would require States to conduct appropriate human rights diligence before granting an export licence."

81 *United Nations Committee on the Elimination of Racial Discrimination, Consideration of Reports submitted by States Parties under Article 9 of the Convention: Concluding observations of the Committee on the Elimination of Racial Discrimination,* 72nd Session, 18 February–7 March 2008, UN Doc. CERD/C/USA/CO/6, para. 30.

82 630 F.2d 876 (2nd Cir, 1980). This was the first civil action filed under the Act which resulted in judgement in favour of the plaintiffs, although the award was not enforced because the perpetrator had already been deported.

ing federal jurisdiction. Just recently, the Supreme Court considered the Act subject to a "presumption against extraterritoriality" and that accordingly it would not usually apply to claims involving alleged human rights abuses or other violations of international law to have possibly occurred in foreign countries.[83]

Critics of exterritoriality point out that it may result in the making national courts an international forum for human rights abuses or violations, while lacking the specialization and practice to that effect.[84] Certain principles and rules of international law are only susceptible to breach by individuals and not by corporations, therefore giving rise only to individual liability. For corporations and generally any non-State actor to become liable for breach of principles and rules of international law, the application of a combination of international and domestic law is required.

The extraterritorial dimension of the duty to protect human rights remains unsettled in international law, partly due to the obligation to respect the national sovereignty of the State where they render their services (the host-State). It is this same concept of sovereignty that also lays the foundation for the responsibility of States to protect human rights, consequently raising the issue of liability of States where services are provided, *i.e.* the flag State. This liability is essentially founded in the duty of flag States to control and exercise due diligence with respect to the rendering of services by PMSCs and their personnel on board ships flying its flag.

Liability of Flag States

There is no strict liability regime applicable to flag States for the actions of PMSCs and their personnel on board ships, even though PMSCs are subject to the control and the law of the flag State when their operate on board flying its flag. This control concerns the protection of national interests when they render their services but also before PMSCs establish themselves in that

83 *Kiobel v. Royal Dutch Petroleum*, No. 10–1491 (U.S. Apr. 17, 2013).

84 Andrea Bianchi underlines the potential political implications of these situations: "By means of particular interpretative and legal argumentation techniques judges have occasionally implemented international human rights at the risk of causing embarrassment to their national state or to foreign countries." Andrea Bianchi, "Globalization of Human Rights: The Role of Non-State Actors," in G. Teubner, ed., *Global Law without a State* (Ashgate Publishers Ltd.: Aldershot, Brookfield USA; Dartmouth reprinted, 2004), 194.

States' jurisdiction, either directly or through local partnerships.[85] Also, when flag States make a decision to allow the presence and use of armed personnel on board ships, they ought to establish the rules for the assessment of PMSCs, namely their human rights track record and the aptitude of their personnel. These procedural duties are connected with the general obligation of all States to protect human rights.

The foundation of the responsibility to protect lays in obligations inherent to the concept of sovereignty, which in turn embraces the responsibilities to prevent and to react. Indeed, the concept of sovereignty must be viewed both as control and responsibility and taken into account in the context of international accountability.[86] For example, the *DRC/Uganda case* at the ICJ found that States' duty to protect from violence extends to a positive obligation to protect people from violence at the hands of that State's military or of non-State actors, even abroad. In this case, the violence was committed by rebel groups acting in the territory of DRC occupied by Uganda.[87] Different question altogether, and one which has been addressed, is if home-States have extraterritorial obligations under IHRL.[88]

Human rights law is not and cannot be without procedural duties. By comparison, international jurisprudence has recognized that in the case of protection and preservation of the environment, an environmental impact assessment is necessary as a procedural obligation of actors undertaking activities that might pose a risk of environmental harm. In the *Pulp Mills on the River Uruguay* case, the ICJ took into consideration the distinction between procedural and substantial duties in its evaluation of States' compliance with the duty of information, particularly regarding the obligation of one State to provide the other with the results of environmental impact assessments of activities that have transboundary effects and the adoption of measures to protect and preserve the environment.[89] The Court considered whether the

85 M. Sornarajah, *The International Law on Foreign Investment* (Cambridge: Cambridge University Press, 2004), 117.

86 ICISS, *The Responsibility to Protect: Report of the International Commission in Intervention and State Sovereignty* (International Development Research Centre, 2011), 12.

87 *Case concerning Armed Activities on the Territory of Congo (Democratic Republic of Congo v. Uganda), Judgment I.C.J., 19 December 2005*, paras. 178–180.

88 Robert McCorquodale and Penelope Simons, "Responsibility Beyond Borders: State Responsibility for Extraterritorial Violations by Corporations of International Human Rights Law," *Modern Law Review* 70 (2007): 605.

89 *Pulp Mills on the River Uruguay (Argentina v. Uruguay), Judgment, I.C.J. Reports 2010*, paras. 113, 115.

State had an obligation to inform its neighbour of activities that may cause significant transboundary harm.[90]

The flag State is the most efficient and immediate legal system for ensuring compliance, the flag State is obligated to adopt the necessary measures to protect those rights or stop potential harm. States remain responsible for a lack of due diligence and for allowing a permissive environment that give rise to human rights abuse.[91]

Likewise, flag States must exercise jurisdiction, adjudicating and enforcing claims against PMSCs and their personnel for human rights abuses or violations on board ships. Flag States must safeguard the right of victims to obtain reparation through civil litigation and take appropriate measures to investigate and offer effective remedies for wrongful conduct of PMSCs and their personnel. Flag States also must recognize that the serious violation of a particular human right and the gross or systematic failure by the company to comply is actionable directly in the national courts on the grounds of breach of international law.[92]

Persistent failure by flag States to comply with these obligations is generally attributed to the lack of means to conduct investigations and for effective adjudication, irrespective of States overwhelming recognizing that the applicable

---

90 *Ibid.*, paras. 119, 120, 158 Also *see Gabčikovo-Nagymaros Project (Hungary/Slovakia), Judgment, I.C.J. Reports 1997*, 78; I.T.L.O.S., *Request for Advisory Opinion Submitted to the Seabed Disputes Chamber*, paras. 145–149. The Seabed Disputes Chamber considered that carrying out an environmental impact assessment was a direct obligation under UNCLOS and a general obligation under customary international law and consequently "*should be included in the system of consultations and prior notifications set out in article 142.*"

91 Chia Lehnardt, "Private Military Companies and State Responsibility," in Simon Chesterman and Chia Lehnardt, eds., *From Mercenaries to Market: The Rise and Regulation of Private Military Companies* (Oxford: Oxford University Press, 2007), 155. In a similar sense, *see* McCorquodale and Simons, "The Changing," 619–620. Also *see* Reinisch, "The Changing," 81: "The policy rationale underlying such 'vicarious' or 'subsidiary' liability is clear: to increase pressure on states by continuing to hold them responsible for 'out-sourced' or 'delegated' activity in order to make sure that they have a direct interest in regulating the behavior of non-state actors to whom they have transferred state tasks."

92 *Ibid.*, p. 569. Also *see* UN Resolution (General Assembly) n.º 60/147, 21 March 2006, *Basic Principles and Guidelines on the Rights to a Remedy and Reparation for Victims of Gross Violations of International Human Rights Law and Serious Violations of International Humanitarian Law*. This resolution emphasis that the Basic Principles and Guidelines do not entail new international or domestic legal obligations but identify mechanisms, modalities, procedures and methods for the implementation of existing legal obligations under international human rights law and international humanitarian law, which the resolution recognizes are complementary though different as to their norms.

international legal framework to the protection of human rights has a high degree of legitimacy. States are frequently reluctant to enforce the law and to act against corporate abuses or to step in when business and human rights are at cross-roads.[93] Indeed, States may grant PMSCs and their personnel immunity from prosecution, or enact other similar procedural impediments such as guaranteeing non-extradition in the case of foreign entities. Such limitations restrict the effective protection of human rights, underlining the complexities resulting from PMSCs and their personnel not being accountable for their actions.[94] Unsuccessful attempts have been made to recognize the existence of corporate responsibility to respect human rights in all situations and independently of States' duties or capabilities.[95] Flag States cannot justify the violation of certain international human rights or their failure to protect them based on

---

93 United Nations, "Promotion and Protection of All Human Rights, Civil, Political, Economic, Social and Cultural Rights, Including the Right to Development. Business and Human Rights: Further Steps towards the Operationalization of the 'Protect, Respect and Remedy' Framework," *Report of the Special Representative of the Secretary-General on the Issue of Human Rights and Transnational Corporations and other Business Enterprises*, John Ruggie, UN Doc. A/HRC/14/27, 9 April 2010, para. 18. Also *see* "Report of the High Commissioner for Human Rights on her Office's Consultation on Operationalizing the Framework for Business and Human Rights," Human Rights Council, 14th Session, UN Doc. A/HRC/14/29, 16 April 2010; and "Report of the United Nations Commissioner on Human Rights on the Sector Consultation Entitled 'Human Rights and the Financial Sector'" 16 February 2007, Human Rights Council, 4th Session, UN Doc. A/HRC/4/99, 6 March 2007.

94 Yäel Ronen underlines that, although there is nothing in human rights theory that precludes the imposition of legal obligations on actors other than States, these are under international law the exclusive holders of human rights obligation. This is turn, Ronen concludes, is an inadequate system of guarantee of human rights. Non-State actors that exercise powers similar to those of States often remain unaccountable for their abuse of that power because their conduct does not amount to international crimes and is not related to an armed conflict. Ronen also draws the attention that State might be willing to recognize the applicability of human rights obligation in situations where such applicability would exempt States themselves from responsibility, thus evading their responsibility for failure to prevent conduct which amounts to human rights violations by non-State actors over which they exercise some control. *See* Yäel Ronen, "Human Rights Obligations of Territorial Non-State Actors," *Cornell International Law Journal* 46 (2013): 21–50.

95 John Ruggie in his report points out that "a large and profitable company operating in a small and poor country could soon find itself called upon to perform ever-expanding social and even governance functions—lacking democratic legitimacy, diminishing the State's incentive to build sustainable capacity and undermining the company's own economic role and possibly its commercial viability," Ruggie, "Promotion," paras. 64–65 Also

the fact that these are not recognized in their national law; yet they are ultimately responsible for the protection of such rights, even when means are not in place to do so,[96] or when perpetrators are of different nationality.[97]

Third State jurisdiction also may be applicable, for example, when a services-contract between a PMSC's and a shipping company is subject to the law of a different State. In these cases, private international law can assist in identifying the correct jurisdiction and applicable law, such as providing means to obtain evidence from abroad or facilitating the recognition and enforcement of any judgment rendered. These disputes are often prolonged and expensive, and may result in hindering or even subsuming the resolution of the underlying substantive dispute.[98]

Nonetheless, when flag States allow the 'externalization' of military and security services they retain their obligations under international law and are responsible for violations of IHL and IHRL, or other rules of international law committed by PMSCs and their personnel when such violations are attributable to the State. In this respect, the Draft Articles on Responsibility of States for Internationally Wrongful Acts[99] provide in article 5 that the conduct of persons or entities that are legally empowered by the law of the State (*i.e.* the flag State) to exercise elements of the governmental authority (*i.e.* security on board ships) shall be considered an act of that State under international law. It also refers in article 8 to the conducts of persons and entities that take place under the control of the State.

---

see United Nations, *The Corporate Responsibility to Respect Human Rights: An Interpretive Guide* (2012).

96 *Barcelona Traction, Light and Power Company, Limited, Judgment, I.C.J. Reports 1970*, para. 33.

97 *The M/V "Saiga" (No. 2) Case*, para. 106: "The provisions referred to in the preceding paragraph indicate that the Convention considers a ship as a unit, as regards the obligations of the flag State with respect to the ship and the right of a flag State to seek reparation for loss or damage caused to the ship by acts of other States and to institute proceedings under article 292 of the Convention. Thus the ship, everything on it, and every person involved or interested in its operations are treated as an entity linked to the flag State. The nationalities of these persons are not relevant."

98 International Law Association, "Private," 568.

99 International Law Commission, *Draft Articles on Responsibility of States for Internationally Wrongful Acts: Report of the International Law Commission on the Work of its Fifty-Third Session, in: Yearbook of the International Law Commission* (2001), *Report of the Commission to the General Assembly on the Work of its Fifty-Third Session, Vol. 2–2*, p. 26.

Furthermore, although one may not refer directly to complicity between flag States and PMSCs,[100] any State, by its actions or omissions may assist or aid such corporations to violate human rights, both within and outside its own territory. On the other hand, it has been argued that if prosecution against a State that violates human rights in its territory may not be possible (such as on board a ship flying its flag), it may be possible to go after non-State actors that are complicit in direct human rights violations or collaborate or profit from such violations.[101]

The IMO has expressly recognized the responsibility of flag-Sates to ensure that ships flying their flags implement measures to protect themselves and the conditions in which they can do so, including with respect to the carriage and use of firearms and ammunition, and the presence of PCASP.[102] The IMO has developed recommendations for flag States regarding PMSCs, although it has also "strongly discouraged" the use of firearms by seafarers for personal protection or for the protection of a ship and has underlined that it "may pose an even greater danger if the ship is carrying flammable cargo or

---

100 *See* Vaughn Lowe, "Corporations as International Actors and International Law Makers," *Italian Yearbook of International Law* 13 (2004): 31. The ICJ considered that complicity included the provision of means to enable or facilitate the commission of the crime, *see Application of the Convention on the Prevention and Punishment of the Crime of Genocide* (*Bosnia and Herzegovina* v. *Serbia and Montenegro*), *Judgment, I.C.J. Reports* 2007, 43, at 217, para. 419. Also, article 16 of the Draft Articles on Responsibility of States for Internationally Wrongful Acts, reads that: "[a] *State which aids or assists another State in the commission of an internationally wrongful act by the latter is internationally responsible for doing so if: a*) [*t*]*he State does so with knowledge of the circumstances of the internationally wrongful act; and b*) [*t*]*he act would be internationally wrongful if committed by that State.*" The Commentary to art. 16 states in para. 5 that it will be sufficient if the aid or assistance contributed significantly to the act. In the case of PMSCs, if a corporation assists a States to commit an international crime, it will be responsible as an accomplice under international criminal law if it supports the crime or had knowledge that its actions would assist the commission of the crime; whereas, if a corporation assists a State in violating customary international law without committing an international crime, it be found liable for breach of its own obligations. *Contra, see* McCorquodale and Simons, "Responsibility," 613–615.

101 Reinisch, "The Changing," 65. Underlines that corporate complicity is a complex issue and one should differentiate between various degrees of involvement and culpability.

102 IMO Resolution MSC.324(89). Also *see* IMO MSC.1/Circ.1046/Rev.2, 25 May 2012, *Revised Interim Recommendations for Flag States Regarding the Use of Privately Contracted Armed Security Personnel on Board Ships in the High Risk Area*; IMO Resolution A.1044(27), para. 8.

similar types of dangerous goods."[103] The IMO further referred that if armed security personnel are allowed on board ships, the master, shipowner, operator and company should take into account the possible escalation of violence and other risks.[104]

Simultaneously, the IMO has underlined the importance of complying with both port and coastal States' law on the carriage and use of firearms on board. Ships in the territorial sea and ports are subject to port and coastal States' legislation, including with respect to the presence of firearms and ammunition on board ships and the use of force for self defense, and that, therefore, masters, shipowners, operators and companies should seek clarity of the applicable national regulations. Consequently, port and coastal States must make their national policy known regarding the embarkation, disembarkation and carriage of armed personnel, firearms and ammunitions and security-related equipment.[105]

The IMO has provided interim guidance to PMSC on placing personnel on board ships in high risk areas[106] and has called upon States to adopt the Best

103 IMO MSC.1/Circ.1334, paras. 59 to 61: "Seafarers are civilians and the use of firearms requires special training and aptitudes and the risk of accidents with firearms carried on board ship is great. Carriage of arms on board ship may encourage attackers to carry firearms or even more dangerous weapons, thereby escalating an already dangerous situation. Any firearm on board may itself become an attractive target for an attacker. It should also be borne in mind that shooting at suspected pirates may impose a legal risk for the master, shipowner or company, such as collateral damages." Also *see* IMO MSC.1/Circ.1333, 26 June 2009, paras. 5 to 7. The European Union has too made recommendations on these matters, *see Commission Recommendation of 11 March 2010 on Measures for Self-Protection and the Prevention of Piracy and Armed Robbery against Ships*, 2010/159/EU, *Official Journal of the European Union*, L 67/13, 17.3.2010. These recommendations are intended to complement the guidance provided by IMO MSC.1/Circ.1334.

104 IMO MSC.1/Circ.1046/Rev.2.

105 IMO Resolution A.1044(27), para. 8(d). Also *see* IMO MSC.1/Circ.1408/Rev.1, 25 May 2012, *Revised Interim Recommendations for Port and Coastal States Regarding the Use of Privately Contracted Armed Security Personnel on Board Ships in the High Risk Area.*

106 IMO MSC.1/Circ.1443. 'High Risk Area' is defined in IMO MSC31/Circ.1339, para. 2.4, section 2, 4: "The High Risk Area defines itself by where pirate activity and/or attacks have taken place. For the purpose of BMP the High Risk Area is an area bounded by Suez and the Strait of Hormuz to the North, 10°S and 78°E." Unless otherwise defined by the flag-State. Also *see* IMO MSC.1/Circ.1405/Rev.2. The IMO has also issued recommendations in the event of hijacking, *see* IMO MSC.1/Circ.1390, 9 December 2010, *Guidance for Company Security Officers (CSOS)—Preparation of a Company and Crew for the Contingency of Hijack by Pirates in the Western Indian Ocean and the Gulf of Aden.*

Management Practices.[107] Despite the fact that the interim guidance is not legally binding, it provides a 'minimum denominator' level of recommendations on the competencies and abilities of PMSCs' personnel, as well assists in the development of an international standards and certification process for PMSCs.[108] The recommendations followed a questionnaire issued by the IMO, which underlined this organization's understanding that "the carriage of armed personnel was a matter to be decided by the shipowner within the framework of the national legislation and policies of the flag State concerned, that there was an urgent need for a consistent approach to the issue and the avoidance of an escalation of violence due to the inappropriate use of force."[109] The questionnaire emphasized the importance of the shipping industry and PMSCs knowing the conditions and requirements regarding privately contracted armed security personnel on board, firearms and security-related equipment. Moreover, States should adopt the legislation, policies and procedures relating to the carriage, embarkation and disembarkation of firearms and security related equipment through their territory and the movement of such armed personnel.[110] States must also amend weapons legislation and provide for efficient sanctions and compliance procedures that also serve as reasonable deterrent for abuses and violations to occur. Germany and the United Kingdom are two of the countries that have taken significant steps to regulate maritime security providers, including of companies and individuals.[111]

---

107 Circular letter No. 3164, 14 February 2011, *Responding to the Scourge of Piracy*. For an overview of the regulation of PMSCs and the position of States, the IMO and the shipping industry regarding the use of private contracted armed security personnel on board ships, *see* Kraska, "International," 219–249.

108 IMO MSC.1/Circ.1443, paras. 1.5 and 2.3.

109 IMO MSC-FAL.1/Circ.2, 22 September 2011, *Questionnaire on Information on Port and Coastal State Requirements Related to Privately Contracted Armed Security Personnel on Board Ships*.

110 *Ibid.*, paras. 6 and 7.

111 On the United Kingdom's change in policy to allow the use of privately contracted armed guards on board ships flying its flag, *see* Andrew Murdoch, "Piracy and the UK," in Panos Koutrakos and Achilles Skordas, eds., *The Law and Practice of Piracy at Sea: European and International Perspectives* (Oxford and Portland, Oregon: Hart Publishing, 2014), 218–220. On the German law applicable to maritime security providers, *see* Doris König and Tim René Salmon, "Fighting Piracy—The German Perspective," in Panos Koutrakos and Achilles Skordas, eds., *The Law and Practice of Piracy at Sea: European and International Perspectives* (Oxford and Portland, Oregon: Hart Publishing, 2014), 241–247. For a comparative study of different jurisdictions, *see* International Chamber of Shipping, *Comparison of Flag State Laws on Armed Guards and Arms on Board* (16 January 2013) [online: http://

Companies and shipowners seeking to hire PMSC must exercise due diligence.[112] This may be reflected in key aspects of the contracting process, such as determining the necessary insurance coverage for PMSCs and their personnel. Such coverage takes into account the security threats affecting ships and port facilities, personal accident, medical care and hospitalization and repatriation of injured persons. Insurance coverage should cover loss or damage as a result of criminal activity, similar to insurance against acts of piracy, war (which may also include piracy) and strike risks.[113] In particular, damages caused by the actions of armed personnel on board ships, including death, should be covered.

In these cases, assessment of the actions or omissions and compliance with rules of command and control must be assessed,[114] including the on the use of force, and the contribution of such actions and omissions towards the damage or risk and of potential concurrent causes. In the case of piracy, for example, the subject-matters which insurance is intended to protect are the ship and its cargo.[115] Therefore, in the case of the use of armed personnel on board ships the relevant subject-matters should be substantially broader and as a result the respective insurance coverage would most likely be higher. For example, recalling the facts of the *Enrica Lexie* incident, if armed personnel on board a ship decided to sink a vessel suspected of imminent piracy attack and killed those on board, regardless of the lawfulness of their actions, the costs associated with such an insurance coverage might be exorbitant. These and other similar costs will inevitably have an impact on shipping and maritime trade.[116]

---

www.ics-shipping.org/docs/default-source/Piracy-Docs/comparison-of-flag-state-laws-on-armed-guards-and-arms-on-board3F9814DED68F.pdf.

112 Kraska, "International," 237.

113 Keith Michel, "Piracy and Carriage of Goods by Sea," in Douglas Guilfoyle, ed., *Modern Piracy: Legal Challenges and Responses* (Cheltenham, UK, Northampton, MA, USA: Edward Elgar, 2013), 313.

114 Clear rules for command and control, particularly involving the use of force, should be included in the contract, *see* Kraska, "International," 237. On the implications of the use of armed personnel for the command-and-control procedures, *see* Petrig, "The Use of Force," 695–697.

115 Peter MacDonald Eggers, "Insurance Protection against Piracy," in Douglas Guilfoyle, ed., *Modern Piracy: Legal Challenges and Responses* (Cheltenham, UK, Northampton, MA, USA: Edward Elgar, 2013), 269.

116 On the potential economic impact of terrorist attacks on maritime trade, *see* Philipp Wendel, *State Responsibility for Interferences with the Freedom of Navigation in Public International Law* (Berlin, Heidelberg, New York: Springer Verlag, 2007), 28–29.

Conclusions

The rendering of military and security services by PMSCs and the presence of armed personnel on board ships shall continue to be a key element of maritime security as a result of their impressive levels of efficiency, taken together with the fact that the overwhelming majority of the world's trade will continue to depend on maritime transportation. Likewise, the profitable and complex business that is criminality at sea makes it highly difficult for States to tackle comprehensively the different root causes.

The currently developing forms of ocean governance, whereby non-State actors increasingly undertake *de facto* roles that belong to States, underline the contrast between, on the one hand, the importance of these entities and persons being accountable, and on the other hand, the lack of constraints upon imposed by any international legal framework to protect human rights.

Flag State jurisdiction is currently the only effective means of control available. The exercise of control and due diligence by the flag State of PMSCs and PCASP are essential. Legislation regarding military and security services and the presence and use of firearms on board ships must be enacted, and provide a clear legal framework. Second, flag States also must make civil available recourse for victims of wrongful actions, and safeguard the right of victims to obtain reparation through civil litigation. Third, flag States must take appropriate measures to investigate and offer effective remedies in their criminal laws for wrongful conduct, and they must ensure prosecution of entities and persons for human rights violations and abuses committed on board ships flying their flags. In this regard, the greater cooperation between States' judicial and judiciary bodies is indispensable. Finally, under the general obligation of cooperation, States should continue to develop an international regulatory framework in order to achieve harmonization of standards and best practices. In this respect, the work of the IMO and the efforts of other organizations and the PMSC have been decisive in ensuring progress.

CHAPTER 9

Eight Dimensions of Maritime Security Law and Practice Among Member States of the Indian Ocean Rim Association

*Bimal N. Patel**

Introduction

Maritime security interests of Indian Ocean Rim Association (IORA) member States are significantly influencing the way the UNCLOS is being implemented. Unlike previous years, since 9/11 and more specifically the period since the emergence of piracy off the coast of Somalia, these nations have started giving greater scope to these interests in the making of their larger foreign and security policy. The IORA members face several potential and existing maritime threats such as terrorist attacks, money-laundering (through maritime shipping financing) designated for terrorism purposes, proliferation of weapons of mass destruction, maritime piracy as well as drug trafficking, people smuggling, and armed robbery at sea. Yet these nations are still giving high, and apparently overriding, priority to traditional sovereignty concept in defining their maritime security, even as the scope and intensity of maritime security dangers increase across the whole region. It is therefore no surprising that these States are slow in reconciling new responses to piracy or other threats.

Although at the global and regional levels, especially in Europe, several legal, political and security initiatives have been launched, the IORA is yet to take any meaningful institutionalised response in this regard. This cynical introductory remark is justified because although the Indian Ocean is besieged with the realities of increasing security threats, IORA member States are yet to reach an accord that would reflect a broad understanding of security issues, let alone reflect the critical importance of dealing collectively with them. The least one could have expected is a meaningful response by the IORA members to the initiatives of the Security Council; but even this has not been forthcoming. The maritime security threats identified by the UN Secretary General

* Director and Professor of International Law, Gujarat National Law University, Gandhinagar (Gujarat); Member, 21st Law Commission of India; email: <vc@gnlu.ac.in>.

 | DOI 10.1163/9789004311442_011

in a 2005 Report requires urgent consideration of common policing authority in the maritime areas of the member States.[1]

The practice of IORA does not suggest that member States adhere to the overarching importance of inclusive (that is, collective) interests. An inclusive view means "the law of the sea should be constructed and understood in a way that supports common use of an area, so that a mutual benefit is shared by the majority."[2] Instead, these members, like vast majority of developing countries, believe in the exclusive interests that are held by individual States and are asserted against the rights and responsibilities of all other states.

A large majority of IORA members apparently adhere to the goal of pursuing "creeping jurisdiction," whereby States wish to assert sovereignty or jurisdiction in adjacent waters beyond the rights they strictly enjoy under the UNCLOS. As these members face maritime and non-maritime security threats, they tend to adopt a common approach to ensure that their security interests are upheld. Although the issues of freedom of navigation, overflight and fishing are yet not reaching a high level of concern, it is expected that unless the IORA formulates a common strategy and cooperation policy, a majority of IORA members' actions would run against the inclusive claim of the ocean space. An explicit agreement on collective rights and interests will be essential to achieve an accommodation of divergent laws and policies.

Maritime Security

The state practices of the IORA member States show that maritime security for them is defined by military interests of an individual member defined in the framework of the classic concept of state sovereignty, independence and non-encroachment by other states into internal affairs. The activities of vessels in the ports, internal waters and territorial waters, as well as the uses of naval forces, constitute the major elements of this classic concept. But the exploration and exploitation of oceans' resources have opened up a new set of

1 UN Doc. A/63/174/Corr.1, Report on the work of the United Nations Open-ended Informal Consultative Process on Oceans and the Law of the Sea at its ninth meeting, Sept. 10, 2005, http://daccess-dds-ny.un.org/doc/UNDOC/GEN/N08/503/59/PDF/N0850359.pdf?OpenElement. See also, Secretary-General, United Nations, A More Secure World: Our Shared Responsibility (2005), http://www.un.org/en/peacebuilding/pdf/historical/hlp_more_secure_world.pdf.

2 Natalie Klein, *Maritime Security and the Law of the Sea*, (Oxford: Oxford University Press, 2011), p. 3.

challenges; requiring that these states re-conceptualize their maritime security interests and doctrines. There is no common understanding among IORA member States of the doctrines for pursuit of maritime security. Although a common understanding is not strictly required, the measure of success for IORA success largely turns on how it protects and promotes member states' oceanic interests. A broad understanding is a prerequisite to effective formulation of the necessary policies and strategies.

In discussing the fundamental notion of maritime security, it is essential to distinguish and understand the definition or meaning of maritime security and maritime safety as formulated by the IMO. In the IMO concept, "maritime safety" refers to preventing or minimizing the occurrence of accidents at sea that may be caused by substandard ships, unqualified crew or operator error, whereas "maritime security" is related to protection against unlawful, and deliberate acts. This chapter uses the meaning of maritime security as "the protection of a state's land and maritime territory, infrastructure, economy, environment and society from certain harmful acts occurring at sea."[3]

Military Activities

The state practices of IORA members are divergent with regard to the innocent passage of warships. This divergence gives rise to clashes as demonstrated through several incidents.[4] For example, India practices that the coastal state require prior notification to its designated authority for the passage of foreign warships through its territorial sea.[5] Indonesia requires prior notice of

3 Natalie Klein, Joanna Mossop and Donald Rothwell, "Australia, New Zealand and Maritime Security," in Natalie Klein, Joanna Mossop and Donald R. Rothwell (eds.), *Maritime Security: International Law and Policy Perspectives from Australia and New Zealand* (Routledge, Oxford 2010), 1, 8.

4 For example, KRI Usman Harun warship of Indonesia was not welcomed in Singapore territorial waters in 1965. *The Strait Times*, 8 February 2014. http://www.straitstimes.com/singapore/kri-usman-harun-not-welcome-in-singapore-waters-0.

5 The right of foreign merchant ships to pass through the territorial sea of a coastal state has long been an accepted principle of customary international law. India has subscribed to this principle too. However, with regards to the innocent passage of foreign warships, India wanted to impose a requirement of prior notification. This position had to change as article 23 of the UNCLOS required that costal states must not hamper the innocent passage of foreign ships, either by imposing requirements upon them which would have the practical effect of denying or impairing the right or by discrimination. The issue of whether the passage of warships in peacetime is or is not innocent remained one of the most controversial issues. India rejected the principle of innocent passage for warships in the territorial sea.

intention of warships to transit its territorial sea. With regard to IORA members in the East Asian region, this divergence gives rise to clashes as demonstrated through several incidents.[6] In the Straits of Malacca and Singapore, innocent passage has been deemed the appropriate regime by two IORA members namely Indonesia and Malaysia, with a reservation when foreign ships pose a threat to the marine environment or national security interests.[7] As ITLOS Judge Jin-Hyun Paik observes, the problem of achieving "a commonly accepted interpretation is more acute in (East Asia) than in any other region, mainly because the practice of the major aspects of ocean use diverges substantially among the coastal states of the region."[8] The same applies for the South Asian region and, it may be safely concluded, across the IORA member States. Whether the current practices of these States ensure a proper balance between the demands for national security and control over ocean space may be subject to a detailed scrutiny and readjustments. As there is no common regional mechanism, the imbalance will persist.

The current military, security and conflict environment in the Indian Ocean continues to generate apprehension among coastal states, especially among weaker members, about military activities at sea undertaken by outside powers as well as immediate regions' own large states. Their inability to effectively monitor waters under their jurisdiction, partly arising due to lack of effective internal legal mechanisms, adds into their vulnerability. While some IORA member States will continue to push for inclusive use theory with respect to military activities, a large majority of states in the region will continue to prefer the theory of exclusive use of ocean space in waters under their jurisdiction. Australia's maritime zones, to cite one example, extend well into the Timor Sea

---

India's position was in stark contrast with position of the US, Western Powers and also the Eastern Bloc. UN Doc., A/CONF.62/WP.8/Part II (1975). The reason for receding from this position was India's self-realization as a growing maritime nation interested in the freedom of navigation both for itself and the international community. It shall be noted that Pakistan, Bangladesh, Sri Lanka and Myanmar, by 1977, had all unilaterally promulgated that prior authorization and notification was required for the passage of foreign warships through their respective territorial waters.

6 For example, KRI Usman Harun warship of Indonesia was not welcomed in Singapore territorial waters in 1965. *The Strait Times*, 8 February 2014. http://www.straitstimes.com/singapore/kri-usman-harun-not-welcome-in-singapore-waters-0.

7 See *Indian Ocean: A Sea of Uncertainty*, published by Future Directions International (Australia), February 2012.

8 Jin-Hyun Paik, "Enhancing Maritime Security in Asia," in Dalchoong Kim and Jing-Hoon Lee (eds), *Comprehensive Security—Conceptions and Realities in Asia* (Seoul, Yonsei University Press, 2000), p. 101.

to the northwest and cause significant concern to Timor. With varied interpretations and claims of maritime security zones among the members, the situation is becoming less stable and unpredictable.

Law Enforcement Activities

Maritime law enforcement is the most important tool to deal with maritime security threats. The IORA member States often display an incoherent, unpredictable, and often conflicting approach to the international law of the sea during maritime law enforcement operations, which include maritime surveillance, stopping and boarding vessels, search or inspection, reporting, arrest or seizure of persons and vessels, detention and formal application of law by judicial or administrative processes, such as imposition of sanctions. A majority of these States lack effective government agency structures and operations, as well as lack of funding for police and coast guard forces. Furthermore, maritime law enforcement powers often are crippled by a lack of radar, boats, and effective command and control. Thus, an unfortunate mix of lack of legal regimes, combined with the lack of government authority and funding availability make these states incapable of conducting proper maritime law enforcement activities.

Piracy and Armed Robbery

IORA member States have variety of concerns as to piracy and robbery. After the entry into force of the UNCLOS and the 9/11 terrorist attacks, several of these concerns, such as troubling environmental issues, have become more prominent. India, Kenya, Singapore, Australia, and South Africa have enacted laws to prevent terrorism,[9] but these laws lack an emphasis on environmental, economic, and transportation impacts. For example, Kenya does not have specific laws to address maritime terrorism. The Prevention of Terrorism Act 2012 and the Maritime Authority Act 2006 provide jurisdiction and enforcement in dealing only with terrorism in general and to some extent within its offshore maritime zones.[10] Likewise, Australia does not have specific legislation dealing with maritime terrorism. The Criminal Code Act 1995, however, provides

9 For example Kenya enacted the Prevention of Terrorism Act 2012 and the Maritime Authority Act 2006, however, no changes were made to the Maritime Zones Act 1989.

10 Kenya: Prevention of Terrorism Act 2012 and the Maritime Authority Act 2006.

a clear definition of terrorism more generally, and it prescribes specified penalties. Thus, Australia's Maritime Transport and Offshore Facilities Act 2003 defines unlawful legal activity on maritime transport and offshore facilities, and it demarcates the extent of the jurisdiction, explains the different maritime security levels, empowers authorities, and sets forth security plans and penalties for non-compliance.[11]

Singapore's laws are more precise. The Maritime Terrorism Act 2002 deals with the hijacking of ships and penalties.[12] Similarly, the Terrorism (Suppression of Financing) Act 2002 caters to deal with concerns arising after the 9/11 terrorist attack situations.[13] The Protection of Constitutional Democracy against Terrorist and Related Activities Act 2004 of South Africa is similarly precise and specific in nature and has comprehensive provisions relating to the hijacking of aircrafts and ships.[14] It should be noted, however, that the South Africa Maritime Safety Authority Act 1998 needs to be amended to bring the provisions in consonance with the 2004 Act.

India does not have a specific law dealing with maritime terrorism *per se*. The Prevention of Terrorism Act 2002 (PoTA), however, does apply to persons on ships.[15] The Maritime Piracy Bill 2012 has lapsed with the dissolution of the 15th Lok Sabha which laid down far more effective provisions concerning jurisdiction and enforcement in dealing with various maritime security threats.[16] Indian statutes lack the necessary legal provisions to deal with the maritime piracy. The famous *Alondra Rainbow* case has left much to be desired. Until now, the Indian Penal Code, the Indian Passport Act and the Indian Firearm Act occasionally[17] were applied to deal with crime at sea.[18] However, the

11 Australia: The Maritime Transport and Offshore Facilities Act 2003.

12 Singapore: The Maritime Terrorism Act 2002.

13 Singapore: The Terrorism (Suppression of Financing) Act 2002 2004.

14 South Africa: The Protection of Constitutional Democracy against Terrorist and Related Activities Act.

15 India: Prevention of Terrorism Act 2002 (PoTA).

16 India: The Maritime Piracy Bill 2012.

17 Diana Chang, "Piracy Laws and the Effective Prosecution of Pirates," *Boston College International and Comparative Law Review* 33(2) (2010): 286. A classic illustration of the universality principle in relation to piracy is offered by the Alondra Rainbow incident and prosecutions. The Alondra Rainbow was a Japanese owned tanker with a Filipino crew under the command of two Japanese officers. The tanker was sailing from Indonesia to Japan when pirates hijacked the ship. The Indian Navy later captured the pirates and towed the vessel to India. The pirates were tried and convicted by an Indian court.

18 Indian Penal Code, 1860, the Indian Passport Act, 1967 and the Indian Firearm Act, 1959 and amendments thereto.

Maritime Piracy Bill provides a comprehensive regime to deal with jurisdiction and enforcement issues concerning the piracy problem.[19] The Bill is quite rigorous, but it has certain in-built weaknesses. The most fundamental concern is the scope of the definition of piracy.[20] Although the Bill provides a definition and brings it into conformity with the UNCLOS, it should have broadened the scope to capture elements of ship hijacking, crew kidnapping, and extortion, all of which have posed more severe threats than traditional piracy.

The problem is especially intense since the Bill emphasizes the requirement of two ships to manifest the crime of piracy. This requirement, however, omits situations of internal attacks, thefts, extortion, and armed robbery. Section 7 of the Bill shifts the burden of proof to the accused. It envisages the *in absentia* criminal prosecution of persons, and prosecution of stateless persons. By extending the jurisdiction to the EEZ, the Bill enables the government to prosecute pirates within areas beyond the territorial sea that otherwise are under India's jurisdiction.

The Bill is not free from certain potential legal challenges. For example, the settled principle that accused are innocent until proven guilty is tested by way the Bill has shifted the responsibility on the accused to prove that he is not guilty. Similarly, the Bill enables the Indian Navy and the Indian Coast Guard to make arrests and seize pirated vessels or cargo outside of India's territorial jurisdiction, raising potential jurisdictional challenges. In other words, the Bill seeks to exercise universal criminal jurisdiction.[21] The Bill enables any gazetted officer to arrest, investigate and criminally prosecute apprehended suspects, and it seeks to establish fast track courts for speedy trials.[22] The solution could be to establish a specially empowered tribunal to monitor govern practice, application and interpretation of the provisions. There is ample precedent for this course of action. The Indian government has been establishing

19 India: The Maritime Piracy Bill 2012.

20 Although this definition aspect definitely needs comparative analysis, the Piracy Act 2001 of Sri Lanka does accommodate modern piracy concerns in its definition. According to the Act, "any person who dishonestly takes or appropriates any ship, by means of theft, force, intimidation, deception, fraud or by other similar means shall be guilty of the offence of piracy and subject to the provisions of sections 8, 9, 10 of the Act with imprisonment... whichever is the greater."

21 This has been already practiced by Indian Courts. See *Pride Foramer v. Union of India (UOI) And Ors and Aban Loyd Chiles Offshore Ltd. v. Union of India & Ors*, Supreme Court of India, Appeal (civil), 2236 of 2002.

22 S. Ruhal argues that the jurisdiction of piracy cases should be given to Higher/Constitutional Courts since case involves the various aspects of international law. Sanjeet Ruhal, *The Piracy Bill 2012: A Solution or Mere Paper Tiger?* Unpublished paper.

subject-specific tribunals in the areas of the natural environment, taxation, and competition laws. It has been suggested that the IORA should consider establishing such a dedicated tribunal to try prosecution of all persons apprehended in a particular member State. India can consider taking the lead and establishing such a tribunal under a dedicated regional agreement since the country enjoys a solid legal and judicial infrastructure, operational capacities, and inherent interests in the region.

The Bill also could have incorporated lessons learned from the prosecution of pirates in Kenya in this regard. The Bill could have incorporated best practices for the domestic prosecution of piracy. It is unclear whether the Bill can reconcile the requirement that any solution to piracy be legitimate on the one hand, and efficient and effective on the other. The most problematic part of the Bill is the death penalty in cases of maritime piracy that result in death during the commission or attempt to commit piracy. This punishment is not in conformity with UNCLOS and is presumably not acceptable to many countries to which the accused may be extradited.

The Bill also should have been drafted with a broader focus than maritime piracy. It would have been preferable for the Bill to recognize piracy as a threat to national maritime security, and to provide the government with much wider authority under Indian laws. Indian vessels also should have been given a much wider role and self-defense authorizations to repel maritime pirates in Indian and high seas waters. The Bill also fails to address numerous commercial, security and industry concerns of the Indian maritime trade community. In view of the proliferation of maritime piracy attacks, the shipping community has developed new policies for self-defense, and upgraded both hard and soft measures of on-board deterrent. Shipping companies have also been stuck with higher insurance premiums and the costs of armed security. If the Bill had dealt effectively these concerns, it would have been more effective in meeting the long-term economic, security concerns and interests of India.

The Bill also could have authorized India to accept offenders subject to the Convention for the Suppression of Unlawful Acts against the Safety of Maritime Navigation (SUA) for criminal prosecution unless they could argue successfully that the Convention does not apply.[23] Similarly, taking provisions from the UN Security Council resolutions against Somali piracy,[24] the Bill could have expanded navy and coast guard power to pursue pirates including their

23 India: Suppression of Unlawful Acts against safety of Maritime Navigation And Fixed Platforms on Continental Shelf, 2002.

24 UN Security Council Resolutions 1816 (2008), 1838 (2008), 1846 (2008), 1851 (2008), 1897 (2009), 1918 (2010), 1950 (2010), 1976 (2011), 2015 (2011) and 2020 (2011).

areas of staging and refuge on land. Similarly, the Bill lacks clarity on practical issues concerning the prosecution, such as the balance between human rights duties and obligations on the part of India for the protection of interests of innocent parties, consideration of the gravity of the crime committed by the pirates, preservation and transportation of evidence, practical mechanisms to expeditiously extradite the accused, and the need for authorizing the government to conclude bilateral agreements with partner states. Several Western European nations have used foreign courts to process pirates, and it is unclear whether India should retain this option for the future. A detailed examination of state practice concerning these rules would be helpful to make the proper assessment or reach the optimal balance.

Marine Environment

Enforcement of international environmental law on the open ocean is a big challenge.[25] Since 1970, no fewer than 17 major oil spills have occurred in the waters of IORA members.[26] The discharge of oil, oily residues, and other noxious or harmful substances in these waters has resulted in great economic and environmental impacts. These spills should be the subject of a separate research study, which might spur IORA members to enact more stringent environmental laws. The MARPOL Convention 73/78, which covers not only accidental and operational oil pollution but also pollution by chemicals, goods in

25 On the various environmental challenges discussed in the remaining sections of this chapter, see the complementary analysis offered in Ximena H. Oyarce, "Current Ocean Law Issues in the Indian Ocean Region," in Harry N. Scheiber and Jin-Hyun Paik (eds.), *Regions, Institutions, and the Law of the Sea* (Leiden and Boston: Brill/Nijhoff, 2013), pp. 359–381.

26 Thailand (Rayong Oil Spill, 2013), India (Mumbai–Uran pipeline Spill–Arabian sea, 21 January 2011), Mumbai Oil spill, MV MSC Chitra and MV Khalija 3 (August 2010), MT Bunga Kelana 3 (Singapore Strait, May 2010), Great Barrier Reef / MV Shen Neng 1 (Australia, Great Keppel Island, April 2010), Montara oil Spill (Australia, Timor Sea, August 2009), Queensland (Australia, March 2009), Limburg (Yemen, Gulf of Aden, October 2002), Treasure (South Africa, Cape Town, June 2000), Apollo sea (South Africa, Cape Town, June 1994), Port Bonython Oil Spill (Australia, Spencer Gulf, August 1992), Kirki (Australia, Indian Ocean, July 1991), Gulf War oil Spill (Iran, January 1991), Nova (Iran, Gulf of Iran, December 85), Castillo de Beliver (South Africa, Saldanha Bay, Aug. 83), Nowruz Field Platform (Iran, Persian Gulf, Feb. 83), Sea Star (Iran, Gulf of Oman, December 72), Oswego-guardian / Texanita Collison (South Africa, Stillbaai, Aug. 72), Wafra (South Africa, Cape Agulhas, February 71).

packaged form, sewage, garbage and air pollution, has been ratified by IORA members. These states, however, lack adequate implementing legislation concerning the construction, design equipment, and manning (CDEM) of vessels. These rules provide an additional mechanism for marine environment protection because ship disasters due to CDEM deficiencies often lead to significant marine pollution.[27] IORA members and, perhaps more specifically IORA as an institution, can be considered an agent or trustee to oversee the facilitation of compliance of marine environmental rules in the Indian Ocean. Although the instances of flags of convenience among IORA members are hardly reported or identified as a problem, the IORA members have an additional obligation to ensure that their flag state administrations do not approve flags of convenience. Otherwise their collective ability to enforce and protect the Indian Ocean from marine pollution will erode.

Due to prevailing ambiguity in international environmental law concerning maritime areas, both port states and coastal states need to exercise more control and jurisdiction over their various maritime zones. Wherever treaty law is silent, these states need to initiate state practices which ensure long term protection and promotion of their interests, especially to prevent ships operating under flags of convenience from endangering the environment.

UNCLOS enables port states to undertake investigations and, where the evidence so warrants, to institute proceedings in respect of any discharge from a vessel that has voluntarily entered the port when the discharge is in violation of international standards, and has occurred outside the internal waters, territorial sea and EEZ of that port state.[28] It is clear that any violation can trigger coastal state denial of the right of innocent passage and, depending upon the gravity of the discharge, enable the exercise of criminal jurisdiction.[29] Such authority and the enforcement powers and authorities of coastal states to control vessel-source pollution and other sources of marine pollution are clearly stipulated in the UNCLOS. It is important, however, to note that these controls and jurisdiction powers bestowed upon the coastal states do not encroach on the freedom of navigation in any way. Rather, it is incumbent upon flag state vessels to ensure and comply with internationally accepted standards in the exercise of their rights of freedom of navigation. Likewise, coastal states have a legitimate role concerning maritime environmental control and jurisdiction.

27 It would be interesting to see how many of the 17 reported incidents mentioned above could have been avoided with the proper implementation of the CDEM standards.

28 UNCLOS, Art. 218(1).

29 UNCLOS, Art. 19(2)(h) and 27.

The law is silent concerning marine pollution on the high seas, except for some generalized obligations. Nevertheless, states may resort to soft law instruments such as declarations, guidelines, etc.,[30] depending upon the gravity of marine pollution and the effect on the EEZ and subsequently contiguous zone and territorial waters over a period of time. No international law prohibits states from undertaking necessary means and measures at their disposal to prevent pollution from reaching its waters or land territory. The 1988 SUA Convention and 2005 SUA Protocol recognize a right of visit by coastal states in case the source of damage to the marine environment falls within the list prescribed by these instruments. One would agree with Klein, however, that "there has been insufficient concern regarding damage to the marine environment outside national jurisdiction to warrant any reallocation of policing powers on the high seas at this point."[31]

Marine Biodiversity

Asian and African nations possess high concentrations of marine biodiversity; this aspect of their ecosystems can play a crucial role in achieving sustainable development. However, due to a lack of scientific knowledge, technological capacity, financial resources, and/or the political will and adequate assistance from developed nations, these nations are unable to assess the individual and cumulative impacts of various human activities in the Indian Ocean—impacts that put at serious risk their biodiversity. The "loss in marine diversity and productivity of marine ecosystems," the UN official Stephen Mathias has pointed out, "will [also] hamper efforts to meet development goals."[32]

In recent years, the United Nations has emphasized the conservation and management of marine biodiversity and ecosystems.[33] In this regard, Mathias notes, UN member States have agreed on significant steps, including the initiation of a process aimed at ensuring that the legal framework for the conservation and sustainable use of marine biodiversity in areas beyond national jurisdiction (ABNJ) effectively addresses those issues. To that end, such a

30 UN Doc. UNGA Res. 37/7 of 28 October 1982.

31 Klein, *Maritime Security*, p. 319.

32 Stephen Mathias, Assistant Secretary-General for Legal Affairs, UN, Inaugural address at the AALCO Legal Experts Meeting to commemorate the 30th anniversary of the UN Convention on the Law of the Sea, 5 Marcy 2013, p. 18.

33 See also UN General Assembly Resolution, 66/288 "The Future We Want" and 67/78 "Oceans and the Law of the Sea." *Ibid.* at 18.

process will identify gaps and ways forward, including through the implementation of existing instruments and the possible development of a multilateral agreement under the UNCLOS.[34]

This analysis suggests a need for IORA members to enact new legislation or amend existing laws and to develop a coordinated instrument, at the multilateral IORA level, to achieve a regional approach. Just as the United Nations has deliberated on these issues over the past three years, seeking to include international cooperation and coordination as well as capacity-building and the transfer of marine technology, the IORA should actively realize the benefits offered under these approaches.

IORA needs to formulate a clear response on the issue of jurisdiction of marine and genetic resources, as there is a sharp division currently among governments.[35] While discussing the issue of marine biodiversity and ecosystem, it is important to realize that there is no multilaterally agreed legal regime for the establishment of marine protected areas (MPAs) in areas beyond national jurisdiction. There also is disagreement over whether MPAs could be established unilaterally or must be done by a group of States only, or through a multilateral organization, such as the International Maritime Organization (IMO). Although these ambiguities persist with regard to the legitimacy and legality of the establishment of MPAs in areas beyond national jurisdiction, an IORA initiative to produce at least an informal understanding is essential. Such an understanding will form a framework for IORA members to protect and promote their interests on an interim basis while a multilateral instrument under the UN auspices or through some other institution takes shape. Such an approach will help to fill gaps in governance and regulation.[36] In the ongoing discussion of a new and more comprehensive, regime for marine biodiversity, set in motion in late 2015 by action of the UN, an IORA- specific approach is

---

34 Mathias, Inaugural address, cited note 31 above.

35 As Luther Rangreji identifies, the "positions of governments are sharply divided between those who consider that they are regulated by Part VI of UNCLOS relating to the High Seas, and those who contend that they are governed by Part XI relating to "the Area," namely the deep-seabed beyond areas of national jurisdiction. The former view is based essentially on the literal interpretation of UNCLOS provisions. According to this view, marine resources in areas beyond national jurisdiction are living resources, and thus fall under the region of the high seas. Part XI of the regime applies only to resources in the Area, which is explicitly defined in Article 133 as mineral resources and thus excludes living resources. The activities relating to genetic resources are therefore governed by the freedom of the high seas." Luther Rangreji, "Issues for Developing Countries under the Nagoya Protocol," in *AALCO UNCLOS Verbatim Records 2013*, p. 32.

36 Rangreji, *ibid.*, at p. 34.

essential since it will also help to give the IORA member States vital input in the current negotiation and the drafting process.

While this chapter suggests creation of an IORA-specific approach in the marine biodiversity area, it is important to also underline that IORA should develop a position on better benefit sharing provisions under the Nagoya Protocol, which deals with genetic resources and access and benefit sharing of genetic resources. Individuals and institutions dealing with the regulation of genetic resources in IORA member States also need an "awareness regime" that can identify legal rights and obligations flowing from national jurisdiction and different treaties that regulate these resources beyond national jurisdiction—a way, in sum, to manage the common heritage of mankind. An associated issue is the terms and conditions known and available for exploitation of marine flora and fauna available in IORA members. Finally, IORA members need to study the application of an Access and Benefit Sharing Protocol, not only for areas within national jurisdiction, but also for areas under other international treaty regimes and processes, and how to build capacity of IORA members in this area.[37]

As far as IORA is concerned, one can discern that the region needs legal instruments to protect all types of marine areas within the region. An effective institutionalized regional compliance mechanism is, therefore, needed for IORA members.[38] IORA member States need to develop environmental law protecting the marine environment, especially given the additional problems of extraterritorial enforcement on the high seas.

IUU Fishing

UNCLOS provides that enforcement rights against foreign vessels on the high seas are available, subject to agreements with the flag state either bilaterally or through regional fisheries management organizations. With regard to IORA state practice, there are no such bilateral agreements in the region. In this regard, it may be noted that among IORA member States, Australia, Bangladesh, India, Indonesia, Iran, Kenya, Oman, Seychelles, South Africa, and Sri Lanka have ratified the 1995 Fish Stocks Agreement (FSA), which requires that States Parties are only to authorize their vessels to fish on the high seas

37 Rangreji, "Issues for Developing Countries," at p. 39.

38 It is worth suggesting that IORA should carry out empirical research on pollution from land-based sources, seabed activities subject to national jurisdiction, activities in the Area, by dumping, by vessels and from or through the atmosphere.

only where those states are able to exercise effectively their responsibilities. These responsibilities include controlling fishing through issuing of licences and authorizations, establishing regulations to address the conduct of fishing, and undertaking monitoring, control, and surveillance of fishing vessels that fly the flag of the member state. As per the FSA, the States Parties must permit access by duly authorized inspectors from other states, consistent with sub-regional and regional schemes for cooperation. However, the current laws of the IORA member States do not provide such explicit provisions, and therefore need to be updated.

The Indian Ocean is known for its straddling fish stocks within EEZs. Similarly, highly migratory fish stocks swim long distances across the high seas and often across numerous EEZs in the Indian Ocean area. These migratory stocks, occurring in more than one maritime zones, are subjected to fishing by vessels flying any number of flags; hence, cooperation among IORA members is essential for effective conservation and management of such stocks. IORA members need to consider on an urgent basis several strategies to integrate ecosystem considerations into fisheries management, including:

- Urgent reduction of fishing capacity to levels commensurate with the sustainability of fish stocks;
- Strengthening the mandates of RFMOs to implement modern approaches to fisheries;
- Conduct RFMO performance reviews, develop a legally binding instrument on minimum standards for port State measures and a comprehensive global register of fishing vessels;
- Provide expanded assistance to developing countries; and,
- Continue a dialogue to address concerns raised by non-parties.[39]

One of the recommendations of the FSA Review Conference 2006 was to eliminate subsidies that contribute to IUU fishing, overfishing and fishing overcapacity.[40] With the exception of a few major IORA members, other members are little involved in overfishing, owing to technological, financial and other limitations. The question of subsidies therefore hardly arises. The IORA needs to focus more on sanctions, development of a legally binding instrument

39 See United Nations Report on Review of Fish Stocks Agreement 2006.

40 Review Conference on the Agreement for the Implementation of the Provisions of the United Nations Convention on the Law of the Sea of 10 December 1982 relating to the Conservation and Management of Straddling Fish Stocks and Highly Migratory Fish Stocks, New York, 22 to 26 May 2006 and 24 to 28 May 2010.

on port State controls, regulation of transhipment, inter-regional assistance monitoring, and control and surveillance and alternative mechanisms. The IORA also needs to develop regional guidelines for fisheries sanctions to ensure adequate disincentives to non-compliance; to initiate a process within FAO to develop an instrument on minimum standards for port State measures, and to develop a comprehensive IORA Register of fishing vessels.

Some IORA members will be in need of financial assistance, capacity building, specialist training, human resource development, technology transfer, and the exchange of knowledge and expertise to ensure that they are able to deal with the IUU fishing threat. IORA needs to identify and assess the reasons for the lack of ratification of the Fish Stocks Agreement (FSA) by the IORA member States and to propose mechanisms and assistance to increase higher adherence to the FSA. IORA provides an excellent forum to achieve, eventually, high seas fisheries management in the region. In this regard, it is an important point of concern that the application of the Precautionary Approach and the establishment of precautionary requirements in state practice among IORA are highly variable, falling substantially short of what the Agreement envisages. To address this concern, the IORA may consider adopting a Precautionary Approach Framework such a NAFO PAF 2004.[41]

IORA needs a stringent approach because IUU fishing does not only damage the marine environment, especially marine biodiversity, it also represents the theft of common fisheries resources and results in considerable losses to those fishermen who do abide by the law.[42] With the globalization of the fisheries sectors, IORA will also face challenges with regard to increased trade flows of fisheries products whose legality is difficult to ascertain. IORA needs to ascertain the utility of the FAO IUU Action Plan and adopt an approach suitable for the IORA. According to one estimate, Sub-Saharan Africa losses € 800 million a year due to IUU fishing.[43] If the losses across the IORA are ascertained, the revenue lost to the IORA members due to IUU likely is staggering. Some firms that practice illegal fishing, including EU operators, operate substandard

---

41 W. Brodie, P. Shelton, E. Couture, K. Dwyre, *A Discussion of the NAFO Precautionary Approach Framework*, Northwest Atlantic Fisheries Organisation, Serial No. N6177, June 2013.

42 Communication from the Commission to the European Parliament, the Council, the European Economic and Social Committee and the Committee of the Regions on a new strategy for the Community to prevent, deter and eliminate Illegal, Unreported and Unregulated fishing /* COM/2007/0601 final *.

43 See *Grain, Fish, Money: Financing Africa's Green and Blue Revolutions*, Africa Progress Report 2014, p. 89.

vessels flying the flag of states which apply no or very low standards of social protection. As a result, their crews have to endure unacceptable living and working conditions, and even a form of slavery. This situation undermines efforts to achieve international progress on social standards for fishermen, reflected in the consolidated convention on work in the fishing sector adopted by the International Labour Organisation (ILO) in June 2007.[44]

In view of these issues, a few initiatives taken by IORA deserve special mention. First, IORA has established a Fisheries Support Unit (FSU) in Oman that identifies and discusses key fisheries issues and aims to study proposals and facilitates research for the IORA members. The role of the FSU as an Advisory Committee for Fisheries and Aquaculture is being explored, especially, in view of the fact that the future global demand for fish consumption have to be met by aquaculture.

While IORA outlines a networking mechanism, it is uncertain whether it will have any impact on the fisheries conservation and management. What it needs to take on board is that the IORA should consider IUU fishing as a maritime security threat and deal accordingly instead of focussing just on conservation and management. Seminars and workshops that have taken place in Oman in 2013, India in 2013 and Tanzania 2014 have left much to be desired, for only a more security-oriented strategic approach can yield desired results. In this regard, the current initiative of the IORA to develop a regional strategy to enable access to EEZ fisheries within the region will partially meet the IUU fishing concerns.

It is essential that IORA develop a framework which can, *inter alia*, advance the development of an effective enforcement regime in respect of particular fisheries or fish stocks. Such a mechanism needs to include traditional maritime enforcement measures such as stopping, inspecting and potentially arresting a vessel. The IORA can also develop a catch documentation scheme that can monitor landings of fish and the trade flow of particular species. The member States also may cooperate in vessel monitoring mechanisms to track the location of vessels. It is equally important that IORA member States adopt the 2009 Port State Measures Agreement, which would provide added ability for them to respond to unlawful fishing on the high seas.

Concluding Remarks

Based on above analysis, this chapter draws few important conclusions. First, there exists a low-level of critical awareness among policy-makers about the

44 ILO Convention No. 188 (2007).

scope and size of each of the maritime security challenges. The awareness level varies from one member state of IORA to another. This calls for a continuous and regular dialogue among them. As IORA states have higher preference for less-formal or less-institutional approaches to international cooperation, it is imperative that a mind-set gradually evolves towards a more institutional response system. This suggestion is made, also because, the current challenges require formalistic and institutional response. Second, the development of legal and juridical framework ranges from a very robust system to a weak system. While the states machinery recognize the importance of well-uniform and cohesive framework, their legal capacity in terms of well-trained human resources in the area of international law and law of the sea in particular, do not enable them to meet the new complex challenges. Third, the IORA states also need to identify and systematically assess the impact of each of these threats on their varied level of political and economic development including the future growth potential. As Indian Ocean waters are becoming more than ever important, these member States are advised to undertake such exercises earlier than later. Obviously, this calls for international cooperation and cooperation especially from the major seafaring nations. While their economic and technology cooperation is more important, their political mature understanding of the region is even more important. Fourth, legislative and judicial capacity building is extremely important which works as a major bottleneck to countries in the region despite their existing and potential political will. Lastly, the Indian Ocean has emerged and will continue to remain the most important water body for the IORA states and external states, hence, it is essential that a regime of legal cooperation is established to ensure long-term stability and peace in this region.

CHAPTER 10

Black Sea Security under the 1936 Montreux Treaty

*Nilufer Oral**

Introduction

The Convention Regarding the Regime of the Straits was adopted on 20 July 1936 in the Swiss city of Montreux, entered into force on 9 November 1936 and has remained in force.[1] The Convention regulates the passage of merchant and military vessels through the Turkish Straits. The Montreux Convention created a unique regime of passage not only for the Turkish Straits but also for the Black Sea. It is the only international treaty that imposed limits on the total number and tonnage of military vessels as well the duration of their stay in a sea that would otherwise be open to international navigation. Negotiated against the backdrop of World War II, the Convention sought to safeguard the security of Turkey as well as that of the Black Sea, as expressly stated in the Preamble of the Convention.

The 1982 Law of the Sea Convention (UNCLOS)[2] created a new regime of transit passage for both merchant vessels and war ships engaged in passage through straits used for international navigation. It maintained and codified the customary international law regime of non-suspendable innocent passage for a limited category of straits. Furthermore, having in mind long-standing treaties such as the 1936 Montreux Convention, article 35(c) excluded a small number of these straits from the scope of Part III. Instead, these straits are subject to the conditions of passage as established under the applicable long-standing convention. Among the long-standing treaties governing straits, however, the 1936 Montreux Convention is the only one to regulate the passage of military vessels in detail.

The Montreux Convention, having withstood the pressures of World War II and the Cold War period, faces new challenges from the post-Cold War period that struggled in redefining the role of NATO, the emergence of regional and global conflicts, and new energy corridors critical for energy security. This

* Member of the Istanbul Bilgi University Law Faculty.

1 173 *League of Nations Treaty Series* 215 (1936).

2 Adopted 10 December 1982, entered into force 16 November 1994, 1833 *United Nations Treaty Series* 3.

 | DOI 10.1163/9789004311442_012

chapter begins by an examination of the historical background to the negotiation and adoption of the 1936 Montreux Convention and then explores recent events raising security issues in the Black Sea, and the role of the Montreux Convention and why it continues to be a viable and important agreement. Despite changes in the international political order, the Montreux Convention has maintained the delicate balance of regional and, by extension, global powers consistent with international law and the law of the sea.

A Legal History of the Black Sea

The Ottoman conquest of Constantinople (Istanbul) in 1453 gave it complete control over the Turkish Straits and access to the Black Sea. Until the demise of the Ottoman Empire in 1922, foreign warships were not allowed passage through the Straits without the consent of the Sultan. The absolute right of the Ottoman Sultan to prohibit the passage of foreign warships through the Straits was known as the *ancient rule of the Ottoman Empire*. In other words, foreign warships could only enter the Turkish Straits with the express permission of the Ottoman Sultan.

The Ottoman Empire's absolute control over the Black Sea was first successfully challenged in 1696 when Peter the Great captured the Fort of Azov on the Crimea coast and gave Russia its first strategic foothold in the Black Sea.[3] In 1699, after losing a battle against the Austro-Hungarian Empire, the Ottoman Empire was forced under the Peace Treaty of Carlowitz to cede to Russia the Fort of Azov and eighty miles of coast along the Crimean coastline. The agreement was signed on 26 January 1699.[4] Following a defeat against the Ottomans in 1711, however, Russia was forced to return these gains under the Treaty of Pruth, signed on 21 July 1711. The Treaty of Pruth is the first international treaty to regulate the status of the Black Sea, and it prohibited Russia from maintaining a naval fleet in the Black Sea. Catherine the Great, however, took Crimea from the Ottomans in 1783, and she succeeded in creating a permanent Russian

3 In 1699 after losing a battle against the Austro-Hungarian Empire the Ottoman Empire was forced to give to Russia the fort in Azov and eighty miles of coast along the Crimean coastline under the Peace Treaty of Carlowitz, signed 26 January 1699. However, following a defeat against the Ottomans in 1711 Russia was forced to return these gains under the Treaty of Pruth, signed 21 July 1711.

4 For an overview of the legal history of the Turkish Straits and Black Sea *see* Nilufer Oral, *Regional Co-operation and Protection of the Marine Environment Under International Law* (Martinus Nijhoff, 2013), 17–28.

presence in the Black Sea. In 1794, she developed a city and port at Odessa on the Crimean coast, which was Russia's first Black Sea port.

The Ottoman Empire jealously guarded its control over the Straits as protected by the *ancient rule of the Ottoman Empire*. There were only few instances when the Sultan allowed foreign warships passage through the Straits. The first time was as part of a mutual security agreement between the Russian and Ottoman Empires against the threat of Napoleon Bonaparte. Consent was granted in the Treaty of 1798 and later in the 1805 Treaty of Mutual Defense. These agreements granted Russian naval ships passage through the Straits for the first time.[5] In 1833, once more, the Ottoman Empire granted passage rights to Russian warships under the Treaty of Defensive Alliance (Hünkar İskelesi Treaty) for protection against the rebellious Ottoman governor in Egypt.[6] The treaty included secret provisions that gave Russia the control over the Black Sea, including an agreement that the Ottoman Empire would prevent the passage of all foreign warships through the Çanakkale Strait (Dardanelles).

In the 1809 Treaty of Çanak (Dardanelles), Great Britain gave official recognition and support to the ancient rule of the Ottoman as means to contain the expansionist desires of Russia.[7] Subsequently, in the 1840 Convention of London and the 1841 London Straits Convention, Britain once again acceded to Ottoman control. The treaties recognized the authority of the Sultan under the *ancient rule of the Ottoman Empire* to grant or deny passage rights to foreign warships through the Straits under.[8] These two treaties effectively codified the *ancient rule* as a principle of international law governing the Straits applicable to the Contracting Parties.[9]

In 1853 when the Crimean War erupted, the Ottoman Empire was supported by the British and French against Russia, which was ultimately defeated. The 1856 Treaty of Paris ended the Crimean War and for the first time in history neutralized an entire sea.[10] In addition to the prohibition on the access of

---

5 Sergei Gorianow, *Le Bosphore et les Dardanelles Etude Historique Sur la Question des Detroits* (1910) 4. Available online at https://archive.org/stream/lebosphoreetlesdoogoriuoft#page/4/mode/2up.

6 20 *British & Foreign State Papers* (James Ridgway & Sons, London, 1842) 1176.

7 Signed 5 January 1908, 1 *British & Foreign State Papers*, 768.

8 Reprinted in Clive Parry, 92 *Consolidated Treaty Series* (Oxford, 1969) 1.

9 See, H.M. Wood, "The Treaty of Paris and Turkey's Status in International Law," *American Journal of International Law* 37 (1943): 262, 273 [citing Wheaton, *Elements of International Law*, 3rd ed. (1846)].

10 Wood, "The Treaty of Paris and Turkey's Status in International Law," 262–274; C.G. Fenwick, "The New Status of the Dardanelles," *American Journal of International Law* 30 (1936): 701–706, 702.

all foreign warships into the Black Sea, both Russia and the Ottomans were prohibited from keeping naval forces in the Black Sea. The Treaty of Paris was signed in 1871 and modified the 1856 Treaty of Paris by permitting both the Ottomans and the Russians to maintain warships in the Black Sea. The Treaty also reaffirmed the authority of the Sultan to prohibit the passage of foreign warships through the Straits.[11]

World War I, however, drastically upset the balance of power established under the Treaty of Paris. Great Britain, an ally of the Ottoman Empire, did an about-face and launched the disastrous 1915 Gallipoli offensive against the Ottomans, who had sought new alliance with Germany. The Ottomans closed the Straits to all vessels belonging to hostile states, including Russian vessels, merchant ships and warships.

The successful conclusion of the Turkish War of Independence in 1922 ended the six hundred year old Ottoman Empire and created the new Republic of Turkey. In Russia, the Czarist regime had been replaced with the revolutionary socialist regime of Lenin in 1917. The political landscape of the Black Sea, and Europe, had changed radically. But the strategic importance of the Straits and Black Sea remained unchanged as revealed during the negotiations over the legal regime for the Turkish Straits and Black Sea that would be part of the 1923 Lausanne Convention on the Regime of the Straits and, later replacing it, the 1936 Montreux Convention on the Regime of the Straits. The latter agreement remains in effect nearly eighty years after it was negotiated, making it one of the oldest treaties still being implemented.

The 1923 Convention Respecting the Regime of the Straits and Other Instruments signed at Lausanne

The Lausanne Convention Relating to the Regime of the Straits signed 24 July 1923,[12] was concluded by the new Republic of Turkey, the United Kingdom, France, Italy, Japan, Greece, Bulgaria, Serbo Croat Slovenes State, Russia and Romania. This Convention demilitarized coastal areas of the Straits, islands in the Sea of Marmara and the islands of Samothrace, Lemnos, Imbros, and

11 Francis Henry Skrine, *The Expansion of Russia* (Cambridge University Press, 1915), 226.

12 The 1923 Convention on the Regime of the Straits was part of the Treaty of Peace signed with Turkey and the British Empire, France, Italy, Japan, Greece, Romania and the Serb-Croat-Slovene State. The Soviet Union, which was a Party to the Convention on the Regime of the Straits, did not participate in the Lausanne Peace Treaty.

Tenedos and Rabbit Islands in the Aegean Sea.[13] Article 1 of the Lausanne Convention recognized "the principle of freedom of transit and of navigation by sea and by air in the Strait of the Dardanelles, the Sea of Marmora and the Bosphorus..." While freedom of navigation was recognized for warships, article 2 imposed a tonnage limit based on the total tonnage of the most powerful naval fleet of a Black Sea littoral state. Each of the Contracting Powers reserved the right to send to the Black Sea *at all times and under all circumstances, a force of not more than three ships, of which no individual ship shall exceed 10,000 tons.* Subject to this limitation all warships were allowed passage during peace time and war if Turkey was a neutral power.[14] The only restriction was that warships and military aircraft of belligerent powers were prohibited from making "any capture, to exercise the right of visit and search, or to carry out any other hostile act in the Straits." According to the Convention in time of war, with Turkey as a belligerent, neutral warships continued to enjoy freedom of passage. Although Turkey could take measures to prevent the passage of enemy warships and aircraft, such measures could not result in preventing the passage of neutral ships and aircraft. The Lausanne Convention only allowed for surface passage by submarines of Powers at peace with Turkey.[15]

A significant aspect of the 1923 Lausanne Convention was internationalizing the Straits, which had been under the domination of the Ottoman Empire for centuries. An International Straits Commission was established composed of representatives of the Contracting Powers and chaired by Turkey[16] and oversaw the administration of the Straits. The total tonnage of foreign war ships allowed into the Black Sea was determined from reports supplied to the Commission by the Black Sea littoral countries twice each year as to the strongest naval force in the Black Sea (Article 10). The Commission was required to report these immediately to the Contracting Powers. This requirement enabled the non-Black Sea Powers to follow closely the regional naval developments.

Turkey, however, was not happy to have delegated administrative power over the Straits to an international commission.[17] By 1935 with the prospects for peace in Europe quickly diminishing as the military forces of Mussolini and Hitler transgressed borders, the Turkish Government, in a rare invocation of the international doctrine of *rebus sic stantibus*, asserted self-defense needs

13 Article 4.

14 Article 2(a).

15 Article 3(a).

16 Articles 10 and 12.

17 Anthony De Luca, *Great Power Rivalry at the Turkish Straits: The Montreux Conference and Convention of 1936* (New York: Columbia University Press, 1981).

to re-militarize the Straits. Consequently, Turkey convinced the Contracting Parties to meet in the Swiss city of Montreux in July 1936 to renegotiate the Straits Convention.

After one month of negotiations that included all the Contracting Parties, the 1923 Lausanne Convention on the Straits was superseded by the new 1936 Convention on the Regime of the Straits, which was signed on 20 July 1936. The Parties to the new agreement were the same, except Australia attended independently of the United Kingdom, the State of Serbo-Croate-Slovene was now Yugoslavia, and Italy, as a belligerent power, was not invited to participate.

The new Convention brought significant changes. First, the International Straits Commission was disbanded. Full Turkish sovereignty, including all aspects of administration over the Straits, was reinstated, and the shores were remilitarized.[18] Further, the Soviet Union, reiterating its position on the Black Sea voiced in Lausanne in 1923, once more was concerned with preserving Black Sea security against foreign encroachment. The security concern for both Turkey and the Black Sea coastal States was expressly included in the preamble of the Montreux Convention, which states that the agreement is designed to safeguard Turkish security and the security of the Black Sea region. Both Conventions distinguished between "time of war" and "time of peace" and "Black Sea powers" and "non-Black Sea powers" imposing different rules of passage accordingly for merchant vessels and warships.

Passage of Warships under the Montreux Convention

The Montreux Convention revisions to the 1923 Lausanne Convention provided a more detailed regime that offered greater security than the 1923 Lausanne Convention. Most importantly, Turkey was afforded control over the Straits. Indeed, one author at the time noted that "[i]n one sense the new convention merely restores old rights formerly possessed and exercised by Turkey, but taken from it in 1923 by the Treaty of Lausanne."[19]

Under the Montreux Convention light surface vessels, minor war vessels and auxiliary vessels are not subject to restrictions and benefit from the right of free passage. Article 2 recognizes the customary international rule of non-suspendable innocent passage for merchant vessels.[20] All major warships are

18 Harry N. Howard, "The Straits After the Montreux Convention," *Foreign Affairs* 15 (October, 1936).

19 Fenwick, "The New Status of the Dardanelles," 702.

20 See Nihan Ünlü, *The Legal Regime of the Turkish Straits* (Martinus Nijhoff, 2002).

subject to "special conditions" of passage. These conditions include a 15,000 ton maximum aggregate tonnage limit for passage of all foreign-flagged naval forces, including Black Sea States, through the Straits at the same time, and a nine vessel numerical limit. Warships that are paying a visit to a port in the Turkish Straits or paying a courtesy visit to the Turkish Government or were damaged during passage, however, are excluded from the tonnage limitation. The passage of aircraft carriers through the Straits is prohibited. Turkey further has the right to regulate civil aircraft passage and establish forbidden flight zones over the Straits.

In addition, the Montreux Convention imposes restrictions for entry of non-Black Sea naval forces into the Black Sea. During the negotiations of the Convention both Turkey and the Soviet Union stressed the importance of providing for the security of the Black Sea. The Soviet representative, Foreign Minister Maxim Litvinov had asserted that allowing foreign warships into the Black Sea would be a departure from the long-standing rule under the Ottomans that closed the Straits and the Black Sea to all foreign warships. Litvinov argued forcibly in favor of keeping the Straits closed to foreign warships for the security of the entire Black Sea, further pointing out that the Black Sea was a closed sea and did not provide a transit route to other countries.[21]

The Montreux Convention, like the Lausanne Treaty, distinguishes between Black Sea powers and non-Black Sea powers. The latter are only permitted to have a maximum aggregate warship tonnage of 30,000 tons in times of peace, which may be increased to 45,000 tons under specific conditions. At no time, however, may the aggregate warship tonnage of a single non-Black Sea Power exceed two-thirds of the allowable aggregate tonnage for the Black Sea. In order to enforce these strict limitations, Black Sea powers are required to report their total fleet tonnage in the Black Sea to the Turkish Government, which transmits the information to the other Contracting Parties. Passage of submarines through the Straits is prohibited, except for boats from Black Sea States that are underway to rejoin their base or for repair, but passage must be on the surface during daytime and in single file.

The Montreux Convention requires Black Sea States to provide an 8-day notice to the Turkish Government before passage, a 15-day notice for non-Black Sea States. Warships sent for humanitarian purposes into the Black Sea are exempt from the 15-day notice requirement. During the Russian-Georgian crisis of 2008, reminiscent of the current Crimean crisis, political tensions arose when the United States sent warships carrying humanitarian aid to Georgia.

21 Seha L. Meray and Osman Olcay (transl.), *Montreux Bogazlar Konferansi Tutanaklar Belgeler* (Ankara University, 1976), 40.

Russia protested and accused Turkey of violating the Montreux Convention by allowing passage into the Black Sea.

The Montreux Convention allows for an aggregate 8,000 tons for the passage of naval vessels sent for humanitarian purposes. While such vessels are exempt from the 15-day notice, prior authorization is still required from Turkey. Furthermore, authorization also needs to be obtained from the other Black Sea States if warships transiting for humanitarian purpose results in either reaching or exceeding the total tonnage of non-Black sea warships allowed in the Black Sea. The Black Sea States have 24 hours to respond. The Convention is silent on what happens if there is a disagreement among the Black Sea States. Presumably, as gatekeeper, Turkey has responsibility as well as legal competence to make the final decision.

Finally, during peacetime, non-Black Sea warships cannot remain in the Black Sea over 21 days. In times of war, the security and interests of Turkey take precedence under the Convention. If Turkey is not a belligerent, war vessels that do not belong to belligerent Powers continue to enjoy peacetime passage rights. War vessels of belligerent powers, however, cannot transit through the Straits except if they are returning to their base or if Turkey is obligated under a treaty of mutual security.[22] This rule applies to both Black Sea and non-Black Sea powers.

If Turkey is a belligerent, the Turkish Government has complete discretion to determine the passage rights of all war vessels. Moreover, the Turkish Government also retains complete discretion to prohibit the passage of warships in cases where Turkey feels threatened with imminent danger of war. This unique provision allows Turkey to take pre-emptory protective measures against the threat of war.

Passage of Warships under UNCLOS

The Montreux Convention provisions for the passage of warships through the Turkish Straits and into the Black Sea are unique under international law. The right of innocent passage through straits for warships was recognized as a rule of customary international law in the historic *Corfu Channel* case.[23] The

---

22 Article 19. However, the Convention concluded when the League of Nations was still in existence only referred to mutual security treaties concluded within the framework of the Covenant of the League of Nations and duly registered and published accordingly.

23 *Corfu Channel Case* (*UK v Albania*) (Merits) [1949] ICJ Rep 4., at p. 29.

1982 United Nations Law of the Sea Convention (UNCLOS)[24] established the right of transit passage for warships through straits used for international navigation that did not otherwise fall under one of the enumerated exceptions.[25] The Turkish Straits were one of the straits subject to a long-standing convention that informed the drafting of article 35(c) of UNCLOS.[26]

Warships are required to (1) comply with the laws and regulations of a coastal State during passage through its territorial sea (art. 30); and (2) the flag State bears international responsibility for any damage caused to the coastal State resulting from non-compliance of a warship, or other governmental ships operated for non-commercial purposes, with the laws and regulations of a coastal State during passage through its territorial sea or with other provisions of the Convention or other rules of international law (art. 31). In other words, warships are subject to the same rights and duties of other vessels, including transit passage, non-suspendable innocent and archipelagic-sea-lane passage.

Article 31(1) mandates coastal States provide for unimpeded passage of all vessels engaged in transit passage, defined in the article as the exercise … of the freedom of navigation and overflight solely for the purpose of continuous

---

24 John N. Moore, "The Regime of Straits and the Third United Nations Conference on the Law of the Sea," *American Journal of International Law* 74 (1980): 77–121; William L. Schachte, Jr, and J.P.A. Bernhardt, "International Straits and Navigational Freedoms," *Virginia Journal of International Law* 33 (1993): 503, 544–45; Jon M. Van Dyke, "Transit Passage Through International Straits," in A. Chircop, T.L. Dorman & S.J. Rolston (eds.), *The Future of Ocean Regime-Building: Essays in Tribute to Douglas M. Johnston* (Brill, 2009), 177–232.

25 The transit passage" regime is set out in article 38(1) of UNCLOS. The exception to the transit passage regime are (1) Straits which include a route through the high seas or an exclusive economic zone, if this route is of similar convenience with respect to navigational and hydrographical characteristics, are subject to the applicable provisions of the other relevant parts of UNCLOS including freedom of navigation and overflight (article 36); Straits located between an island and the mainland of the same State, used for international navigation between one part of the high seas or an EEZ and another part of the high seas or an EEZ, in cases where, seaward of the island, another route of similar convenience, including hydrographical characteristics, exists through the high seas or exclusive economic zone. By virtue of article 38(1) and article 45(1)(a) and (2), the traditional non-suspendable regime of innocent passage applies; Straits used for international navigation between a part of the high seas or an exclusive economic zone and the territorial sea of a foreign State. The traditional non-suspendable innocent passage regime also applies here (article 45(1)(b); Straits that are governed in whole or in part by long-standing international conventions.

26 Satya N. Nandan, Shabtai Rosenne and Neal R. Grandy (eds.) *United Nations Convention on the Law of the Sea 1982: A Commentary*, Vol. II (Martinus Nijhoff, 1993) 301–308, 308.

and expeditious transit of the strait between one part of the high seas or an exclusive economic zone and another part of the high seas or an exclusive economic zone. Ships are to refrain from any threat or use of force against the sovereignty, territorial integrity or political independence of States bordering the strait, or in any other manner in violation of the principles of international law embodied in the Charter of the United Nations (39(1)(b)).

In those straits subject to non-suspendable innocent passage, according to Article 19(1), passage is innocent so long as it is not prejudicial to the peace, good order or security of the coastal State and in conformity with the Convention and with other rules of international law. Non-innocent activities are enumerated in Article 19(2), and include: any threat or use of force against the sovereignty, territorial integrity or political independence of the coastal State, or in any other manner in violation of the principles of international law embodied in the Charter of the United Nations (art. 19(2)(a)); any exercise or practice with weapons of any kind (art. 19(2)(b)); any act aimed at collecting information to the prejudice of the defense or security of the coastal State (art. 19(2)(d)); the launching, landing or taking on board of any aircraft (art. 19(2)(e)); and the launching, landing or taking on board of any military device (art. 19(2)(f)).

The differences between the rules of passage for warships under the Montreux Convention and under UNCLOS are significant. Under the latter regime, coastal States must allow for transit passage and cannot impose restrictions on passage, such as tonnage, numerical limitations or type of ship or class of warship. Furthermore, UNCLOS recognizes the right of submerged transit passage for submarines in straits used for international navigation. Likewise, aircraft enjoy transit passage rights over straits, whereas military aircraft overflight of the Turkish Straits is prohibited. The 21-day time restriction for warships staying in the Black Sea is another unique aspect of the Montreux Convention, as is the requirement that all warships give Turkey prior notification before passage. Some coastal States have purported to impose restrictions on military vessel activities in their territorial seas, such as requiring prior notification, but these restrictions have been challenged.[27]

A specific incident involving the Black Sea took place in 1988, albeit prior to the entry into force of UNCLOS. The Soviet Union, in addition to the restrictions imposed under the Montreux Convention, had also imposed prior notification requirements for foreign warships seeking passage through its territorial sea.

---

27 See James Kraska, *Maritime Power and the Law of the Sea* (Oxford University Press, 2011), 117–118.

The United States, under the Freedom of Navigation program (FON),[28] created in 1979 to challenge what it considered to be excessive maritime claims, sent two naval warships, the *USS Yorktown* and the *USS Caron* to the Black Sea. On 12 February 1988, these two vessels, which were exercising their "innocent passage" rights but without having given prior notification to the Soviet Union, were bumped or shouldered by Soviet warships.[29]

There also remains some controversy over the rights of the coastal State to regulate warship activities in the EEZ.[30]

The Montreux Convention and the 1923 Lausanne Treaty expressly differentiate between peacetime and war time. Whereas UNCLOS does not expressly limit its application to time of peace or indicate its status in time of war or armed conflict, except for the last paragraph of the preamble, which provides that "matters not regulated by this Convention continue to be governed by the principles of general international law." The general view is that the law of naval warfare is implicitly excluded by the language "matters not regulated" by the Convention, and would presumably continue to be regulated by other sources of international law, such as the United Nations Charter and right of self-defense under article 51, the Geneva Conventions and their protocols, as well as customary rules of international law.[31] The Montreux Convention is quite clear, as it was concluded when the League of Nations still existed and it refers to the organization. The notions of war and peace changed after World War II, with the replacement of the League of Nations by the UN Charter, an issue that had been raised by Stalin during the Yalta Conference as grounds to revise the Montreux Convention.[32] Nonetheless, over the decades since and

28 See US State Department explanation of the FON at http://www.state.gov/e/oes/ocns/opa/maritimesecurity/.

29 William J. Aceves, "Diplomacy At Sea: U.S. Freedom of Navigation Operations in the Black Sea," in John Norton Moore and Robert F. Turner (eds.), *International Law Studies, Readings on International Law From the Naval War College Review*, Vol. 68 (1978–1994), Chapter 18.

30 For a detailed analysis of military activities and security issues in the EEZ see Kraska, *id.*, at 221–289; Jon M. Van Dyke, "Military Ships and Planes Operating in the Exclusive Economic Zone of Another Country," *Marine Policy* 28 (2004): 29–39.

31 See Nandan *et al.*, *United Nations Convention on the Law of the Sea 1982: A Commentary*, 89; Bernard H. Oxman, "The Regime of Warships Under the United Nations Convention on the Law of the Sea," *Virginia Journal of International Law* 24 (1983–1984): 811–863; Brian Wilson and James Kraska, "American Security and Law of the Sea," *Ocean Development & International Law* 40 (2009): 268–290, 277–278.

32 Stephen G. Xydis, "New Light on the Big Three Crisis over Turkey 1945," *Middle East Journal* 14 (1960): 416–432, 419.

despite the occasional grumblings for revision, the Montreux Convention has remained intact and operative.

NATO and the Black Sea

During the Cold War the Turkish Straits once again would be a testing ground for regional power politics. During the forty-five year span of the Cold War (1947–1991) the Montreux Convention provided an important control mechanism over Soviet warship access into the Mediterranean Sea. Ironically, where once the USSR stood firmly to protect the restrictive provisions on warship passage of the Montreux Convention, under the expansionist policies of Stalin and with the emergence of the USSR as a world naval power, Moscow saw the treaty as a hindrance to its maritime power projection.

In 1944 Stalin had told Churchill that he wanted the Montreux Convention modified to allow Soviet warships free passage at all times—wartime and peacetime.[33] The next year during the Yalta Conference (4–11 February 1945), Stalin expressed his view that the Montreux Convention was outmoded and needed revision. More importantly, Stalin disagreed with the broad authority given to Turkey to close the Straits in wartime, or if Turkey felt threatened with war. Stalin could not accept a situation with "Turkey's hand on the throat of the USSR."[34] The USSR also made demands for bases along the Straits.[35]

The Straits were raised again during the Potsdam Conference (1945), where the Soviet Union once again voiced its discontent with the Montreux Convention, and communicated its desire to replace it with a bilateral agreement with Turkey that excluded other parties, on the grounds that both Turkey and the USSR had a common security interest in preventing use of the Straits by other countries.[36] The Soviet request for bases along the Straits was again invoked, but Churchill rejected the idea. Discussions held among Stalin, Churchill and Truman during the Potsdam Conference, with the notable absence of Turkey, exposed divergent views: on the one hand, the Soviets desired to share control of the Straits and close off the Black Sea entirely to non-Black Sea Powers, whereas the United States sought to open the Straits and the Black Sea to complete freedom of passage for all warships.[37]

---

33 Xydis, "New Light on the Big Three Crisis Over Turkey 1945," 418.

34 *Ibid.*

35 *Ibid.*

36 *Ibid.*, at 428.

37 *Ibid.*

The Cold War was launched in earnest following President Truman's historic speech to Congress on March 12, 1947 speech to Congress calling for aid to Turkey and Greece over concerns of a growing Soviet threat. The move was partly sparked by Stalin's demand to control the Turkish Straits and build bases in the region.[38] The historic visit on 5–9 April 1946 of the battleship *USS Missouri* to Istanbul marked the first time a U.S. warship anchored in the Istanbul Strait. The voyage sent a strong signal to the Soviet Union that the Straits would not be abandoned to Soviet control. On August 7, 1946 the Soviet Charge, Orekhov presented U.S. Secretary of Dean Acheson the following Soviet demands for the Straits:

1. Straits should be open to all merchant ships at all times
2. Straits should be open to warships ships of Black Sea states at all times
3. No passage to non-Black Sea warships unless specifically provided for
4. Straits regime should be under competence of Turkey and Black sea powers
5. Turkey and USSR to provide joint means of defense of the Straits[39]

The United States saw these demands as a means for the USSR to eventually take control over the Straits. The demands brought a change to U.S. policy that now viewed the Straits and Turkey's independence as part of U.S. vital interests.[40]

Until the fall of the USSR in 1991, the Turkish Straits and the Black Sea would serve as a backdrop to the Cold War. Turkey served at the frontline, and the Montreux Convention provided the legal framework to control and maintain the delicate balance of power and peace in both the Black Sea and the Mediterranean Sea.

The Treaty of Washington,[41] establishing the North Atlantic Treaty Organization (NATO) was adopted in 1949, in response to the new Soviet threat and the spread of communism. Belgium, Canada, Denmark, France, Iceland, Italy, Luxembourg, the Netherlands, Norway, Portugal, the United Kingdom, and the United States joined the alliance, which provided for the collective security of the member countries in accordance with the UN Charter. Turkey joined NATO in 1952.

38 Knight, "American Statecraft and the 1946 Black Sea Straits Controversy," 459.

39 *Ibid.*

40 *Ibid.*, at 464.

41 The North Atlantic Treaty, signed 4 April 1949, Washington DC.

During the Cold War, the Black Sea served as the Eastern border between NATO and the USSR. One of the great political challenges for Turkey, as the only NATO country bordering the Black Sea, was to maintain the balance of power in the region through judicious application of the 1936 Montreux Convention for the passage of warships and military aircraft.

The first challenge for Turkey arose in 1964 when the first permanent Soviet naval base was established in the Mediterranean Sea, resulting in an increase in the number of Soviet warships deployed through the Straits.[42] Turkey's decision to allow the passage of the *Kiev*, a 40,000 ton Soviet naval vessel with a 600-foot flight deck that carried helicopters and aircraft raised questions of whether the vessel transit was a violation of the Convention.[43] Annex II of the Montreux Convention prohibits the passage of vessels designed or adapted primarily for the purpose of carrying and operating aircraft at sea. Turkey decided to grant passage to the vessel, which the USSR classified as an anti-submarine cruiser and not an aircraft carrier.[44] Nonetheless, Turkey succeeded in maintaining the integrity of the Montreux Convention throughout the Cold War.

In 1991, however, the foundation of NATO was shaken with the dissolution of the USSR. The historic downfall of the USSR also brought significant changes to the security of the Black Sea region. NATO's principal antagonist had disappeared from the pages of history. The Russian Federation emerged as a new state and initially enjoyed close ties with the west.[45] In 1994, the Russian Federation joined NATO's Partnership for Peace Program. Other former members of the Soviet Union were being integrated into the international system. In 1999, the Czech Republic, Hungary, and Poland became NATO members. Thereafter, Bulgaria, Estonia, Latvia, Lithuania, Romania, Slovakia, and Slovenia joined the organization. NATO, which was established as a transatlantic alliance to protect Europe against the potential threat of a military intervention from the Soviet Union, seemed on the brink of irrelevance. NATO actively engaged with Black Sea States following the dissolution of the USSR. In addition to the 1994 Partnership Program with Russia and the NATO-Russia

42 Stephen S. Roberts, "The Turkish Straits and the Soviet Navy in the Mediterranean," Professional Paper 331, Center for Naval Analysis, (March 1982), https://www.cna.org/sites/default/files/research/5500033100.pdf.

43 H. Gary Knight, "The Kiev and the Turkish Straits," *American Journal of International Law* 71 (1977): 125–129.

44 David. F. Froman, "Kiev and the Montreux Convention: The Aircraft Carrier that Became a Cruiser to Squeeze Though the Turkish Straits," *San Diego Law Review* 14 (1977): 681–717.

45 Luca Ratti, "Back to the Future? International Relations Theory and NATO-Russia Relations since the End of the Cold War," *International Journal* 64 (2009): 399–422.

Founding Act of 27 May 1997, NATO established the 1997 Partnership Program with Ukraine. In 1992 Georgia joined the North Atlantic Cooperation Council. Bulgaria and Romania were also being integrated into the European Union and by 2005 had become full members of NATO. Warm relations between the West and Russia became increasingly brittle, and Russia looked at NATO expansion with increasing suspicion.[46]

Black Sea relations following the end of the USSR smoldered for a number of years. In 1997 the economically weak Russian Federation ceded its prized Crimea to Ukraine, including the port of Sevastopol and with it the once great pride of the USSR Navy—the Black Sea fleet. Russia and Ukraine signed the Partition Treaty on the Status and Conditions of the Black Sea Fleet on 28 May 1997, and later renewed the agreement in 2010. The arrangement remained fragile however, and relations strained under unresolved issues, including delimitation of the Kerch Straits in the Sea of Azov between Ukraine and Russia.[47] The 1992–1993 war between Georgia the independence-seeking Abkhazia destabilized the region, as Russia supported the Abkhazian secessionists. These early seeds of conflict were exacerbated as NATO openly sought to strengthen its Black Sea presence. By 2004 Romania and Bulgaria already had been admitted to NATO. In 2008 NATO considered admitting Georgia and Ukraine. France and Germany, both concerned about Russian reaction, did not agree, and the issue was left to a future date.[48]

As NATO's presence in the Black Sea was expanding, a somewhat unexpected push-back came from Turkey, which together with Russia, objected to allowing the NATO—led 2003 Operation Active Endeavor anti-terror operation into the Black Sea.[49] Turkey instead established Operation Black Sea Harmony as a Black Sea corollary to Operation Active Endeavor, and extended

46 J.L. Black, "Russia and NATO Expansion Eastward: Red-Lining the Baltic States," *International Journal* 54 (1999): 249–266; Ronald D. Asmus, "Europe's Eastern Promise: Rethinking NATO and EU Enlargement," *Foreign Affairs* 87 (2008): 95–106.

47 Alexander Skaridov, "The Sea of Azov and Kerch Straits," in David D. Caron and Nilufer Oral, eds., *Navigating Straits: Challenges for International Law* (Brill, 2014) 220–237.

48 John J. Meirsheimer, 'Why the Ukraine Crisis is the West's Fault," *Foreign Affairs*. (Sept–Oct. 2014), available on-line at <http://www.foreignaffairs.com/articles/141769/john-j-mearsheimer/why-the-ukraine-crisis-is-the-wests-fault.

49 Operation Active Endeavour was established under NATO in response to the terror attacks of 11 September 2001 to patrol the Mediterranean Sea for terrorists and ships carrying weapons of mass destruction. See http://www.nato.int/cps/en/natolive/topics_7932.htm.

an invitation to participate to other Black Sea states, including Russia.[50] Other Black Sea regional security initiatives include the "The Black Sea Naval Co-Operation Task Group-BLACKSEAFOR,"[51] the Black Sea Coast and Border Guards Cooperation Forum (BSCF) established in 2006[52] and Confidence- and Security-Building Measures in the Naval Field (CSBM) in the Black Sea.[53]

Tensions came to crisis with the Russian invasion of Georgia on 8 August 2008. Following the conflict, NATO established the NATO-Georgia Commission (NGC). Georgia had aspired for NATO membership and over the years had supported NATO operations.[54] Ukraine, a partner country since 1997 and actively involved with NATO, on 9 July 2009 signed the Declaration to Complement the Charter on Distinctive Partnership between NATO and Ukraine.[55] NATO—Ukraine relations are cited as a factor leading to the annexation of Crimea by Russia in 2014.[56] On 1 April 2014 NATO-Russian relations were suspended following the Crimea and Eastern Ukraine crises.[57]

---

50 Operation Black Sea Harmony became operational on 1 March 2004. See Deborah Sanders, *Maritime Power in the Black Sea* (Ashgate, 2014), 111.

51 The BLACKSEAFOR was initiated by Turkey in 1998. The BLACKSEAFOR establishment agreement was signed by Bulgaria, Georgia, Romania, Russian Federation, Turkey and Ukraine on 2 April 2001 in Istanbul. See http://www.dzkk.tsk.tr/icerik.php?icerik_id=229&dil=0&blackseafor=1.

52 The Black Sea Cooperation Forum (BSCF) agreement was signed in Istanbul during the Meeting of the Leaders of the Black Sea Littoral States Border/Coast Guard Agencies in Istanbul on November 07–09, 2006. See http://www.sgk.tsk.tr/baskanliklar/genel_sekreterlik/ingilizce/international_relations.asp; See also, Panagioti Manoli, "Cross-border Cooperation and Regional Policy Coordination: The Role of BSEC," in A. Ergun and H. Isaxanli, eds., *Security and Cross-Border Cooperation in the EU, Black Sea Region and Southern Caucasus* (IOS Press, 2013), 70–78.

53 OSCE, Joint Declaration of the Republic of Bulgaria, Georgia, Romania, the Russian Federation, the Republic of Turkey, and Ukraine, signed 25 April 2002, OSCE document FSC.DEL/298/02, 27 May 2002. See Zdzislaw Lachowski, Confidence-And Security-Building Measures in the New Europe, SIPRI Research Report No. 18 (Oxford University Press, 2004).

54 http://www.nato.int/cps/en/natohq/topics_38988.htm?selectedLocale=en#.

55 http://www.nato.int/cps/en/natohq/official_texts_57045.htm.

56 Meirsheimer, "Why the Ukraine Crisis is the West's Fault," *Foreign Affairs*, http://www.foreignaffairs.com/articles/141769/john-j-mearsheimer/why-the-ukraine-crisis-is-the-wests-fault.

57 Adrian Croft and Sabine Siebold, "NATO Suspends Cooperation with Russia over Ukraine Crisis," *Reuters*, 1 April 2014, available at http://www.reuters.com/article/2014/04/01/us-ukraine-crisis-nato-idUSBREA2U1UF20140401.

Montreux and the Black Sea Crises

The 2008 Georgia-Russia conflict arose following the hostilities between Georgia and the separatists forces of South Ossetia. Russia sent armed forces to support South Ossetia and Abkhazia. The Russian Black Sea Fleet also established a maritime blockade of Georgian ports Turkey's refusal to allow the passage through the Turkish Straits of two U.S. hospital ships, USNS *Mercy* (69,552 tons)[58] and the USNS *Comfort* (69,552 tons),[59] was somewhat of a surprise. The vessels were sent to Georgia for humanitarian purposes. Turkey denied passage on the grounds that the tonnage of the two ships exceeded the maximum total displacement 45,000 tonnage limitations under the Montreux Convention for all non-Black Sea powers present in the Black Sea at the same time, and 30,000 tons for a single Black Sea power. Turkey, however, was also concerned about escalation of the conflict.[60] On 22 August 2008 the United States sent three smaller ships through the Turkish Straits to Georgia—the guided missile destroyer, USS *McFaul* (8,915 tons), the U.S. Coast Guard cutter *Dallas* (3,250 tons) and the USS *Mount Whitney* (18,400 tons). These three vessels had an aggregate tonnage of slightly more than the 30,000 ton limit allowed under the Convention for a single non-Black Sea State to have in the Black Sea at one time.

On 21 August, however, Turkey granted passage to four NATO frigates from non-Black Sea States (Spain, Germany, Poland and the United States) with an aggregate displacement over 17,500 tons, to conduct "routine port visits and exercises with NATO member nations bordering the Black Sea."[61]

One question that was raised was whether the two hospital vessels should have been granted passage as auxiliary vessels under Article 10, and thereby be exempt from the tonnage limitation.[62] The exception under article 10 is only granted "in times of peace." It should be noted that unlike other provisions in the Convention, which expressly specify Turkey's status as being a belligerent[63]

58 http://www.msc.navy.mil/inventory/ships.asp?ship=124.

59 http://www.msc.navy.mil/inventory/ships.asp?ship=124.

60 Selçuk Çolakoğlu, "Turkey and Russia to determine the future of the Black Sea region," *The Journal of Turkish Weekly*, 6 March 2013.

61 "NATO Ships in Black Sea on Routine Visit, Unrelated to Georgia Crisis," NATO Press release (28 August 2008), available at http://www.nato.int/docu/pr/2008/p08-110e.html.

62 Annex II describes auxiliary vessels as "naval surface vessels the standard displacement of which exceeds 100 tons (102 metric tons), which are normally employed on fleet duties or as troop transports, or in some other way than as fighting ships, and which are not specifically built as fighting ships" and do have any of a set of enumerated characteristics.

63 Article 5.

or non-belligerent,[64] article 10 does not. Thus, given the state of hostilities between Georgia and Russia, Turkey could in principle refuse passage.

The next challenge came with the 2014 Russia-Ukraine-Crimea crises. As a result of mass public protests, the pro-Russian Ukrainian President Viktor Yanukovych fled Ukraine in February 2014. In March, Russia annexed Crimea with the stated purpose of protecting Russians living there, and Moscow brought the region under its control.[65] The successful annexation of Crimea returned to Russia the port of Sevastopol and the entire Black Sea Fleet. As a result of the annexation, Russia now posesses the largest Black Sea naval fleet. Shortly after the annexation, the Russian Navy Commander-in-chief issued a statement that thirty new warships would be added to the Black Sea Fleet over the next six years. Under the 1997 agreement with Ukraine, as extended in 2010, Russia could only do ship-for-ship swap that replace older warships with similar new ones of the same size.[66] The clear signals for expansion of the Russian Black Sea Fleet will have significant impact on the balance of naval power in the Black Sea and regional maritime security.

On 7 March 2014, just days before the Crimean referendum and as Russian military build-up along the Ukrainian border increased, the United States deployed the USS *Truxtun*, a guided missile destroyer, through the Turkish Straits into the Black Sea. The United States reported that the ship was on a routine and pre-planned operation.[67] Two days earlier, Turkey had also granted the USS *Taylor* passage into the Black Sea. Tensions in the Black Sea also spread to the skies above it as Turkey sent up F-16 fighter jet aircraft after a Russian surveillance jet aircraft flew close to Turkish airspace in the Black Sea.[68] The

64 Article 4.

65 Following a referendum held on 16 March 2014, President Putin signed on 17 March 2014 the Executive Order *On Recognising the Republic of Crimea* as a "sovereign and independent state, whose city of Sevastopol has a special status." See http://eng.kremlin.ru/acts/6884. On 18 March President Putin signed another executive order that officially admitted the newly created Republic of Crimea into the Russia Federation. See Executive Order on executing Agreement on Admission, available in English at http://eng.kremlin.ru/acts/6888.

66 John C. Daly, "Hot Issue: After Crimea: The Future of the Black Sea Fleet," 22 May 2014, *The Jamestown Foundation*, available at http://www.jamestown.org/programs/hotissues/single-hot-issues/?tx_ttnews%5Btt_news%5D=42411&tx_ttnews%5BbackPid%5D=61&cHash=18fb1cd8a0f31ab19fc65b39ef83d021.

67 http://www.stripes.com/news/destroyer-uss-truxtun-heads-for-black-sea-amid-heightened-tensions-over-crimea-1.271401.

68 "Turkish fighters scramble against Russian aircraft," *Global Times*, 4 March 2014, available at http://www.globaltimes.cn/content/846222.shtml.

second U.S. warship, the USS *Donald Cook*, entered the Black Sea on 10 April 2014, and was closely observed by a Russian fighter jet aircraft.[69]

This time, however, the Black Sea crisis brought protests from Russia, which claimed that Turkey had violated the Montreux Convention by allowing the USS *Taylor* to stay 11 days over the 21-day limit on the grounds of a mechanical breakdown.[70] Over the following weeks Russia continued to observe with apprehension a notable NATO naval presence in the Black Sea.[71]

In July 2014 Russian news agencies reported a record number of nine NATO vessels in the Black Sea.[72] Russia warned that it was extremely dangerous to add the "NATO-factor" to the "complicated situation in Ukraine."[73] During the same month, standing NATO Mine Counter-Measures Group TWO, one of four standing NATO maritime forces, composed of four ships from Italy, the United States, Turkey and the United Kingdom, entered the Black Sea in a "regularly scheduled deployment to enhance NATO's solidarity and readiness in the region."[74] In response, the Russian Black Sea Fleet conducted 20 military vessel drills in the Black Sea. The following month, supposedly unconnected to the Ukraine crisis, Azerbaijan, Georgia, and Turkey made a tripartite agreement for conducting joint military exercises, education and cooperation for protection of energy pipelines.[75] Between 8–10 September 2014, a multinational naval exercise in the Black Sea took place with the participation of Ukraine, Georgia, Romania, the United States, Turkey and three NATO warships [Canadian frigate HMCS *Toronto*; Spanish frigate ESPS *Almirante Juan de Borbón*; and Romanian frigate ROS *Regele Ferdinand*].

69 "Russian fighter jet made repeated passes close to US warship in Black Sea," *The Guardian*, 14 Apr. 2014, available at http://www.theguardian.com/world/2014/apr/14/russian-fighter-jet-repeated-passes-us-warship-black-sea.

70 http://www.hurriyetdailynews.com/turkey-urges-russia-to-drop-montreux-off-the-agenda.aspx?pageID=238&nid=64988.

71 http://rt.com/news/uss-donald-cook-montreux-716/.

72 "NATO Sets Record By Sending Nine Ships Into Black Sea," *Sputnik News*, 8 July 2014. Available at http://sputniknews.com/world/20140708/190842033/NATO-Sets-Record-By-Sending-Nine-Ships-Into-Black-Sea.html#ixzz3QbV8geyW.

73 "NATO factor to create additional tensions in Ukraine–Moscow (Part 2)," *Interfax*: (Russia & CIS Military Newswire), 3 Jul. 2014.

74 "Allied Maritime Command News Release," (3 July 2014) available at http://www.nato.int/cps/en/natohq/news_111570.htm?selectedLocale=fr.

75 Joshua Kuchera, "Georgia, Azerbaijan, Turkey Agree On Joint Military Exercises," *EURASIANET*, 21 Aug. 2014. Available at http://www.eurasianet.org/node/69646.

The significance of the Crimean–Ukraine–Russia crisis as a broad security threat was underscored in NATO's Wales Declaration (September 2014), which placed Russian aggression against Ukraine in its first item:

> We, the Heads of State and Government of the member countries of the North Atlantic Alliance, have gathered in Wales at a pivotal moment in Euro-Atlantic security. Russia's aggressive actions against Ukraine have fundamentally challenged our vision of a Europe whole, free, and at peace. Growing instability in our southern neighborhood, from the Middle East to North Africa, as well as transnational and multi-dimensional threats, are also challenging our security. These can all have long-term consequences for peace and security in the Euro-Atlantic region and stability across the globe.[76]

Conclusion

The passage of warships through the Turkish Straits and the Black Sea has a long history of restrictions dating back to the *ancient rule of the Ottoman Empire* and later the Turkish Republic. In this regard, there is a universally recognized and long-standing international customary practice of limiting foreign warship access through the Straits and the Black Sea. Since the time of the Great Game in the Nineteenth Century and throughout the Cold War period, the restrictions placed on Black Sea Powers as well as the non-Black Sea Powers, created a balance that served the interests of regional and global security, and was generally accepted despite occasional diplomatic grumbling.

The 1936 Montreux Convention has been shaped by the historic dynamics of the Black Sea region, which in many aspects remain unchanged. The Contracting Powers consciously negotiated a regime for the passage of warships different from general rules and practice of international law, such as the regime of innocent passage of warships in the territorial seas, as recognized in the *Corfu Channel* case. Article 35(c) of UNCLOS preserved the regime of passage for straits used in international navigation regulated in whole or in part by a long-standing international convention. In a sense, the Parties to UNCLOS, in adopting Article 35(c), implicitly recognized and reconfirmed the unique regime of the Montreux Convention. No attempts were made during the nine-year negotiation period of UNCLOS to revise the Montreux Convention.

---

76 Wales Summit Declaration, 5 September 2014. Available at http://www.nato.int/cps/en/natohq/official_texts_112964.htm.

The recent conflicts in the Black Sea, including the Russia-Georgia conflict and the Crimean conflict, demonstrate the continued volatility of the region. Turkey has the responsibility to administer the provisions of Montreux Convention, while preserving regional and Turkish security. The potential is significant for escalation of sub-regional conflicts in an enclosed sea such as the Black Sea. While some may have the political view that as a NATO member Turkey should give NATO free access into the Black Sea, on the other hand, this approach could provoke a wider-spread conflict within and beyond the Black Sea. By limiting non-Black Sea forces access into the Black Sea, there is less likelihood that the conflict will spread.

As provided for under the Montreux Convention, Turkey is the ultimate guardian of Black Sea security. In addition to ensuring faithful implementation of the various conditions of passage for warships, Turkey also holds the keys to the Straits and Black Sea in times of war or during periods in which Turkey feels threatened with the imminent danger of war. This latter provision does not require actual war but only the danger of war. In these cases, Turkey has discretion to either prevent passage warships or remove the tonnage and numerical restrictions on military vessels. Turkey can apply this authority to all foreign warships. However, given Turkey's long-standing membership in NATO and duties under the Washington Treaty, this discretionary measure could give Russia greater concern. If Russia were to engage in further hostile acts in the Black Sea, Turkey could feel threatened with the danger of imminent war in the region and could exercise discretion and allow a greater number of NATO warships into the Black Sea as a measure of its self-defense. Turkey, however, has never exercised this discretionary right and is unlikely to do so, given the importance of Turkish-Russian relations.

Nearly eighty years after its adoption, the later Montreux Convention remains a robust international agreement that has played an important role in maintaining regional security in the Black Sea through the vicissitudes of World War II, the tense days of the Cold War, and the upheavals of the post-Cold War period. While some revisions could be made to the Convention, such as replacing reference to the League of Nations with the United Nations and making reference to the UN Charter, clearly there is a reluctance open up the Convention and risk unsettling a security framework that has served the interests of both Black Sea and non-Black Sea Powers in maintaining the greater balance of peace.

CHAPTER 11

The Conceptualization and Construction of a Northeast Asian Maritime Security Architecture: Might Europe Serve as a Model?

Seokwoo Lee and Hee Eun Lee***

Introduction

Northeast Asia (NEA) remains the last vestige of the Cold War. The Cold War balance of power paradigm in the region had the United States, Japan, and South Korea pitted against the Soviet Union, China, and North Korea. The divided Korean peninsula was the focus of tensions in the region. Yet, despite the fall of the Soviet Union and the subsequent end of the Cold War, security issues in NEA continue to endure around the Korean peninsula. The Demilitarized Zone (DMZ), still under UN Command on the southern side, remains a flashpoint for conflict and is a constant reminder of the enduring standoff between North and South Korea. Beyond just Korea, the situation has become immensely more complicated. The region has seen the realization of North Korea's nuclear weapons ambitions, the ascendancy of China as a legitimate global power, and Japan's determination to maintain its standing as a world leader, all while the Obama administration pivots the United States toward Asia. The security stakes in NEA have been raised as China has significantly increased its military budget, making it the second largest spender behind the United States.[1] Japan appears intent on revising its "pacifist Constitution," worrying both China and South Korea.[2] North Korea has conducted more nuclear weapons tests and instigated armed conflict with South Korea along their border. Finally, South Korea pushes Japan to address the legacy of Japanese

* Professor, Inha University Law School; Co-Editor-in-Chief, Asia-Pacific Journal of Ocean Law and Policy (APJOLP).

** Professor, Handong International Law School.

1 "China's Military Budget Increasing 10% for 2015, Official Says," N.Y. Times, Mar. 5, 2015, accessed June 5, 2015, http://www.nytimes.com/2015/03/05/world/asia/chinas-military-budget-increasing-10-for-2015-official-says.html.

2 "Abe Is Said to Have Plans to Revise Pacifist Charter," accessed June 5, 2015, http://www.nytimes.com/2015/02/06/world/asia/abe-wants-to-revise-pacifist-constitution-as-early-as-2016-ally-says.html?_r=0.

 | DOI 10.1163/9789004311442_013

colonialism from issues surrounding Korean "comfort women" and forced labor, and contentious territorial claims over the Dokdo islets.

Accordingly, in recent years the potential for conflict in the region has reached dangerous levels as the uneasy and relative stability experienced during the Cold War era has given way to an unsettling reconfiguration of power and influence. How the countries of NEA deal with security risks under these conditions is less clear than their approach during the Cold War bipolar balance of power system. Today, the old system has been replaced by a more complex set of political, economic, and military realities. The changed security dynamics of the region has resulted in sharper political rhetoric and the tendency for states in the region to take bolder stances over islands and islets that were muted during the Cold War.[3]

The disputes over these islands and islets also reveal the additional feature of the NEA security dilemma—geography. As large bodies of water separate the countries in the region, any concerted effort to manage regional security must take into account maritime issues. In this area, the countries of NEA have not made any meaningful collective effort to address this particular subject of concern. Though increased economic ties among China, Japan, and Korea are being pursued through the creation of a free trade agreement, politically, there is a mutual distrust that emanates from perceived historical injustices.[4] The circumstances in NEA are strikingly similar to Europe right after World War II, which saw the creation of the European Coal and Steel Community in 1951 despite the destructive history experienced by the member states. While there are fundamental differences between NEA and Europe, the European experience in the creation and establishment of collective security mechanisms might prove to be instructive in highlighting the relevant issues to be dealt with when imagining the possible ways NEA states could constructively engage each other in the area of maritime security. Accordingly, the goal of this chapter is to highlight the Northeast Asian maritime security dilemma and the risks of military conflict between the states in the region with an emphasis on the Korean peninsula. Given the relative success of the North Atlantic Treaty Organization and collective security mechanisms in Europe, this chapter

---

3 For example, in April 2006, Japan and South Korea had a standoff in the waters around Dokdo with South Korea patrol ships intent on blocking two Japanese vessels that were attempting to conduct an underwater survey. Such an incident in NEA would have been difficult to imagine during the Cold War era when the relationships between the countries were more clearly defined.

4 "Editorial: Time to end the cycle of mutual distrust in East Asia," accessed June 5, 2015, http://ajw.asahi.com/article/views/editorial/AJ201412090028.

will briefly examine the presence of collective security in Europe and its lack thereof in NEA and then utilize the European context as a point of reference to imagine how a collective security system could be constructed in NEA to promote stability and predictability while noting the unique challenges facing such a project in NEA.

Collective Security in Europe and NEA

Essentially, the term "security" has two main focuses for the state. The first is the goal of deterring a hostile force and the second revolves around assuring the integrity and political independence of the state. When applying such a notion of security among and between states through a collective mechanism in the NEA context, critical issues are raised such as: what are the functions and roles for collective security and how does the mechanism of collective security achieve such purposes and goals? If there is no collective security mechanism, is the risk of armed conflict increased? Can a collective security mechanism, such as those found in Europe and including the North Atlantic Treaty Organization, European Union, Organization for Security and Co-operation in Europe, Council of Europe, and other types of regional cooperation, be utilized effectively in NEA? What would be the implications of such a mechanism in the region?

NATO, created through The North Atlantic Treaty of 1949, was originally established to cement a strong American presence in Europe to check Soviet expansion, halt militant European nationalism, and encourage European political unity.[5] Given the momentous changes in the security landscape in Europe caused by the end of the Cold War, NATO's mission has adapted to the times to address "complex new risks to Euro-Atlantic peace and security, including oppression, ethnic conflict, economic distress, the collapse of political order, and the proliferation of weapons of mass destruction."[6] In the maritime sector, NATO has set forth its strategy in four distinct areas which include: (1) deterrence and collective defense; (2) crisis management; (3) cooperative security; and (4) maritime security.

Another important aspect of European collective security can be found in the European Union's Common Foreign and Security Policy (CFSP). The conclusion of the Cold War in the early 1990s saw the development of NATO-EU

5 "A short history of NATO," accessed July 1, 2015, <http://www.nato.int/history/nato-history.html>.

6 *Id.*

relations with NATO negotiations with the Western European Union and the eventual establishment of the CFSP and the Common Security and Defense Policy (CSDP). An important feature of the CSDP is the EU's commitment to a unified maritime security strategy.[7]

It is evident that NATO has had some success in Afghanistan and Libya, but Russian incursion in the Crimea has instigated a new concern for European security. Presently, there is a general understanding that developments in the Crimea do not necessarily mean a return to the dynamics of the Cold War era, and that the economic and financial sanctions on Russia will be ineffective. Because of the Russian actions in Crimea, NATO is now undergoing a test of whether it can address a crisis in an area bordering a NATO member. It also appears that there is no guarantee that Russia will not invade a NATO member state. If that is the case, what then are the implications of this type of security arrangement for NEA? In this regard, among the three main functions of NATO—collective security, cooperative security, and crisis management—what would be the most suitable and effective model for NEA?

This new development in Europe as to the application of Article 5 of the NATO Treaty (an attack against one member shall be considered an attack against all NATO members) raises another question: does NATO, or for that matter, the EU, have sufficient counter-mechanisms in place to deal with conflicts or crisis between member states? This issue compels consideration of appropriate countermeasures and continuing cooperation in conflicts between NATO and EU members and other states in the region. Furthermore, if the development of the collective security mechanism in Europe is a model for other non-European states, then there should be preparation for hybrid threats such as international terrorism, failing states, regional and sub-regional conflicts, nuclear proliferation, and bad governance.

There are significant differences between the security situation in NEA and Europe. In Europe, for example, the deliberate institution of a post-WWII arrangement, the memorial of historical injustices, the effort to build good neighbourly relations with former enemy states, and the diminution of sovereignty through a monetary union are substantially different responses to potential regional conflicts compared to what has occurred in NEA. It is notable that the examples of the development of supranational regional cooperative bodies in the pan-Asian region are marginal. In particular, regional security and military cooperation is virtually non-existent. The possibility of an

7 The Council of the European Union, European Union Maritime Security Strategy, accessed July 1, 2015, http://register.consilium.europa.eu/doc/srv?l=EN&f=ST%2011205%202014%20INIT.

outbreak of a regional war in NEA and the possibility of U.S. military intervention is a topic of continued interest in the region, particularly in relation to maritime conflicts, which include island and maritime boundary disputes among powerful states.

Unlike the collective security mechanism that NATO provides for Europe, NEA's collective security mechanism remains primitive. Currently, the system of collective security consists mostly of bilateral agreements that function more as a conflict avoidance or confidence building structure rather than a cooperative effort to preserve peace and security in the region. The effort to establish a NEA maritime collective security system is hampered primarily by several on-going conflicts among the regional states involving territorial disputes over islands and islets in the region.

The most internationally notable conflict in NEA is the territorial dispute and corresponding disagreement over maritime zones between China/Taiwan and Japan over the Diaoyu/Senkaku Islands. The islands currently remain under the control of Japan, although the ownership of the islands has been fiercely contested as evidenced by protests from China and Taiwan. Tensions have been exacerbated by incidents around the islands resulting from the states exerting their sovereignty or jurisdiction. Japan argues that it discovered the islands in 1895 when they were *terra nullius*. China and Taiwan both assert that the islands were discovered and incorporated into the Ming and Qing Dynasties before Japan's purported discovery and incorporation. China and Taiwan also argue that when Japan won the Sino-Japanese War, it annexed the islands along with Taiwan and all its appertaining islands. China claims the islands should have been returned to China at the conclusion of World War II.[8]

Another on-going conflict within the region is the conflict between the Republic of Korea and Japan over the ownership of the Dokdo/Takeshima Islands. While the islands currently remain under the control of South Korea, Japan has argued its claim over the islands. South Korea asserts its claims through historical records that the islands belonged to Korea prior to Japanese occupation. Japan, on the other hand, argues that the island was *terra nullius* when it annexed them, and that the Allied powers granted the islands to Japan after World War II.[9]

8 Guifang Xue and Lei Zhang, "East Asian Sea—Conflicts, Strategies for Peaceful Resolutions and Accomplishments (Panel 1): Maritime Disputes in Northeast Asian and Escalation of the Sino-Japan Islands Dispute: Implications and Prospects," *University of Hawai'i Law Review* 35 (2013): 465–466.

9 *Ibid.* at 464–465.

An additional conflict over the ownership of islands in NEA involves Japan and Russia over the Northern Territories/Kurile Islands. Japan and Russia both claim sovereignty over the small group of islands, Etorofu/Iturup, Kunashiri/Kunashir, Shikotan, and the Habomais. The islands are currently occupied by Russia, and lie to the north of Hokkaido, the northernmost of Japan's four main islands.

Another hindrance to the effort for a collective maritime security mechanism is the uncompromising attitude of each state in exuding their full claim for exclusive economic zones (EEZ) around their maritime borders. The extension of maritime zones that resulted from the adoption of the United Nations Convention on the Law of the Sea (UNCLOS) have sparked substantial tension in the region. With UNCLOS' entry into force on November 16, 1994, some coastal states have made unilateral assertions about their maritime borders that substantially overlap with those of neighbouring states due to the confined geographical space between them. Currently, most regional states claim the full 200 nautical miles while engaging in negotiations regarding maritime boundaries with neighbouring states.[10]

Finally, China's new policy for a stronger military force, especially a stronger navy, has led to heightened tensions within the region. Through initiatives to modernize its military force, China has impressively achieved great success, as exemplified by China's production of its first aircraft carrier, *Liaoning*. Also, China is modernizing its fleet of submarines as well as taking a more aggressive approach in strengthening its defenses to deter against incursions in waters over which it claims sovereignty. China has modernized and manufactured patrol ships, larger class destroyers, and anti-access/area-denial capabilities to heavily fortify its maritime border. This effort is motivated by the goal of catching up with the United States and its naval prowess in order to balance military power in the region in its favor. While many agree that China is still behind the United States in terms of sheer military capacity, Chinese accomplishments have caught the attention of its neighbors.[11] China's military modernization program and build-up has understandably caused a sense of anxiety among neighboring states such as Japan, South Korea, and Taiwan. Consequently, these tensions necessitate a more restrained approach towards collaboration for a collective NEA maritime security system.

In the midst of these tensions, regional states need to build upon existing foundations to create a similar collective security mechanism in the European

10 *Ibid.* at 460–461.

11 Ting Wai, "China's Naval Development and the Evolving Security Dilemma in East Asia," *KAS Journal on Contemporary Korean Affairs* 1 (2013): 23–42.

mold. For instance, there are several regional maritime security regimes that already have been established to avoid conflicts in the region. Some agreements are of direct security significance, such as the 1972 Incidents at Sea (INCSEA) agreement between the United States and Russia, Russia and South Korea (1994), and Russia and Japan (1993).[12] Others are bilateral confidence-building agreements strengthen regional peace, and include the exchange of military officers among national defence universities and cooperation in maritime search and rescue missions.[13] Building upon these bilateral agreements, the next natural step is to achieve consensus through multilateral agreements on measures to secure common security and peace in the region.

While bilateral agreements have been successful in avoiding the potential for armed conflict within the region, only a few multilateral agreements have shared the same level of success. The most successful of these in multilateral maritime affairs is the North Pacific Coast Guard Forum (NPCGF), which includes the United States, China, Russia, Japan, and South Korea.[14] The relative lack of successes in establishing multilateral agreements in the region testifies to the difficulty in implementing collective security mechanisms similar to those in Europe.

The Role of the United States and a NEA Collective Security Mechanism

Since World War II, the United States has had a strong military presence in the Pacific. The U.S. Seventh Fleet, the largest numbered fleet in the United States Navy, includes a nuclear-powered aircraft carrier, close to eighty naval ships, more than one hundred aircraft, and forty thousand personnel.[15] While some have argued that the United States has tried to diminish its role in the Pacific by collaborating more with its allies in the region, the conflicts that exist between its allies has enticed the United States to recommit to a strong regional presence.

The Obama Administration announced its approach to national security policy as well as its stance on preserving global security in the 2010 *National*

12 Mark J. Valencia, "A Maritime Security Regime for Northeast Asia," *Asian Perspective* 33 (2008): 161.

13 *Ibid.* at 162.

14 *Ibid.*

15 Commander, U.S. Seventh Fleet, *United States Navy*, accessed April 8, 2015, http://www.c7f.navy.mil/forces.htm.

Security Strategy. While the *Strategy* did not emphasize a particular region, it highlighted an intention to engage with states and international institutions in a more comprehensive manner to promote a sustainable international order.[16] Then, in the span of just two years, the Obama Administration shifted its focus from Europe and the Middle East to the Asia-Pacific region. The primary reason for the transition from a region-neutral strategy to a region-focused strategy is to prioritize the options available to the United States.[17] Thus, by 2015, the language of the *National Security Strategy* changed drastically, as the Obama Administration shifted from a region-neutral approach toward highlighting the strength of its military presence in the Pacific Ocean, and referring to its approach as "Advanc[ing] Our Rebalance to Asia and the Pacific."[18] The Obama Administration made it known that the United States would not only diversify and strengthen its cooperation with allies in the region, such as Japan, South Korea, Australia, and the Philippines, but also that it would promote further cooperation among the allies themselves.[19]

Despite an effort to promote cooperation, the United States has pointed out the failure of Japan to adequately settle its legacy of colonialism. The failure to fully resolve the issues emanating from colonialism in NEA puts forward a zero-sum gain among regional allies. This situation continues to serve as a barrier to the states in the region to maximize their mutual interests. There are numerous recommendations and suggestions for the major states of NEA to draw them away from a self-centred view of the world and adopt a comprehensive and holistic approach to the salient issues facing the region, but the prospects for the implementation of such recommendations and suggestions are unclear. Many of the current issues and problems in the region require active engagement of all the relevant countries. Given the current political climate, however, it is very difficult to estimate how these will be handled. A question that emerges is what is the possibility of an outbreak of armed conflict and what level of interaction will the United States take in its regional foreign policy. In particular, the strategic reduction of the U.S. role in Europe, together

16 U.S. President Barack Obama, *National Security Strategy*, May 2010 (The White House: Washington, DC, 2010).

17 Jonathan G. Odom, "What Does a 'Pivot' or 'Rebalance' Look Like? Elements of the U.S. Strategic Turn Towards Security in the Asia-Pacific Region and Its Waters," *Asian Pacific Law & Policy Journal* 14 (2012): 4.

18 U.S. President Barack Obama, *National Security Strategy*, February 2015 (The White House: Washington, DC, 2015).

19 *Ibid.*

with the growing power of China in NEA, requires an approach to these issues that is comprehensive and holistic.

In light of such a complex situation among the states in the region, the United States plays the key role in securing collective cooperation to better secure peace in the region. Currently, three primary relationships for the United States define its existing approach to security in the region: South Korea, Japan, and Taiwan.[20]

The relationship between the United States and South Korea was forged from U.S. involvement in the Korean War. When the two Koreas signed the Armistice Agreement, the United States and South Korea developed a complex command structure that allowed for the United States to retain a strong presence on the Korean peninsula.[21] This cooperation has evolved into various joint military efforts between the two states.

The United States-Japan relationship also bolsters the United States role in NEA. Though the United States' presence and initial involvement arose as a precautionary measure and U.S. occupation of Japan after World War II, cooperation has further developed into a joint military effort to maintain peace in the region. With the Security Treaty of 1951, which later was revised into the Treaty of Mutual Cooperation and Security of 1960, the United States was able to strengthen its presence with the use of bases in exchange for security guarantees.[22] The United States considers that its alliance with Japan is the "cornerstone of US security interests in Asia."[23]

The final relationship that empowers the United States as the key figure for a collective security mechanism is its relationship with Taiwan. While its relationship with Taiwan has diminished into a more symbolic form because of China's growing influence in the global arena, the relationship nevertheless is maintained from a previous security commitment that existed between United States and Taiwan through the United States-China Mutual Defence Treaty. The United States and Taiwan have since continued their relationship, despite the United States rescinding the bilateral defense treaty when it recognized Beijing as the legal government of China.[24]

In addition to cooperation between the United States with its existing allies within the region, the new rebalance to the Pacific will see the United States

---

20 Michael J. Kelly and Sean Watts, "Rethinking the Security Architecture of North East Asia," *Victoria University of Wellington Law Review* 41 (2010): 275.

21 *Ibid.*

22 *Ibid.* at 276.

23 U.S. Dept. of State, Bureau of East Asian and Pacific Affairs, *U.S. Relations with Japan*, accessed April 30, 2016, http://www.state.gov/r/pa/ei/bgn/4142.htm.

24 Kelley and Watts, "Rethinking," at 277.

expand its cooperative efforts with new emerging powers. Recent strategic documents indicate that the United States has a great interest in expanding the number of its allies in the region, and is especially interested in the potential for closer strategic partnership with Vietnam and India. Like its partnerships with existing allies, such new partnerships will see the United State take into consideration the unique interest of its new partner states, which has successfully served the United States in building strong cooperative measures with its allies in the past. Eventually, it is hoped that such collaborative efforts would develop into the building of a regional collective security architecture in which the United States plays the key role of shaping collective security.[25]

Understanding the U.S. role in the Asia-Pacific region can bring into perspective the reason why, despite the lack of an existing collective security measure, the NEA region has retained stability despite the possibility of an actual eruption of armed conflict. A robust U.S. presence in the region has largely prevented low-level conflicts from spiraling out of control; and particularly in the context of Korea-Japan relations, the physical presence of U.S. troops in both countries serves as a safety valve for any major dispute.

From the perspective of U.S. interests, a maritime collective security mechanism in NEA would serve the purpose of combining the divided efforts of the United States with its individual allies into a single effort. This would require not just the involvement of the United States working with each individual player in the security of the region, but a common understanding and commitment among the individual states that would outweigh their disputes that deter cooperation.

The Role of Germany in Europe: Lessons for a NEA Collective Security Mechanism?

Germany has played a key role in the creation and maintenance of Europe's modern collective security system. German integration into Europe was a key ingredient for maintaining European security after World War II. For its own part, Germany is in a strategically advantageous environment since it is nestled in the middle of Europe where NATO and EU member states surround it. The conditions Germany faced during the post-World War II era of demilitarization, economic crisis, and the reduction of defense budgets can be differen-

25 Odom, "What Does a 'Pivot' or 'Rebalance' Look Like? Elements of the U.S. Strategic Turn Towards Security in the Asia-Pacific Region and Its Waters," 9–11.

tiated with the situation in NEA. These circumstances helped to contribute to the relative success of European collective security.

Another contributing factor for the development of Europe's collective security system is that there was no war between East and West Germany to achieve German unification. In NEA, however, the Korean War was an attempt at coerced Korean unification, so the simple adoption of the European model is not necessarily the answer given the stark differences in the political and military environment. Korean unification will certainly be one of the important factors to be considered when formulating a collective security mechanism in NEA. The essential issue of NEA's strategic dilemma is based on the dynamics of the divided Korean peninsula and the prospects for unification, rather than a confrontation with Japan regarding the settlement of the legacy of Japanese colonialism.

Another critical distinction between Europe and NEA is Germany's role in Europe is clearly different than that of the role of Japan in NEA. Given the experience of integration in Europe after World War II, the national policies of Germany are influenced by its position within the EU system, and Germany recognizes the importance of multi-national mechanisms. For Germany, multilateral relations are perhaps more important than bilateral relations. In NEA, however, the politics of the Cold War make any kind of meaningful integration unlikely. From that point of view, without any impetus for greater engagement among NEA states, it is unlikely that Germany can be a model for NEA in terms of resolving the salient issues affecting the region.

There is also a very distinctive point that should be noted when comparisons are made between Germany and Japan in terms of the settlement of history. The resolution of historical injustices can be a very positive force to enhance mutual understanding and trust between the states, but in NEA, China and Korea continue to struggle with Japan over historical recognition of Japanese actions in the first half of the twentieth century.

A Collective Security Mechanism in NEA and Korea

The collective security mechanism in Europe has overcome various obstacles and barriers that emerged from World War II through the leadership of important European powers. Basically, the establishment of an effective collective security mechanism is based on the political will of the states that recognize the long-term importance of such a system. The states in NEA, however, including Korea, lack the desire to adopt a European-style collective security mechanism. Korea and Japan have given the weight of their cooperation to the United

States separately, while Korea has also made efforts to constructively engage China. Though China and Japan are now competing for regional hegemony, the old Cold War paradigm no longer has the same resonance as the prospects for Korean unification may disrupt the balance of power in the region. Therefore, if any collective security mechanism is to be established, there must be a common understanding on the part of China, Japan, and Russia regarding a unified Korea.

Nevertheless, Cold War tensions continue to exist in NEA on the Korean peninsula. While the two Koreas signed an armistice agreement in 1953, tensions have continuously existed between them along both coasts—the East Sea (Sea of Japan) and the Yellow Sea. The tension has further been fuelled by direct and indirect maritime incidents between the two states. Some of these incidents include North Korean kidnapping of fishery boats, infiltration of the maritime border using submarines and special operations forces, and attacks against South Korean naval vessels and islands near the maritime border, such as the torpedo attack that sank the *Cheonan* corvette in 2010 and the shelling of the Yeonpyeong Islands that same year.[26]

On March 26, 2010, ROK Navy warship *Cheonan* was conducting a routine patrol near Baengnyeon Island of South Korea. Without warning, the *Cheonan* corvette was struck by a torpedo and sank about 1.8 km southwest of the island. The news of the *Cheonan's* sinking caused an uproar among the South Korean people who requested that the government uncover the truth behind the ship's sinking. The South Korean government commissioned a Joint Civil-Military Investigation Group of investigators from South Korea, the United States, the United Kingdom, Australia and the Kingdom of Sweden. After a thorough investigation, the group released a final report on May 20 that concluded the ship was hit by an external underwater explosion from a torpedo made by North Korea, thereby placing the responsibility solely on North Korea.[27] While South Korea was still experiencing the aftermath of the *Cheonan* sinking, North Korean artillery shelled Yeonpyeong Island on November 23, 2010, with South Korean marines stationed on the island executing an immediate counterattack. Yeonpyeong Island is located not far from the very site where the *Cheonan* sinking took place. As a result of the North Korean attack, four South Koreans died and nineteen were injured.[28]

---

26 Kyengho Son, "Maritime Security Environment in East Asian Waters," *KAS Journal on Contemporary Korean Affairs* 1 (2013): 18.

27 Seokwoo Lee and Hee Eun Lee, "Lethal Maritime Conflicts between North and South Korea," *KAS Journal on Contemporary Korean Affairs* 1 (2013): 106–107.

28 *Ibid.* at 107.

To fully understand the recent incidents between North Korea and South Korea, an appreciation of the Northern Limit Line ("NLL") is required. The NLL originated from the Inter-Korean Armistice Agreement on May 27, 1953. At the time, the United Nations Command and the army of North Korea established a Military Demarcation Line on land, but failed to extend it to establish a maritime border. United Nations Commander General Mark Clarke drew a seaward extension of the Military Demarcation Line that separated five islands of South Korea (Paengyong, Taechong, Sochong, Yeonpyeong, and Woo Islands) and North Korea. Ever since the drawing of this seaward extension, South Korea has argued that this line marks the maritime border between South Korea and North Korea despite its absence in the Armistice Agreement.[29]

Constant tension between the two states and the heavy military presence on both the Military Demarcation Line and the NLL inevitably resulted in unfortunate incidents and exchanges of gunfire. Though neither side would go as far as instigate a heightened form of armed conflict, this constant sense of military anxiety has heavily shaped the security horizon for both states. This state of affairs is caused by a dispute between North and South Korea over the validity of the NLL. While the NLL has served as the *de facto* maritime border between the two sides, North Korea does not recognize its legitimacy.

If UNCLOS were applied to the NLL, it "would probably not stand as a legitimate maritime boundary under the 'equitable principles' that have evolved from the decisions based on Articles 74 and 83 of the Law of the Sea Convention, because it denies North Korea access to adjacent sea areas."[30] Article 7(6) of UNCLOS provides that no state can employ a system of straight baselines "in such a manner as to cut off the territorial sea of another State from the high seas or an exclusive economic zone." Moreover, according to legal precedents and international practice, islands do not have an equal capacity with land masses to create maritime zones, nor do they command equal strength with an opposing continental area or land mass.[31] This conclusion has been reached

---

29 *Ibid.* at 100–105.

30 Van Dyke, Jon M. "Disputes Over Islands and Maritime Boundaries in East Asia." in *Maritime Boundary Disputes, Settlement Processes, and the Law of the Sea*. Eds. Seoung-Yong Hong and Jon M. Van Dyke (Leiden: Martinus Nijoff, 2009), 43.

31 Islands can generate maritime zones (Article 121 of the Law of the Sea Convention), but they do not generate full zones when they are opposite or adjacent to continental land areas or substantially larger islands. This conclusion has been reached consistently by the International Court of Justice and arbitral tribunals, *e.g., North Sea Continental Shelf Case (Fed. Rep. of Germany v. Denmark & Netherlands)*, 1969 I.C.J. 3, para. 101(d); *Arbitration Between the United Kingdom of Great Britain and Northern Ireland and the French Republic on the Delimitation of the Continental Shelf* (1977–78), 18 I.L.M. 397 (1979);

most dramatically in the 2009 decision between Ukraine and Romania, where the ICJ completely ignored Ukraine's Serpents' Island in the delimitation of the maritime boundary between the two countries in the Black Sea.[32]

Therefore, using the "enclave" approach introduced in the *Anglo-French Arbitration*,[33] a territorial sea would be drawn around the five South Korean islands referred to above, but the islands themselves would be ignored in drawing the main maritime boundary. In sum, all recent decisions of the ICJ and arbitral tribunals have reflected two fundamental principles of international law: (1) large land masses trump smaller islands in terms of the generation of maritime zones and (2) each state must have at least some access to adjacent coastal waters. The NLL appears inconsistent with these principles.

The conflict between North and South Korea creates a problem for the establishment of a collective maritime architecture in the region. South Korea's primary security concern is to protect itself from a possible armed attack by its old enemy. Seoul's essential security interest to protect its territory from armed conflict by North Korea hinders possibility of a common collective security measure, much less a dialogue, within the region. Without a resolution between the two Koreas, a collective maritime security measure may prove to be a substantial challenge since the conflict involves other states such as the United States, China, Russia, and Japan.

Beyond the tense relationship between the Koreas, South Korea continues to experience disputes with states in the region that hinder the establishment of a maritime collective security system. The first of these is Korea's disagreements with Japan. Having experienced centuries of conflict, the history between these two states has hindered substantial security and diplomatic cooperation. The failure to set out a straightforward historical understanding of what happened in the past has continued to spark great tension between the two neighbors, exemplified by the territorial dispute between the two states over Dokdo/Takeshima.

Case Concerning the Continental Shelf (*Tunisia v. Libya*), 1982 I.C.J. 18; *Continental Shelf* (*Libya v. Malta*), 1985 I.C.J. 13; *Case Concerning Delimitation of the Maritime Boundary in Gulf of Maine Area* (*US v. Canada*), 1984 I.C.J. 246; *Delimitation of the Maritime Boundary Between Guinea and Guinea-Bissau*, 25 I.L.M. 252 (1986); *Case Concerning Maritime Delimitation in the Area Between Greenland and Jan Mayen* (*Denmark v. Norway*), 1993 I.C.J. 38; *Delimitation of the Maritime Areas Between Canada and France* (*St. Pierre and Miquelon*), 31 I.L.M. 1149 (1992); the *Eritrea-Yemen Arbitration*, http://www.pca-cpa.org (1998–99); *Qatar-Bahrain Maritime Delimitation and Territorial Questions*, 2001 I.C.J. 40.

32 *Maritime Delimitation in the Black Sea* (*Romania v. Ukraine*), I.C.J., Feb. 3, 2009, accessed July 6, 2015, http://www.icj-cij.org/docket/files/132/14987.pdf.

33 *Arbitration Between the United Kingdom of Great Britain and Northern Ireland and the French Republic on the Delimitation of the Continental Shelf* (1977–78), 18 I.L.M. 397 (1979).

Dokdo/Takeshima is comprised of two islets in the East Sea (Sea of Japan) where South Korea claims that it has rightful ownership that dates back as far as the sixth century when the Silla dynasty conquered a smaller kingdom that governed the present-day Ulleung Island and the Dokdo islets. Japan, on the other hand, argues that the islets were *terra nullius* when it occupied them in 1905. The conflict dates back to Japan's forced annexation and occupation of the Korean peninsula and is directly related to the emotional feelings tied with the suffering of the Korean people during the period of Japan's imperial rule of Korea, which included forced military service, forced labor, and the sex slaves or "comfort women" for the Imperial Japanese Army.[34]

The conflict between South Korea and Japan, two of the more powerful military forces in the region, has hindered the formulation of collective maritime security. While continued efforts exist to seek greater cooperation between the two states, a collective military effort seems implausible to both sides because of the lack of bilateral trust.

While the relationship between South Korea and Japan has deteriorated over the past several years, South Korea and China have continued to strengthen their cooperation since they established diplomatic ties in 1992. Despite the close cooperative relations between Korea and China recently, it is still very doubtful to pin down the exact nature of Chinese policies regarding Korean unification. The Chinese strategic vagueness and its view of North Korea are also very questionable. Furthermore, China's lack of support on the *Cheonan* sinking was unhelpful. There may be a misunderstanding in the West about the recent engagement between Korea and China. Korea understands the importance of a relationship with China, but Korea prioritizes its relationship with the United States. Korea must deal with China because of its influence over North Korea, and the role China likely would play during Korean unification (especially in the event of a complete North Korean collapse). The prospect of Chinese intervention in North Korea after collapse must be factored into South Korea's position on China.

The relationship that South Korea shares with China truly tests its ability to perform diplomacy between the United States and China. South Korea's relationship with China must be understood in the context of China's historical mentality for dominance in East Asia, as well as South Korea's relationship with the United States. First, the situation must be seen in light of China's policy regarding maritime security. Some have argued that China's pursuit of military modernization is an attempt to conduct a Sino-centric reordering

34 Suk Kyoon Kim, "Understanding Maritime Disputes in Northeast Asia: Issues and Nature," *The International Journal of Marine and Coastal Law* 23 (2008): 230–231.

of the region. This perspective is supported by China's attempt to settle regional territorial issues, especially in Southeast Asia, as well as the use of its economic power to become a dominant presence in the region. For China, the ocean presents the potential threat of invasion because of its long coastline. Additionally, within China there is a belief that the United States, through its relationship with Taiwan, may continue to use its maritime dominance to contain China from expanding its influence in East Asia.[35]

Between the United States, which wants to retain its position as a dominant maritime power in NEA, and China, which wishes to recover its former glory as the Middle Kingdom, South Korea is faced with the challenge of playing a middle-power role as it balances its treaty obligations with the United States and its economic relationship with China. Though the current situation may pose a challenge that tests South Korea's ability to diplomatically resolve tensions between the two superpowers that wish to strengthen their ties with Korea, Korea has the potential to play a bridging role as the middle-power, using its economic and geo-political influence within the region to ensure a measure of regional security.[36]

Conclusion

The rise of nationalism in NEA has put pressure on improved relations in the region. However, strained relations in NEA have not come about only because of nationalism, but also due to the economic deflation and domestic political uncertainty in Japan, the rise of China's economic and military power, and the reduction of the U.S. defence budget and the pivot to Asia. In addition, North Korea plays a unique and separate role in creating tensions in the region.

The kind of collective security mechanism that is present in Europe does not exist in NEA currently because it would require closer cooperation among member states. Cooperation between NATO and Japan in particular, i.e., the role of Japan in Afghanistan, has significance because the mechanism of regional stability can be potentially exported to NEA through Japan. A lack of a common understanding of history and what was derived from World War II in NEA, however, has shown that cooperative security or crisis management cannot be at the same level in Asia as it is in Europe. Such a mechanism requires trusted

35 Sukjoon Yoon, "The Emergence and Consolidation of a Chinese Version of the Monroe Doctrine: Implications for Korea's China Policy," *KAS Journal on Contemporary Korean Affairs* 1 (2013): 43–72.

36 *Ibid.*

partners, and these are lacking in NEA. Furthermore, there is a low probability of the sincerity of any program planning and execution among NEA states. China has a grand perspective for East Asia under its tutelage, but it lacks an ability to spread and maintain that view. The troubling economic, social and environmental problems that it faces means that China may be unable to effectively manage these issues, hampering its ability to establish a legitimate consensus to spur on the creation of a collective security mechanism.

What then are the prospects for the adoption of a collective security mechanism in NEA? The region needs a diagnosis for its security situation that has an outlook of ten to fifteen years. For example, the regional states would need to carefully study the timing of the adjustment of the defence budgets for each country. The United States is expected to choose and focus rather than take a holistic and comprehensive approach. China will undoubtedly pay more attention to the Asia-Pacific region. The next twenty-five years likely will reveal radical changes in the region and the concept of a collective security mechanism inevitably will be modified. Nevertheless, the fact that there is no dialogue among NEA countries is a tremendous obstacle. It is at this point where the European experience can inspire the states in the region to seek out a dialogue while looking at the achievements within the European collective security mechanism. Then, concrete steps can begin to be taken towards a measure of a more durable and peaceful coexistence.

PART 3

Marine Scientific Research and the Marine Environment

CHAPTER 12

International Governance of Ocean Fertilization and other Marine Geoengineering Activities

*Sherry P. Broder**

Introduction

The human created Climate Crisis is here. The earth has warmed and is continuing to grow warmer.[1] The weather is changing and extreme droughts, flooding, super storms, torrential rains, heat waves and wildfires are becoming more frequent and severe. The oceans are rising and acidification is destroying the coral reefs. Sea ice in the Arctic Sea reached its lowest maximum extent ever in 2015.[2] The west coast of the Antarctic Peninsula has experienced a temperature rise of 2.5°C over the last 50 years and is thus one of the fastest warming places in the world.[3] The warmer atmosphere and warmer ocean currents

* Lecturer in Law at the William S. Richardson School of Law at the University of Hawai'i, Adjunct Research Fellow at the East-West Center, and long-time Hawai'i attorney with expertise in complex civil litigation. The author would like to thank Jordan Fletcher Davis, William S. Richardson School of Law, for her assistance in preparing this article. This article is a sequel to Sherry P. Broder and Marcus Haward, "Geoengineeering: Ocean Iron Fertilization and the Challenges for International Regulatory Action," in *Regions, Institutions, and Law of the Sea: Studies in Ocean Governance*, edited by Harry N. Scheiber and Jin-Hyun Paik (Martinus Nijhoff 2013).

1 "2014 Earth's Warmest Year on Record; December 2014 Record Warm; Global Oceans Also Record Warm For 2014," *Summary Information, State of the Climate: Global Analysis for December 2014*, U.S. NOAA National Centers for Environmental Information, published online January 2015, http://www.ncdc.noaa.gov/sotc/global/201412. ("The globally averaged temperature over land and ocean surfaces for 2014 was the highest among all years since record keeping began in 1880. The December combined global land and ocean average surface temperature was also the highest on record.")

2 "Arctic Sea Ice Reaches Lowest Maximum Extent Ever," *Arctic Sea Ice News and Analysis*, National Snow and Ice Data Center, March 19, 2015, http://nsidc.org/news/newsroom/arctic-sea-ice-maximum-reaches-lowest-extent-record.

3 Press Release, "New study shows Antarctic ice shelf is thinning from above and below," *British Antarctic Survey*, May 13, 2013, http://www.antarctica.ac.uk/press/press_releases/press_release.php?id=2989.

 | DOI 10.1163/9789004311442_014

of Antarctica contribute to the accelerating loss of ice shelves.[4] Dangerous secondary effects from the Climate Crisis, including international and national conflicts, are predicted. Some societies are already losing their homelands, water supplies or food supplies, and biodiversity, including the coral reef systems, and experiencing unprecedented heat waves and increasingly severe storms.

As global concern about climate change impacts has increased, attention has turned to increasing the ocean's natural function as a carbon sink. The Fifth Assessment Report of the Intergovernmental Panel on Climate Change (IPCC), is clear that warming of the climate system is unequivocal, driven primarily by anthropogenic greenhouse gas emissions, and many aspects of climate change would continue even if greenhouse gas emissions stopped immediately.[5] Methods that aim to deliberately alter the climate system to counter climate change, termed geoengineering, have been proposed. It is in this context of the climate crisis that the discussion on climate-geoengineering continues to grow.[6]

Thirteen open ocean experiments of ocean iron fertilization (OIF) have been initiated since 1993, and most of them have been on the high seas in areas beyond national jurisdiction.[7] Since 2007, OIF has been the subject of discussions in international forums such as the Conference of Parties to the Convention on Biological Diversity (CBD) and the Consultative Meetings of the Parties to the London Convention and Protocol on the Prevention of Marine Pollution by Dumping of Wastes and Other Matter and Protocol (LC-LP).[8] Had the parties to the LC-LP not taken action, experiments would

---

4 *Ibid.*

5 Subsidiary Body on Scientific, Technical and Technological Advice, *supra* note 28 at 3.

6 *Ibid.*

7 The open ocean experiments have included Ironex in 1995, SOIREE (Southern Ocean Iron Release Experiment) in 1999, EisenEX in 2000, SEEDS (Subantarctic Pacific Iron Experiment for Ecosystem Dynamics Study) in 2001, SOFeX (Southern Ocean Iron Experiments—North and South) in 2002, SERIES (Subarctic Ecosystem Response to Iron Enrichment Study) in 2002, SEEDS II in 2004, EIFEX (European Iron Fertilization Experiment) in 2004, CROZEX (CROZet Natural Iron Bloom and Export Experiment) in 2005, and LOHAFEX in the South Atlantic in 2009 (LOHA is Hindi for iron, FEX stands for Fertilization Experiment). Williamson, "Ocean Fertilization for Geoengineering," 477; Philip Boyd, *et al.*, *Mesoscale Iron Enrichment Experiments 1993–2005: Synthesis and Future Directions*, 315 Science 612 (2007).

8 Convention on the Prevention of Marine Pollution by Dumping of Wastes and Other Matter, London, Dec. 29, 1972, 1046 U.N.T.S. 120 (London Convention); Protocol to the Convention on the Prevention of Marine Pollution by Dumping of Wastes and Other Matter, Nov. 7, 1996, 36 I.L.M. 1 (London Protocol). The Protocol embodies a complete revision of the London Convention. Until the Protocol eventually replaces the Convention as Contracting Parties

probably have continued on the high seas without any regulation. Beyond the limits of coastal state jurisdiction, where OIF activities generally take place, the principle of freedom of the high seas and the corresponding principle of exclusive flag state jurisdiction apply.[9] Organized international action on this issue is critical and the LC-LP parties have reasoned that the jurisdiction to address ocean fertilization lies within their authority and have taken substantive steps to implement regulations.

In October 2013, the LC-LP parties adopted a resolution to amend the Protocol to provide a global regulatory mechanism for OIF and other marine geoengineering activities. This amendment includes marine geoengineering on the high seas.[10] It is significant that an international legal framework has been developed for activities on the high seas and demonstrates that the concept of "freedom of the high seas" and unregulated activities is gradually being eroded. The Protocol Amendment incorporates the important principles of international environmental law such as prevention and precaution as expressed in the 1992 Rio Declaration, and the recommendations contained in Agenda 21, chapter 17 § 30(b)(i). It requires the parties to adopt national legislation to implement and be in accordance with the amendment.

The Climate Crisis

The most recent Intergovernmental Panel on Climate Change ("IPCC Report Fifth Assessment Report ['AR5']") is clear that the earth is reacting and changing due to the excessive amounts of carbon dioxide in the atmosphere. "Our assessment finds that the atmosphere and oceans have warmed, the amount of snow and ice has diminished, sea level has risen and the concentration

---

ratify the Protocol, the two instruments exist in parallel. Eighty-seven Parties have ratified the Convention and forty-five, the Protocol. Status of the Convention and Its Protocol, June 24, 2015. International Maritime Organization, Status of Conventions, http://www.imo.org/en/About/Conventions/StatusOfConventions/Pages/Default.aspx.

9 United Nations Convention on the Law of the Sea, Dec. 10, 1982, 21 I.L.M. 1261 (entered into force Nov. 16, 1994). See UNCLOS, art. 87 on Freedom of the High Seas.

10 See "Resolution LC/LP 4(8) on the amendment to the London Protocol to regulate the placement of matter for ocean fertilization and other marine geoengineering activities," Report of the Thirty-Fifth Consultative Meeting and the Eighth Meeting of Contracting Parties, IMO Doc. LC 35/15, Annex 4 (18 October 2013). The amendment will enter into force for those Parties which have accepted it on the 60th day after two-third of the Parties that have deposited their instrument of acceptance with the International Maritime Organization, Art 21(3). (The US is not yet a Party to the Protocol). As of November 2014, no instrument of acceptance had been deposited with the IMO.

of carbon dioxide has increased to a level unprecedented in at least the last 800,000 years."[11] IPCC AR5 is clear that warming of the climate system is unequivocal and this warming is driven primarily by anthropogenic greenhouse gas emissions. Moreover, many aspects of climate change will continue even if greenhouse gas emissions are stopped immediately.

We know that we have to reduce our carbon emissions, but it is unlikely that we will achieve the goal of a cap of a 2 degree increase as set forth in the Copenhagen Accord. Those most affected immediately by climate change have rejected this 2 degree goal as inadequate. Small Island Developing states and Least Developed Countries have identified global warming of 1.5°C as warming above which there would be serious threats to their own development and, in some cases, survival.[12]

The IPCC AR5 concludes that "[c]limate change poses a moderate threat to current sustainable development and a severe threat to future sustainable development. (high confidence; medium evidence, high agreement),"[13] The IPCC predicts that at the current rate of emissions, global temperature will increase in the range of 3.7–4.8°C by 2100.[14] The first report in the *Turn Down the Heat* series of reports, prepared for the Word Bank by the Potsdam Institute for Climate Impact Research and Climate Analytics Potsdam, warns that the world is on track to warm by 4°C if the global community fails to take action on carbon emissions and that a world that is 4°C warmer could be "devastating"

---

11 Statement of Thomas Stocker, Co-Chair of IPCC Working Group I ("WGI"), "'Leaders must act', urges Ban, as new UN report warns climate change may soon be 'irreversible,'" *The United Nations and Climate Change, via UN News Centre*, Nov. 3, 2014, http://www.un.org/climatechange/blog/2014/11/leaders-must-act-urges-ban-new-un-report-warns-climate-change-may-soon-irreversible/.

12 The Alliance of Small Island States and the Least Developed Country group called for warming to be limited to 1.5°C at the 14th Conference of the Parties to the United Nations Framework Convention on Climate Change in Poznan, Poland, December 2008. ENB (2008) COP 14 Highlights: Thursday, 11 December 2008; www.iisd.ca/download/pdf/enb12394e.pdf.

13 Fatima Denton, *et al.*, "IPCC 2014: Climate-resilient pathways: adaptation, mitigation, and sustainable development," in Christopher B. Field, *et al.*, eds., *Climate Change 2014: Impacts, Adaptation, and Vulnerability. Part A: Global and Sectoral Aspects. Contribution of Working Group II to the Fifth Assessment Report of the Intergovernmental Panel on Climate Change*, Cambridge University Press, chap. 20, 1104, https://ipcc-wg2.gov/AR5/images/uploads/WGIIAR5-Chap20_FINAL.pdf.

14 Ottmar Edenhofer, *et al.*, "IPCC 2014: Summary for Policymakers," in Ramon Pichs-Madruga, *et al.*, eds., *Climate Change 2014, Mitigation of Climate Change, Contribution of Working Group III to the Fifth Assessment Report of the Intergovernmental Panel on Climate Change*, Cambridge University Press, 13–14, http://report.mitigation2014.org/spm/ipcc_wg3_ar5_summary-for-policymakers_ approved.pdf.

to human society everywhere.[15] The poor and most vulnerable will suffer the most.[16] The World Bank report warns that "a 4°C world is so different from the current one that it comes with high uncertainty and new risks that threaten our ability to anticipate and plan for future adaptation needs."[17]

Ocean Fertilization as a Geoengineering Climate Mitigation Tool?

There has been a lack of success in curbing greenhouse gas emissions, and geoengineering is seriously considered as a supplement to the implementation of conventional mitigation practices of reducing CO_2 emissions. Geoengineering is now on national and international agendas. International environmental law will have an important role in the governance of scientific research of climate engineering, as well as any large-scale deployment. As a minimum, climate engineering techniques will have the potential to cause transboundary harm, create damage in areas beyond national jurisdiction, and alter biodiversity.

Methods that aim to deliberately alter the climate system to counter climate change, termed "geoengineering," have been proposed. Limited evidence precludes a comprehensive quantitative assessment of the two most prominent methods—Solar Radiation Management (SRM) and Carbon Dioxide Removal (CDR)—and their impact on the climate system. CDR methods have biogeochemical and technological limitations to their potential on a global scale. There are two distinctly different types of geoengineering technologies that are being investigated and assessed at the present time. First, carbon dioxide removal (CDR), which is intended to reduce the level of atmospheric carbon dioxide; and, second, solar radiation management (SRM), which is theorized to reduce the incidence and absorption of incoming solar radiation so as to cool the atmosphere at ground level. Solar radiation management, however, provides no remedy to counteract further damage to the environment caused by greenhouse gas concentrations, such as ocean acidification.

Due to its colossal volume, the oceans have a great capability for CO_2 storage, although this ability to serve as a vast greenhouse gas sink has been greatly reduced by the amount of CO_2 already stored in the oceans. The oceans

---

15 Hans Joachim Schellnhuber, *et al.*, "Why A 4 Degree Celsius Warmer World Must Be Avoided," *Turn Down The Heat Series*, World Bank, Executive Summary (2014), Forward ix, http://www.worldbank.org/en/topic/climatechange/publication/turn-down-the-heat.

16 *Ibid.*

17 *Ibid.* 1–2.

comprise over 70 percent of the earth's surface with a total volume of 1.37 billion cubic kilometers.[18]

Ocean fertilization is one CDR technique being considered.[19] Several experiments have been completed involving the intentional introduction of iron to the upper ocean to stimulate phytoplankton[20] growth in iron-deficient areas.[21] Phytoplankton photosynthesize CO_2 and retain the carbon in their cells. The theory is that the resulting increase in phytoplankton blooms from iron fertilization increase the ability of the oceans' biological pump to draw down CO_2 from the atmosphere and consequently expand the food supply for marine life. The carbon contained in dead plankton and fish tissue may eventually sink and be sequestered in the ocean depths.[22] There have been a small number of experiments to add phosphorus in low nutrient waters to spur phytoplankton growth.[23] Adding nitrogen has also been proposed as ocean fertilization.[24]

In 2005, IPCC Working Group III issued a Special Report on Carbon Dioxide Capture and storage that set forth the standards for determining whether an OIF operation can be successful in disposing carbon from the atmosphere. The scientists found that carbon must be exported below the main thermocline, typically 1000 meters, to remain out of contact with the atmosphere for a

---

18 Alyn C. Duxbury, *An Introduction to The World's Oceans* (6th ed. 2000).

19 Philip Williamson, *et al.*, "Ocean Fertilization for Geoengineering: A Review of Effectiveness, Environmental Impacts and Emerging Governance," *Process Safety and Environmental Protection* 90 (2012): 475–488.

20 Phytoplankton are microscopic plants microscopic organisms that inhabit the upper sunlit layer of almost all oceans and bodies of fresh water of the sea, are responsible for roughly half of all the carbon that is fixed on Earth's surface and are the primary component of the biological pump that draws carbon dioxide out of the atmosphere into the deep ocean. Sallie W. Chisholm, Paul G. Falkowski & John J. Cullen, "Dis-Crediting Ocean Fertilization," *Science* 294 (2001): 309; Christopher. L. Sabine, *et al.*, "The Ocean Sink for Anthropogenic Co_2," *Science* 305 (2004): 367.

21 Oceanographer John Martin advanced the theory known as the iron hypothesis. Martin posited that by using a relatively small amount of iron on certain areas of the ocean, known as high-nutrient, low-chlorophyll zones (HNLCs), large blooms of unicellular aquatic plants commonly known as algae would result. If enough of these HNLC zones were fertilized with iron, it was his opinion that the growth in algae could take in enough carbon from the atmosphere that the greenhouse effect would be reversed and he went so far as to claim that the Earth would be cooled. John H. Martin & S.E. Fitzwater, "Iron Deficiency Limits Phytoplankton Growth in the North-East Pacific Subarctic," *Nature* 331 (1988): 341–343. John H. Martin, "Glacial-Interglacial CO_2 Change: The Iron Hypothesis," *Paleoceanography* 5 (1990).

22 Williamson, "Ocean Fertilization for Geoengineering," 477.

23 *Ibid.* 478.

24 *Ibid.*

prolonged period, the standard 100-year long timescale, to be considered successful.[25] It is not actually known, however, whether this depth is enough or the 100-year timescale is sufficient. Knowing whether the transport of carbon to the deep ocean at least meets these standards is one the first steps to evaluating the viability of fertilization as a carbon reduction strategy.

Prior to the work of the IPCC AR5, neither geoengineering science nor technology had been addressed in any official negotiation texts or reports by the United Nations Framework Convention on Climate Change (UNFCCC) and IPCC. In the 2013 IPCC AR5 *Summary for Policy Members*, geoengineering was specifically addressed, and serious concerns were stated about its limitations, side effects and long-term consequences.

There is insufficient knowledge to quantify how much CO_2 emissions could be partially offset by CDR on a century timescale. Modelling indicates that SRM methods, if realizable, have the potential to substantially offset a global temperature rise, but they would also modify the global water cycle, and would not reduce ocean acidification. If SRM were terminated for any reason, there is high confidence that global surface temperatures would rise very rapidly to values consistent with the greenhouse gas forcing. Furthermore, CDR and SRM methods carry side effects and long-term consequences on a global scale.[26]

In addition, in the chapter on Carbon and Other Biogeochemical Cycles, the Panel analyzed the potential effect of CDR methods,[27] and in particular addressed the impacts of enhanced carbon sequestration in the ocean.[28] The Panel concluded in its Synthesis that "[i]n contrast to SRM methods, CDR methods that manage the carbon cycle are unlikely to present an option for rapidly preventing climate change.... The level of confidence on the effects

---

25 Special Report on Carbon Dioxide Capture and Storage (SRCCS) has been prepared under the auspices of Working Group III (Mitigation of Climate Change) of the Intergovernmental Panel on Climate Change (IPCC).

26 Lisa V. Alexander, *et al.*, "IPCC: 2013, Summary for Policy Makers," in Thomas F. Stocker, *et al.*, eds., *Climate Change 2013: The Physical Science Basis, Summary for Policy Makers, Working Group I Contribution to the Fifth Assessment Report of the Intergovernmental Panel on Climate Change*, Cambridge University Press 29, http://www.ipcc.ch/pdf/assessment-report/ar5/wg1/WG1AR5_SPM_FINAL.pdf.

27 *Ibid.* Philippe Ciais, *et al.*, "IPCC: 2013, Carbon and Other Biogeochemical Cycles," in Thomas F. Stocker, *et al.*, eds., *Climate Change 2013: The Physical Science Basis, Summary for Policy Makers, Working Group I Contribution to the Fifth Assessment Report of the Intergovernmental Panel on Climate Change*, Cambridge University Press, chap. 6 546–552, http://www.ipcc.ch/pdf/assessment-report/ar5/wg1/WG1AR5_Chapter06_FINAL.pdf.

28 *Ibid.* 6.5.3.2 "Impacts of Enhanced Carbon Sequestration in the Ocean," at 551.

of both CDR and SRM methods on carbon and other biogeochemical cycles is very low."[29]

The United Nations General Assembly has encouraged states to support further research and study of iron fertilization,[30] and four U.N. bodies and associated secretariats have done so: the Intergovernmental Oceanographic Commission of UNESCO (IOC),[31] the Convention on Biological Diversity (CBD),[32] the International Maritime Organization (IMO) (as secretariat for the London Convention, 1972,[33] and its related 1996 London Protocol),[34] and the meeting of the contacting parties of the U.N. Convention on the Law of the Sea.[35] In addition, other treaties and international instruments are implicated such as the U.N. Framework Convention on Climate Change (UNFCCC),[36] the Environmental Modification Treaty,[37] the Straddling Fish

29 *Ibid.* 6.5.5 "Synthesis," at 552.

30 G.A. Res. 67/78, U.N. Doc. XXX (Dec. 11, 2012); G.A. Res. 62/215, U.N. Doc. A/62/PV.79 (Dec. 22, 2007).

31 Intergovernmental Oceanographic Commission (IOC) of the U.N. Educational Scientific & Cultural Organization (UNESCO) (hereafter cited as IOC/UNESCO).

32 Convention on Biological Diversity, June 5, 1992, Rio de Janeiro, 1760 U.N.T.S. 79, 31 I.L.M. 818 (1992), *entered into force on* Dec. 29, 1993 [hereafter cited as CBD].

33 Convention on the Prevention of Marine Pollution by Dumping of Wastes and Other Matter, Nov. 13, 1972, London, 1046 U.N.T.S. 120, 11 I.L.M. 1294 (1972), *entered into force on* Aug. 30, 1975 (hereafter cited as London Convention).

34 1996 Protocol to the London Convention 1972, Nov. 7, 1996, 36 I.L.M. 1 (1997), *entered into force on* March 24, 2006 [hereinafter London Protocol].

35 United Nations Convention on the Law of the Sea, Dec. 10, 1982, Montego Bay, Jamaica, 1833 U.N.T.S. 397, 2 I.L.M. 1261 (1982), *entered into force on* Nov. 16, 1994 [hereafter cited as Law of the Sea Convention].

36 United Nations Framework Convention on Climate Change, May 9, 1992, New York, 1771 U.N.T.S. 107, 31 I.L.M. 849 (1992), *entered into force on* March 24, 1994 [hereafter cited as UNFCCC].

37 Convention on the Prohibition of Military or Any Other Hostile Use of Environmental Modification Techniques, May 18, 1977, Geneva, 1108 U.N.T.S. 152, *entered into force on* Oct. 5, 1978 [hereafter cited as Environmental Modification Treaty]. Militaries have utilized geo-engineering to manipulate the weather. During the Cold War, the United States and the Soviet Union pursued weather control for military applications. The United States conducted weather warfare during the Vietnam War. During both the Johnson and Nixon administrations, the military seeded clouds with silver iodide flares along parts of the Ho Chi Minh Trail to induce greater rainfall, in an attempt to limit enemy movement, in "Operation POPEYE." J.R. Fleming, "The Climate Engineers," *Wilson Quarterly* 31(2) (2007): 46–60. Approximately 2,600 cloud seeding sorties were flown over the course of five years. *Ibid.*

Stocks Agreement,[38] the Antarctic Treaty,[39] the Madrid Protocol,[40] and the Convention on the Conservation of Antarctic Marine Living Resources.[41]

Governmental agencies have been exploring and analyzing geoengineering technological and governance issues in the United States[42] and the United Kingdom.[43] In 2009, the UK Royal Society published a report entitled *Geoengineering the Climate: Science, Governance and Uncertainty*,[44] which introduced the concept of geoengineering the climate for mitigation purposes. The Royal Society defined geoengineering as: "deliberate large-scale

---

38 United Nations Agreement for the Implementation of the Provisions of the United Nations Convention on the Law of the Sea of 10 December 1982 relating to the Conservation and Management of Straddling Fish Stocks and Highly Migratory Fish Stocks, New York, Aug. 4, 1995, 2167 U.N.T.S. 88, 34 I.L.M. 1542 (1995), *entered into force on* Dec. 11, 2001 [hereafter cited as Straddling Fish Stocks Agreement].

39 Antarctic Treaty, Dec. 1, 1959, Washington D.C., 402 U.N.T.S. 71, 19 I.L.M. 860 (1980), *entered into force* June 23, 1961.

40 Protocol on Environmental Protection to the Antarctic Treaty, Oct. 4, 1991, Madrid, Spain, 30 I.L.M. 1455 (1991), *entered into force on* Jan. 14, 1998 ("Madrid Protocol").

41 Convention on the Conservation of Antarctic Marine Living Resources, May 20, 1980, Canberra, 1329 U.N.T.S. 48, 19 I.L.M. 841 (1980), *entered into force on* April 7, 1982.

42 *See* Cong. Research Serv., RL 41371, *Geoengineering: Governance and Technology Policy* by Kelsi Bracmort and Richard K. Lattanzio, R41371, 2, n. 11 (2011), https://www.fas.org/sgp/crs/misc/R41371.pdf; Staff of H. Comm. on Sci. & Tech., 111th Cong., Engineering the Climate: Research Needs and Strategies for International Coordination, at ii–iii (Comm. Print 2010), http://www.gpo.gov/fdsys/pkg/CHRG-111hhrg62619/pdf/CHRG-111hhrg62619.pdf; In 2009–2010, the Science and Technology Committee of the U.S. House of Representatives conducted hearings to consider the benefits and risks of a variety of climate geo-engineering techniques, including iron fertilization. *See* U.S. House of Representatives, 111th Congress, Hearing Before the Committee on Science and Technology, *Geoengineering: Parts I, II, and III* (November 5, 2009–March 18, 2010), http://www.gpo.gov/fdsys/pkg/CHRG-111hhrg53007/pdf/CHRG-111hhrg53007.pdf.

43 See Department of Energy and Climate Change, Government Response to the House of Commons Science and Technology Committee 5th Report of Session 2009–10: The Regulation of Geoengineering, 2010, Cm. 7936 (U.K.); Science and Technology Committee, The Regulation of Geoengineering, 2009–10, H.C. 221, at 6 (U.K.); Innovation, Universities, Science and Skills Committee, Engineering: Turning Ideas into Reality: Government Response to the Committee's Fourth Report, 2008–9, H.C. 759, at 11 (U.K.); Innovation, Universities, Science and Skills Committee, Engineering: Turning Ideas into Reality, 2008–9, H.C. 50-I, P195 (U.K.).

44 The Royal Society, "Geoengineering the Climate: Science, Governance and Uncertainty" (2009), http://royalsociety.org/Geoengineering-the-climate/.

interventions in the earth's climate system to diminish climate change or its impacts."[45]

The National Academy of Science, a private, nonprofit institution chartered by the U.S. Congress in 1863 to provide science, technology, and health policy advice to the government, has recently published a new two-volume report that concludes that proposed climate intervention technologies of CDR and SRM are not ready for wide-scale deployment and should not be considered substitutes for reductions in carbon dioxide emissions and adaptation efforts to combat climate change.[46] However, it recommends that the Federal research and development investment increase to improve technologies on a larger scale.[47] The panel concludes that "if society ultimately decides to intervene in Earth's climate, the Committee most strongly recommends any such actions be informed by a far more substantive body of scientific research—encompassing climate science and economic, political, ethical, and other dimensions—than is available at present."[48] The panel proposed small-scale experiments but admitted that governance and legal standards were still to be developed.[49] "The committee felt that the need for information at this point outweighs the need for shoving this topic under the rug."[50]

The debate among scientists and policy-makers can be expected to continue. Chemical oceanographers Long Cao and Ken Caldeira have, for instance,

---

45 Geo-engineering is also referred to as climate engineering, climate remediation, and climate intervention. Geo-engineering encompasses a range of techniques to remove CO_2 from the atmosphere or to block incoming sunlight.

46 "Report in Brief, Climate Intervention Carbon Dioxide Removal and Reliable Sequestration Reflecting Sunlight to Cool Earth," National Academies of Sciences (2015), http://dels.nas.edu/resources/static-assets/materials-based-on-reports/reports-in-brief/climate-intervention-brief-final.pdf.

47 Press Release, "Climate Intervention Is Not a Replacement for Reducing Carbon Emissions; Proposed Intervention Techniques Not Ready for Wide-Scale Deployment," *News*, National Academies of Sciences (Feb. 10, 2015), http://www8.nationalacademies.org/onpinews/newsitem.aspx?RecordID=02102015.

48 NAS, "Report in Brief, Climate Intervention," 4.

49 Henry Fountain, "Panel Urges Research on Geoengineering as a Tool Against Climate Change," *New York Times* (Feb. 10, 2015), http://www.nytimes.com/2015/02/11/science/panel-urges-more-research-on-geoengineering-as-a-tool-against-climate-change.html?_r=0.

50 *Ibid.* Statement of Marcia K. McNutt, chairwoman of the panel and the editor in chief of the journal Science. Chemical oceanographers Long Cao and Ken Caldeira, members of the NAS panel, had earlier maintained that "[s]mall-scale fertilization experiments can be useful to help understand the role of nutrients in marine biogeochemical cycles."

suggested that "[s]mall-scale fertilization experiments can be useful to help understand the role of nutrients in marine biogeochemical cycles."[51]

The effectiveness of ocean fertilization in actually sequestering carbon and its safety and effect on the marine and human environments has not been established. No ocean fertilization experiments large enough to test the idea of geoengineering the climate have been conducted. Scientists in the field, however, conclude that currently the research results from experiments do not substantiate that ocean fertilization on a large-scale would provide a particularly effective approach to reduce CO_2 in the atmosphere.[52] A study on iron seeding authored by 47 ocean scientists concluded that such attempts to seed the ocean artificially were unlikely to sequester much carbon dioxide.[53]

The efficiency of large-scale ocean fertilization is unknown. Some scientists have suggested that one ton of iron added to iron deficient parts of the ocean could draw down 30,000 to 110,000 tons of carbon from the air.[54] Other studies have estimated that ocean fertilization might cost approximately $30 to $300 per ton of carbon sequestered.[55]

Iron Fertilization Experiments

None of the thirteen open ocean experiments have been longitudinal (involving repeated observations of the same variables over time), nor have they been conducted over a large area of the ocean.[56] The long-term risks of heightened iron levels, the amount of CO_2 captured and retained, and whether capture and retention would make a difference in the level of atmospheric carbon, are still unknown. Moreover, the removal of carbon dioxide from the atmosphere does just that—remove CO_2. All other greenhouse gases—such as methane, nitrous oxide, and fluorinated gases—are not reduced by the technique.

Model studies have also shown that the efficiency of ocean fertilization is poor in the Southern Ocean and that most of the carbon is re-released into

51 Long Cao & Kenneth Caldeira, "Can Ocean Iron Fertilization Mitigate Ocean Acidification? A Letter," *Climate Change* 99 (2010): 303, 310.

52 *Ibid.* 484.

53 S. Blain, *et al.*, "Effect of Natural Iron Fertilization on Carbon Sequestration in the Southern Ocean," *Nature* 446 (April 26, 2007): 1070–74.

54 Hugh Powell, "Fertilizing the Ocean with Iron: Should We Add Iron to the Sea to Help Reduce Greenhouse Gases in the Air," *Oceanus* (Nov. 13, 2007).

55 Philip W. Boyd, "Implications of large-scale iron fertilization of the oceans," *Marine Ecology Progress Series* 364 (July 29, 2008): 216–217.

56 *See*, fn. 7.

the atmosphere.[57] Earlier computer modelling simulations of iron fertilization on a global and cumulative basis showed an increased uptake of CO_2 after 100 years.[58] But the results of these models also suggest that iron fertilization may create conditions in which additional powerful greenhouse gases, such as nitrous oxide or methane, are released to the atmosphere, and thus iron fertilization may actually increase the concentration of other climate warming gases in the atmosphere.[59]

Adverse ecological effects on the marine environment are likely. Other computer simulations show that long-term storage of carbon in an artificially fertilized part of the ocean will significantly reduce the productivity of other parts of the ocean that rely on nutrients from the altered location, and will produce negative effects on fisheries in these other locations.[60] Ocean fertilization will impact and modify the highly complex structure of ocean food chains. There could be eutrophication (excessive nutrient enrichment) and toxic algae blooms. A study on an actual open ocean fertilization experiment in the subarctic Pacific showed an increase in diatom blooms producing a strong neurotoxin, which could have a detrimental effect on ocean ecosystems.[61]

---

57 Josie Robinson, *et al.*, "How deep is deep enough? Ocean iron fertilization and carbon sequestration in the Southern Ocean," *Geophysical Research Letters*, 41 (7), (April 11, 2014): 2489–2495.

58 Jorge L. Sarmiento and James C. Orr, "Three-Dimensional Simulations of the Impact of Southern Ocean Nutrient Depletion on Atmospheric CO_2 and Ocean Chemistry, 1928–1950," *Limnology and Oceanography*, 36 (8), (1991): 1928–1950; Katharina D. Kurz and Ernst Maier-Reimer, "Iron Fertilization of the Austral Ocean—The Hamburg Model Assessment," *Global Biogeochemical Cycles* 7 (1) (1993): 229–244; Richard E. Zeebe and David Archer, "*Feasibility of Ocean Fertilization and Its Impact on Future Atmospheric CO_2 Levels*," *Geophysical Research Letters* 32 (2005): 1–5; Olivier Aumont and Laurent Bopp, "Globalizing Results from Ocean in Situ Iron Fertilization Studies," *Global Biogeochemical Cycles* 20 (2006): 2; Konstantin Zahariev, James R. Christian, and Kenneth L. Denman, "Preindustrial, Historical, and Fertilization Simulations Using a Global Ocean Carbon Model with New Parameterizations of Iron Limitation, Calcification, and N_2 Fixation," *Progress In Oceanography* 77(1) (2008): 56–82.

59 Robinson, "How Deep is Deep Enough?" Andreas Oschlies, *et al.*, "Side Effects and Accounting Aspects of Hypothetical Large-Scale Southern Ocean Iron Fertilization, Biogeosciences," *Marine Ecology Progress Series* 7 (2010): 4017–4035, 4027. *See also* Xin Jin & Nicolas Gruber, "Offsetting the Radiative Benefit of Ocean Iron Fertilization by Enhancing N_2O Emissions," *Geophysical Research Letters* 230 (24) (2003): 249.

60 John J. Cullen and Phillip Boyd, "Predicting and Verifying the Intended and Unintended Consequences of Large-Scale Ocean Fertilization,". *Marine Ecology Progress Series* 364 (2008): 295.

61 The Silver study has demonstrated that iron fertilization leads to increased biomass of diatom species from the genus *Pseudo-nitzschia*. This species has been found to produce

In addition there is concern that the sinking of large quantities of dead plankton will increase acidification of the deep ocean and also produce other greenhouse gases such as methane.[62] These studies have led some scientists to conclude that "We propose that until the side-effects of widespread OIF can be shown to be verifiable—and there is good reason to believe that they cannot—OIF should not be considered a viable technology for climate mitigation."[63]

Focus on the Southern Ocean

Iron-deficient waters in the Southern Ocean around Antarctica are currently the prime places of interest for testing the usefulness of ocean fertilization as a geoengineering tool, and the area has been the site of several ocean fertilization experiments.[64] Ocean iron fertilization is predicted to artificially stimulate

---

the neurotoxin domoic acid. The Silver study surveyed 34 stations ranging from the Pacific subarctic to the Southern Ocean, some of which were experimental iron fertilization sites, collecting and analyzing near-surface-water samples and sediment samples, including eleven *Pseudo-nitzschia* species from these samples, and measuring domoic acid concentrations in the cells and water of each sample. Although it has not yet been shown that these elevated toxin levels impact higher trophic levels in oceanic ecosystems, as seen in coastal regions, it has been proven that domoic acid has reached levels that pose a threat to oceanic ecosystems. The data from this study also suggest that domoic-acid neurotoxins are delivered to deeper depths in the intact cells of these *Pseudo-nitzschia* species and at higher rates with iron fertilization. M. Silver, *et al.*, "Toxic Diatoms and Domoic Acid in Natural and Iron Enriched Waters of the Oceanic Pacific," *Proc. Nat'l Acad. Sci.*, 107(48) (Nov. 30, 2010): 20762–20767.

62 Another emerging concern of ocean scientists is the potential for acidification of parts of the world's oceans due to increased absorption of carbon into the sea. Scientists have concluded that "[i]f the amount of net carbon storage in the deep ocean by iron fertilization produces an equivalent amount of emission credits, ocean iron fertilization further acidifies the deep ocean without conferring any chemical benefit to the surface ocean." Long Cao & Kenneth Caldeira, "Can Ocean Iron Fertilization Mitigate Ocean Acidification? A Letter," *Climate Change* 99 (2010): 303, 310. *See generally* David Freestone & Rosemary Rayfuse," Ocean Fertilization and International Law," *Marine Ecology Progress Series* 364 (2008) 227–233. *See also* L. Cao & K. Caldeira, "*Ocean Fertilization and Ocean Acidification*," Abstract, American Geophysical Union, Fall Meeting (2008), http://adsabs.harvard.edu/abs/2008AGUFM.B22C.06. Richard S. Lampitt, *et al.*, "Ocean fertilization: a potential means of Geo-Engineering?" *Philosophical Transactions A*, The Royal Society Publishing, (Nov. 18, 2008): 3919–3945, http://rsta.royalsocietypublishing.org/content/366/1882/3919.

63 John J. Cullen and Philip W. Boyd, "Predicting and Verifying the Intended and Unintended Consequences of Large-Scale Ocean Fertilization," *Marine Ecology Progress Series* 364 (July 29, 2008): 295–301.

64 Williamson, "Ocean Fertilization for Geoengineering," 483.

the biological pump by exploiting the surplus macronutrients in high-nutrient low-chlorophyll (HNLC) regions where biological activity is restricted by the availability of the micronutrient iron. The Southern Ocean contains the largest expanse of HNLC waters, and is specifically iron deficient.[65] Scientists have investigated the hypothesis or carbon sinking 1000 meters and found that natural ocean circulation patterns in the Southern Ocean work in such a way that most of the carbon probably will not stay submerged long enough for a significant effect on the climate.[66] "More than 66 percent of sequestered carbon returns into contact with the atmosphere within 100 years, with a mean time scale of 37.8 years.… Carbon exported to 2000 m experiences lower leakage to the atmosphere (29%), suggesting that more stringent depth criteria may facilitate more accurate carbon credits systems."[67] The pristine ecosystems of Antarctica and its ocean have resulted in special conservation treaty protections for the Southern Ocean south of 60°, in the Antarctic Treaty, the Madrid Protocol, and the Convention on the Conservation of Antarctic Marine Living Resources, and these treaties will also have to be complied with in the event of an interest in ocean fertilization experiments in areas governed by them.[68]

2007 Controversial Russ George Failed Plans

In 2007, Russ George, a California businessman and CEO of a U.S. company named Planktos, had plans to dump iron in the ocean in different places without obtaining any permits or following any protocols. His 2007 project helped to increase interest in regulation of ocean fertilization. Several international organizations and agencies expressed great concern and objections to the Planktos plan.

The year before, in 2006, the parties to LP had taken steps to regulate CO_2 sequestration in the sub-seabed.[69] At the following LP meeting in 2007,

65 Robinson, "How Deep is Deep Enough?"

66 *Ibid.*

67 *Ibid.*

68 *See*, fn. 39–41.

69 At the 2006 meeting, Annex 1 to the L/P was amended in order to include the storage of CO_2 in sub-seabed geological formations in the list of permitted dumping-related activities. *See*, "Resolution Relating to Risk Assessment and Management Framework on carbon sequestration," 1st Meeting of Contracting Parties to the Protocol, (Nov. 3, 2006), IMO Doc. LP.1(1). This amendment entered into force on February 10, 2007 for all contracting parties to the Protocol, as no objection was presented to the IMO by the deadline provided by Art. 22(4) of the Protocol. *See also*, discussion on the actions of the IMO

CO_2 Specific Guidelines for the assessment, conditions and monitoring required for a permit were developed and adopted.[70] It seemed appropriate for the IMO to take the lead in developing international regulations, given its experience with an assessment framework and application of the precautionary principle incorporated in the Protocol.[71] In 2012, the Parties adopted revised *Specific Guidelines for Assessment of Carbon Dioxide Streams for Disposal into Sub-seabed Geological Formations*, to provide for possible transboundary migration of CO_2 waste streams within sub-seabed geological formations after injection, and to implement the 2009 amendment to article 6 of the LP.[72] In 2013, the Parties adopted guidance on the implementation of Article 6.2 on the export before injection of carbon dioxide streams for disposal in sub-seabed geological formations for the purpose of sequestration.[73]

Russ George attempted to dump 45 tons of iron near the Galapagos and Canary Islands.[74] George proposed to sell carbon offsets from this 2007

---

to regulate incineration. Bettina Boschen, "The Regulation of Ocean Fertilization and Marine Geoengineering under the London Protocol," *Climate Change Impacts on Ocean and Coastal Law: U.S. and International Perspectives*, Randall S. Abate, ed. (New York: OUP, 2014).

70 The Amendments to Annex 1 were supplemented by the "Specific Guidelines for the Assessment of Carbon Dioxide for Disposal into Sub-seabed Geological Formations," IMO Doc. LC 29/17 (2007). At the 2009 Meeting, the parties voted to propose a LP amendment sponsored by Norway to enable the export of CO_2 for sequestration purposes, which otherwise would have been prohibited under Article 6 of the Protocol. *See*, "Resolution On the Amendment to Article 6 of the London Protocol," 4th Meeting of the Contracting Parties to the London Protocol, LP.3 (4) (Oct. 30, 2009). The proposed amendment to article 6 will enter into force for Parties that have accepted it on the 60th day after two-thirds of all Contracting Parties have adopted it. It appears unlikely that a sufficient number of contracting parties will ratify this amendment in the near future.

71 The precautionary principle is applied the frameworks for both CCS and ocean fertilization: a strong version of the precautionary principle for CCS with the burden of proof falling on the operator, and a much stronger version for ocean fertilization with explicit prohibition except for research. Both technologies demonstrate the need for good scientific and technical evidence-base to support assessments, decision-making and regulatory progress on new climate change mitigation technologies.

72 "Specific Guidelines for the Assessment of Carbon Dioxide for Disposal into Sub-seabed Geological Formations," IMO Doc. LP.7. LC 34/15, Annex 8 (Nov. 2, 2012) [aka Revised CO2 Specific Guidelines or Revised CO2 Sequestration Guidelines].

73 "Guidance on the Implementation of Article 6.2 on the Export of CO2 Streams for Disposal in Sub-seabed Geological Formations for the Purpose of Sequestration," IMO Doc. LC 35/15, Annex 6 (2013).

74 *Ibid.*

commercial mission based on alleged results from prior small-scale research conducted off the coast of Hawaii.[75] The Spanish and Ecuadorian governments both halted his plans by barring his ships from their ports. Legal objections were raised by the U.S. Environmental Protection Agency and others, and the United States made an official submission to the Scientific Groups of the London Convention and Protocol expressing concerns about Planktos's proposed actions using a U.S.-flagged vessel.[76] To no one's surprise, Planktos had earlier tried to avoid regulation and accountability by stating that it would remove itself as a U.S.-flagged vessel and leave from a non-U.S. port.[77]

IUCN and Greenpeace also notified the London Convention and Protocol parties of the Planktos plans and the need to develop regulations on marine geoengineering.[78] Planktos closed down the project in 2008, claiming that "[a] highly effective disinformation campaign waged by anti-offset crusaders" made it difficult to raise the required funds.[79]

Commencement of an International Governance Regime

OIF was first discussed within an international intergovernmental framework at the 2007 meeting of the LC/LP Scientific Groups when Greenpeace, IUCN, and the United States made submissions to consider the matter particularly in response to imminent plans to conduct an OIF activity by the U.S.-based for-profit company, Planktos Inc.[80] Immediately there was a statement of concern issued by the LC/LP Scientific Working Groups emphasizing the lack of knowledge about the environmental effects of ocean fertilization on the marine environment and human health and asserting the jurisdiction of London

75 Quiren Schiermeier, "Climate Change: The Oresmen," *Nature* 421 (Jan. 9, 2003): 109–110.

76 United States, "Planktos, Inc., Large-Scale Ocean Iron Addition Projects," United States, IMO Doc. LC/SG 30/INF.28 (1 June 2007).

77 *Ibid.*

78 World Conservation Union (IUCN), "Regulation of CO_2 sequestration," IMO Doc. LC/SG 30/12 (May 8, 2007); Greenpeace International, "Challenging 'geo-engineering solutions' to climate change: The urgent need for detailed scientific scrutiny and international regulations to protect the oceans from large-scale iron fertilization programmes," IMO Doc. LC/SG 30/12/1 (8 May 2007).

79 Rachel Courtland, "Planktos dead in the water," *Nature* (Feb. 15, 2008), http://www.nature.com/news/2008/080215/full/news.2008.604.html.

80 *See*, "Report of the 30th Meeting of the Scientific Group of the London Convention and the First Meeting of the Scientific Group of the London Protocol," IMO Doc. LC/SG 30/14 (July 25, 2007).

Convention and Protocol by stating that any such projects be "evaluated carefully to ensure they were not contrary to the aims of the London Convention and Protocol."[81]

The statement by the Scientific Groups was adopted by the subsequent Consultative Meetings of Contracting Parties to the LC/LP, and expanded to specifically assert that the jurisdiction of the LC/LP includes ocean iron fertilization relying on the general objective to "protect and preserve the environment from all sources of pollution" contained in article I of the Convention and article 2 of the Protocol. At the same time, however, the Parties pointed out that was up to each state to assess ocean iron fertilization proposals on a case-by-case basis.[82]

The Ninth Conference of Parties to the CBD in 2008 also reacted to the risks posed by Russ George's plans. CBD Decision IX/16 is often described as a moratorium on ocean fertilization on ocean fertilization activities beyond national jurisdiction, including those involving legitimate scientific research, because it states that the Parties and governments should act "to ensure that ocean fertilization activities do not take place until there is an adequate scientific basis on which to justify such activities."[83] By its terms, however, CBD does not have authority for binding decisions. The decision is clear that any activities must be in accord with the "precautionary approach." A limited exception is provided for "small-scale" scientific research studies undertaken within "coastal waters" but only if those experiments do not involve generating or selling carbon offsets and are subject to strict control and a prior environmental impact assessment. The decision also provides that there should a global, transparent, and effective regulatory mechanism for any such activities.[84] Finally, the CBD recognized in the decision that the scientific and legal analysis of parties to the 1972 London Convention and its 1996 Protocol relating to iron fertilization are ongoing.[85] Other international bodies responded to the CBD, and, in particular, the UNESCO Intergovernmental Oceanographic Commission (IOC)

---

81 "Statement of Concern," Report of the 30th Meeting of the Scientific Group of the London Convention and the First Meeting of the Scientific Group of the London Protocol, IMO Doc. LC/SG 30/14 (25 July 2007): ¶¶ 2.23–2.25.

82 Report of the 29th Consultative Meeting of the Contracting Parties of the London Convention and the Second Meeting of Contracting Parties of the Protocol, IMO Doc. LC 29/17, ¶ 4.23.

83 "Decision IX/16, Biodiversity and Climate Change §C. Ocean Fertilization," Conference of the Parties to the CBD at Its Ninth Meeting, UNEP/CBD/COP/IX/16 (Oct. 9, 2008).

84 *Ibid.*

85 *Ibid.* Article 22 of the CBD states that other international treaties already in place at ratification, such as the Law of the Sea Convention, should be respected.

criticized the CBD parties for "plac[ing] unnecessary and undue restriction on legitimate scientific activities."[86]

The IOC scientists recognized the London Convention (but not the London Protocol) as the governing body of law concerning iron fertilization, and generally disagreed that the *size* of the iron fertilization project should be relevant in deciding whether a project constituted a scientific experiment.[87] "An ocean fertilization activity might be damaging even if conducted over one square kilometer (for example, over a coral reef) just as another ocean fertilization activity might be benign even though conducted over many thousands of square kilometers."[88] The parameters of scientific experiments should be decided by scientists.[89] The IOC scientists also criticized the concept of making a distinction between coastal waters and the open ocean.[90]

Further, the IOC Statement encourages iron fertilization experiments as "important tools" for learning about the ocean environment,[91] criticizes the "bureaucratic burden" the London Protocol parties have placed on iron fertilization research,[92] and denounces the CBD parties for "plac[ing] unnecessary and undue restriction on legitimate scientific activities."[93] The IOC scientists argue that there will not be a true assessment of risks associated with iron fertilization until scientists can conduct valid experiments.[94]

After the 2008 CBD meeting, the parties to the London Convention and Protocol adopted Resolution LC/LP.1 on the Regulation of Ocean Fertilization, which assets the jurisdiction of these treaties by specifically stating that ocean fertilization is within the "scope" of the London Convention and Protocol,

---

86 "Report on the IMO London Convention Scientific Group Meeting on Ocean Fertilization," Ad Hoc Consultative Group on Ocean Fertilization, Intergovernmental Ocean Commission, (IOC) of UNESCO Doc. IOC/INF-1247 (June 15, 2008).

87 "There is no well-established meaning to 'large scale' that would allow it to usefully distinguish between activities that would and activities that would not damage the ocean environment…. [i]t is impossible to assess the impacts of experiments through information on spatial scale alone." *Ibid.* ¶ IV.1 (1) and 2(A)–(C).

88 *Ibid.* ¶ IV.1 (iv).

89 *Ibid.* ¶ IV.1 (v).

90 *Ibid.* ¶ IV.3 (2).

91 *Ibid.* ¶ IV.1 (5). ("We should promote better scientific understanding of the ocean. Manipulative experiments, including ocean fertilization, are important tools that scientists use to develop a better understanding of the marine environment. Such scientific research should be promoted with a minimum of additional bureaucratic burden.")

92 *Ibid.* ¶ IV (5).

93 *Ibid.* ¶ IV (3).

94 *Ibid.* ¶ IV.1 (5).

"*any* activity undertaken by humans with the principal intention of stimulating primary productivity in the oceans," and any activities outside an LC/LP assessment framework are contrary to the London Convention and Protocol.[95] The Resolution confirms that, in light of the insufficient knowledge on its effectiveness and potential environmental impacts, no ocean fertilization activities other than legitimate scientific research should be allowed, using an assessment framework to be developed by the LC/LP Scientific Groups[96] and "utmost caution" should be exercised.[97] The Resolution clarifies that iron fertilization is the "placement of matter for a purpose other than mere disposal." Large-scale or commercial iron fertilization projects are defined as contrary to the Convention and Protocol and are not exempt from their definitions of dumping.[98] In seeming response to UNESCO IOC scientists, the size of the "legitimate scientific experiment" is not quantified.[99] The LC/LP Scientific Groups were requested to draft an "assessment framework" for submission at a later meeting.[100]

Nonetheless, despite the strong language, the 2008 LC/LP.1 is a non-binding resolution. But its language is quite strong in allowing for scientific research while restricting other OIF activities, and committing to the basic principles for regulation of ocean fertilization for research studies through a specific assessment framework. Some Parties, led by Australia again, advocated for more aggressive regulation in the form of an amendment to the London Protocol to provide a "global, transparent and effective regulatory and control

95 "Resolution LC/LP.1 on the Regulation of Ocean Fertilization," 30th Meeting of the Contracting Parties to the London Convention and Third Meeting of the Contracting Parties to the London Protocol, IMO Doc. LC 30/16, Annex 6 (Oct. 31, 2008). See ¶ 1 (The parties "[agree] that the scope of the London Convention and Protocol includes ocean fertilization activities"). "Report of the 31st Meeting of the Scientific Group of the London Convention and the 2nd Meeting of the Scientific Group of the London Protocol," LC/SG 31/16, annex 2 (May 19–23, 2008). 30th Meeting of the Contracting Parties to the London Convention and the Third Meeting of the Contracting Parties to the London Protocol, *Ocean Fertilization: Report of the Working Group on Ocean Fertilization*, LC 30/WP.6 (Oct. 27–31, 2008). The Parties debated a draft resolution prepared by the Ocean Fertilization Working Group.

96 *Ibid.* fn. 3, clarifying that the resolution excludes the processes of conventional aquaculture, or mariculture, or the creation of artificial reefs.

97 *Ibid.* ¶ 6.

98 *Ibid.* ¶ 3.

99 IMO "Resolution LC/LP.1."

100 *Ibid.* The development of an Assessment Framework for adoption is included and anticipated to be worked on until 2012. IMO, Joint Long-Term Programme [JLTP] for the London Convention and Protocol (2010–2012), LC-LP.1/Circ.32, Annex, ¶ 3.2.2 (Jan. 12, 2010).

mechanism" in line with CBD Decision IX/16.[101] The Parties did agree to continue to discuss a potentially legally binding measure to be adopted under the LC/LP framework.[102]

Controversial Joint German-Indian LOHAFEX Experiment

In 2009, a joint OIF scientific experiment project, the LOHAFEX, was carried out by National Institute of Oceanography (NIO) of the Council of Scientific and Industrial Research (India) and the Alfred Wegener Institute for Polar and Marine Research (Germany). There was a lot of controversy about this experiment.[103] The Polarstern, a German-flagged vessels travelled from South Africa (Cape Town) to Chile (Punta Arenas) and conducted OIF tests in the Southern ocean.[104]

The experiment initially was authorized by the German Federal Research Ministry (BMBF) without requiring a specific environmental impact assessment. However, later the Federal Environment Ministry did not approve the authorization, finding that the experiment violated the 2008 CDB COP Decision IX/16 and made a request to BMBF to cancel the project. There was concern that the experiment would violate the CDB and LC/LP 2008 decisions. The experiment was relatively large-scale compared to previous OIF experiments.[105] Due to public pressure, efforts by non-governmental organizations, and disputes between government departments, the expedition was initially suspended by BMBF, but eventually allowed after all.

---

101 "Report of the 30th Consultative Meeting, Third Meeting of Contracting Parties," IMO Doc. LC 30/16 (Dec. 9, 2008): ¶ 4.

102 *Ibid.* ¶ 4.14.

103 For background on the research collaboration between the Alfred-Wegener-Institute of Germany and India's National Institute of Oceanography (NIO), *see* http://www.awi.de/de/aktuelles_und_presse/selected_news/2009/lohafex/experiment/.

104 *See* details at http://www.awi.de/en/infrastructure/ships/polarstern/weekly_reports/all_expeditions/ant_xxv/ant_xxv3/. (LOHA is the Hindi for "iron" joint with Fertilization Experiment).

105 Harald Ginzky, "Ocean Fertilization as Climate Change Mitigation Measure—Consideration under International Law," *Journal of European Environmental & Planning Law* 7 (2010): 57, 71–73. *See also*, Randall S. Abate & Andrew B. Greenlee, "Sowing Seeds Uncertain: Ocean Iron Fertilization, Climate Change, and the International Environmental Law Framework," *Pace Environmental Law Review* 27 (2012): 555, http://digitalcommons.pace.edu/pelr/vol27/iss2/550.

Prof. Alexander Proelss, who advised the LOHAFEX project on its legality under international law, concluded that as "CDB COP decisions are, as a basic principle, legally non-binding [...the] Federal Republic of Germany is therefore not obliged under Decision IX/16 to prohibit the performance of the LOHAFEX marine research experiment." He found that the experiment was included under the exception of small scale experiments within coastal waters, and could not thus constitute dumping under the LC-LP regime. Finally, he concluded that "marine research experiments on the high seas which are permissible under international law and are conducted by German flag vessels are not subject to [national] authorisation."[106]

The LOHAFEX experiment and rogue commercial ocean fertilization activities of Russ George acted in part as catalysts for international organizations to continue and push forward with their work on the management and regulation of such activities.

The Development of Non-Binding International Governance

The Development of an Assessment Framework for Iron Fertilization and the Adoption of LC/LP.2.

Work commenced on an Assessment Framework for the London Convention and Protocol during intersessional scientific meetings in 2009.[107] At the 2009 LC/LP Meeting of the Parties, there was extensive discussion on how to make an assessment framework legally binding. Eight options were debated, including amendments to the Protocol.[108] Australia and New Zealand, joined by other Pacific Island countries, argued for the establishment of a clear legal

106 Prof. Dr. Alexander Proelss, "Legal Opinion on the Legality of the LOHAFEX Marine Research Experiment under International Law," (Kiel, 22 January 2009), Alfred-Wegener-Institut of Germany and India's National Institute of Oceanography (NIO), see http://www.awi.de/de/aktuelles_und_presse/ selected_news/2009/lohafex/experiment/.

107 "Report of the 32nd Meeting of the Scientific Group of the London Convention and the Third Meeting of the Scientific Group of the London Protocol," IMO Doc. LC/SG 32/15, ¶ 2 & annex 2 (May 25–29, 2009): "Report of the 31st Meeting of the Contracting Parties to the London Convention and 4th Meeting of the Contracting Parties to the London Protocol," LC 31/15, ¶ 5.18 (Oct. 26–30, 2009).

108 The options included a statement of concern, a simple resolution, a resolution building upon resolution LC-LP.1 (2008), an interpretative resolution, an amendment to Annex 1 of the London Protocol, an amendment to the definition of dumping and exclusions for dumping, and a new, stand-alone article to the London Protocol. IMO Doc. LC 31/15, ¶¶ 4.15.

regime, because ocean fertilization experiments were being carried out and the Southern Ocean was a focus.[109] No decision was made at that time.

In October 2010, the CBD parties adopted Decision X/33[110] to expand its limitations on even scientific experiments to all forms of geo-engineering, not just iron fertilization.[111] Decision X/33 applies the precautionary approach, limits scientific research to "small scale" studies in a "controlled setting," and requires a "thorough prior assessment of the potential impacts on the environment."[112] It appears that by "a controlled setting," the CBD was referring again to research studies and experiments in coastal waters that would be subject to national jurisdiction.

In 2010, IOC/UNESCO published its report by scientists on ocean fertilization, which concluded that it is unclear whether findings from small scale experiments can even be extrapolated to apply to larger applications. This report also underscored the extreme difficulty of assessing the long-term effectiveness of any carbon capture or the unintended consequences.[113]

---

109 "Ocean Fertilization: Regulating Ocean Fertilization Experiments Under the London Protocol and Convention," 31st Meeting of the Contracting Parties to the London Convention and 4th Meeting of the Contracting Parties to the London Protocol, submitted by Austl. & N.Z., IMO Doc. LC 31/4/1 (Oct. 26–30, 2009); "Report of the 31st Meeting of the Contracting Parties to the London Convention and 4th Meeting of the Contracting Parties to the London Protocol," IMO Doc. LC 31/15, ¶ 5.18.

110 "Decision X/33, Biodiversity and Climate Change 3," Convention on Biological Diversity, Tenth Conference of the Parties, (Oct. 18–29, 2010). At this meeting, Tuvalu, whose very survival is immediately threatened by climate change, on behalf of the Pacific Island parties, supported a moratorium pertaining to ocean fertilization and other forms of geo-engineering activities to counteract the effects of climate change. Soseala Tinilau and Solomona Lotoala, from Tuvalu's Department of Environment, Ministry of Foreign Affairs, Environment and Labour, spoke on behalf of the Pacific Island parties: "Unfortunately for our Pacific Island countries both climate change and biodiversity are integral recognized as parts of our daily lives. One represents an opportunity and supports our daily lives the other is imposed on our daily lives through global processes beyond our control and threatens our survival." "Pacific Supports Ocean Moratorium on Ocean Fertilization at Nagoya Biodiversity Conference," *Island Business International* (Oct. 22, 2010), http://www.islandsbusiness.com/news/index_dynamic/containerNameToReplace=MiddleMiddle/focusModuleID=130/focusContentID=21187/tableName=mediaRelease/overideSkinName=newsArticle-full.tpl?PHPSESSID=66118ca2e179e9e12e02e3ef583752cc.

111 "CBD Decision X/33."

112 *Ibid.* ¶ 8 (w).

113 Doug Wallace, *et al.*, "Ocean Fertilization, A Scientific Summary for Policy Makers," UNESCO-IOC Doc., IOC/BRO/2010/2 (2010): 1, 3. "[E]ven using the highest estimates for both carbon export ratios and atmospheric uptake efficiencies, *the overall potential for*

The LC/LP Ocean Fertilization Correspondence Group could not reach a consensus on whether the CDB report adequately summarized the current state of scientific knowledge on iron fertilization. The Correspondence Group identified two gaps in the CBD report: (1) the report focused only on the potential impacts iron fertilization may cause to marine diversity and did not provide a comprehensive summary of the current state of knowledge on iron fertilization, and (2) the report did not provide sufficient guidance for determining what level of impact to marine diversity is acceptable.[114]

The LP Intersessional Legal and Related Issues Working Group on Ocean Fertilization, met in February 2009 to start the process of analyzing the different options for making a potential legally binding resolution or an amendment to the London Protocol on ocean fertilization.[115]

The LC/LP Scientific Groups worked further on a draft Assessment Framework for Scientific Research Involving Ocean Fertilization, which was presented to the Parties at the November 2010 meeting. Some delegations sought to completely prohibit ocean fertilization, including research experiments.[116] Others wanted a liability and compensation regime for ocean fertilization including experiments. Despite these concerns, the parties proceeded and narrowed the number of options to explore further to three different types of Resolutions. In addition, Canada proposed (1) a Protocol amendment to create a permitting authority that could be adapted to other marine geoengineering activities to be accompanied and (2) an interpretive resolution for the interim until the amendment could enter into force and the Convention could also be amended.[117]

In 2010, while recognizing the lack of a liability and compensation regime and continuing concerns about the lack of a binding mandate, the parties to the

---

ocean fertilization to remove CO2 from the atmosphere is relatively small. Thus recent calculations of cumulative sequestration for massive fertilization effort over 100 years are in the range 25–75 Gt (gigatonnes) of carbon [], in comparison to cumulative emissions of around 1,500 Gt carbon from fossil fuel burning for the same period under business-as-usual scenarios." (Emphasis added), *Ibid.* 12.

114 "Ocean Fertilization: Development of Science Overviews on Ocean Fertilization, Report of the Ocean Fertilization Correspondence Group," 33rd Meeting of the Scientific Group of the London Convention and 4th Meeting of the Scientific Group of the London Protocol, IMO Doc. LC/SG 33/2/1 (April 19–23, 2009).

115 "Report of the 1st Meeting of the LP Intersessional Legal and Related Issues Working Group on Ocean Fertilization," London Protocol Intersessional Legal and Related Issues Working Group on Ocean Fertilization, IMO Doc. LP/CO2 2/5 (20 February 2009).

116 IMO Doc. LC/LP 32/15 ¶ 4.4.

117 *Ibid.* ¶¶ 4.2–4.9.

London Convention and Protocol took the next step and adopted Resolution LC/LP.2 on the Assessment Framework for Scientific Research Involving Ocean Fertilization, which was referred to as the next step in the original Resolution LC/LP.1.[118] Parties are to follow the Assessment Framework to determine whether the proposed activity constitutes legitimate scientific research that is not contrary to the aims of the Convention and Protocol and is not a commercial activity. It lists detailed steps for an environmental assessment including risk management and monitoring. It also requires the establishment of a baseline as a condition precedent for issuance of a permit. Furthermore, it provides for the sharing of research results. The Assessment Framework prohibits economic gain from qualifying experiments, which includes carbon credits.[119] Finally, it provides guidance for assessing iron fertilization proposals for scientific experiments and imposes the precautionary approach.[120] Under this framework, researchers apply to the appropriate regulatory agency of their home state for approval to conduct scientific research.

In Resolution LC/LP.2, the parties commit to a system of "global, transparent, and effective control and regulatory mechanism for ocean fertilization activities," environmental assessment, including risk management and monitoring, sharing of research results, The parties also expanded the scope to "other activities that fall within the scope of the London Convention and London Protocol and have the potential to cause harm to the marine environment."[121] The Assessment Framework is to be considered a "living document" and thus subject to future amendment in light of new and relevant scientific information

118 "Resolution LC/LP.2 on the Assessment Framework for Scientific Research Involving Ocean Fertilization," Report of the 32nd Meeting of the Contracting Parties to the London Convention and 5th Meeting of the Contracting Parties to the London Protocol, IMO Doc. LC 32/15, Annex 5 (2010). *See also* "Assessment Framework for Scientific Research Involving Ocean Fertilization," Report of the 32nd Meeting of the Contracting Parties to the London Convention and 5th Meeting of the Contracting Parties to the London Protocol, IMO Doc. LC 32/15, Annex 6 (2010). It should be noted that the assessment framework has been modelled after the 'Risk Assessment and Management Framework for CO2 Sequestration in Sub-seabed geological Structures' (CS-SSGS), adopted in 2006. *See*, Report of the 1st Meeting of the LP Intersessional Legal and Related Issues Working Group on Ocean Fertilization, London Protocol Intersessional Legal and Related Issues Working Group on Ocean Fertilization, IMO Doc. LP/CO2 2/5 (20 February 2009): ¶ 2.2.

119 "Assessment Framework," IMO Doc. LC 32/15, Annex 6.

120 *See, Ibid.* ¶¶ 1.3.2.6, 4.3 (explaining that risk management procedures are precautionary and that the decision to reject a proposal should take a precautionary approach into account).

121 "Resolution LC/LP.2 on the Assessment Framework," IMO Doc. LC 32/15, Annex 5.

and experience.[122] Objections were raised about the proposed requirement of obtaining "prior informed consent" from parties in the Region of Potential Impact, and as a result the final version merely required consultation, notification and reporting.[123]

In anticipation of another regulatory gap, some Parties to the LC/LP sought to broaden the scope of the regulatory framework to include other forms of geoengineering and Resolution 2 extends the commitment of Contracting Parties to include "other activities that fall within the scope of the London Convention and Protocol and have the potential to cause harm to the marine environment."[124] Similarly, CBD Decision X/33 reaffirms Decision IX/16 C and extends its purview to geoengineering in general.[125] Intersessional groups were assigned to develop the science and technical evaluation and permitting process and to solve the thorny issue of how to make sure the LC/LP permitting process would be legally binding and enforceable.

Russ George Again—2012 Commercial Rogue Experiment

Totally ignoring the adoption of an assessment framework by the LC/LP, in July 2012, Russ George was finally able to accomplish his goal of fertilizing the ocean. He used a fishing charter to dump 100 tons of iron-sulfate. A few weeks later, another 20 tons was dumped.[126] The dump was in the migration routes of pink and sockeye salmon in the Pacific Ocean and an eddy approximately 200 nautical miles west of the islands of Haida Gwaii, a remote island archipelago located 70 kilometers off the northwest coast of British Columbia, Canada, and just south of Alaska.[127] This commercialized dump is estimated to be 100 times larger than any previous iron fertilization experiment.[128]

---

122 *Ibid.*

123 *Ibid.*

124 *Ibid.*

125 Conference of the Parties to the CBD at Its Tenth Meeting, Nagoya, Japan, Oct. 18–29, 2010, X/33.

Biodiversity and Climate Change, 8(w), UNEP/CBD/COP/DEC/X/33 (Oct. 29, 2010).

126 Bruce Falconer, "Can Anyone Stop the Man Who Will Try Just About Anything to Put an End to Climate Change," *Pacific Standard*, May 5, 2014, http://www.psmag.com/navigation/nature-and-technology/battlefield-earth-can-anyone-stop-man-will-try-just-anything-fix-climate-78957/.

127 *Ibid.*

128 Greg Rasmussen and Marissa Harvey, "Iron Fertilization Project Stirs West Coast Controversy," *CBCNEWS/British Columbia*, Oct. 16, 2012 6:31 AM PT Last Updated: Oct. 16, 2012 11:21 AM PT, *see* Statement of Maite Maldonado, Biological Oceanographer,

George found funding by persuading the Old Massett Village Council comprised of Haida Indians in Haida Gwaii archipelago that his iron fertilization project would revive the steeply declined salmon population, and with it the village economy, and in addition there would be a second payoff from the sale of profitable carbon credits.[129] The Haida Salmon Restoration Corporation raised over two million dollars from the community for this venture.[130] Russ George became the CEO, director and officer of the HSRC but was subsequently fired.[131]

After the 2012 ocean fertilization, once again George was the subject of criticism from scientists, environmental groups, and international organizations that govern climate engineering.[132] In October 2012, after an investigation, the *Guardian* exposed George's expedition and interviewed international scholars who said it violated the UN convention on biological diversity (CBD) and the London convention on the dumping of wastes at sea, both of which prohibit for-profit and as well as large-scale ocean fertilization experiments.[133] A few days later, the Canadian Environment Minister stated that Environment Canada did not approve this non-scientific event,[134] and labelled it "rogue science."[135] The dumping led not only to a criminal investigation by the

---

at University of British Columbia, http://www.cbc.ca/news/canada/british-columbia/iron-fertilization-project-stirs-west-coast-controversy-1.1172422.

129 Falconer, "Can Anyone Stop the Man Who Will Try Just About Anything?".

130 Andrew Revkin, "A Fresh Look at Iron, Plankton, Carbon, Salmon and Ocean Engineering," Dot Earth (blog), *New York Times* (July 18, 2014 7:55 a.m.), http://dotearth.blogs.nytimes.com/2014/07/18/a-fresh-look-at-iron-plankton-carbon-salmon-and-ocean-engineering/, Zoe McKnight, "Why was Iron Dumping a Surprise?" Vancouver Sun (Sept. 3, 2013), http://www.vancouversun.com/technology/iron+dumping+surprise/8865130/story.html.

131 L. Jim Thomas, "Don't Dump Iron—Dump Rogue Climate Schemes," *Huffington Post Canada Blog* (Posted: 05/27/2013 2:07 am EDT Updated: 07/24/2013 5:12 am EDT), http://www.huffingtonpost.ca/l-jim-thomas/geoengineering-russ-george-iron-dumping_b_3332235.html.

132 Neil Craik, Jason Blackstock and Anna-Maria Hubert, "Regulating Geoengineering Research through Domestic Environmental Frameworks: Reflections on the Recent Canadian Ocean Fertilization Case," 7 *Carbon & Climate Law Review* (2013): 117, 118.

133 Martin Lukas, "World's Biggest Geoengineering Experiment 'Violates' UN Rules," *Guardian* (London), Oct. 15, 2012, http://www.theguardian.com/environment/2012/oct/15/pacific-iron-fertilisation-geoengineering.

134 Mark Hume and Ian Bailey, "Businessman Russ George Defends Experiment Seeding Pacific with Iron Sulphate," *The Globe and Mail*, Friday, Oct. 19, 2012 1:55 AM EDT, Last updated Friday, Oct. 19, 2012 4:18 PM EDT, http://www.theglobeandmail.com/news/british-columbia/businessman-russ-george-defends-experiment-seeding-pacific-with-iron-sulphate/article4622528/.

135 Falconer, "Can Anyone Stop the Man Who Will Try Just About Anything?".

Canadian Government,[136] but also to the publication of an official statement of concern by the London Convention and London Protocol parties. The LC-LP Contracting Parties issued a statement deeming the project to be in fact "ocean fertilization."[137] The UNESCO Intergovernmental Oceanic Committee (IOC) issued a statement stating its concern for ocean fertilization activities like George's, which are "carried out in the absence of transparency and in violation of international conventions."[138]

Satellite imagery a few months later showed a 10,000 square kilometer phytoplankton bloom in the Gulf of Alaska, centered in the area where the iron was dumped.[139] In 2014, Haida Salmon Restoration Corporation's lawyer stated that there were massive increases in pink salmon in Washington State and the Squamish River, in addition to increased populations of Chinook and Coho salmon in the Pacific Ocean and the fertilization was successful.[140] Scientists dismissed these claims as speculative, given the lack of concrete evidence and the complexity of determining the reasons for any increases in salmon population.[141]

The Haida tribe that invested money into Russ' project was interested in increasing the salmon stock in their coastal waters to revive a damaged economy, while George himself hoped to replenish the oceans and earn valuable carbon credits.[142] It was clear that Russ George did not have any legitimate scientific reason for conducting his ocean fertilization experiment, and that

---

136 "Canadian Statement Report," 34th Consultative Meeting of the Contracting Parties to the London Convention and the Seventh Meeting of the Contracting Parties to the London Protocol, LC 34/15 (Nov. 23, 2012), Annex 3. Canada stated that its government did not authorize the activity nor did it receive any details from the proponent about the proposed activity in advance that would have allowed an assessment of the project.

137 "Report of the 34th Consultative Meeting of the Contracting Parties to the Convention on the Prevention of Marine Pollution by Dumping of Wastes and Other Matter, 1972, and 7th Meeting of Contracting Parties to the 1996 Protocol," IMO Doc. LC 34/15, Nov. 23, 2012, Annex 3 (expressing "grave concern" at the ocean fertilization activity that took place in the Pacific Ocean off of the coast of Canada). *See also, Ibid.* Annex 7.

138 "Statement by the Intergovernmental Oceanographic Commission of UNESCO Regarding Ocean Fertilization," IOC Executive Secretary Wendy Watson-Wright, UNESCO (Oct. 19, 2012), http://www.unesco.org/new/fileadmin/MULTIMEDIA/HQ/SC/pdf/IOC_statement_Ocean_fertilization.pdf.

139 Lukas, "World's Biggest Geoengineering Experiment."

140 "Haida Gwaii Ocean Iron Dump Investigation to Proceed," *Canadian Broadcasting News British Columbia* (Feb. 5, 2014), http://www.cbc.ca/news/canada/british-columbia/haibida-gwaii-ocean-iron-dump-investigation-to-proceed-1.2523649.

141 *Ibid.*

142 *Ibid.*

he violated the provisions LC/LP.1 and LC-LP.2 of the London Convention and Protocol. George also violated the Convention on Biological Diversity, which at one time had established a complete moratorium on all ocean fertilization projects.[143]

Haida Salmon Restoration Corporation is under investigation by Environment Canada for allegedly committing ten counts of ocean disposal in violation of the provisions of the Canadian Environment Protection Act (CEPA) governing dumping[144] during the 2012 incident. Canada seems to have made some effort to incorporate LC-LP resolutions pertaining to ocean fertilization into its domestic law. A March 2011 information bulletin from Environment Canada stated, for example, that that "[o]cean fertilization activities that fall within the definition of disposal under CEPA 1999 are considered to be disposal at sea and are not allowed without a permit." The bulletin does not elaborate about which ocean-fertilization activities fall within the definition of disposal under CEPA.[145] The bulletin also establishes that, per the 2008 LC-LP Resolution, only ocean fertilization that constitutes legitimate scientific research may proceed, and such research must be assessed per the 2010 *Assessment Framework*, after which Environment Canada will "render a decision" (but not issue a permit, which is not required).[146] This text suggests, but does not explicitly state, that Canada considers all other ocean fertilization to be contrary to the disposal requirements of CEPA. It appears that the Canadian Environment Agency continues to investigate this incident.[147] Whether the Haida Salmon Restoration Corporation will be charged with violation of the environmental laws of Canada is unclear.

Development of LC/LP International Governance

Haida Salmon Restoration Corporation's 2012 ocean fertilization experiment gave considerable impetus to the development of a binding legal regime to regulate ocean iron fertilization. The LC/LP parties issued another statement of concern at the 2012 annual meeting: "The Parties … express grave concern

143 "Statement of Executive Secretary IOC Wendy Watson-Wright" (Oct. 19, 2012).

144 Canadian Environmental Protection Act, S.C. 1999, c. 33, §§ 122–137.

145 Environment Canada, Government of Canada, "Information Bulletin—Ocean Fertilization Activities Are Currently Not Allowed Except for Qualified Research," (2011).

146 *Ibid.*

147 *See*, email response from Canadian Environment Agency on file with author (Feb. 26, 2015).

regarding the deliberate ocean fertilization activity that was recently reported to have been carried out in July of 2012 in waters off the Canadian west coast. This activity involved the deliberate introduction into surface waters of 100 metric tonnes of iron sulfate. The Parties recognize the actions of the Government of Canada in investigating this incident"[148] At this meeting, it was considered to expand their scope to regulation of all marine geo-engineering activities.

On October 18, 2013, the Parties to the London Protocol (but not those to the London Convention) following a proposal from Australia, Nigeria and South Korea and adopted by consensus amendments to the Protocol to establish a global regulatory mechanism for ocean iron fertilization and other marine geoengineering activities.[149] The Parties adopted Resolution LP.4(8), *On the Amendment to the London Protocol to Regulate the Placement of Matter for Ocean Fertilization and Other Marine Geoengineering Activities*,[150] to "create a legally binding regime for ocean fertilization, incorporating an assessment and permit system for legitimate scientific research and a prohibition on any other ocean fertilization activities."[151] Once accepted by two-thirds of the Parties, the amendment will implement a binding regulatory framework[152] and the Contracting Parties will have agreed to be governed by an international legal framework and have conceded their right to sponsor activities that were arguably previously unregulated and within the scope of the "freedom of the high

148 "Statement of Concern Regarding the Iron Fertilization on Ocean Waters West of Canada," Report of the 34th Consultative Meeting of the Contracting Parties to the Convention on the Prevention of Marine Pollution by Dumping of Wastes and Other Matter, 1972, and 7th Meeting of Contracting Parties to the 1996 Protocol thereto, IMO Doc. LP 7. LC 34/15, Annex 7 (Nov. 23, 2012). *See also*, "Statement of Concern," Report of the 30th Meeting of the Scientific Group of the London Convention and the First Meeting of the Scientific Group of the London Protocol, IMO Doc. LC/SG 30/14 (July 25, 2007): ¶¶ 2.23–2.25.

149 "Report of the 35th Consultative Meeting of Contracting Parties to the London Convention and 8th Meeting of Contracting Parties to the London Protocol, International Maritime Organization," LC 35/15 (Oct.10, 2013), http://www.umweltbundesamt.de/sites/default/files/medien/376/dokumente/report_of_the_thirty-fifth_consultative_meeting_london_convention_2013_10_21.pdf.

150 *Ibid.* Annex 4.

151 *Ibid.* Annex 5.

152 *See* Resolution. LP.4(8) on the Amendment to the London Protocol to Regulate the Placement of Matter for Ocean Fertilization and Other Marine Geoengineering Activities, Oct. 18, 2013, "Report of the 35th Consultative Meeting and the 8th Meeting of Contracting Parties," IMO Doc. LC 35/15, Annex 4 (Oct. 21, 2013) (adopting an amendment to the London Protocol to regulate marine geoengineering).

seas" principle.[153] Additionally, Resolution LP.4 (8) emphasizes that marine geoengineering should not be considered a viable substitute for reducing humanity's output of greenhouse gases into the atmosphere.[154]

Parties to the London Protocol intended for LP.4 (8) to provide a basis through which marine geoengineering activities other than simply ocean fertilization could be regulated in the future.[155] The definition of marine geoengineering is broad: "A deliberate intervention in the marine environment to manipulate natural processes, including to counteract anthropogenic climate change and/or its impacts, and that has the potential to result in deleterious effects, especially where those effects may be widespread, long last or severe."[156] If new marine geoengineering techniques begin to threaten the ocean environment, parties to the London Protocol can employ Resolution LP.4 (8) and the LC-LP.2 framework assessment to provide a regulatory framework for those activities.[157]

Annex 4.1 of the amendment defines ocean fertilization as "any activity undertaken by humans with the prime intention of stimulating primary productivity in the oceans," but not including conventional aquaculture, mariculture, or the creation of artificial reefs.[158] Annex 4.2 provides that ocean fertilization experiments, which are not conducted for the purpose of legitimate scientific research, will not be permitted.[159] Annex 4.3 requires that ocean fertilization proposals intended for scientific research will have to be evaluated under the LC-LP.2 framework assessment.[160] The LC-LP.2 framework is based on the precautionary principle and does takes into consideration Decisions X/33 and XI/20 of the Convention on Biological Diversity which is clear that climate geoengineering projects are not to take place "in the

153 *Ibid.* The amendment will enter into force for those Parties which have accepted it on the 60th day after two-third of the Parties that have deposited their instrument of acceptance with the International Maritime Organization (London Protocol, Art 21 (3). The US is not a Party to the Protocol). As of February 2014 no instrument of acceptance had been deposited with the IMO.

154 *Ibid.*

155 *Ibid.*

156 *Ibid.* Annex 4.

157 *Ibid.* Annex 5, "Assessment framework for matter that can be considered for placement under Annex 4." This generic framework is to be complemented by the 2010 specific Assessment Framework on Scientific Research on Ocean Fertilization.

158 *Ibid.* Annex 4.

159 *Ibid.*

160 *Ibid.* Annex 4.

absence of science based, global, transparent and effective control and regulatory mechanisms."[161]

New LP article 6*bis* establishes the basic structure of the regulatory mechanism for marine geoengineering. Marine geoengineering activities that involve the placement of matter into the sea from vessels, aircraft, platforms, or other man-made structures fall within the scope of the LP and may be considered for inclusion in new Annex 4. Once listed, such activities are prima facie prohibited unless the listing in Annex 4 provides that the activity, or the subcategory of an activity, may be considered for a permit. Such permits are issued by the Contracting Party with jurisdiction to regulate the activity and thus article 6(2)*bis* requires Contracting Parties to adopt administrative or legislative measures that ensure permits comply with the provisions of Annex 5 and the requirements of any specific assessment framework developed for the relevant activity under the LP. As a minimum, any permit issued must determine that: (1) the activity is not contrary to the aims of the Protocol, and (2) that pollution of the marine environment is, as far as practicable, prevented or reduced to a minimum.

Annex 5 sets out a "General assessment framework for matter that may be considered for placement under Annex 4" ("GAF"). The purpose of the GAF is to provide a legally binding minimum standard of assessment requirements for all marine geoengineering activities listed in Annex 4.[162] The GAF sets out a consultation procedure where the activity may affect maritime areas under the jurisdiction of another state, or where the activity affects areas beyond national jurisdiction. The Parties are required to establish a consultation process with all stakeholders nationally or internationally when a proposal is submitted and during the assessment process. A peer review of the scientific and technical quality of the information and data provided by the applicant is to be conducted.[163]

The amendments are a landmark for the international control of climate engineering activities; when they enter into force, they will be the first legally binding regulation of such activities in international law.[164] The amendments have two substantial parts. First, ocean fertilization is now prohibited, except for ocean fertilization for research purposes only, and the assessment

161 *Ibid.* Annex 4.

162 "Report of the 4th Meeting of the Intersessional Working Group on Ocean Fertilization," IMO Doc. LC 34/4 (July 27, 2012) ¶¶ 3.4–3.6.

163 "Draft Guidance on a Procedure for Considering the Inclusion of New Activities in Annex 4 to the London Protocol," IMO Doc. 35/15, Annex 5.

164 Harold Ginzky and R. Frost, 'Marine Geo-Engineering: Legally Binding Regulation under the London Protocol', *Carbon and Climate Law Review* 2 (2014): 82–96.

framework and permitting conditions are to be followed. Second, the amendments provide for the regulation of additional emerging marine geoengineering activities not yet specified which have the potential to impact adversely on the marine environment.[165]

At the 2014 LP/LC Consultative Meeting of Contracting Parties, further work was accomplished on marine geoengineering. Guidelines were accepted to be utilized in considering whether and, if so, in what form and context, marine geoengineering activities of potential concern should be addressed. The Parties also moved ahead with arrangements for establishing a roster of international independent experts who would expertise on assessing marine geoengineering activities listed under LP Annex 4 and providing advice.[166]

This new regime includes two parts: a non-binding element focused on OIF that addresses all LC/LP Parties, and a binding element that primarily reinforces the non-binding measures once the LP amendment enters into force. An important benefit of the amendment to the Protocol is that it creates a legally binding permitting regime making the regulation of OIF and other marine geoengineering activities more effective. Since the amendment requires that national legislation regulating marine geoengineering activities in accordance with the amendment be adopted, hopefully the legality of projects will have a process to be vetted in advance and ambiguities encountered in the 2009 German-Indian LOHAFEX ocean fertilization experiment will be avoided. Moreover, there will be a lack of a clear legal regime to prosecute those that refuse to adhere to international standards such as the 2012 Russ George project.

With these amendments, the LC/LP parties have made significant progress on developing a system to regulate legitimate ocean fertilization experiments and to include other possible climate engineering. The legal effect of these regulations and standards should include not only the LC/LP parties, but all parties to UNCLOS. Under UNCLOS, parties are required to cooperate through competent international organizations, including the IMO. Art 210(6) of UNCLOS regulates dumping activities, and specifies that national laws and measures shall be "no less effective" than the global rules and standards. These regulations provide more detail for governance of marine scientific research

165 Report of the 35th Consultative Meeting of Contracting Parties to the London Convention and 8th Meeting of Contracting Parties to the London Protocol (Oct. 21, 2013), IMO Doc. LC 35/15, Section 4, Annex 4.

166 IMO Press Release (Nov. 7, 2014), 36th Consultative Meeting of Contracting Parties to the London Convention and 9th Meeting of Contracting Parties to the London Protocol (3–7 November 2014), http://www.imo.org/en/MediaCentre/MeetingSummaries/LCLP/Pages/LC-36-LP-9.aspx.

under UNCLOS article 240(d), which requires that marine scientific research be conducted "in compliance with regulations ... for the protection and preservation of the environment."[167] Thus, the international rules, standards, and recommended practices and procedures developed under the London Convention and Protocol may be applicable to states as a result of their obligations under UNCLOS even if they are not a party to either the LC or LP.[168]

Unfortunately, there are only 87 parties to the 1972 London Convention, comprising 61.76% of world tonnage, and only 45 parties to the 1996 Protocol to the London Convention comprising 36.6% of world tonnage.[169] There are few parties (52) to the Antarctic Treaty (29 consultative and 23 nonconsultative), and only 34 parties to the Madrid Protocol.[170] Thus, vessels not flagged by state parties to these conventions may ignore the legal obligations to limit activities to scientific research and proceed through the IMO process of obtaining a permit. The limited participation in these treaties potentially detracts from their ability to be the source of comprehensive global regulation.

Professors Freestone and Rayfuse have observed that "[n]o matter how strict an approach is taken by the parties to the LC/LP, the very real potential exists for proponents of ocean fertilization to undermine the LC/LP regulatory efforts by simply incorporating their companies, flagging their vessels, and loading their fertilizer in non-LC/LP party states."[171] This potential problem is a serious one. However, if the flagging and loading were done in a country that had ratified UNCLOS, article 210 would require adherence to the rules promulgated under the LC/LP. To completely avoid such rules, the flagging and loading would have to be done in a country that had not ratified either UNCLOS,

---

167 UNCLOS is clear that states have marine environmental protection responsibilities including among other things the obligation to protect and preserve the marine environment (Art. 192) and the duty to minimize the release of toxic, harmful, or noxious substances into the marine environment (Art. 194, ¶ 3(a)). UNCLOS also requires environmental impact assessment for planned activities under the jurisdiction or control of states that may cause substantial pollution or significant and harmful changes to the marine environment (Art. 206) and reporting of results (Art. 205).

168 Tullio Treves, "The Law of the Sea 'System' of Institutions," Max Planck Yb. *United Nations Law* (1998): 325; Bernard H. Oxman, "The Duty to Respect Generally Accepted International Standards," *New York University Journal of International Law & Policy* 24 (1991): 109.

169 IMO, "Summary of Status of Conventions," http://www.imo.org/About/Conventions/StatusOfConventions/Pages/Default.aspx (accessed April 10, 2015).

170 Secretariat of the Antarctic Treaty, "Parties," http://www.ats.aq/devAS/ats_parties.aspx?lang=e (accessed April 10, 2015).

171 David Freestone & Rosemary Rayfuse, "Ocean Fertilization and International Law," *Marine Ecology Progress Series* 364 (2008): 227–233.

or the LP. Even under that scenario, it can be said that the regulations established by the parties to the LC-LP have become binding customary international law because they were adopted by the countries most concerned with these issues and enjoy widespread adherence.

The parties to the LC-LP face challenges similar to those confronted by the International Seabed Authority in dealing with seabed mining, namely that commercial entities have and will continue to look to the exclusive economic zones of coastal countries to conduct activities, where the regulations may be more lax.

CBD's Continued Involvement

In 2014 the CBD's Subsidiary Body on Scientific, Technical, and Technological Advice prepared a report to examine the potential impacts of geoengineering on biodiversity.[172] This interim update was intended to inform parties to the CBD on recent developments regarding climate-engineering technologies and the international regulatory framework governing climate engineering.[173] The CBD update contains an extensive bibliography of sources examining the 2012–2014 science, ethics, and governance issues surrounding climate geoengineering and their relation to biodiversity.[174] The CBD Interim Update considers the portions of the IPCC's AR5 that were available and points out that climate change is progressing while policy needed to confront climate change lags far behind.[175]

Conclusion

There is debate over the London Convention and Protocol approach to ocean iron fertilization and other marine geoengineering technologies. There continue to be many serious unresolved questions and issues that demand additional analysis and work.

---

172 "Interim Update of Information on the Potential Impacts of Climate Geoengineering on Biodiversity and the Regulatory Framework Relevant to the Convention on Biological Diversity," Subsidiary Body on Scientific, Technical, and Technological Advise, Convention on Biological Diversity, UNEP/CBD/SBSTTA/18/INF/5 (May 22, 2014), http://www.cbd.int/doc/meetings/sbstta/sbstta-18/information/sbstta-18-inf-05-en.pdf.

173 *Ibid.* 1.

174 *Ibid.*

175 *Ibid.* 7–8.

Some view the limitation to legitimate scientific research as too restrictive.[176] Critics claim that the LC/LP approach will interfere with the ability to use science to effectively confront a coming climate crisis.[177] Proponents of geoengineering argue that ocean iron fertilization could be a useful carbon removal technology because it is relatively cheap and easily carried out in comparison with other geoengineering schemes.[178] On the other hand, there are serious questions about the effectiveness of ocean iron fertilization and concern about the uncertainty of impacts on the environment and human health. It is clear that ocean fertilization and other climate engineering ideas will alter the ocean ecosystems and will have other known and unforeseen effects on other ecosystems.

Nonetheless, marine geoengineering is on the agenda. One section of the IPCC AR5 report generated controversy because it discussed the use of geoengineering technology as an actual response to continuing climate change.[179] Although the report considered the fact that geoengineering is risky and should be used with great caution, the report's mere coverage of geoengineering as a possible solution drew criticism and concern.[180] The IPCC report did not focus its attention on ocean iron fertilization, but instead discussed Bioenergy with Carbon Capture and Sequestration (BECCS) as a possible climate-engineering solution.[181] The IPCC's mention of BECCS in its report is yet another demonstration that scientists and policy makers are considering and evaluating geoengineering as a possible response to the problem of climate change.

The continuing analysis and development of international regulatory regimes to ensure climate-engineering technologies are employed carefully is

---

176 Michael C. Branson, Comment, "A Green Herring: How Current Ocean Fertilization Regulation Distracts from Geoengineering Research," *Santa Clara Law Review* 54 (2014): 163, http://digitalcommons.law.scu.edu/cgi/viewcontent.cgi?article=2771&context=lawreview.

177 Edward A. Parson & Lia N. Ernst, "International Governance of Climate Engineering," *Theoretical Inquiries in Law, 2013, UCLA School of Law Research Paper No. 12–23* (Oct. 5, 2012): 309.

178 Kerstin Güssow, *et al.*, "Ocean Iron Fertilization: Time to Lift the Research Taboo," *Climate Change Geoengineering: Philosophical Perspectives, Legal Issues, and Governance Frameworks* (Cambridge University Press 2013): 246.

179 Emma Woods, "New Kid on the Block: Geoengineering and the IPCC," The Royal Society, in *Verba* (April 14, 2014), http://blogs.royalsociety.org/in-verba/2014/04/14/.

180 Nafeez Ahmed, "Leaked IPCC Climate Plan to Worsen Global Warming- Ecologists," *Guardian* (April, 7, 2014) http://www.theguardian.com/environment/earth-insight/2014/apr/07/ipcc-un-climate-change-mitigation-wg3-worsen-geoengineering.

181 Woods, "New Kid on the Block: Geoengineering and the IPCC."

increasingly necessary. The LC/LP framework offers an advantage with regard to marine geoengineering because its marine geoengineering provisions will apply to all marine areas, including areas beyond national jurisdiction.

There are ethical and moral issues relating to ocean fertilization, as well as other marine issues involved in utilizing geo-engineering.[182] The populations of developing countries who have had a minor role in creating the problem may find the risks of geoengineering unacceptable. Most important, it is critical that geoengineering not be pursued as an alternative to the necessary reduction of greenhouse gases. Governance should include work to ensure informed consent of those affected by geoengineering research experimentation. The principle of intergenerational equity should be observed as a governing norm as well.

LP Article 3(2) establishes the principle of liability:

> Taking into account the approach that the polluter should, in principle, bear the cost of pollution, each Contracting Party shall endeavor to promote practices whereby those it has authorized to engage in dumping or incineration at sea bear the cost of meeting the pollution prevention and control requirements for the authorized activities, having due regard to the public interest.

There are still many unresolved issues relating to liability. Even when geoengineering measures might conceivably have a positive influence on the climate in one region, they can at the same time have considerable adverse consequences for the climatic conditions of other regions. But there are no specific provisions in the LC/LP regulations governing liability between states. Like for other regulatory instruments, the inclusion of liability rules will also depend on national implementation establishing procedures for determining, adjudicating, or settling disputes. At least, while there is no specific compliance mechanism under the Convention, the Protocol has established a set of rudimentary compliance procedures and mechanisms.[183] These include a subsidiary Compliance Group that meets and provides advice on compliance issues.

---

182 Albert C. Lin, "Does Geoengineering Present A Moral Hazard?," *Ecology Law Quarterly* 40 (2013): 673; Steve Rayner, *et al.*, "The Oxford Principles," *Climatic Change* 121 (2013): 399, 501–02; *see also*, Stephen M. Gardiner, "Some Early Ethics of Geoengineering the Climate: A Commentary on the Values of the Royal Society Report," *Environmental. Values* 20 (2011): 163, 183–84.

183 "Report of the 29th Consultative Meeting of Contracting Parties to the Convention on the Prevention of Marine Pollution by Dumping of Wastes and Other Matter and the Second Meeting of Contracting Parties to the 1996 Protocol to the Convention on the Prevention

Liability regimes are especially important because they require a consideration of internalizing the costs of pollution into the polluter's financial equation. They provide incentives for compliance with environmental protection standards and presumably help to prevent environmental harm from occurring in the first place. They provide an important role should environmental harm occur notwithstanding the regulatory efforts of the underlying protective regime.

Ocean fertilization and geoengineering proposals raise issues beyond technical and scientific assessments, including social and ethical impacts.[184] First and foremost, it is critical that geoengineering not be pursued as an alternative to the necessary reduction of greenhouse gases. Furthermore, the principle of intergenerational equity should be an operative and governing norm. Finally, geoengineering research governance should have a human focus and work to ensure informed consent of those affected by geoengineering research experimentation.[185] Clearly there is more work required to make the LC/LP regulations fair and complete.

---

of Marine Pollution by Dumping of Wastes and Other Matter", IMO Doc. LC 29/17 (Dec. 14, 2007).

184 Albert C. Lin, "Does Geoengineering Present A Moral Hazard?," *Ecology Law Quarterly* 40 (2013): 673; Steve Rayner, *et al.*, "The Oxford Principles," *Climatic Change* 121 (2013): 499, 501–02; *see also* Stephen M. Gardiner, "Some Early Ethics of Geoengineering the Climate: A Commentary on the Values of the Royal Society Report," *Environmental Values* 20 (2011): 163, 183–84.

185 Governance foundation with a human focus should utilize the distinguished Oxford Principles and Asilomar recommendations. Both call for public participation and consent of those affected by geoengineering experimentation and the need to include the principle of intergenerational equity. Oxford Principle 2 calls for "public participation in geoengineering decision-making" and states: "Wherever possible, those conducting geoengineering research should be required to notify, consult, and ideally obtain the prior informed consent of, those affected by the research activities." The fifth Asilomar recommendation calls for "public involvement and consent," and elaborates, "Public participation and consultation in research planning and oversight, assessments, and development of decision-making mechanisms and processes must be provided. Approaches are needed to ensure consideration of the international and intergenerational implications of climate engineering." "Oxford Principles: History, Oxford Geoengineering Programme," http://www.geoengineering.ox.ac.uk/oxford-principles/history/ (accessed Mar. 3, 2014); Asilomar Scientific Organizing Committee, "The Asilomar Conference Recommendations on Principles for Research into Climate Engineering Techniques," (Nov. 2010): 7. *See also*, Eli Kintisch, "'Asilomar 2' Takes Small Steps Toward Rules for Geoengineering," *Science* 328 (2010): 22.

CHAPTER 13

Institutional Interplay between Marine Biodiversity beyond National Jurisdiction—A New Agreement?

*Yasuko Tsuru**

Introduction

The interdisciplinary relationship between international law and international politics has become particularly salient.[1] In recent years, the increased density of international norms and proliferation of international institutions has not necessarily created a more cooperative world. Although it is counterintuitive, an increasing number of international laws and a proliferation of institutions may produce intermingling, overlap, and conflict in ways that complicate rather than stabilize world order. For example, states may seek to cooperate with regimes and institutions based solely on their state interests and on what they regard as the likely judgments of international tribunals. This phenomenon of "forum shopping" tends to fragment the fabric of international order by setting up institutional conflict. It is uncertain, however, what happens when institutions approach the same issue from different angles and how this overlap is best managed. It may become even more difficult to make institutions work efficiently and to realize the purposes for which they were constructed. Furthermore, the power or strength of norms influences this process.

The development of technology contributed to negotiation of the Third United Nations Conference of the Law of the Sea, which adopted the United Nations Convention on the Law of the Sea (UNCLOS) in 1982. UNCLOS is often said to be the framework convention dealing as a "package" with all of the

* Professor of International Politics at Faculty of Global Studies, Sophia University, Tokyo, Japan.

1 See Jeffrey L. Dunoff & Mark A. Pollack, *Interdisciplinary Perspectives on International Law and International Relations-The State of the Art* (Cambridge University Press, 2013) as one of the latest books from the point of international law. The soft law debate from the 1970s on pay attention to international politics. In international politics the constructivist approach much concerned with norms became popular in the 1980s and "legalization" is considered to bridge two areas. As for institutional overlap, see R. Keohane & D. Victor, "The Regime Complex for Climate Change," *Perspectives on Politics* 9 (2011): pp. 7–23.

 | DOI 10.1163/9789004311442_015

various marine issues, including fisheries, marine pollution, navigational rights, and maritime boundaries. In each area, functional organizations, such as the International Maritime Organization (IMO) and the Food and Agricultural Organization (FAO), regional fishery management organizations (RFMOs), and the UN General Assembly, all have contributed to the development of the law of the sea by developing additional standards and norms. Oceans governance emerges from this process of institutional interplay.[2]

This chapter analyzes the remarkable issue of marine biodiversity beyond areas of national jurisdiction (BBNJ) from the perspective of institutional interplay. In particular it considers why, states must come of agreement on the efficacy of a proposed implementation agreement within the structure of UNCLOS to address BBNJ; and, of course, the specific substantive terms of such an agreement, however structured, must be negotiated. There already exist two UNCLOS implementation agreements: the 1994 the UN General Assembly agreement relating to part XI of UNCLOS concerning amendments to the seabed mining regime;[3] and the 1995 Fish Stocks agreement relating to straddling fish stocks and highly migratory stocks, which further develops articles 63 and 64 of UNCLOS.[4]

The Rio+20 outcome document in 2010 and the 2015 Ad Hoc Open-ended Informal Working Group to study issues relating to the Conservation and Sustainable Use of Marine Biological Diversity beyond Areas of National Jurisdiction (WGBBNJ) demonstrate progress already made by 2015 toward a new implementation agreement. What is the background of these issues and how and why did this become a new agenda of the law of the sea? This chapter provides an account of the political process of negotiations on this issue to date; and it concludes that developing states comprise a large and powerful bloc of states that may seek to "forum shop" and will tend to gravitate toward creating institutions that promote their interests. In other words, these states could become the leaders in institutional interplay, as negotiations go forward, on the specifics of both new rules and new procedures regarding resource uses and protections in the area beyond national jurisdiction.

---

2 T. Henriksen, G. Honneland and A. Sydnes, *Law and Politics in Ocean Governance: the UN Fish Stocks Agreement and Regional Fisheries Management Regimes* (Nijhoff, 2006). It is focused on only fisheries aspects using governance.

3 Agreement relating to the Implementation of Part XI of the UNCLOS, adopted on 28 July 1994 and entered into force on 28 July 1996, UNTS vol. 1836, p. 42. As of October 2014, 167 states and entities were members of UNCLOS and 147 on this 1994 implementation agreement. Among the major maritime States, only US had not ratified UNCLOS.

4 Signed on 4 December 1995 and entered into force on 11 December 2001, UNTS vol. 2167, p. 88.

Institutional Effectiveness for BBNJ

Environmental issues in UNCLOS are not as well-focused or fully defined as some are numerous other issues. Moreover, marine "biodiversity" is not explicitly given explicit attention by the UNCLOS. Principle 2 of the 1972 Stockholm Declaration (issued a full decade before signing of UNCLOS) referred to the protection of the marine ecosystem.[5] Two decades later, the Convention on Biological Diversity (CBD), adopted at the Rio Earth Summit in 1992, essentially a "framework" instrument, was of course concerned with biodiversity.[6] The three objectives of the CBD are expressed in Article 1:

- the conservation of biological diversity;
- the sustainable use of its components; and
- the fair and equitable sharing of the benefits arising out of the utilization of genetic resources.

Furthermore, "genetic resources" means genetic material of actual or potential value. And "genetic material" means any material of plant, animal, microbial or other origin containing functional units of heredity.[7] Stated differently, every living resource is also a genetic resource. Even though quantity matters with regard to living resources, information on genesis is what matters for genetic resources; in practical terms quantity is not so important, because even a handful of resources could produce substantial capital.

In the early stages of discussing this process, the United States was the leader of negotiations of CBD. The U.S. sought to influence developing countries to conserve resources in order to provide a sustainable, stable supply for developed countries. But the negotiations that produced the CBD produced

5 Principle 2 says that "The natural resources of the Earth including the air, water, land, flora and fauna and especially representative samples of natural ecosystems must be safeguarded for the benefit of present and future generations through careful planning or management, as appropriate," in P.W. Birnie and A. Boyle, *Basic Documents on International Law and Environment* (Oxford University Press, 1995).

6 1760 U.N.T.S. 79. Entered into force on December 29, 1993. As for January 25, 2015, 193 states are parties to the convention and this treaty is one of the most universal treaties like the UN Charter. See the CBD website at: http://www.cbd.int/information/parties.shtml.

7 Article 2 of the CBD defines "biological diversity" as "the variability among living organisms from all sources including inter alia, terrestrial, marine and other aquatic ecosystems and the ecological complexities of which they are a part of; this includes diversity within species, between species and ecosystems."

a result beyond the initial U.S. objective; it took the form, instead, of essentially a benefit-sharing treaty. Consequently, the United States was one of the few countries that have not ratified the CBD. Developing countries see CBD as an opportunity to take back structural exploitation by developed countries. CBD concerns genetic resources on land and in particular resources derived from plants, and not marine or deep seabed resources beyond coastal states' extended continental shelf. CBD does not specify the identity of the "components" themselves, nor did it address the problem of access or benefit from such resources beyond national jurisdiction.[8] By contrast, UNCLOS—which adopted a zonal approach to rights and duties—covers all the areas of the ocean, even though it does not specifically address issues relating to biodiversity. The treaty established the exclusive economic zone (EEZ) extending up to 200 miles from the baseline from which the breadth of the territorial sea. UNCLOS also established a new distinct regime that applies in the "Area," or the deep seabed beyond the limits of national jurisdiction.[9] At the time of the UNCLOS negotiations, genetic technology was only in its initial phase; and the authors of the Convention did not foresee the importance of possible exploitation of genetic resources on the Area. Hence, Part XI of the Convention on the Area, referred only to mineral resources of the seabed as comprising the "common heritage of mankind."

The irony is that genetic resources beyond national jurisdictions, which could possibly soon be exploited and become much more profitable than mineral resources, lack specific provisions in UNCLOS. States disagree, as do some specialist scholars, over whether such resources may be addressed within the regime of high seas freedoms or whether they might fall under the regime of seabed mining in Part XI.[10] How to fill this gap institutional design and

8 See generally Harry N. Scheiber, "The Biodiversity Convention and Access to Marine Genetic Materials in Ocean Law," in Davor Vidas and Willy Østreng (eds.), *Order for the Oceans at the Turn of the Century* (Brill/Kluwer Law, 1998), 187–202; and Alexander Proelss, "Marine Genetic Resources under UNCLOS and the CBD," in *German Yearbook of International Law* 51 (2008): 417. In this situation all states respect the interests of other states and it does not mean they can do whatever they like.

9 It is defined as the "seabed and ocean floor and subsoil thereof beyond national jurisdiction" (UNCLOS Art. 1), which would be the outer edge of the continental margin, or to the distance of 200 miles from the baseline where the margin does not extend up to that distance.

10 For another area of disagreement with regard to the intended (or properly interpreted) scope and application of the terms of Article XI, see Alexander Proelss, "The Role of the Authority in Ocean Governance," in Harry N. Scheiber and Jin-Hyun Paik, eds., *Regions,*

regulatory measures, in other words this "deficit," is the hottest issue in BBNJ, and it overlaps the recurring negotiations of CBD Conference of the Parties (COP) and forums under UNCLOS.

Development of Negotiation: The Jakarta Mandate

This process may be traced back to the 1995, called the "Jakarta Mandate" (CBD COP 2, Decision 10) on marine and coastal biodiversity. The lack of institutional measures for regulation of marine biodiversity beyond national jurisdiction was added to the agenda. Decision 10 pointed out the necessity for the conservation of marine diversity by incorporation of CBD objectives. This movement produced two main streams of the debate: marine protected areas and genetic resources.

Marine protected areas have the potential to raise separate controversies,[11] but this chapter focuses on genetic resources specifically. The Jakarta Mandate, COP 2, Decision 10, para. 12 states that signatories agree that:

> ... the Executive Secretary, in consultation with the United Nations Office for Ocean Affairs and the Law of the Sea, [will] undertake a study of the relationship between the Convention on Biological Diversity and the United Nations Convention on the Law of the Sea with regard to the conservation and sustainable use of genetic resources on the deep seabed, with a view to enabling the Subsidiary Body on Scientific, Technical and Technological Advice (SBSTTA) to address at future meetings, as appropriate, the scientific, technical, and technological issues relating to bio-prospecting of genetic resources on the deep seabed.

Institutions, and the Law of the Sea: Studies in Ocean Governance (Leiden and Boston: Brill/Martinus Nijhoff, 2013), Chap. 10.

11 See further M.M. Goote, "The Jakarta Mandate on Marine and Coastal Biological Diversity," *International Journal of Marine and Coastal Law* 12 (1997): 377–89. The Jakarta Mandate is a global consensus on the importance of marine and coastal biological diversity. It is part of the Ministerial statement of the COP meeting and Annex I, para. (iv) says as follows: In relation to paragraph 11, critical habitats for living marine resources should be an important criterion for the selection of marine and coastal protected areas, within the framework of integrated marine and coastal area management, taking into consideration the objectives of the Convention on Biological Diversity. Conservation measures should emphasize the protection of ecosystem functioning, in addition to protecting specific stocks. UNEP/CBD/COP/DEC II /10 Annex I available at the CBD website.

The final report of the Jakarta meeting was forwarded by the CBD Secretary during the eighth SBSTTA in 2003. This document was the starting point for considering whether the CBD would address marine biodiversity beyond national jurisdiction. In CBD COP 7 in 2004, marine protected areas entered the discussion. The COP requested that the Executive Secretary, in consultation with states and the International Seabed Authority—established as the only competent organization for marine science within the UN system—undertake to compile information on the methods for the identification, assessment and monitoring of genetic resources of the seabed and ocean floor and subsoil thereof in areas beyond the limits of national jurisdiction, and also to synthesize information on their status and trends, including identification of threats to such genetic resources and the technical options available for their protection.[12]

COP 8 in 2008 sorted options for utilizing for the protection of deep seabed genetic resources beyond national jurisdiction as follows: (1) the use of codes of conduct, guidelines and principles; and (2) reduction and management of threats through environmental regulation environmental impact assessments, establishment of marine protected areas, and prohibition of detrimental and destructive practices in vulnerable areas. These options and others were considered in the specific context that this work would unfold within the UN framework.[13]

Discussion under the UN Framework

The UN Open-ended Informal Consultative Process on Oceans and Law of the Sea (UNICPOLOS) was established in 1999 to facilitate the annual review by the General Assembly of the international ocean affairs and law of the sea by considering the UN Secretary-General's report on Oceans and the Law of the Sea, and by suggesting the particular issues to be considered by the General Assembly. To follow up on its annual agenda topics, which are decided by the General Assembly each year illustrates the most important marine issues (Table 13.1).

12 UNEP/CBD/COP/DEC/VII/5, para. 54.

13 UNEP/CBD/COP/DEC/VIII/21, para. 5.

TABLE 13.1 *Topics of UNICPOLOS*

Year	Session	Topics
2000	1st	
2001	2nd	
2002	3rd	(a) Protection and preservation of the marine environment (b) Capacity-building, regional cooperation and coordination and integrated ocean management
2003	4th	(a) Safety of Navigation; for example, Capacity-building for the Production of Nautical Charts (b) Protecting Vulnerable Marine Ecosystems
2004	5th	New Sustainable Uses of the Oceans, Including the Conservation and Management of the Biological Diversity of the Seabed in Areas Beyond National Jurisdiction ⇒*WG on the Marine biodiversity beyond national jurisdiction*
2005	6th	(a) Fisheries and their contribution to sustainable development (b) Marine Debris
2006	7th	Ecosystem approaches and Oceans
2007	8th	Marine Genetic Resources
2008	9th	Maritime Security and Safety
2009	10th	Implementation of the outcomes of the Consultative Process, including a review of its achievements and shortcomings in its first nine meetings
2010	11th	Capacity-building in Ocean affairs and the Law of the sea, including Marine Science
2011	12th	Contributing to the assessment, in the context of the UN Conference on SD, of progress to date and the remaining gaps in the implementation of the outcomes of the major summits on sustainable development and addressing new and merging challenges.
2012	13th	Marine Renewable Energies
2013	14th	Impact on Ocean Acidification on the Marine Renewable Energies
2014	15th	Role of Seafood in Food Security

The topic "New Sustainable Uses of the Oceans, Including the Conservation and Management of the Biological Diversity of the Seabed in Areas beyond National Jurisdiction"[14] was discussed at the 5th UNICPOLOS in 2004, and it became evident that there was an interest gap between developed countries and developing states.[15] The WGMBBNJ working group emerged from the meeting.[16] The main topic for the 8th UNICPOLOS in 2007 was marine genetic resources, but states did not reach consensus on how marine genetic resources were to be treated legally.[17] The G77 states and China argued that marine genetic resources should be regarded as a common heritage of mankind, much like mineral resources in the Area, and they sought to get agreement on a sharing of benefits, even if successful exploitation of a genetic resource should take fifteen years from discovery to production.[18] On the other hand, some developed countries, chiefly the United States, Russia, and Japan, resisted including marine resources within Part XI. These nations assert that the freedom of high seas regime for living resources applies to living resources in the Area and should preclude a restrictive new regulatory regime.[19]

With this impasse, all eyes turned to the WGMBBNJ, which started in 2006. In the beginning, the WG was an informal meeting regarded as a kind of study group to encourage productive and open discussions on marine biodiversity

14 U.N. Dec. A/59/122 Report on the work of the UNICPOLOS at its 5th meeting (7–11 June 2004), esp. paras. 83–94.

15 Summary of the 5th meeting of the UNICPOLOS, 2004, *Earth Negotiations Bulletin* (*ENB*), Vol. 25 No. 12, p. 8. The handful of states that invest in the technology to explore and exploit deep seabed genetic resources (developed countries) would not renounce the economic benefits by benefit-sharing measures, on the other hand, G77/China argued that the common heritage of mankind encompasses living resources and called for benefit sharing.

16 U.N. Doc. A/59/24 (November 17, 2004).

17 U.N. Doc. A/62/169 Report on the work of the UNICPOLOS at its 8th meeting (25–29 July 2007). This UNICPOLOS was the first time with no negotiated outcome at all.

18 Summary of the 8th meeting of the UNICPOLOS, 25–29 June 2007, *ENB*, Vol. 25, No. 43, p. 10. At the time of UNCLOS negotiation, mineral resources of the Area thought to be profitable soon.

19 Art 87 (1) of UNCLOS exemplifies the freedom of high seas as follows:... It comprises, *inter alia*, both for coastal and land-locked States: (a) Freedom of navigation; (b) Freedom of overflight;...(f) Freedom of scientific research, subject to Parts IV and V. The term '*inter alia*' suggests that the freedom of the high seas may comprise other freedoms which are not provided in this Article 87(1). Yet it is unclear what activities may be embraced in this concept. See Yoshifumi Tanaka, *The International Law of the Sea* (Cambridge University Press, 2012), p. 151.

beyond national jurisdiction in a comprehensive and integrated way.[20] The 4th WGMBBNJ in 2011, however, was the turning point. The mantra that "the status quo is not an option" was promoted by numerous delegates, who were ready to engage in a discussion on the concepts and processes required for effective conservation and sustainable use of WGMBBNJ.[21]

The 2010 Nagoya Protocol on Access and Benefit-sharing under CBD in began to gain traction and influence development of WGMBBNJ. The Nagoya Protocol makes clear that the CBD did not cover marine biological resources beyond national jurisdiction because of the inherent limits in the CBD framework itself.

In leading the G77, China, the European Union and Mexico agreed to take all the issues as a package, including questions of benefits-sharing, marine protected areas, environmental impact assessments, and technology. During this 4th WGMBBNJ that these proponents clearly asked for an approach that includes benefit sharing, and that mentioned property rights for the first time.[22] Finally, the 4th WG adopted by consensus a set of recommendations to initiate a process on the legal framework for the conservation and sustainable use of marine biodiversity beyond national jurisdiction by identifying gaps and ways forward, including through the implementation of existing instruments and the possible development of a multilateral agreement under UNCLOS.[23] Of course, benefit sharing is no exception of this package.

The main priority for developing countries is protection of their interests in marine genetic resources, whereas developed states give their priority to the development of marine protected areas and environmental impact assessments. While MPAS, EIA and marine genetic resources remain the pillars of the agenda for the 5th WG in 2012, property rights and fishery subsidiaries also were considered.[24] The package of issues swelled with so many issues that the 5th WG failed to develop a final document. Thereafter, negotiations shifted to Rio+20.

---

20 Over 250 participants representing governments, UN agencies, NGOs and industry attended the meeting. Summary of WGMBBNJ, 13–17 February 2006, *ENB*, Vol. 25, p. 1. U.N. Doc. A61/65, pp. 3–4.

21 U.N. Doc.A66/119, Letter dated 30 June 2011 from the Co-Chairs of the Ad Hoc Open-Ended Informal Working Group to the President of the General Assembly, para. 42.

22 *Ibid.*, paras. 15–22. See also summary of the 4th WGMBBNJ, 31 May, 3 June 2011, *ENB*, Vol. 25, 70, pp. 5–7.

23 U.N. Doc., A66/119, Annex. I Recommendations of the WGMBBNJ and Co-Chairs' summary of discussions.

24 Summary of the 5th WGMBBNJ, 7–11 May 2012, *ENB*, Vol. 25, No. 83, p. 9.

Negotiation at UNCSD

The UN Conference on Sustainable Development (UNCSD), known as "Rio+20", was held in June 2012, and it marked the twentieth anniversary of the 1992 UN "Rio Conference" (the Conference on Environment and Development (UNCED). The objectives of the Rio+20 meeting was to secure a renewed (and strengthened) political commitment for sustainable development, assessing the progress to date, and to address the remaining gaps in the implementation of the outcomes of the major summits on sustainable development and addressing new and emerging challenges.[25]

Preparatory negotiations of UNCSD[26] began in 2010. The 2nd session of the Preparatory Committee convened in March 2011, and ocean-related issues were mentioned strongly by many delegates. The complexity, fragmentation and incompleteness of the high seas regime as a governance gap was one of the problems to be addressed at Rio+20. At the end of this 2nd session, a decision was adopted on the process for preparing the draft of the outcome document for Rio+20 by no later than early January 2012.[27] There were no concrete negotiations on the details of any issues, but the deliberations on oceans seemed to send a strong message to the ongoing WGMBBNJ.[28]

In January 2012 the Co-Chairs on behalf of the bureau submitted the zero-draft, which was accepted as the basis of negotiations to UNCSD in the initial informal consultations held at the end of the same month.[29] Para. 80 of the zero draft in 2012 stated:

> We note the establishment by the UN General Assembly of an Ad Hoc Open-ended Informal Working Group to study issues relating to the conservation and sustainable use of marine biological diversity beyond areas

25 U.N. Doc. A/Res/64/236, para. 20.

26 Other than three Preparatory Committee, inter-sessional meetings and Informal Consultations were convened and there existed totally 10 preparatory conventions before Rio+20.

27 A/CONF/216/PC/9: Report of the Preparatory Committee for the UNSC, 2nd session, 7 and 8 March 2011, para. 75.

28 See U.N. Doc. A/Conf.216/PC/9: report of the Preparatory Committee for the UNCSD, 2nd session, 7–8 March 2011 and summary of the 2nd session of the PC for the UNCSD, *ENB*, vol. 27, no. 3, p. 7.

29 The Future We Want—Zero draft of the outcome document was 19 pages. Available on the Rio+20 website at http://www.unced2012.org.content/documents/370... (Dated August 2012). As for the informal meeting in December 2011, the secretariat received approximately 6,000 pages of input from states and other stakeholders.

> of national jurisdiction, and we agree to initiate, *as soon as possible*, the negotiation of an *implementing agreement to UNCLOS* that would address the conservation and sustainable use of marine biodiversity in areas beyond national jurisdiction (italics added).

At informal consultations less than three weeks before the Rio+20 revised draft there remained disagreements on main issues, and the oceans were one of the most controversial. On the final day of Rio+20 the outcome document, "The Future We Want" was adopted by consensus. The document was possible because of the effective management of the host country, Brazil. The Rio meeting was not permitted to end the conference without successful adoption of an outcome document. Rio successfully produced a consensus while preserving the quality and integrity of the main outcome.[30] As for the implementation agreement, The Future We Want, para. 162 states:

> We recognize the importance of the conservation and sustainable use of marine biodiversity beyond areas of national jurisdiction ... Building on the work of the ad hoc working group and ***before the end of the 69th session of the General Assembly*** we commit to address, on an urgent basis, the issue of the conservation and sustainable use of marine biological diversity of areas beyond national jurisdictions, including by taking a decision on the development of an ***international instrument*** under the United Nations Convention on the Law of the Sea.[31]

In comparing the zero draft and the "Future We Want" document, the strong opposition of some states against the zero draft is very evident.

After Rio+20

After Rio+20 the negotiations moved to WGMBBNJ once again. The next iteration was held four times, and the 9th and last WGMBBNJ was conducted on January 23, 2015. The group reached consensus on the recommendations for a decision to be taken at the 69th session of the UN General Assembly to develop

30 Summary of the UNCSD 13–22 June 2012, *ENB* vol. 27, no. 51, p. 22.

31 U.N. Doc. A/Res66/288 (2012). Emphasis added. The Future We Want is organized into six sections with totally 283 paragraphs. Oceans and Seas are paras. 158–177.

a new and legally binding instrument on biodiversity beyond national jurisdiction under UNCLOS.[32]

This document refers to the international instrument connoting the international agreement with a time schedule; the door was now open to a new and third implementation agreement for UNCLOS.

Looking back at deliberations of WGMBBNJs, from the beginning the reasons why the WG had been constructed were apparent. With the progress of the WG atmosphere that the status quo was insufficient entered the main stream, and there was no consensus on how to reach to this goal. The final WGMBBNJ recommendations contained two dimensions—the nature of the final document and the issues to be addressed in the preparatory committee. China/G77 favored adoption of a new agreement. The EU sought a third implementation agreement under UNCLOS. The differences arose from dissatisfaction of developing countries with existing ocean law and intepretations of UNCLOS that had been made by developed countries to their own advantage. Several developed countries including the United States, Japan and Russia did not support a legally binding agreement. The United States was strongly opposed to a new agreement, and it suggested that the existing regime of freedom of high seas was sufficient to address the conservation and sustainable use of biodiversity beyond national jurisdiction.

The idea of a legally-binding instrument emerged as a potential compromise between those favoring legally binding agreements and those who are skeptical about such instruments.[33] The negotiation process and time schedule of the instrument was established.

Nearly all delegations from developing countries and the EU were disappointed with the time schedule because it delays the decision of whether the intergovernmental conference would be convened by the General Assembly in 2018. The timetable depends on the progress of the Preparatory Committee.

32 Outcome of the Ad Hoc Open-ended Informal Working Group to study issues relating to the conservation and sustainable use of marine biological diversity beyond areas of national jurisdiction and Co-Chairs' summary of discussions, in a letter dated 13 February 2015 from the Co-Chairs of the Ad Hoc Open-ended Informal Working Group to the President of the General Assembly, 20–23 January 2015, UN.Doc.A/68/780.

33 Summary of the 9th meeting of WGMBBNJ, 20–23 January 2015, *ENB*, vol. 25, no. 94, p. 8. Para. (d) of recommendations stress the need for the comprehensive global regime and having considered the feasibility of developing an international instrument under UNCLOS.

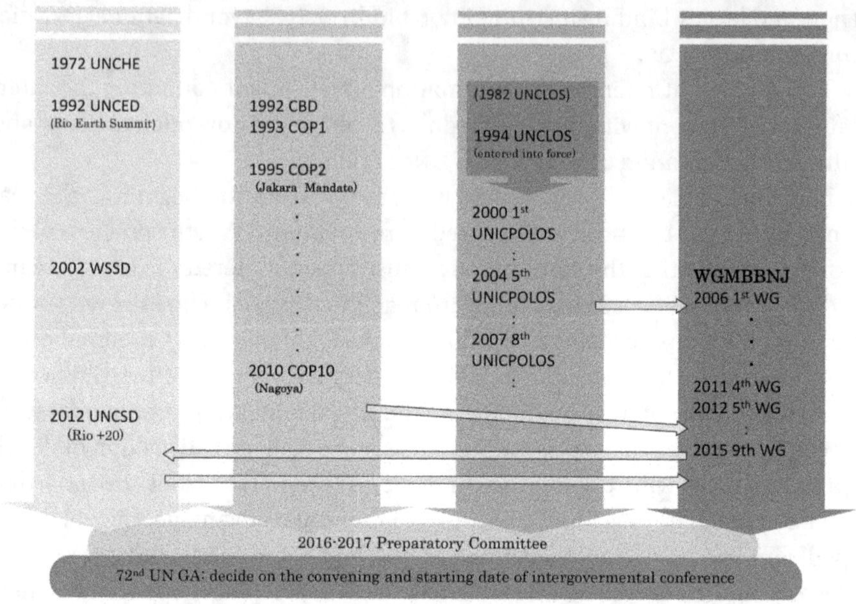

FIGURE 13.1 *Institutional Interplay: Four Pillars toward Implementation Agreement.*

Rough Sketch of Negotiation Structure

This chapter traced how the door was opened to a legally binding instrument to adopt an implementation agreement for the marine biodiversity beyond national jurisdiction, a process that runs through many venues and forums set forth in Figure 13.1.

Figure 13.2 arranges the negotiating positions of several groups involved in the process.

States are in agreement that conservation of biodiversity within or beyond national jurisdiction is necessary, but they are divided on the nature of a future document resulted in the intergovernmental conference. Furthermore, even states that tend to be sympathetic toward other positions are sometimes in disagreement. China/G77 and the EU, for example, do not target the same goal. The target of the EU was to increase MPAs and from the 4th WGBBNJ, the EU opposed China/G77, which sought a legally binding implementation agreement. The United States, Japan and Canada strongly opposed an implementation agreement on genetic resources, and they were also concerned that the MPA approach would unduly burden the freedom of fisheries.

On the other hand, China/G77 presented its position as a single bloc, at first insisting that genetic resources were covered by the UNCLOS deep seabed

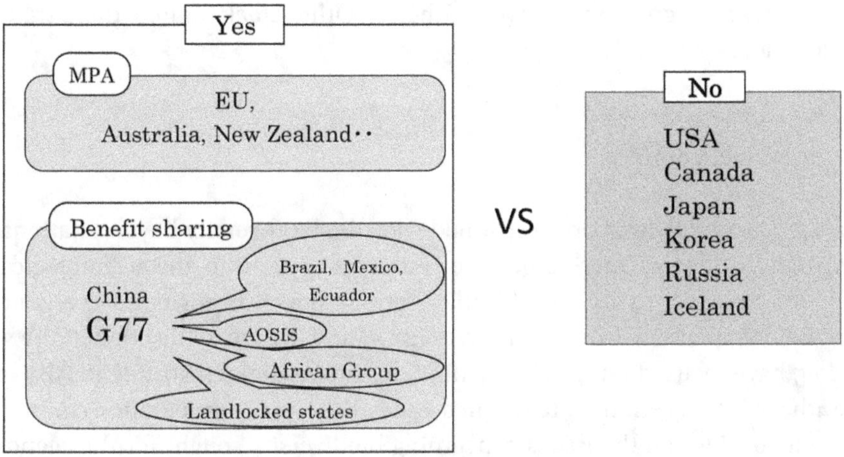

FIGURE 13.2 *Positions on Implementation Agreement.*

regime as a "common heritage of mankind," eventually they opposed the increase of MPAs because they were concerned that marine protected areas would become an obstacle to development.

Developing countries believe developed countries have exploited wealth from their natural resources, and impede their ability to break away from poverty. On this issue, these states are searching for common but differentiated responsibilities, and their ultimate goal is a framework for benefit sharing because developing countries lack technology needed for exploitation of marine genetic resources.

China regarded itself as a developing country and joined with G77. With this sense of entitlement, these states sought action toward big, but vague goals with no specificity. They have resulted in using every related forum where deliberations and decisions by consensus were made—a form of forum shopping. The seemingly monolithic bloc of China and the developing countries have differentiated positions dependent on whether they are rich or poor in biodiversity. At the 9th WGMBBNJ there appeared some cracks among them supporting a new legally binding instrument, with the African Group on one side and Asian countries and Latin American states on the other.[34] As the process shifts into the mode of formal negotiations the possibility of a new agreement will likely affect a whole range of activities in the oceans in which

34 Some countries in Latin America are not party to UNCLOS itself.

some states have greater stakes than others.[35] Other cracks among these states could emerge.

Conclusion

The CBD originally was not concerned with the biodiversity beyond national jurisdictions, but as marine genetic resources came into the technological reach of humans CBD COP and other forums began to focus on the issue. Discussions spanning from the 4th WGMBBNJ and Rio+20 to the 9th WGMBBNJ in 2015 have focused on development of an international instrument. Also as the 4th WGMBBNJ enumerated related issues, the coming preparation committee will consider all the issues concerning biodiversity sought for the agenda by developing countries. To this point, developing countries often have found themselves at cross purposes. Their advantage in numbers, however, may propel them toward a new and legally binding instrument. The door is open, and it is difficult to predict what will happen in 2018. Even at this point, it is clear that existing institutions are not clear on how to address the biodiversity problem, especially the "institutional deficit" regarding marine genetic resources. Their exploitation, with a potential impact for future generations, will surely accelerate; and if the management of biodiversity is left unsettled without control, it would be devastating and a tragedy for the Earth.

---

35 *ENB*, vol. 25. no. 94, p. 11.

PART 4

Ongoing Issues in Ocean Law and Policy

CHAPTER 14

Standard of Review in the Law of the Sea: Reflections from the Bench

*Jin-Hyun Paik**

Over the past five years or so, I have served as Judge of the International Tribunal for the Law of the Sea (the Tribunal or ITLOS). I joined the Tribunal when the Tribunal started to have more and more cases. I have been fortunate in this respect, because, unlike some of my colleagues, I did not have to suffer from the lack of cases to work on.

During this period, I have dealt with several important and interesting cases and been faced with many difficult legal issues. One of those issues is the standard of review in assessing the actions of national authorities. By the standard of review, I refer to the standard applicable to judicial review of national decisions or measures by an international judicial body. Before proceeding further, let me clarify one or two things.

Judicial Review

An international court or tribunal, in discharging its judicial function, is frequently faced with whether the action taken by national authorities, including a national court, is in conformity with international law. In such a case, the international court undertakes to "review" the action of national authorities. When an action of national authorities under review is that of a national court, the review may look like a judicial review in the sense widely used in domestic law that an appellate court reviews a lower court's ruling. However, the review by an international court is not a judicial review in that sense. It goes without saying that an international court or tribunal is not an appellate court against the decisions of a national court. What an international court does is to consider not whether national authorities applied its law correctly in making its decision, but rather whether national authorities, in applying its national law, acted in conformity with international law. As the Permanent

* Judge, International Tribunal for the Law of the Sea.

 | DOI 10.1163/9789004311442_016

Court of International Justice famously held in the 1926 decision of *Certain German Interests in Polish Upper Silesia case*, "[T]here is nothing to prevent the Court's giving judgment on the question whether or not, in applying that law, Poland is acting in conformity with its obligations towards Germany under the Geneva Convention."[1] The review of the decisions of national authorities by an international judicial body is essentially the exercise of its own judicial function to decide an international legal question, that is to say, whether national authorities acted in conformity with international law. Here I use the term "judicial review" in this sense.

Standard of Review in International Law

Those who are international trade law experts are familiar with the term "standard of review." However, while this term is widely used in the context of WTO dispute settlement, it is not unique to the WTO or even to the field of international economic law. The question of the standard of review frequently arises also in international human rights law under the doctrine of margin of appreciation. This question is not unknown to the law of the sea either.

The Tribunal was faced with this question in the past in the context of assessing the reasonableness of bond in the prompt release proceedings under article 292 of the United Nations Convention on the Law of the Sea (the Convention or UNCLOS). The Tribunal again came across this question in the latest *M/V "Virginia G" Case*,[2] which I will touch upon later. The recent decision of the International Court of Justice also addressed this question in the context of the consistency of special permits granted by Japan with the terms of Article VIII of the International Convention for the Regulation of Whaling.[3]

The question of the standard of review rarely arises, however, when an international norm is sufficiently clear and specific, without leaving much room for the discretion on the part of national authorities in interpreting and applying it. This question arises usually when an international norm or international obligation with which a State must comply is flexible, open-ended, or indeterminate, thus leaving a substantial room for discretion on the part of the State. Such norms are called "standard-type rules" or "discretional rules", and

1 *Case concerning certain German interests in Polish Upper Silesia, Merits, Judgment, PCIJ Series A no 7*, 1926, p. 19.

2 *The M/V "Virginia G" Case (Panama/Guinea-Bissau), ITLOS Report* 2014.

3 *Whaling in the Antarctic (Australia v. Japan: New Zealand intervening), Judgment, ICJ Reports* 2014, p. 226.

there are many of those in international law. For example, an international rule often requires States to take "necessary and appropriate" measures or "reasonable" measures, or may allow States to take an exceptional measure if such a measure is "necessary" to achieve certain policy goals or if there are "sufficiently serious" injuries. When an international court reviews whether the measures taken by national authorities were necessary, appropriate or reasonable, the first question faced by the court is what should be the level or standard of review. In other words, how intensely should the international court review the decision of national authorities and how much discretion, if any, should the international court grant to national decision-makers? In essence, this issue comes down to the question of who should interpret and apply such norms—national authorities or an international court.

The standard of review thus is a sensitive issue, as it touches upon the fundamental balance of power between States and an international judicial institution. The standard of review is also an issue that is relevant to the efficiency and legitimacy of the international court. An international court applying what appears to be a stringent standard is often criticized for being too intrusive and undermining the sovereignty of States. Those critics argue that an international court, in so doing, tries to substitute its own judgment for that of national authorities. Thus, the question merits some serious reflection.

Different Standards

There are two different standards for judicial review: a low or deferential standard and a high or stringent standard. A low or deferential standard gives deference to national decisions unless they are manifestly unreasonable or arbitrary. On the other hand, a high or stringent standard reviews all matters, both legal and factual, afresh or *de novo*.

The rationale for a low standard is well known. According to the proponents of such a standard, the application of standard-type norms or obligations, such as allowing States to take measures necessary to achieve certain policy goals, necessarily requires the understanding of complexities of local conditions, which an international court is not well positioned to assess. National authorities, it is contended, know more and better about their own circumstances and have greater resources for making rational assessment. Therefore, an international court should acknowledge these relative strengths of national authorities and give deference to their decisions.

On the other hand, the proponents of a high standard point out that national authorities are more likely to be biased, as they are an interested party to the dispute before an international court. While an international court can be relied upon to be neutral, national authorities may not always be so. This concern as to bias, according to this view, requires an international court to adopt a more stringent standard. Furthermore, if too much discretion is granted to national decision-makers, it would be difficult to achieve a uniform interpretation and application of international norms.

I believe that these relative strengths and weaknesses of an international court and of national authorities should be taken into account in determining an appropriate standard of review. The appropriate standard depends on several factors, especially but not limited to the nature of international norm or obligation in question. It may vary from case to case, and from issue to issue.

Standard of Review in the Law of the Sea

Now let me turn to the decision of the *M/V "Virginia G" Case*, in which ITLOS was faced with the question of standard. The case was concerned with a dispute between Panama and Guinea-Bissau over the bunkering of fishing vessels in the Exclusive Economic Zone (EEZ). *M/V "Virginia G"*, an oil tanker flying a Panamanian flag (and beneficially owned by a Spanish company), was arrested and later confiscated by the authorities of Guinea-Bissau for providing gas oil to fishing vessels in the EEZ of Guinea-Bissau without obtaining a valid written authorization. Panama argued in pertinent part that Guinea-Bissau violated article 58(1) of the Convention by regulating the bunkering of *M/V "Virginia G"*. It further argued that Guinea-Bissau violated article 73(1) of the Convention by arresting, detaining, and confiscating its vessel.

On the other hand, Guinea-Bissau argued that the coastal State has competence to regulate bunkering of fishing vessels as a fishing-related activity under article 56(1) of the Convention. It further argued that since *M/V "Virginia G"* failed to obtain a valid authorization for bunkering services as was required under the law of Guinea-Bissau, its arrest and detention, and eventual confiscation was not a violation of the Convention.

The Tribunal agreed with Guinea-Bissau on the point that the sovereign rights of coastal State in the EEZ under article 56(1) of the Convention encompass the competence to regulate the bunkering of foreign vessels fishing in the zone.[4] It also found that Guinea-Bissau did not violate article 73(1) of the

4 *M/V "Virginia G" Case (Panama/Guinea-Bissau)*, *ITLOS Report* 2014, paras. 225–236.

Convention by arresting and detaining the *M/V "Virginia G"*, as the vessel had engaged in unauthorized bunkering in the EEZ of Guinea-Bissau.[5]

However, on the confiscation measure taken by the authorities of Guinea-Bissau under article 73(1) of the Convention, the Tribunal found that Guinea-Bissau violated that provision.[6] Article 73(1) reads: "The coastal State may, in the exercise of its sovereign rights to explore, exploit, conserve and manage the living resources in the EEZ, take such measures, including boarding, inspection, arrest and judicial proceedings, *as may be necessary* to ensure compliance with the laws and regulations adopted by it in conformity with this Convention"(italics added). The key issue here was the legality of the confiscation of the *M/V Virginia G* under article 73(1), and specifically whether the confiscation was "necessary" to ensure compliance with the laws and regulations of Guinea-Bissau.

In assessing the conformity with UNCLOS of the action of the authorities of Guinea-Bissau, the Tribunal was faced with the question of the standard of review: how much deference or respect should be given to the decision of Guinea-Bissau to confiscate the vessel; or how intensively should the Tribunal review the decision of Guinea-Bissau. Should the Tribunal respect the decision unless it was manifestly arbitrary or unreasonable, or should the Tribunal scrutinize this decision, both legally and factually?

The Tribunal, I believe, adopted a relatively high standard of review in this case, although it did not elaborate on this point clearly in its judgment. The Tribunal reviewed both the legal and factual aspect of the decision of Guinea-Bissau. After examining the circumstances and manner in which the confiscation measure was imposed, as well as the legal factors relevant to the case, it concluded that the confiscation of the vessel and the gas oil on board was not necessary within the meaning of article 73(1) of the Convention, and that Guinea-Bissau was thus in violation of that provision. This approach of the Tribunal was in line with its previous jurisprudence, among others the decision of the *"Camouco" Case*.[7] In that case, the Tribunal, in determining whether the bond imposed by the French court of 20 million French Francs was reasonable for the purpose of the prompt release proceedings. To reach its judgment, the Tribunal considered a number of factors, including the gravity of the alleged offenses, the range of penalties that could be imposed, value of detained vessels and cargos seized, the amount of the bond imposed by the

5 *Ibid.*, para. 265.

6 *Ibid.*, paras. 266–271.

7 *The "Camouco" Case* (*Panama v. France*), *ITLOS Report* (2000).

detaining State and its form, rather than simply giving deference to the decision of the detaining State.[8]

Conclusion

Now let me come to the conclusion. The standard of review matters in the law of the sea, as the Convention contains several provisions of general standard-type or discretionary norms. When the Tribunal is asked to interpret and apply those provisions, it needs to determine what should be the appropriate standard in assessing the action of national authorities. While it is difficult to speak of such standard in general terms, I think that a relatively high standard of review may be called for.

There are several reasons for such a standard. Apart from the general consideration of the relative strengths and weaknesses of an international judicial institution and national authorities, there is a consideration particularly relevant to the interpretation and application of the Convention. The Convention was adopted to establish an international legal order for the oceans applicable to all States. Indeed the Preamble of the Convention recognizes "the desirability of establishing through this Convention ... a legal order for the seas and oceans." The uniform interpretation and application of the provisions of the Convention is indispensable to establishing a stable legal order for the seas and oceans. If each State is allowed to interpret and apply the provisions of the Convention as it sees appropriate, uniformity would be impossible and obligations under the Convention would mean different things to different States. This outcome would upset the delicate balance achieved in the Convention through painstaking negotiation and render the multilateral ocean regime created by the Convention ineffective. I believe that the Tribunal, as a judicial institution created by the Convention to safeguard the integrity of the regime of the law of the sea, has an obligation to prevent this from taking place.

I am sure that the question of standard of review will come back to the Tribunal over and again in the future. Although the appropriate standard may vary from case to case, this last point should be particularly borne in mind in responding to that question.

---

8 *Ibid.*, paras. 65–67.

CHAPTER 15

UNCLOS and Non-Party States before the International Court of Justice

*Tullio Treves**

About three quarters of the existing States in the global community of nations are parties to UNCLOS. The non-parties are a minority of States that includes the United States, Colombia, Israel, Peru, Turkey and Venezuela. Non-parties live in a world in which the vast majority of States is a party to the Convention. They consequently must take into account the fact that the basic rules concerning the law of the sea binding the majority of other States and regulating their relationships are the rules set out in UNCLOS. As a matter of fact, non-parties normally rely on the UNCLOS in determining which rules apply to them.

Non-parties are bound by customary law and not by UNCLOS. It becomes, consequently, important to assess whether certain provisions of UNCLOS correspond to customary international law. When it is so, it may be held that the rules set out in UNCLOS are binding also for non-parties.

In examining the relationship between UNCLOS and customary law at the Oslo Conference of the Law of the Sea Institute a few years after the opening to signature of the Convention, the present writer came to the conclusion that the correspondence between UNCLOS and customary law had to be assessed case-by-case, and that there was no presumption in favor of such correspondence.[1] Re-examining the matter on the occasion of the thirtieth anniversary of the Convention, I expressed the view that the legal situation had radically changed, stating that: "At present it can be said that there is a presumption

* University of Milan, former Judge of the International Tribunal for the Law of the Sea, Senior International Law Consultant, Curtis, Mallet-Prevost, Colt & Mosle LL.P.

1 T. Treves, "The UN Convention on the Law of the Sea as a Non-Universally Accepted Instrument: Notes on the Convention and Customary Law," in A.W. Koers and B.H. Oxman, eds., "The 1982 Convention on the Law of the Sea," *Proceedings, Law of the Sea Institute, Seventeenth Annual Conference*, Co-Sponsored by the Fridtjof Nansen Institute, July 13–16, 1983 (Oslo, Norway, *and* Honolulu, 1984), pp. 685–690. I maintained the same position, with more detailed analysis of practice and scholarly writing in my Hague lectures of 1990, *Codification du Droit International et Pratique des Etats dans le Droit de la Mer,* 223 (1990), pp. 9–302, at 43–50.

 | DOI 10.1163/9789004311442_017

that the provisions of the Convention correspond to customary law. It is, however, a rebuttable presumption as, again on a case-by-case basis, evidence can be submitted to argue that a specific provision has a merely treaty character. Of course, when institutions and mechanisms for the settlement of disputes are concerned, the UNCLOS provisions only apply as conventional rules."[2]

To assess whether a rule set out in UNCLOS corresponds to a customary rule is of particular interest when an international judge has to settle a maritime dispute involving a State that is not a party to UNCLOS.

A first instance of such situation presented itself before the International Court of Justice (ICJ) in the *Qatar v. Bahrain* case. In its judgment on the merits of 2001, the Court observed:

> Neither Bahrain nor Qatar is party to the Geneva Conventions on the Law of the Sea of 29 April 1958; Bahrain has ratified the United Nations Convention on the Law of the Sea of 10 December 1982 but Qatar is only a signatory to it. Customary international law, therefore, is the applicable law. Both Parties, however, agree that most of the provisions of the 1982 Convention which are relevant for the present case reflect customary law.[3]

Two recent instances are the cases decided in two judgments rendered by the ICJ in 2012 and 2014 involving, respectively, Colombia and Peru, both non-party States. These are the *Nicaragua v. Colombia* judgment of 19 November 2012 and the *Peru v. Chile* Judgment of 27 January 2014. These judgments contain pronouncements on the correspondence to customary law of several articles of the Convention. Both judgments consider articles concerning delimitation of the exclusive economic zone (EEZ) and of the continental shelf. The *Nicaragua v. Colombia* judgment also concerns article 121 on the regime of islands and article 76 on the continental shelf—including a mention of the role of the Commission on the Limits of the Continental Shelf (CLCS) in a case involving a non-party. The *Peru v. Chile* judgment contains a passage showing that disputes before the ICJ may influence non-parties to conform to the Convention.

This analysis focuses on these two recent judgments, although it must be noted that two new cases were brought to the ICJ by Nicaragua against Colombia at the end of 2013. Both of these cases involve questions in which the fact that Nicaragua is a party to UNCLOS and Colombia is not is relevant.

---

2 T. Treves, "UNCLOS at Thirty: Open Challenges," *Ocean Yearbook* 27 (2013): pp. 49–66 at 51.

3 Case Concerning Maritime Delimitation and Territorial Questions between Qatar and Bahrain (*Qatar v. Bahrain*), Merits, [2001] ICJ Reports 40, at para. 167.

As the cases are pending and as I serve as counsel in both, I will abstain from commenting on these.[4]

As regards the UNCLOS identically drafted articles concerning delimitation of the EEZ (art. 74) and of the continental shelf (art. 83), both the *Nicaragua v. Colombia* and the *Peru v. Chile* judgments confirm the Court's jurisprudence on the three-step methodology to be followed, and state that the above-mentioned articles correspond to customary international law.

As far as methodology is concerned, the Court summarizes its position as follows in *Peru v. Chile*:

> 180. The methodology which the Court usually employs in seeking an equitable solution involves three stages. In the first, it constructs a provisional equidistance line unless there are compelling reasons preventing that. At the second stage, it considers whether there are relevant circumstances which may call for an adjustment of that line to achieve an equitable result. At the third stage, the Court conducts a disproportionality test in which it assesses whether the effect of the line, as adjusted, is such that the Parties' respective shares of the relevant area are markedly disproportionate to the lengths of their relevant coasts (Maritime Delimitation in the Black Sea (Romania v. Ukraine), Judgment, I.C.J. Reports 2009, pp. 101–103, paras. 115–122; Territorial and Maritime Dispute (Nicaragua v. Colombia), Judgment, I.C.J. Reports 2012 (II), pp. 695–696, paras. 190–193).[5]

As regards the applicable law, the relationship between the UNCLOS articles and customary international law, is presented as follows in *Nicaragua v. Colombia*:

> The Court has recognized that the principles of maritime delimitation enshrined in Articles 74 and 83 reflect customary international law (Maritime Delimitation and Territorial Questions between Qatar and

4 Question of the Delimitation of the Continental Shelf between Nicaragua and Colombia beyond 200 nautical miles from the Nicaraguan Coast (*Nicaragua v. Colombia*) submitted on 16 September 2013 (ICJ Press Release 2013/21); Alleged Violations of Sovereign Rights and Maritime Spaces in the Caribbean Sea (*Nicaragua v. Colombia*), submitted on 26 November 2013 (ICJ Press Release 2013/36).

5 Maritime dispute case (*Peru v. Chile*), Judgment of 14 January 2014, available at www.icj-cij.org, para. 180.

> Bahrain (Qatar v. Bahrain), Merits, Judgment, I.C.J. Reports 2001, p. 91, paras. 167 et seq.)[6]

Similarly, in *Peru v. Chile* the ICJ states:

> The Court proceeds on the basis of the provisions of Articles 74, paragraph 1, and 83, paragraph 1, of UNCLOS which, as the Court has recognized, reflect customary international law (Maritime Delimitation and Territorial Questions between Qatar and Bahrain (Qatar v. Bahrain), Merits, Judgment, I.C.J. Reports 2001, p. 91, para. 167; Territorial and Maritime Dispute (Nicaragua v. Colombia), Judgment, I.C.J. Reports 2012 (II), p. 674, para. 139).[7]

The demonstration in the two judgments of the customary character of the rules in articles 74 and 83 is not very satisfactory. It is based on a reference to the *Qatar v. Bahrain* judgment of 2001.[8] This judgment, however, is explicit only as regards the correspondence to customary law of UNCLOS Article 15 on the delimitation of the territorial sea.[9] As regards Articles 74 and 83, such correspondence can, however, be read only by inference. The Court explicitly says only that the parties agreed that most of the provisions relevant to their case reflect customary international law.[10] The reference to such agreement of the parties is also present in the *Nicaragua v. Colombia* judgment.[11] The applicability of Articles 74 and 83 was also assumed by Peru (not a party to UNCLOS) in the maritime dispute case against Chile.[12]

It is interesting to note that in *Nicaragua v. Colombia* the Court found that there were relevant circumstances in delimitation, and proceeded to a modification of the equidistance line. Contrary to opinions to the effect that in the second stage of the delimitation process only modest adjustments are allowed, the Court moved the equidistance line substantially closer to the Colombian islands than to the Nicaraguan coast. It explained that:

---

6 Territorial and Maritime Dispute (*Nicaragua v. Colombia*), Judgment, I.C.J. Reports 2012 (II), p. 674, para. 139.

7 *Peru v. Chile* quoted at note 4, para. 179.

8 *Qatar v. Bahrein* quoted at note 3.

9 *Ibid.*, at para. 175.

10 *Ibid.*, at para. 167.

11 *Nicaragua v Colombia, supra* note 4, at para. 138.

12 Case Concerning Maritime Dispute (*Peru v. Chile*), available at: <http://www.icj-cij.org/docket/index.php?p1=3&p2=3&code=pc&case=137&k=88>.

> Following this approach does not preclude very substantial adjustment to, or shifting of, the provisional line in an appropriate case, nor does it preclude the use of enclaving in those areas where the use of such a technique is needed to achieve an equitable result.[13]

In *Peru v. Chile* the Court, after finding that there was an (albeit tacit) agreement between the parties accepting a delimitation line following a parallel up to a distance of 80 miles, utilized the three-step methodology to draw the delimitation line beyond the end point of the 80-mile delimitation line. It found that no relevant circumstances existed and that the disproportionality test did not require any modification to the equidistance line.[14]

A delicate question in which it may make a difference whether a rule set out in UNCLOS corresponds to customary law concerns article 121. This article provides for the legal regime of islands and specifies, in paragraph 3, that islands which are "rocks that cannot sustain human habitation or economic life of their own," are entitled only to a territorial sea and a contiguous zone. In *Qatar v. Bahrain* the ICJ had stated that paragraphs 1 and 2 of article 121 corresponded to a customary rule,[15] leaving open the question of the "rocks" considered in paragraph 3. In *Nicaragua v. Colombia* the Court stated that paragraph 3 also reflects customary law. Its reasoning is as follows:

> The Judgment in the *Qatar v. Bahrain* case did not specifically address paragraph 3 of Article 121. The Court observes, however, that the entitlement to maritime rights accorded to an island by the provisions of paragraph 2 is expressly limited by reference to the provisions of paragraph 3. By denying an exclusive economic zone and a continental shelf to rocks which cannot sustain human habitation or economic life of their own, paragraph 3 provides an essential link between the long-established principle that "islands, regardless of their size, … enjoy the same status,

13 *Nicaragua v Colombia, supra* note 5 at para. 197. In his Separate Opinion appended to the Judgment, Judge Abraham argues that the three-step method should not have been followed by the Court because "the construction of a provisional median line as a starting-point for the delimitation is not only highly inappropriate in this case, but that it is even virtually impossible. … " See Separate Opinion of Judge Abraham, at para. 24).

14 Case Concerning Maritime Dispute (*Peru v. Chile*), paras 190–195.

15 The *Nicaragua v. Colombia* 2012 judgment in referring to Case Concerning Maritime Delimitation and Territorial Questions between Qatar and Bahrain (*Qatar v. Bahrain*), Merits, [2001] ICJ Reports, quotes paras 167 and 195 as regards art. 121 paras. 1, and 185 as regards art. 121 para. 2. Only this last paragraph is explicit in stating that the provision of UNCLOS considered reflects customary international law.

and therefore generate the same maritime rights, as other land territory" (*ibid.*) and the more extensive maritime entitlements recognized in UNCLOS and which the Court has found to have become part of customary international law. The Court therefore considers that the legal regime of islands set out in UNCLOS Article 121 forms an indivisible regime, all of which (as Colombia and Nicaragua recognize) has the status of customary international law.[16]

Article 76 of UNCLOS is a complex provision of ten paragraphs, complemented by the 9-articles Annex II. It sets out a new definition of the continental shelf. It determines that its external limit is a line 200 nautical miles (nm) from the baseline, and it provides for its extension beyond such limit if certain requirements are satisfied. These requirements are defined using legal-scientific terms and involve the intervention of a specific technical body, the Commission for the Limits of the Continental Shelf (CLCS). The complexity of the rules and the role assigned to the CLCS, an independent international body established under UNCLOS, makes it difficult, at first sight at least, to argue that article 76 corresponds to customary law in each and every detail, although it may be debated whether some aspects of the article reflect a customary rule.

The question was envisaged in *Nicaragua v. Colombia*. In this case, a question put by Judge Bennouna requested the parties to take a position on it. The answers of the parties converged in agreeing that art. 76, para. 1 corresponds to customary law. They disagreed, however, as to whether this finding also applies to other paragraphs. The Court stated the positions of the parties as follows:

> Nicaragua states that the provisions of Article 76, paragraphs 1 to 7, relating to the definition of the continental shelf and to the determination of the outer limits of the continental shelf beyond 200 nautical miles, have the status of customary international law.
>
> While Colombia accepts that paragraph 1 of Article 76 reflects customary international law, it asserts that "there is no evidence of State practice indicating that the provisions of paragraphs 4 to 9 of Article 76 [of UNCLOS] are considered to be rules of customary international law."[17]

16 *Nicaragua v. Colombia* quoted at note 5, para. 139.

17 *Nicaragua v. Colombia* quoted at note 5, paras. 116–117.

The Court then drew its conclusions:

> The Court notes that Colombia is not a State party to UNCLOS and that, therefore, the law applicable in the case is customary international law. The Court considers that the definition of the continental shelf set out in Article 76, paragraph 1, of UNCLOS forms part of customary international law. At this stage, in view of the fact that the Court's task is limited to the examination of whether it is in a position to carry out continental shelf delimitation as requested by Nicaragua, it does not need to decide whether other provisions of Article 76 of UNCLOS form part of customary international law.[18]

Thus the Court left open the customary status of the paragraphs of article 76 following paragraph 1. Crawford has observed that, in so doing, the Court "etched an awkward line."[19] In fact, it is a prudent line. True, it may be argued that some elements of article 76, beyond paragraph 1, are such that it is very likely that non-parties will follow them when claiming external limits for their continental shelves beyond 200 nm: for instance, the maximum limits of 350 nm from the baselines and of 100 nm from the 2000 meters isobath, and maybe even criteria for determining the outer edge of the continental margin. Practice has not, however, reached this stage. It will be interesting to see whether the United States will rely on these criteria in delineating the outer limit of its continental shelf beyond 200 nm, provided it does so while still being a non-party to UNCLOS.

Even such reliance will not be sufficient to state that article 76 (at least up to para. 7, or the basic principles set out in these paragraphs) corresponds to customary law. It will be necessary that the outer limit so delineated be opposable to all States. Under UNCLOS, such opposability (the fact that the delineation is "final and binding" according to the language of article 76, para. 8) is the effect of the conformity of the delineation to the recommendations of the Commission, namely of an act of a creature of UNCLOS. Will there be substantial opposition of parties to UNCLOS to a delineation of the external limit of the continental shelf beyond 200 nm effected without following the procedure set out in UNCLOS?

---

18 *Nicaragua v. Colombia* quoted at note 5, para. 118.

19 J. Crawford, *Chance, Order, Change: The Course of International Law*, General Course of Public International Law, 365 Recueil des Cours, (2013), pp. 9–381, at p. 103 (in the edition in the "Pocketbooks of the Hague Academy of International Law," 2014, at p. 133).

The conclusion that article 76 as a whole (at least in its basic principles) reflects—or, more likely, has generated—customary law depends, in my view, on the answer to this question and to the above proposed question as to whether the United States (and possibly other non-parties) will follow the criteria set out in that article.

Crawford observes that "there is currently no alternative definition of the outer edge of the continental margin in play, and the application of the multilateral definition in UNCLOS Article 76 has not attracted criticism from third States." On this basis, he draws the conclusion that not only the general rule of paragraph 1, but also "the detailed and evidently rather arbitrary provisions in the rest of the Article" are "an accurate reflection of a practice accepted by law."[20] This view may be an accurate forecast of the answers State practice will provide to the questions proposed above. For the time being it is, in my view, too early to speak of a practice accepted as law.

The position put forward by Crawford may be seen as an application of the above mentioned presumption that UNCLOS, unless the contrary is proven, corresponds to customary international law, or a reflection of the attitude frequently taken by the Court to rely on International Law Commission (ILC) based conventions, or as seen above, on UNCLOS, as a basis for stating that a given rule has customary status.[21] This would, nonetheless, be a rather extreme application, as it would certainly not apply to aspects of article 76 involving the CLCS, which as an institution is a creature of conventional law. The provisions concerning the CLCS are in fact the least amenable to becoming or generating customary rules. It may be noted, however, that Crawford, although speaking of article 76 as a whole (art. 76 (1) and "the rest of the Article") does not dwell on the role of the CLCS.

In *Nicaragua v. Colombia* the ICJ has underlined the importance of the procedures to be followed before the Commission stating that these apply to States parties to UNCLOS even in the context of a dispute with a non-party. After recalling its statement in the 2007 judgment in *Nicaragua v. Honduras* that:

> any claim of continental shelf rights beyond 200 miles [by a State party to UNCLOS] must be in accordance with Article 76 of UNCLOS and reviewed by the Commission on the Limits of the Continental Shelf established thereunder (I.C.J. Reports 2007 (II), p. 759, para. 319),

20 Crawford, *Chance, Order, Change*, p. 103 (133 f. of the pocketbook edition).

21 See my remarks in T. Treves, *Codification* quoted *supra* at note 1.

... the Court states:

> Given the object and purpose of UNCLOS, as stipulated in its Preamble, the fact that Colombia is not a party thereto does not relieve Nicaragua of its obligations under Article 76 of that Convention.[22]

This statement does not imply any customary status for the procedures under article 76. It reinforces nonetheless their status, as they are seen as *erga omnes partes* obligations independent of the bilateral context in which the question of their application may arise.[23]

This peculiar status of the CLCS procedures is reinforced by article 50 of the CLCS Rules of Procedure. It provides that all United Nations member States, thus including non-parties to UNCLOS, are to be notified of the submissions deposited with the CLCS. Further, under Annex III to the CLCS Rules of Procedure, all States receiving this notification may present comments on which the submitting State may then comment.[24] Paragraph 5(a) of Annex I to the CLCS Rules of Procedure only permits the CLCS to consider submissions unless prior consent is denied by "any of the States" considered in that provision, thus including non-parties to UNCLOS. As Nicaragua remarked in answering the question put to the parties by Judge Bennouna in the case concluded with the judgment of 19 November 2012, non-parties to the UNCLOS "have a role in the work of the Commission,"[25] and have taken advantage of the possibility to comment relatively often. In particular, on various occasions, the United States[26] and other non-party States[27] have done so.

---

22 *Nicaragua v. Colombia* quoted at note 5, para. 126.

23 A strong criticism is in the dissent of Judge ad hoc Mensah.

24 CLCS Rules of Procedure, section II 2a (v).

25 Written reply of the Republic of Nicaragua to the question put by Judge Bennouna at the public sitting held on the afternoon of 4 May 2012, para. 18.

26 See CLCS.01.2001.LOS/USA of 18 March 2002, CLCS.02.2004.LOS/USA of 9 September 2004 containing the reactions of the United States to the submissions of the Russian Federation and Brazil, available at: http://www.un.org/Depts/los/clcs_new/commission_submissions.htm.

27 See CLCS.01.2001.LOS/CAN of 26 February 2002 and CLCS.01.2001.LOS/DNK of 26 February 2002 containing respectively the reactions of Canada and Denmark—when they had not yet become bound by the Convention—to the submission made by the Russian Federation, available at: http://www.un.org/Depts/los/clcs_new/submissions_files/submission_rus.htm. Note verbale dated 6 February 2014 from the Permanent Mission of Colombia to the United Nations addressed to the Secretary-General (UN document A/68/743 of 11 February 2014).

A dispute concerning delimitation of maritime areas between States parties to UNCLOS falls within the categories set out in articles 15, 74 and 83 of the Convention. It may concern the territorial sea, the EEZ and the continental shelf. When the parties request the judge to draw a "single" or an "all purposes" line, it may concern all these zones and also other zones in which the States (or a State) involved in the dispute claim rights, at least when the rights claimed are those a State is permitted to have under UNCLOS. When the dispute arises between a State party and a non-party State, it may happen that the non-party State has adopted maritime zones that do not correspond to those envisaged in the Convention.

Do the provisions of the delimitation articles of UNCLOS, which the ICJ considers to correspond to customary international law, also apply in a dispute in which delimitation must be effected between a zone envisaged by the Convention, belonging to a State party, and a zone, belonging to the non-party State, different in name and in legal nature? The answer to the question is not easy legally. It is also politically sensitive. An affirmative answer by a court might be read as recognition by the judge of the conformity with international law of the zone—which may be incompatible with the Convention—adopted by the non-party State.

The *Peru v. Chile* case is an example of this kind of problem. While Chile has a 12 nm territorial sea and a 200 nm EEZ, Peru has, since 1947, claimed a 200 nm "maritime domain" in which, according to the formulation set out in article 54 of the Political Constitution of 1993, it exercises "sovereignty and jurisdiction, without prejudice to the freedom of international communications, pursuant to the law and treaties ratified by the State." According to the same article, such sovereignty and jurisdiction also includes the airspace within 200 nm "without prejudice to the freedom of international communications, pursuant to the law and treaties ratified by the State." There is a discussion in Peru as to whether this is a sovereignty claim or a jurisdictional claim, and practice may be read in different ways. It appeared to Peru necessary to avoid that the Court be confronted with the difficult legal question mentioned above, and to avoid the political difficulty that could arise if Chile argued that, in accepting Peru's claims, the Court would be supporting an extremist position, evidenced by a maritime zone in which the coastal State's claims arguably go beyond those permitted by the Convention. So Peru decided to "place on record" formally before the Court "Peru's commitment to the modern law of the sea as reflected in UNCLOS," and that that the term "maritime domain" was applied "in a manner consistent with the maritime zones set out in the 1982 Convention.[28]

28 CR 2012/27, p. 22 para. 26, (Wagner) ICJ sitting of 3 December 2012 at 3 pm.

This situation, the declaration of Peru, and the consequences drawn by the Court are tersely set out in the Judgment:

> While Chile has signed and ratified UNCLOS, Peru is not a party to this instrument. Both Parties claim 200-nautical-mile maritime entitlements. Neither Party claims an extended continental shelf in the area with which this case is concerned. Chile's claim consists of a 12-nautical-mile territorial sea and an exclusive economic zone and continental shelf extending to 200 nautical miles from the coast. Peru claims a 200-nautical-mile "maritime domain". Peru's Agent formally declared on behalf of his Government that "[t]he term 'maritime domain' used in [Peru's] Constitution is applied in a manner consistent with the maritime zones set out in the 1982 Convention". The Court takes note of this declaration which expresses a formal undertaking by Peru.[29]

The *Peru v. Chile* case shows that disputes before the ICJ involving parties and non-parties to UNCLOS may be occasions for non-parties to give interpretations of their law and practice as conforming with the Convention. Although Peru has not acceded to UNCLOS, it has now accepted, because of the necessities of the case, by its unilateral declaration of which the Court has taken note as a formal undertaking, a key consequence of such accession.

The Chilean Judge ad hoc, Francisco Orrego Vicuña, in his partly concurring and partly dissenting opinion, sets out in detail the legal and practical implications that, in his view, are entailed by the Peruvian formal undertaking. According to Judge Orrego Vicuña's Opinion, from Peru's undertaking, as noted by the Court,

> [i]t follows that Peru is entitled to exercise jurisdiction over its maritime areas up to 12 nautical miles for the territorial sea, 24 nautical miles for the contiguous zone and 200 nautical miles for the exclusive economic zone and the continental shelf.[30]

Moreover, according to Judge Orrego Vicuña,

> A more important consequence of this finding is to the benefit of the international community as a whole. Vessels flying the flag of all nations,

29 *Peru v. Chile*, quoted at note 3., para. 178.

30 Judge Ad hoc Orrego Vicuña, partially concurring and partially dissenting Opinion, para. 9.

> including Chile, whether merchant or military, can now have freedom of navigation beyond the 12 nautical-mile territorial sea of Peru, just as submarines will be able to navigate submerged. Aircraft will also have the right of unauthorized overflight, Restrictions applied to such activities will now have to be lifted.[31]

It is too early to say whether Peru will interpret its commitment as including all points made by Judge Orrego Vicuña. It is also difficult to say whether Peru's declaration was important for bringing the Court to the solution, quite satisfactory for Peru, it reached in its judgment. What can be surmised is that it helped the Court to remove the above-mentioned obstacles, legal and political, which would have made more difficult to decide the case.

---

31 *Ibid.*, para. 11.

CHAPTER 16

One Cannot Change the Wind, but One Can Always Adjust the Sail: The Role of Legal Framework in Developing a Blue Economy

*Assunção Cristas**

Introduction

The blue economy is in the global political agenda for various reasons and its development must be supported by scientific research and political commitment, and go hand in hand with food security, job creation, quality of life, equity and sustainability. Portugal has one of the largest exclusive economic zones (EEZs) in Europe and it is investing in its blue economy by designing a "smart, sustainable and inclusive" blue agenda based on three pillars: strategic, legal and financial.

This chapter focuses mainly on the legal pillar, which sets forth an ocean governance framework that accepts maritime spatial planning (MSP) as its main axis and efficiency, simplification, dematerialization, transparency and coordination as its main principles. Portugal is developing this legal framework to ensure that our blue agenda provides jobs, economic benefits, reduced costs for investors and operators, streamlining, legal predictability and certainty, equal opportunities for all maritime sectors and environmental sustainability. Many challenges lay ahead for all of us, academia, policy and polity, private sector and civil society. This is a unique time to be involved with the ocean, and the Portuguese experience with this legal framing of creating, facilitating and promoting the development of the blue economy within a sustainability framing is certainly an interesting case study of ocean governance.

* Portuguese lawyer, professor, politician and current president of the Centro Democrático e Social—Partido Popular/CDS—People's Party. Served from 2011 to 2015 as Minister of Agriculture, Sea, Environment and Territorial Planning, Portugal.

 | DOI 10.1163/9789004311442_018

Developing a Blue Agenda

We all are dependent on the Ocean. Its importance—at environmental, economic and social levels—has become increasingly evident and many challenges lay ahead for all: academia, public and private sectors, and civil society. Having knowledge for forward-looking, sustainable ocean governance becomes crucial in developing a transition to blue sustainable economy. In past years, Portugal decided to invest on the development of its blue agenda with a political commitment aiming at sustainability, job and wealth creation, food security and quality of life. So this is where we are now: on the verge of having a political and economic relevant blue agenda.

Portugal has the ambitious goal of increasing its ocean economy through a smart, sustainable and inclusive growth, which are keywords in the European Strategy—Europe 2020. Therefore we have to adjust the sail to these new winds in order to take the boat efficiently and safely forward. The political role, supported by a solid legal framing—clear, transparent and user friendly—is crucial for the facilitation and coordination of the different forces and interests at play.

Portugal has one of the largest exclusive economic zones in Europe, with a sea area of 1.7 million square km—an area that amounts to 18 times our land area. Moreover Portugal submitted the extension of its continental shelf; once established the area of Portugal's blue economy will be 40 times larger than our terrestrial area. This process will reinforce our role as one of the most prominent Atlantic countries, meaning we have to constantly be at the forefront, and conferring significant responsibility. For example, within this process of our proposal for the extension of our continental shelf, its first result was the recognition, in 2006, of the Rainbow hydrothermal vent field as the very first marine protected area (MPA) beyond 200 nautical miles under the jurisdiction of a country. In accordance with the United Nations Convention on the Law of the Sea (UNCLOS), Portugal already has established five MPAS under its jurisdiction, located in the extended continental shelf beyond 200 nm, which amount to more than 110.000 km^2. This is a trend we want to continue.

The role of the Ocean for a sustainable future world is unquestionable. This key assumption must guide how we foster and strengthen mechanisms aimed at good governance and sound maritime policies. A sustainable ocean development is dominating the discourse all over the world, and Portugal is committed that such vision is more than mere words. For that, Portugal designed the architecture for a "smart, sustainable and inclusive" blue agenda based on three pillars: strategic, legal, and financial. One of the main objectives of Portuguese maritime policy is to develop these three pillars in a way that allows the transition to a sustainable blue economy, ensuring that practical and realistic solutions contribute to a compromise between economic and social development and environmental sustainability.

Strategic

Portugal historically has been a maritime nation and our culture is marked by our relation to the sea, in many areas of Portuguese culture, from gastronomy to poetry, added by the use and enjoyment of the sea in most of our leisure activities. A revealing curiosity is that Portugal has the second highest per capita consumption of seafood in the world (after Iceland). Especially after the ratification by Portugal of UNCLOS in 1997, a series of successive and incremental activities and measures gave Portugal again a prominent place in the maritime narrative of the world. In 1998, in the International Year of the Ocean, Lisbon hosted the World Exhibition dedicated to "The Ocean, a Heritage for the Future" which was a stepping stone for rediscovering how important the sea was for the development of Portugal. This led to a more strategic thinking about the Ocean.

In 2006, one year before the creation of the European Integrated Maritime Policy, that would provide a more coherent approach to maritime issues, and with increased coordination among different policy areas, Portugal published its first National Ocean Strategy 2006–2016, after a long process of a better understanding the sea and of how Portugal relates to it. This Strategy was based on three strategic pillars: knowledge, spatial planning, and an active promotion and defense of national interests. This Strategy promoted a series of measures in the field of maritime identity and international assertion, capacitation and knowledge, as well as maritime governance.

Policy is a dynamic process and there is a need for constant adaptation to new contexts and new timings. So in 2012 on its mid-life review process, new challenges that justified a new Strategy were encountered. These challenges included the need for an Action Plan, the need to make synchronize the timing of the Strategy with the new European Structural and Investment (ESI) Funds 2014–2020, and an adaptation to the changes at European Policy level, namely the adoption of the European Union Maritime Strategy for the Atlantic Area and its respective Action Plan, and the reform of the Common Fisheries Policy. Furthermore, a new emphasis on blue growth, understood as an inter-sectorial perspective based on knowledge, innovation and on promoting a greater efficiency in the use of resources, was considered an important feature for Portuguese Maritime Policy. Building on the previous Strategy, but with a new vision, the National Ocean Strategy (NOS) 2013–2020, approved by the Council of Ministers, was developed to address these challenges.

The NOS presents a new development model of ocean and coastal areas and is the main document guiding our maritime policy. The NOS is structured around the idea of Portugal as essentially a maritime and coastal country and it is based on the nation's territory of reference, its dimension and geography, and on national identity. These are the foundations on which a new view of the sea takes us towards the use of the ocean as a strategic development vector. In fact, the Portuguese sea is geo-strategically central in the Atlantic axis and has the potential of being a gateway to Europe.

The NOS 2013–2020 focuses on five major objectives: (1) recovering our national maritime identity within a modern, proactive and entrepreneurial framework; (2) achieving economic, geostrategic and geopolitical potential by creating conditions to attract investment, national and international, and promoting growth, employment, social cohesion and territorial integrity; (3) increase, by 2020, the direct contribution of the sea sector to national GDP by 50 percent; (4) strengthen national scientific and technological capacity, stimulating the development of new areas of action; (5) consider Portugal, globally, as a maritime nation and as an unavoidable part of the Integrated Maritime Policy and of the Maritime Strategy of the European Union. From these objectives, it might be relevant to single out that knowledge is transverse to all activity areas of the NOS 2013–2020 and it also involves a prevalence of R&D components, as well as the need for a highly qualified workforce on science and technology. Ocean literacy is also acknowledged in NOS 2013–2020 as fundamental to the success of a maritime country like Portugal.

A main component of NOS 2013–2020 is its Action Plan, entitled the Mar-Portugal Plan, aiming economic, social and environmental valorization of the national maritime space through the implementation of sectorial and cross-sectorial projects. The guiding principles for the implementation of the NOS 2013–2020 are integrated management, precaution and effective participation. These principles promote a mindset of a blue growth model based on coordination, sustainability and openness. The implementation of this strategy will allow Portugal to meet the challenges for the promotion, growth and competitiveness of the maritime economy and for that it needs a solid regulatory framework.

Legal Framework

This legislation was conceptualized having in mind the need to promote simplification, coordination, guarantee broad participation, ensure transparency and assure stability, while having in mind as well that the growth of blue economy inevitably will lead to competing activities. Maritime spatial planning (MSP) was the chosen governance mechanism for this legislation. MSP enables adequate managing of different activities, leading to an effective compatibility between competing uses, a better economic use of the marine environment and the minimization of the impacts of human activities.

On April 10, 2014 the Framework Law on Maritime Spatial Planning and Management (Law nº 17/2014) was published. The law establishes the definition of the main principles of maritime spatial planning and the legal regime for the use of national maritime area. Approved in the Parliament by an overwhelming majority of more than two thirds, the law was further developed in a subsequent diploma that was approved in January 2015 by the Council of Ministers. This step took into account the European Union Directive establishing a framework for Maritime Spatial Planning (Directive 2014/89/EU of the European Parliament and of the Council, of 23 July 2014). These two legal acts are articulated with the NOS 2013–2020 and were designed to attract investment, national and international, and to promote growth and employment, while guaranteeing a focus on the protection of the marine environment. The principles enshrined in this legislation include complying with the ecosystem approach to management, adaptive integrated, multidisciplinary and cross-sector management, improvement and promotion of economic activities in a long-term perspective, and regional and cross-border cooperation and coordination.

In order to achieve maritime spatial management, the legal acts envisage two types of spatial planning instruments: situation plan and allocation plans. These two different types of plans are complementary: the situation plan is the snapshot, current and potential, of the national maritime space, identifying the areas for environmental protection and preservation and the distribution in time and space of current and potential uses and activities; the allocation plans assign areas or volumes of the national maritime space to uses and activities not previously identified in the situation plan. Once approved, allocation plans are automatically integrated into the situation plan.

These plans must undergo a public discussion process, guaranteeing stakeholder engagement, and echo the coordination between different public authorities and sectors. With these plans, anyone interested in developing a sea activity can clearly understand what can be done in the ocean and what rules are applied. Therefore this maritime spatial planning provides legal predictability, reduces costs for investors and operators, and is a step forward of the traditional ad-hoc and sectorial approach and provides a more coherent and integrated pattern of sea use. This approach gives a strong answer to the identified concern of investors with the dispersion of licensing institutions and bureaucratic and complex set of procedures for developing any sea activity. Simplification is achieved both by the promotion of a strong articulation between all institutions involved on the required authorizations and by the dematerialization of the access to licensing of the private use of the sea through an electronic platform. Procedures for obtaining the title of private use of the national maritime space will be performed electronically through an electronic one-stop shop.

We are aware that blue economy might lead to competing activities, and the law deals with that by establishing clear rules to deconflict them. First, good environmental stewardship of the marine environment is not negotiable. If competing uses or activities (on-going or intended) are identified, then the criterion used is the greater social and economic benefit to the country, in particular by creating employment and qualification of human resources, by value creation and by the contribution to sustainable development. When this metric fails to meet a conclusive selection of the prevailing use, ensuring maximum coexistence of uses or activities is used as second criterion.

My job for the last three years has been to design this legal framework for governance that would develop a spatial understanding of the ocean. Maritime policy in Portugal is becoming a major area and Portugal cannot afford too many changes of tack since it has one of the largest EEZs in Europe. Once established, the principles and values associated with MSP have to be stable for a long period of time. Participation by all stakeholders, including political negotiations with the opposition party, was a fundamental stage in the process.

Having laws published is only a first step and its proper implementation depends on many factors. A number of institutional, administrative and capacity challenges have to be overcome for this iterative MSP process to become a reality. There are numerous challenges, namely more scientific knowledge and investment in capacity building for which time and resources are needed. These challenges bring us to the third pillar—financial.

Financial

The question of financing is not straightforward, and much work has yet to be developed both at national and international levels to overcome many existing barriers, namely the cost of capital when investing on the ocean. Investment risk is still too high. There is a need for concerted action at scientific, financial and political levels to reduce the perceived intrinsic risk of ocean based activities. The international debate on the need for new financial tools that fit the ocean specific context has an important role in the process and has been very active and promising.

We believe that the approved legal framework will attract private investors as it offers clear and transparent rules, and above all stability. It is only a first step, but a very relevant one. Furthermore, Portugal is entitled to European Funds that were designed to assist countries in its sustainable development. These European funds, namely the recently approved 2014–2020 European Structural and Investment (ESI) Funds and the Horizon 2020 (for research), will be a significant source of capital for projects throughout Portugal and in Europe. The European Maritime and Fisheries Fund (EMFF) is the ESI Fund that promotes the development of a blue economy in Europe. The design of our National Program to leverage the EMFF followed closely the National Ocean Strategy to ensure that these funds will play a very important role in implementation.

Our existing blue economy is still limited and dispersed, but our policy promotes networks, partnerships and co-operatives in order to gain critical mass. The private sector has a crucial role in all areas of blue economy, and scaling up projects is important to ensure financeable, transparent and accountable investments. To develop innovative financial instruments to overcome these hurdles remains not only a national, but also an international challenge for a true transition to blue economy.

Conclusion

There is still a long way to go or a long distance to sail, to use maritime jargon. The transition to a blue economy demands an effort at all levels and from all stakeholders. A blue economy is a priority for Portugal, and we stress the importance of the conservation and sustainable use of the ocean and of its resources. We have been working on enhancing policy, legal and institutional tools for the development of the blue economy as a privileged way to drive sustained, inclusive and equitable economic growth and job creation. This process requires innovative, integrated and multidimensional approaches that take into account not only the conventional uses of the ocean but also the emerging tendencies and visions that have been main objectives of our ocean policy.

Scientific knowledge and increased technological capability to preserve and develop marine resources are paramount, and therefore our legal framework guides stakeholders to promote and develop environmentally friendly maritime activities. Our policy acknowledges that science, technology, innovation and marine literacy are vital to better understand the ocean and essential to support a science-based approach to decision making. Portugal, therefore, is investing in these areas.

The NOS institutes the planning of the national maritime space guaranteeing that the different existing and potential activities are compatible, and establishing the simplification of the respective administrative procedures.

The legal body constructed so far intends to use space more efficiently, more easily and in a more coordinated way in order to promote the rational use of the sea, to improve decision-making and to balance sectorial interests.

A good legal framework is clearly a necessary phase of the process but not sufficient by itself. So this is a unique time to be involved with the ocean, when a lot is needed, from the policy area, from scientific research, from entrepreneurs, from civil society. Only together we can make the blue agenda happen. The role of the legal framework is just as adjusting the sails: it facilitates a better navigation.

CHAPTER 17

Ever More Lines in the Sea: Advances in the Spatial Governance of Marine Space

*Clive Schofield**

Introduction

The oceans dominate the world spatially. These extensive maritime spaces and their associated coastal zones are critically important to sustaining life on Earth. Coastal and ocean areas are also the drivers for the global economy and play a crucial and increasingly important role in global food security. These areas are therefore of critical importance across scales, from the global to the regional, national and sub-national coastal community levels. The oceans are also increasingly under threat, however, notably as a result of enhanced competition for coastal and marine resources, and intense and diverse uses of coastal and marine spaces. All too frequently this competition leads to friction and tension among marine users and incompatible spatial and temporal overlaps between uses of ocean space. The impacts of climate change on the oceans further compound these issues.

Increasing uses of the oceans and escalating threats related to the oceans demand enhanced management, regulation and enforcement. The chapter explores how marine spatial issues are becoming increasingly important to effective oceans governance, leading to the definition of ever more "lines in the sea" or efforts towards ocean zonation. The chapter starts with a short contextual section regarding the importance of the oceans in diverse ways. It then seeks to highlight how lines are being defined offshore at national and sub-national scales, as well as with respect to areas beyond national jurisdiction. The definition of maritime joint development zones and, more recently, the emergence of so-called "grey areas" offshore are also noted. Some

---

* Director of Research, Australian National Centre for Ocean Resources and Security (ANCORS) and Leader of the University of Wollongong's Global Challenge on Sustaining Coastal and Marine Zones. E-mail: <clives@uow.edu.au>. The author wishes to thank Dr Andi Arsana of the University of Gadjah Mada, Indonesia for his assistance with the preparation of figures that accompany this chapter.

 | DOI 10.1163/9789004311442_019

considerations are also offered concerning areas where jurisdictional clarity is lacking, such as in maritime spaces subject to overlapping claims, and problematic issues related to this phenomenon are outlined.

While traditional sector-specific approaches remain, such as concerning shipping, fisheries and marine energy as well as with respect to the protection of the marine environment and biodiversity through the definition of often impressively broad marine protected areas (MPAs), there is an evident trend towards more integrative approaches. This shift is most readily apparent through the advent of marine spatial planning (MSP) initiatives. The definition of maritime limits and boundaries is essential to the exercise of jurisdiction and thus the effectiveness of management measures. Ultimately, however, the definition of lines in the sea, including the application of marine spatial planning techniques, are not enough in isolation—implementation through management, regulation, surveillance and enforcement within designated lines and zones also are essential to the success of such measures.

The Enduring and Increasing Significance of the Ocean

Renowned science-fiction author Arthur C. Clarke once remarked "How inappropriate to call this planet Earth when it is quite clearly Ocean."[1] The oceans dominate the world spatially, encompass about 72 percent of the surface of the planet. These extensive maritime spaces are of enduring importance to the global economy. Sea lines of communication (SLOCs) serve as critical conduits and ports as essential gateways for international trade in a globalized, interdependent world, and 90 percent of global trade by volume is carried by sea.[2]

The vast extent of the oceans and their role in providing the conduit for the majority of global trade only tells part of the story. The oceans are fundamental to life on planet Earth. The oceans produce 50 percent of the planet's oxygen, cycle 93 percent of CO_2 in the atmosphere and play an essential role in driving

1 Quoted in J.E. Lovelock, "Hands up for the Gaia Hypothesis," *Nature*, Vol. 344 (1990): 100–102, at 102.

2 Harry N. Scheiber, "Economic Uses of the Oceans and the Impacts on the Marine Environment," in Davor Vidas and Peter J. Schei (eds.), *The World Ocean in Globalisation: Climate Change, Sustainable Fisheries, Biodiversity, Shipping, Regional Issues* (Leiden and Boston: Brill/Martinus Nijhoff, 2013), 65–97. United Nations Conference on Sustainable Development (UNCSD), *Rio Ocean Declaration*, Co-Chairs Statement of The Ocean Day at Rio+20, 16 June 2012, Annex II, 1, available at, http://www.unesco.org/new/fileadmin/MULTIMEDIA/HQ/SC/pdf/pdf_Rio_Ocean_Declaration_2012.pdf.

the global atmospheric system.[3] Additionally, the oceans serve as a major carbon and heat sink, absorbing 26 percent of CO_2 emitted due to man and fully 90 percent of excess heat in past 200 years. Furthermore, the oceans are vital to global nutrient cycling and represent a key repository and supporter of biological diversity on a world scale, even though the full extent of the oceans cornucopia of biodiversity remains largely unknown. Indeed, the estimated number of species inhabiting the world's oceans is in excess of 10 million—or some fifty times the number of marine species so far reported.[4]

The role of coastal and marine environments in terms of providing resources is also significant and growing. These environments support and sustain and living resources, notably fisheries and aquaculture, and these resources are critical to food security. It is estimated that more than 2.9 billion people depend on fish for almost 20 percent of their intake of animal protein, while fisheries supply over 15 percent of the animal protein consumed by 4.3 billion people globally.[5] Moreover, the fishing industry supports the livelihoods of at least 540 million people worldwide,[6] or, put another way, it is estimated that fisheries and aquaculture are fundamental to the livelihoods of 10–12 percent of the world's population.[7] Additionally, coastal areas within 100 km of the ocean are estimated to contribute 61 percent of gross domestic product (GDP) globally.[8]

The oceans are, furthermore, an increasing source of energy resources. It is estimated, for example, that 45 percent of remaining recoverable conventional oil resources are located in offshore fields.[9] Analogously, oil industry sources have suggested that 30 percent of global oil production and 27 percent of

---

3 Robin Warner and Clive Schofield, "Climate Change and the Oceans: Legal and Policy Portents for the Asia-Pacific Region and Beyond," in Robin Warner and Clive Schofield (eds.), *Climate Change and the Oceans* (Cheltenham: Edward Elgar, 2012), 1.

4 Danielle Skropeta, "Exploring Marine Resources for New Pharmaceutical Applications," in Warwick Gullett, Clive H. Schofield, and Joanna Vince (eds.), *Marine Resources Management* (LexisNexis Butterworths, Australia, 2011), 211–224, at 221.

5 Food and Agriculture Organization (FAO) *The State of World Fisheries and Aquaculture 2014: Opportunities and Challenges*, (Rome: FAO/United Nations, 2014), p. 4.

6 United Nations Conference on Sustainable Development (UNCSD) 2012, *Rio Ocean Declaration*, Co-Chairs Statement of The Ocean Day at Rio+20, 16 June 2012, Annex II, at p. 6, available at, <http://www.unesco.org/new/fileadmin/MULTIMEDIA/HQ/SC/pdf/pdf_Rio_Ocean_Declaration_2012.pdf>.

7 FAO, *The State of World Fisheries*, p. 6.

8 *Rio Declaration*, p. 6.

9 International Energy Agency (IEA) (2013), *Redrawing the Energy-Climate Map*, World Energy Outlook Special Report, (Paris: Organization for Economic Cooperation and Development (OECD) and IEA), 93.

global gas production in 2010 came from offshore sources such that such areas represent "a non-negotiable imperative for oil companies."[10]

The importance of the oceans is growing. Take Australia as an example. It is estimated that the total measurable value of the country's ocean spaces to the national economy in the 2011–2012 period was in excess of A$47 billion in benefits[11]—nearly double (98 percent) the figure of just one decade before.[12] This estimate as to the value of Australia's marine industry is considered to be conservative, as economic data for a number of key activities remains unavailable.[13] Further, the assessment fails to capture the value of ecosystem services, which were estimated to add a further A$25 billion to bring the overall contribution to of marine industry to around A$72 billion.[14]

Such essential, but often not fully acknowledged, ecosystem services include activities such as scientific research and development, marine safety, desalination, marine cable and pipeline services, coastal protection, and aspects of marine tourism as well as more nebulous and intangible cultural and recreational benefits.[15] It was estimated in the late 1990s that marine systems contributed $21 trillion in such ecosystem services as compared with $12 contributed from terrestrial sources.[16] A more recent assessment on the part of the World Wildlife Fund (WWF), employing a conservative assessment approach estimates the total asset value of the oceans to be at least US$24 trillion, with the annual "gross marine product" associated with the oceans being estimated at over US$2.5 trillion.[17] This impressive figure is thrown into sharper relief when it is realized that it is restricted to renewable economic activities directly associated with the oceans, such that offshore oil and gas is excluded from the

---

10 Sylvian Serbutoviez, "Offshore Hydrocarbons," IFP Energies Nouvelles," *Panorama 2012*, p. 1, available at, <http://www.ifpenergiesnouvelles.com/publications/notes-de-synthese-panorama/panorama-2012>.

11 Australian Institute for Marine Science (AIMS), *The AIMS Index of Marine Industry 2014*, (Townsville: AIMS, 2014), p. 1 and 6.

12 *Ibid.*, p. 6.

13 *Ibid.*, p. 3.

14 *Ibid.*

15 *Ibid.*, pp. 4–5.

16 Robert Costanza, R. d'Arge, R. de Groot, S. Farber, M. Grasso, B. Hannon, K. Limburg, S. Naeem, R.V. O'Neill, J. Paruelo, R.G. Raskin, P. Sutton and M. van den Belt, "The Value of the World's Ecosystem Services and their Natural Capital," *Nature*, 387 (1997): 259.

17 World Wildlife Fund (WWF) *Reviving the Ocean Economy: The Case for Action—2015*, (WWF, 2015), at p. 12.

equation. According to these estimates the oceans rank as the world's seventh largest economy.[18]

Increasing Threats to the Oceans

The counterpoint to the significant benefits and opportunities offered by the oceans is that many of the marine resources and services mentioned above are increasingly under threat. Coastal environments are under pressure from population shifts towards the coast and related development pressures, something encapsulated by the term 'coastal squeeze,' as well as pressures induced by climate change, such as sea level rise, as explored in more detail elsewhere in this volume (see Rogers and Schofied, Chapter 18 in this volume).[19] Offshore, growing diversity and intensity of ocean uses has meant that marine environments are similarly under greater stress with it being estimated, for example, that more than 40 percent of oceans already are strongly affected by humans.[20] Notable contemporary oceans governance challenges include: resource depletion; protecting and preserving the marine environment; overfishing including the impacts of illegal, unreported and unregulated (IUU) fishing; biodiversity preservation on the high seas; and, maritime security to protect against threats such as piracy and armed robbery against shipping.

Taking overfishing and IUU fishing as pertinent examples, the FAO estimated that as of 2011, 28.8 percent of fish stocks were being fished at biologically unsustainable levels and therefore overfished, and fully 61.3 percent of stocks were fully fished, with only 9.9 percent of stocks being classified as underfished.[21] Concerning IUU fishing, while it has to be acknowledged that inherent difficulties exist with respect to estimating the precise dimensions of such activities, they are nonetheless understood to be occurring on a grand scale. Estimates suggest that the annual IUU catch in 2011 ranges from 11 to 26 million tonnes or 14–33 percent of the total legal catch globally.[22]

18 *Ibid.*

19 See, for example, Nigel Pontee, "Defining Coastal Squeeze: A Discussion," *Ocean & Coastal Management* 84 (2013): 204–207.

20 B.S. Halpern, S. Wallbridge, K.A. Selkoe, C.V. Kappel, F. Micheli, C. D'Agrosa, J.F. Bruno, K.S. Casey, C. Ebert, H.E. Fox, R. Fujita, D. Heinemann, H.S. Lenihan, E.M.P. Madin, M.T. Perry, E.R. Selig, M. Spalding, R. Steneck and R. Watson, "A Global Map of Human Impact on Marine Ecosystems," *Science* 319 (2008): 5865.

21 FAO, *The State of World Fisheries*, p. 7.

22 See, "The Future of Fish—The Fisheries of the Future," *World Ocean Review* 2013 (Maribus, 2013), p. 70.

The economic losses associated with such activities are likewise enormous, and estimated to be between US$9–24 billion per year.[23] Indeed, as much as 20 percent of seafood globally is derived from IUU fishing,[24] with West Africa alone losing approximately $1.5 billion per annum to IUU fishers.[25]

Further, while the afore-mentioned WWF report assigns hugely impressive dollar figures to offshore activities, the economic value of the ocean is tied to assets that are in steep decline as a consequence of threats such as pollution and over-fishing; and these concerns are compounded by the impacts of climate change, including ocean warming, deoxygenation and acidification.[26] In this context, "impacts on key marine and coastal organisms, ecosystems and services are already detectable" with high risks of severe impacts in the future.[27] Such conclusions are underscored by the well-established fact that many fisheries are fully or over-exploited globally[28] and defaunation of the ocean is accelerating.[29]

Against this backdrop efforts are being made to enhance management and regulation of marine spaces. Such management efforts are often zonal in character, requiring that jurisdictional lines be defined, delineated and delimited. The following sections seek to highlight a number of such approaches across multiple scales and contexts. It is not an exhaustive enumeration or categorization of such initiatives, but is instead intended to illustrate the breadth and diversity of spatial management mechanisms increasingly being applied offshore. The chapter concludes with some considerations of the opportunities and challenges inherent in effecting such line, area or place based jurisdictional

---

23 See, David J. Agnew, John Pearce, Ganapathiraju Pramod, Tom Peatman, Reg Watson, John R. Beddington, and Tony J. Pitcher, "Estimating the Worldwide Extent of Illegal Fishing," *PLOS ONE* 4(2), 2009, available at, http://www.plosone.org/article/info%3Adoi%2F10.1371%2Fjournal.pone.0004570. See also NOAA and European Commission. Joint Statement between the European Commission and the United States Government on Efforts to Combat Illegal, Unreported, and Unregulated (IUU) Fishing. 7 Sept. 2011.

24 Margot L. Stiles, Ariel Kagan, Emily Shaftel, Beth Lowell, *Stolen Seafood: The Impact of Pirate Fishing on Our Oceans*, OCEANA, 2013, at p. 3, available at, http://oceana.org/sites/default/files/reports/Oceana_StolenSeafood.pdf.

25 Richard Valdmanis and Simon Akam, "Illegal Fishing Plunders and Strains West Africa," *Reuters* (March 15, 2012).

26 WWF, *Reviving the Ocean Economy: The Case for Action—2015*, pp. 7–8.

27 J.-P. Gattuso *et al.*, "Contrasting Futures for Ocean and Society from Different Anthropogenic CO_2 Emissions Scenarios," *Science* 349 (6243) (July 2015), at p. 45.

28 FAO, *The State of World Fisheries*, p. 7.

29 Douglas J. McCauley, Malin L. Pinsky, Stephen L. Palumbi, James A. Estes, Francis H. Joyce and Robert R. Warner, "Marine Defaunation: Animal Loss in the Global Ocean," *Science* 347 (6219) (16 January 2015): p. 247.

and management techniques in ocean spaces. In short, the chapter concludes with analysis of how these diverse lines may be drawn together.

A Spatially Sophisticated System of Maritime Zones

The fundamental guiding framework governing claims to maritime jurisdiction is provided by the United Nations Convention on the Law of the Sea (UNCLOS),[30] which has achieved near universal acceptance with, at the time of writing, 166 parties to it comprising 165 States plus the European Union.[31] A key achievement of UNCLOS was the realisation of a clear spatial framework for the limits to national claims to maritime jurisdiction. The Convention provides for a series of national zones of maritime jurisdiction measured offshore from baselines along the coast. It is worth emphasizing here that maritime claims can only be advanced by coastal States on the basis of their sovereignty over land territory. This latter point is in keeping with the longstanding legal maxim that "the land dominates the sea."[32]

These zones include a territorial sea, with consensus on a maximum limit of 12 nautical miles (nm) measured from baselines (something not previously achieved).[33] Seaward of territorial sea limits UNCLOS also provides for coastal States to claim a contiguous zone within which the coastal State may "exercise the control necessary" to prevent or punish infringements of its customs, fiscal, immigration or sanitary laws out to a maximum limit of 24 nm from baselines along the coast.[34] As most States claim a 12 nm breadth territorial sea, the contiguous zone, if claimed, generally extends from the 12 nm to 24 nm limit as measured from baselines along the coast.

30 *United Nations Convention on the Law of the Sea*, adopted 10 December 1982, 1833 UNTS 3 (entered into force 16 November 1994) (UNCLOS).

31 See, United Nations, Status of the United Nations Convention on the Law of the Sea, of the Agreement relating to the implementation of Part XI of the Convention and of the Agreement for the implementation of the Convention relating to the conservation and management of straddling fish stocks and highly migratory fish stocks, New York, updated to 10 January 2014, available at, http://www.un.org/Depts/los/reference_files/status2010.pdf.

32 Proper Weil, *The Law of Maritime Delimitation—Reflections* (Cambridge: Grotius, 1989), 50.

33 UNCLOS, Articles 3 and 4. It is acknowledged that "M" is the technically correct abbreviation for nautical miles. However, "nm" is frequently used in the literature and is employed as the abbreviation for nautical miles in this volume.

34 *Ibid.*, Article 33.

The general acceptance of maritime claims out to 200 nm from baselines through the introduction of the exclusive economic zone (EEZ) was a particularly significant development.[35] The EEZ resulted in an enormous extension of maritime claims offshore. As noted above, most coastal States claim a 12 nm territorial sea, meaning that the actual breadth of the EEZ is usually 188 nm seaward of territorial sea limits. Claims to contiguous zone and EEZ rights therefore overlap—wholly so for the contiguous zone, partially for the EEZ. UNCLOS also includes revised provisions concerning the delineation of the outer limits of the continental shelf of coastal States, including with respect to continental shelf areas seawards of 200 nm EEZ limits (see Figure 17.1).

The introduction of the EEZ concept has had profound implications in terms of rights over marine resources. For example, in 1984, two years after UNCLOS was opened for signature, the United Nations Food and Agriculture Organisation (FAO) estimated that 90 percent of marine fish and shellfish were caught within 200 nm of the coast.[36] Similarly, it was estimated that 87 percent

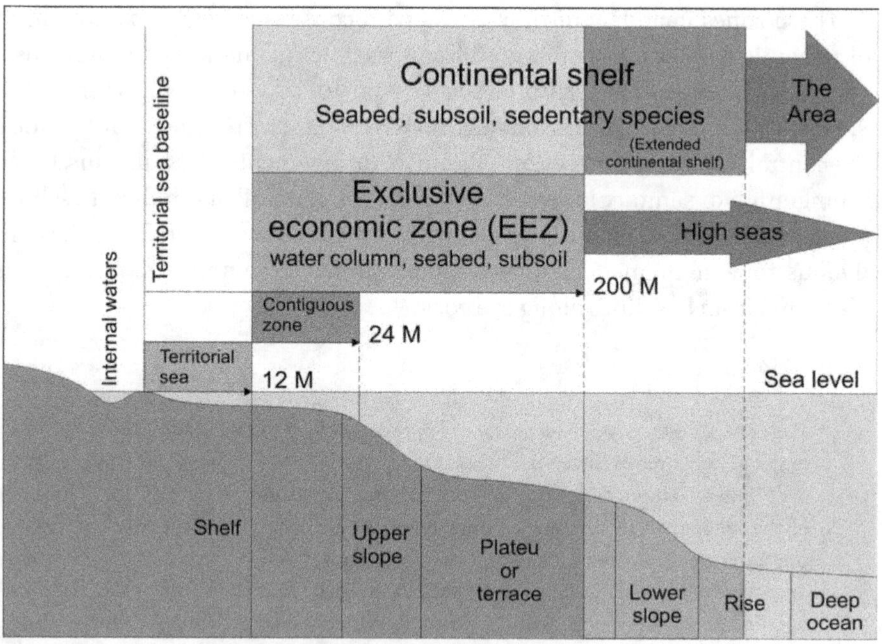

FIGURE 17.1 *Zones of Maritime Jurisdiction under UNCLOS.*

35 *Ibid.*, Article 57.

36 Rachel Schurman, "Tuna Dreams: Resource Nationalism and the Pacific Island's Tuna Industry," *Development and Change* 29 (1998): 107–136, at 107.

of the world's known submarine oil deposits would fall within the 200 nm zone of jurisdiction.[37]

Through its zonal approach UNCLOS provides for a delicate balancing of interests between coastal states and flag states, between the desire for coastal states to access and assert control over marine resources in proximity to their coasts, while simultaneously preserving existing high seas freedoms, especially those related to navigation, within these expanded national maritime areas, notably the EEZ. The consequence is a spatially and jurisdictionally complex, layered and nuanced system of oceans governance where the rights and responsibilities of coastal states and users are reconciled, but may also be in tension. Just as UNCLOS has gained widespread acceptance from the international community of States, so the vast majority of national claims to maritime jurisdiction are consistent with the maritime jurisdictional framework established by the Convention, at least as far as the breadth of maritime claims is concerned.[38]

Ever Creeping Maritime Jurisdiction: Extended Continental Shelf Submissions

UNCLOS established complex criteria regarding how the outer limits of the continental shelf seawards of EEZ limits—the so-called 'outer' or 'extended' continental shelf—may be determined. During the drafting of UNCLOS, States reconciled the already established regime of the continental shelf with a system of maritime zones defined through distance measurements from baselines. By the time of the Third UN Conference on the Law of the Sea it had become clear that the continental shelf of some coastal States would extend considerably beyond the 200 nm limits of the EEZ. The broad continental margin States formed the so-called "marginers" group, and argued successfully for the recognition and preservation of their rights over continental shelf areas

37 Robin R. Churchill and A. Vaughn Lowe, *The Law of the Sea*, 3rd ed. (Manchester, Manchester University Press, 1999), 162.

38 Some outliers do remain in terms of maritime claims which remain inconsistent with the spatial provisions of UNCLOS. For example, a small number of States that still claim territorial seas of greater than 12 nm breadth (Benin, Ecuador, El Salvador and Peru all claim 200 nm territorial seas whilst Togo claims a 30 nm territorial sea). However, numerous formerly excessive maritime claims have been 'rolled back' over time. For instance, Roach and Smith record 26 excessive territorial sea claims rolled back. See, J. Ashley Roach and Robert W. Smith, *United States Responses to Excessive Maritime Claims*, 3rd ed. (Leiden: Martinus Nijhoff, 2012): 139–148.

seawards of 200 nm EEZ limits. This is not to say, however, that there was no change in the rules related to the definition of outer continental shelf limits.

It was well recognized that the existing provisions with respect to the delineation of continental shelf entitlements were flawed. While Article 6 of the Convention on the Continental Shelf of 1958[39] provided for a limit to continental shelf rights based on a depth of 200 meters, it also included, as an alternative, an open-ended definition of continental shelf limits based on exploitability. The latter option for defining continental shelf limits "to a depth beyond that limit where exploitation of resources was possible" was clearly dependent on the technologies available to enable the exploitation of seabed resources leading to the unsatisfactory scenario that continental shelf limits defined on this basis could well change over time.

Consequently, complex criteria were laid down in Article 76 of UNCLOS whereby the outer limits of the continental shelf seaward of EEZ limits—the so-called 'outer' or 'extended' continental shelf—may be determined in partnership with a scientific and technical body established through the Convention—the Commission on the Limits of the Continental Shelf (CLCS).[40] While undoubtedly complex, requiring considerable investment and efforts to acquire information drawn from multiple scientific disciplines, the provisions of Article 76 provide for a definable outer limit to the continental shelf.

Such outer continental shelf limits are delineated by coastal States through a process whereby, in the first instance, proposed outer limits are submitted to the CLCS. The CLCS then provides recommendations related to proposed limits, based on which the coastal State defines its final and binding outer continental shelf limits. The original deadline for such submissions to the CLCS was set at 10 years following the entry into force of the UNCLOS. As the Convention came into force in November 1994, this deadline was set to fall in November 2004. However it became clear that many interested States would be unable to meet this deadline, so it was pushed back to 13 May 2009, ten years after the CLCS adopted scientific and technical guidelines.

The existence of a deadline led to a surge in submissions to the CLCS. From a situation of only 11 submissions one year prior to the deadline, the Commission was faced with 51 full submissions and 41 submissions of preliminary information just after the deadline passed. These figures have subsequently grown further so that by September 2015, over 100 submissions have been made

39 *Convention on the Continental Shelf*, opened for signature 29 April 1958, 499 UNTS 311 (entered into force 10 June 1964).

40 For information on the United Nations Commission on the Limits of the Continental Shelf (CLCS) see, http://www.un.org/Depts/los/clcs_new/clcs_home.htm.

(77 full and 47 preliminary) involving 80 coastal States.[41] These submissions encompass vast areas of continental shelf in excess of 30 million km^2 located seawards of 200 nm EEZ limits (see below).

A regional perspective on coastal State claims to maritime jurisdiction in the South Pacific is set forth in Figure 17.2.

Continuing Jurisdictional Creep: The "Territorialisation" of Maritime Zones

While the breadth of coastal State claims has to a large extent been constrained and contained by the framework established under UNCLOS, States have nonetheless sought to secure additional rights within their national zones and, increasingly, in areas beyond national jurisdiction. Such additional national jurisdictional rights over maritime spaces have been claimed on a variety of grounds including maritime security or the need to ensure environmental

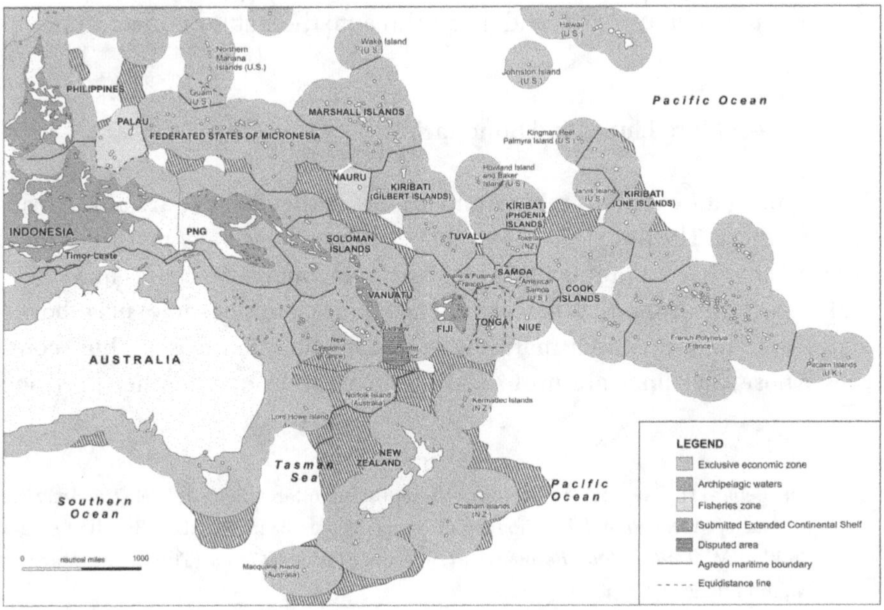

FIGURE 17.2 *Maritime Claims and Boundaries in the South Pacific Region.*

41 Regarding Submissions to the CLCS as well as its Recommendations see the CLCS website at, http://www.un.org/Depts/los/clcs_new/commission_submissions.htm.

protection. With regard to the former concern, it has been noted that there exists a temptation for States to use the 200nm EEZ limit as the basis for claiming broader rights, justified on the basis of contemporary security threats, such as maritime terrorism, leading to "further territorialisation of the EEZ."[42] An example of this type of practice is provided by Iran's 1993 Marine Areas Act, through which Iran claims the right to control the laying of submarine pipelines and cables and to control "any kind of research" within the EEZ.[43] The United States has protested against these provisions in the Iranian legislation. The United States has also objected to Iran's interpretation of the regime of innocent passage, and purported prohibition on military activities of foreign states within its EEZ.[44] Iran is, however, by no means alone in making excessive claims to jurisdiction.[45]

It has further been observed that coastal States have been keen to assert additional controls offshore on environmental grounds, for instance in respect of the transport of hazardous materials,[46] and in the wake of the loss of the *Prestige* oil tanker off the north-western coast of Spain and the resulting oil pollution along those shores, in November 2002.[47] This issue represents a significant challenge to the delicate balance of rights, responsibilities and freedoms contained in UNCLOS, and thus a threat to the stable oceans regime.

Ever More Limits and Boundaries

The significant extensions to maritime claims outlined above are of relatively recent vintage. The practical consequence of the expansion of maritime zones seawards has been a proliferation of overlapping maritime claims and, inevitably, maritime disputes. Wherever the maritime claims of (now) neighboring States overlap, a potential maritime boundary situation exists. Thus, countries whose coastlines are up to 400 nm apart from one another may now

---

42 See, Benard H. Oxman, "The Territorial Temptation: A Siren Song at Sea," *The American Journal of International Law* 100 (2006): 830–851, at 842. See also, Roach, J.A. and Smith, R.W., *United States Responses to Excessive Maritime Claims* (The Hague: Martinus Nijhoff Publishers, 1996), 409–421.

43 United States Department of State, *Iran's Maritime Claims*, Limits in the Seas, No. 114, (Washington D.C.: Office of the Geographer, Bureau of Intelligence and Research, 16 March 1994): 26–33.

44 *Ibid.*, 38–39.

45 See, for example, Roach and Smith, *Excessive Maritime Claims*, 170–178 and 379–398.

46 *Ibid.*, 398–409.

47 Oxman, 2006: 843–849, at 847–848.

require an EEZ and continental shelf boundary to be delimited between them. Furthermore, where extended continental shelf areas are in play the States involved may be considerably further apart. Overall, just over one-half of overlapping maritime entitlements have been resolved through the delimitation of maritime boundaries.

Where each inter-State maritime boundary relationship is as one potential maritime boundary, even if it may be composed of multiple distinct segments, the number of potential maritime boundaries within 200 nm limits has been calculated to be 366.[48] At the time of writing 196 of these 366 potential maritime boundaries or 54 percent of them had at least one agreement in place relating to their delimitation. That said, many of these existing delimitation lines are partial in that they either pre-date the introduction of the EEZ, and therefore relate exclusively to the seabed and subsoil of the continental shelf, rather than the superjacent water or because agreement could be reached on definition of only part of the potential full extent of the boundary line.

Further, these figures do not include potential maritime delimitations with respect to extended continental shelf rights. Submissions to the CLCS exceed 30 million km^2, of which about 3.2 million km^2 is subject to overlapping submissions and therefore constitute large areas located seaward of 200 nm EEZ limits subject to overlapping maritime claims that will ultimately demand resolution, most obviously through the delimitation of extended continental shelf boundaries.[49]

Sub-National Maritime Boundaries and Cooperative Mechanisms

Maritime jurisdictional arrangements within substantially expanded national maritime spaces also have become more complex, especially where efforts have been made towards autonomy and devolution mechanisms. Such domestic political arrangements often have the aim of delivering the benefits to be derived from control over ocean spaces and the marine resources contained

48 Clive H. Schofield, "The Delimitation of Maritime Boundaries: An Incomplete Mosaic," in Wastl-Walter D. (ed.), *The Ashgate Research Companion to Border Studies*, (Farnham: Ashgate, 2011), 665–681, at 670.

49 Clive H. Schofield and Robert van de Poll, "Exploring the Outer Continental Shelf," in Michael Lodge and Kenning Zhang (eds.), *Implementation of Article 82 of the United Nations Convention on the Law of the Sea. Report of the International Workshop convened by the International Seabed Authority in collaboration with the China Institute for Marine Affairs in Beijing, the People's Republic of China, 26–20 November, 2012*, ISA Technical Study No.12, (Kingston, Jamaica: International Seabed Authority, 2013), 72.

therein, as well as responsibility for such areas to sub-national political units. Such a scenario demands further definition of sub-national scale maritime jurisdictional zones and boundaries.

A notable example in this context is provided by the Indonesian experience. Indonesia's Law No.22 of 1999 provides for regional autonomy that extends to the maritime spaces of the largest archipelago in the world. Specifically, Article 10(2) of the legislation authorizes regional authorities to exercise responsibility for exploration, exploitation and management, spatial administration and law enforcement of regulations, albeit in conjunction with national authorities for the enforcement of State security and sovereignty. Under the law, Indonesia's Provincial authorities are granted authority out to 12 nm from their coasts (Article 3) while its lower order Regencies (local government areas) and municipal regions are accorded responsibility over maritime spaces out to one-third of this distance, or 4 nm (Article 10(3)). When it is considered that Indonesia has 34 Provinces, comprising a total of 508 regencies or local government areas, the proliferation of internal maritime boundaries required and the scale of the challenge inherent in this devolutionary enterprise is daunting.

A further example of this type of practice is provided by arrangements that have been drafted as part of the peace process designed to settle the long-running conflict in the southern Philippines. In January 2014, through an *Agreement on the Bangsamoro Waters and Zones of Joint Cooperation* was concluded as an Addendum to the Annex on Revenue Generation and Wealth Sharing and the Annex on Power Sharing between the Government of the Philippines and the Moro Islamic Liberation Front (MILF).[50] This arrangement provides the anticipated autonomous area located in the southern Philippines, the "Bangsamoro political entity," jurisdiction out to 22.224 km (12 nm) from the low-water mark along the coast. The agreement also provides the definition of lines 'equally distant' from the shores of local government units of the Bangsamoro and thoses of the rest of the Philippines, as well as the establishment of "Joint Zones of Cooperation" (see Figure 17.3).[51]

Moreover, even where international maritime boundaries between neighboring States remain disputed or at least undelimited, this does not necessarily preclude cooperation at the sub-national level. Examples of this type of arrangement are provided by a Memorandum of Agreement (MoA) concluded between the authorities of the neighboring coastal provinces of

---

50 See, http://www.gov.ph/downloads/2014/01jan/20140125-On-the-BM-Waters-and-Zones-of-Joint-Cooperation.pdf.

51 *Ibid.*

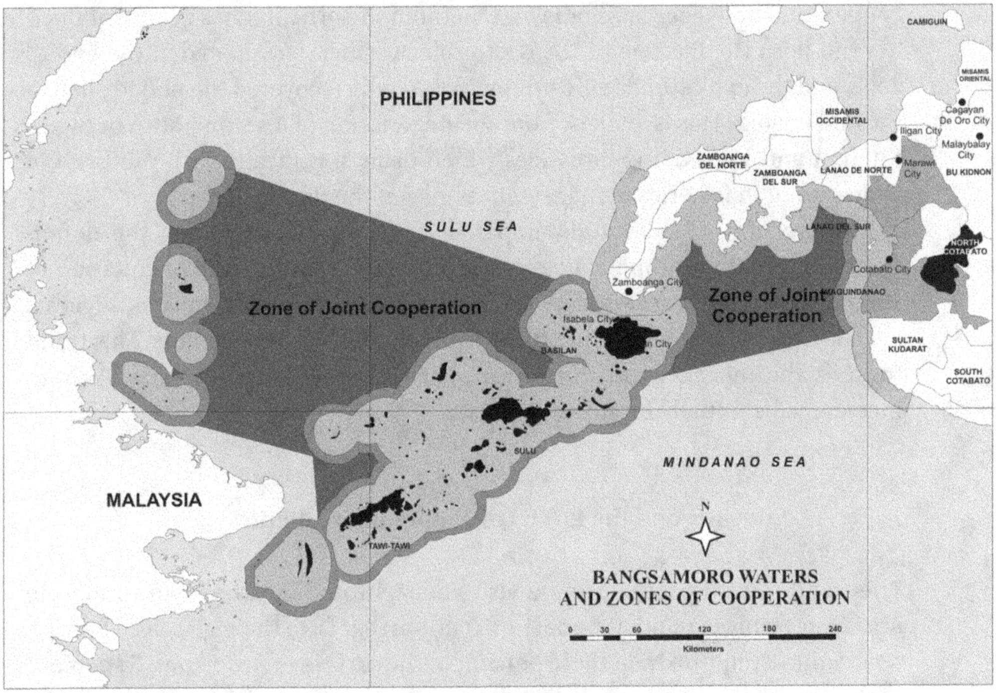

FIGURE 17.3 *Bangsamoro Waters and Areas of Joint Cooperation.*
SOURCE: SCHOFIELD AND ARSANA, 2016.

Cambodia (Kampot Province)[52] and Vietnam (Kien Giang Province) in 2008 and analogously an MoA between the adjoining coastal border provinces of Cambodia (Koh Kong Province) and Thailand (Trat Province), also in 2008.[53] Both of these sub-national maritime cooperative arrangements arose through the United Nations Global Environmental Facility (GEF)[54] project on "Reversing Environmental Degradation Trends in the South China Sea and Gulf of Thailand."[55] It is noteworthy that the sub-national maritime

52 Memorandum of Agreement between The Provincial People's Committee of Kien Giang Province (S.R. Vietnam) and The Governor of Kampot Province (Kingdom of Cambodia), *Policy Framework for Cooperation in the Management of Coastal Ecosystems and Natural Resources*, Kampot, Cambodia, 27 March 2008.

53 MoA included in *Report on The Third Joint Meeting between the Management Teams of the Peam Krasop Wildlife Sanctuary* (PKWS) *and Trat Demonstration Sites*, UNEP/GEF South China Sea Project, Trat Province, Thailand, (18–20 February 2008).

54 See, Global Environment Facility (GEF) webpage at, https://www.thegef.org/gef/home.

55 John C. Pernetta and Si Tuan Vo, "The UNEP/GEF South China Sea Project: Lessons Learnt in Regional Cooperation," *Ocean and Coastal Management* 53 (2010): 589–596.

cooperation between Cambodia and Vietnam in particular has been sustained beyond both the lifetime of the GEF project, which terminated at the end of 2008, and the expiration of the original MoA, which covered the period 2008 to the end of 2012. This is evident from the negotiation of a second MoA between Kampot and Kien Giang Provinces which deals specifically with the issue of cooperation in fisheries management between the parties.[56]

The creation of sub-national spatial arrangements including the definition of internal maritime zones and boundaries as well as the negotiation of sub-national yet transboundary cooperative maritime mechanisms suggests even increasing complexity in offshore jurisdictional arrangements. This trend towards the management of marine space at more local scales is likely to increase substantially in the future.

Zoning Beyond the Limits of National Jurisdiction

The trend towards ocean zoning also appears to be set to extend to marine areas ostensibly beyond national jurisdiction, that is, to the high seas. The high seas comprise approximately 40 percent of the surface of the planet, 64 percent of the area of the oceans and almost 95 percent of their volume.[57] While in the past there were perceptions that the sheer size of areas beyond national jurisdiction (ABNJ) would mean that interactions and conflicts between different activities would be minimized, it is becoming clear that this is unlikely to be the case.[58] For example, in the Clarion-Clipperton Fracture Zone located in the Equatorial North Pacific Ocean, the International Seabed Authority has (ISA) has already designated an extensive system of contract areas as well reserved areas.[59] It has been estimated that the Clarion-Clipperton Zone alone holds

56 Plan of Cooperation in Fisheries Management between Kampot Administrative Committee (Kingdom of Cambodia) and The Provincial People's Committee of Kien Giang Province (S.R. Vietnam), *Implementing the Cooperation in Fisheries between Kampot Fishery Department and Kien Giang Fishery Department*, Kampot, Cambodia, 29 May 2014.

57 Global Environment Facility (GEF) "Areas Beyond National Jurisdiction (ABNJ)," available at https://www.thegef.org/gef/ABNJ.

58 Secretariat of the Pacific environment Programme (SPREP) and the Commonwealth Scientific and Industrial Research Organisation (CSIRO), "Interactions between Sectors in Areas Beyond National Jurisdiction (ABNJ): Deep Sea Mining (DSM), Fishing Vessel Activity, and Ecologically or Biologically Significant Areas (EBSA)," *Information Brief*, 20 May 2015.

59 For maps detailing areas of exploration as well as information on contractors and reserved areas, see *Exploration Areas, International Seabed Authority*, http://www.isa.org.jm/en/scientific/exploration.

more than 27 billion metric tons (mt) of nodules containing on the order of 7 billion mt of manganese, 340 million mt of nickel, 290 million mt of copper and 78 million mt of cobalt, as well as rare earths needed for the production of many high-technology products, such as smart phones.[60] Consequently, this area is highly likely to be the focus for future deep sea mining (DSM) activities. The ISA's Legal Counsel, Michael Lodge, suggested in May 2013 that "[w]e are on the threshold of a new era of deep seabed mining."[61]

The ISA's elaborate system of lease areas for DSM overlap not only with areas of high fishing activity, but also with ecologically or biologically significant areas (EBSAs)[62] relating to the Equatorial High-Productivity Zone and the Clipperton Fracture Zone petrel foraging area.[63] These overlaps give rise to potential conflicts between DSM operations and fishing activities, for example, both of which are liable to have cumulative impacts on the marine environment within the designated areas and beyond.

Recognition of the vulnerability of such essentially ungoverned marine spaces to activities such as overfishing, DSM, dumping and shipping led to the establishment in 2006 of the "Ad Hoc Open-ended Informal Working Group to study issues relating to the conservation and sustainable use of marine biological diversity beyond areas of national jurisdiction" [the Working Group].[64] A major breakthrough on these issues was achieved in 2015. First, in January 2015, the Working group recommended that formal negotiations be launched towards a new implementing agreement under UNCLOS to address these issues. Subsequently, on June 19, 2015, the UN General Assembly adopted a Resolution on the development of a legally binding international instrument

60 David Shukman, "Deep Sea Mining 'Gold Rush' Moves Closer," *BBC News*, 18 May 2013, http://www.bbc.co.uk/news/science-environment-22546875.

61 Vicky Validakis, "Deep Sea Bed Mining Rush a Step Closer to Reality," *Mining Australia*, 21 May 2013, http://www.miningaustralia.com.au/news/deep-sea-mining-rush-a-step-closer-to-reality; see also, Christopher Werth, "Deep Sea Mining: Economic Bonanza or Environmental Boondoggle," 20 May 2013, http://www.theworld.org/2013/05/deep-sea-mining-economic-bonanza-or-environmental-boondoggle/. A metric ton (or tonne) is equivalent to 1,000 kilograms.

62 EBSAs were introduced from 2008 following a meeting of the Conference of the Parties to the Convention on Biological Diversity. See, https://www.cbd.int/ebsa/. For details on the scientific criteria for the identifying EBSAs, known as the "Azores Scientific Criteria and Guidance" see, https://www.cbd.int/doc/meetings/mar/ebsaws-2014-01/other/ebsaws-2014-01-azores-brochure-en.pdf.

63 EBSAs No.16 and No.19 respectively. Details available at, https://www.cbd.int/ebsa/. See also, SREP/CSIRO, "Interactions between sectors in Areas Beyond National Jurisdiction (ABNJ)."

64 See, http://www.un.org/Depts/los/biodiversityworkinggroup/biodiversityworkinggroup.htm.

on the conservation and sustainable use of the marine biological diversity in areas beyond national jurisdiction.[65] In the present context it is notable area based management tools, including marine protected areas are among the four elements to be discussed in the negotiations.[66]

Alternatives to Dividing Lines: Maritime Joint Development Arrangements

Where agreement on the definition of a dividing line offshore through an international maritime boundary has proved difficult to achieve and overlapping claims persist, in some instances an alternative joint or shared management mechanism has been implemented. Such cooperative mechanisms at sea are often termed maritime joint development zones.[67] Although a number of these cooperative mechanisms predate UNCLOS, such joint maritime zones have predominantly been concluded since the Convention was opened for signature. The key provisions within UNCLOS in this respect concern provisional arrangements, and are found in Articles 74(3) and 83(3). These articles provide, in identical text, that:

> Pending agreement as provided for in paragraph 1, the States concerned, in a spirit of understanding and cooperation, shall make every effort to enter into provisional arrangements of a practical nature and, during this transitional period, not to jeopardize or hamper the reaching of the final agreement. Such arrangements shall be without prejudice to the final delimitation.

65 United Nations General Assembly (UNGA), "Development of an International Legally Binding Instrument under the United Nations Convention on the Law of the Sea on the Conservation and Sustainable Use of the Marine Biological Diversity of Areas beyond National Jurisdiction," Resolution 69/292, 19 June 2015, available at, http://daccess-dds-ny.un.org/doc/UNDOC/GEN/N15/187/55/PDF/N1518755.pdf?OpenElement.

66 The other elements under discussion are access and benefit sharing arrangements related to marine genetic resources, environmental impact assessments and capacity building and transfer of marine technology.

67 See, for example, Clive H. Schofield, "Blurring the Lines: Maritime Joint Development and the Cooperative Management of Ocean Resources," *Issues in Legal Scholarship*, Berkeley Electronic Press, Vol. 8, no.1 (Frontier Issues in Ocean Law: Marine Resources, Maritime Boundaries, and the Law of the Sea, 2009), Article 3.

A number of joint zones have been defined in conjunction with agreed maritime boundary lines, such as those between Bahrain and Saudi Arabia (where the entirety of the zone in question is on the Saudi Arabian side of the line),[68] between Norway and Iceland[69] and Denmark and the United Kingdom[70] (where the distribution of the joint zone is uneven across the delimited boundary) and China and Vietnam (where the joint zone straddles the boundary line more evenly.[71]

The majority of maritime joint development zones, however, have been established in the absence rather than in conjunction with the delimitation of a maritime boundary. Examples include the joint zone established between

---

68 *Bahrain-Saudi Arabia Boundary Agreement Dated 22 February 1959* (signed 22 February 1958, entered into force 26 February 1958). Treaty text available from the United Nations Division on the Law of the Sea's (DOALOS) "Maritime Space" database at <http://www.un.org/Depts/los/LEGISLATIONANDTREATIES/index.htm>. See also, Jonathan I. Charney and Lewis M. Alexander (eds.), *International Maritime Boundaries*, Vol. II (Dordrecht: Martinus Nijhoff, 1993), 1489–1497.

69 *Agreement between Norway and Iceland on Fishery and Continental Shelf Questions*, 28 May 1980 (entered into force 13 June 1980). Overall, 61 per cent of the joint zone lies on the Norwegian side and 39 per cent on the Icelandic side of the boundary line. Additionally, while each State is entitled to 25 per cent of revenues deriving from the exploitation of oil and gas on the other side of boundary, hydrocarbon fields straddling the joint zone and Icelandic waters are considered wholly Icelandic.

70 *Agreement between the Government of the Kingdom of Denmark together with the Home Government of the Faroe Islands on the one hand and the Government of the United Kingdom of Great Britain and Northern Ireland on the other hand relating to the Maritime Delimitation in the area between the Faroe Islands and the United Kingdom*, 18 May 1999 (entered into force 21 July 1999) Treaty text available at the DOALOS website, *supra* note 3. See also, Jonathan I. Charney and Robert W. Smith (eds.), *International Maritime Boundaries*, Vol. IV (Dordrecht: Martinus Nijhoff, 2002), 2955–2977. The majority of the 2,337nm^2 (c.8,000 km^2) "Special Area" lies on the UK side of the boundary line.

71 *Agreement between the People's Republic of China and the Socialist Republic of Viet Nam on the Delimitation of the Territorial Sea, the Exclusive Economic Zone and Continental Shelf in Beibu Bay/Gulf of Tonkin* (25 December 2000) and Agreement between the People's Republic of China and the Socialist Republic of Viet Nam on Fisheries Cooperation for the Gulf of Tonkin. Treaty text available at the DOALOS website, *supra* note 3. See also David A. Colson and Robert W. Smith (eds.) *International Maritime Boundaries*, Vol. V (Leiden: Martinus Nijhoff, 2005), 3745–3758. The joint "Common Fishery Zone" established extends 30.5nm on either side of the agreed boundary line and encompasses approximately 30,000km^2 (around 8,747nm^2).

Japan and the Republic of Korea,[72] joint arrangements between Malaysia and Thailand,[73] and between Malaysia and Vietnam[74] in the Gulf of Thailand (see Figure 17.4), and joint mechanisms establishes initially by Australia and Indonesia,[75] and subsequently between Australia and Timor-Leste[76] in the Timor Sea.

Where deadlock in maritime boundary delimitation negotiations exists, maritime joint development arrangements may prove to be attractive, and there arguably has been a growing trend toward their use. That said, such arrangements are not to be entered into lightly, not least because they entail a considerable degree of sustained, ongoing cooperation between the parties involved.[77]

---

72 *Agreement between Japan and the Republic of Korea Concerning Joint Development of the Southern Part of the Continental Shelf Adjacent to the Two Countries*, 30 January 1974 (entered into force 22 June 1978).

73 See, *Memorandum of Understanding between the Kingdom of Thailand and Malaysia on the Establishment of a Joint Authority for the Exploitation of the Resources of the Sea-Bed in a Defined Area of the Continental Shelf of the Two Countries in the Gulf of Thailand*, done on 21 February 1979; and, *Agreement between the Government of Malaysia and the Government of the Kingdom of Thailand on the Constitution and Other Matters Relating to the Establishment of the Malaysia-Thailand Joint Authority.* See, Charney and Alexander, *International Maritime Boundaries*, 1111–1123.

74 *Memorandum of Understanding between Malaysia and the Socialist Republic of Vietnam for the Exploration and Exploitation of Petroleum in a Defined Area of the Continental Shelf Involving the Two Countries* was signed on 5 June 1992 and entered into force on 4 June 1993.

75 *Treaty between Australia and the Republic of Indonesia on the Zone of Cooperation in an Area between the Indonesian Province of East Timor and Northern Australia*, 11 December 1989.

76 Notably the Timor Sea Treaty and Treaty on Certain Maritime Arrangements in the Timor Sea (CMATS). See, *Timor Sea Treaty* (Dili, 20 May 2002, entry into force 2 April 2003) and Treaty on Certain Maritime Arrangements in the Timor Sea (CMATS, (signed 12 January 2006, in force 23 February 2007).

77 See, for example, Clive H. Schofield, "No Panacea?: Challenges in the Application of Provisional Arrangements of a Practical Nature," in Myron H. Nordquist and John N. Moore (eds.), *Maritime Border Diplomacy* (Leiden/Boston: Martinus Nijhoff, 2012), pp. 151–169.

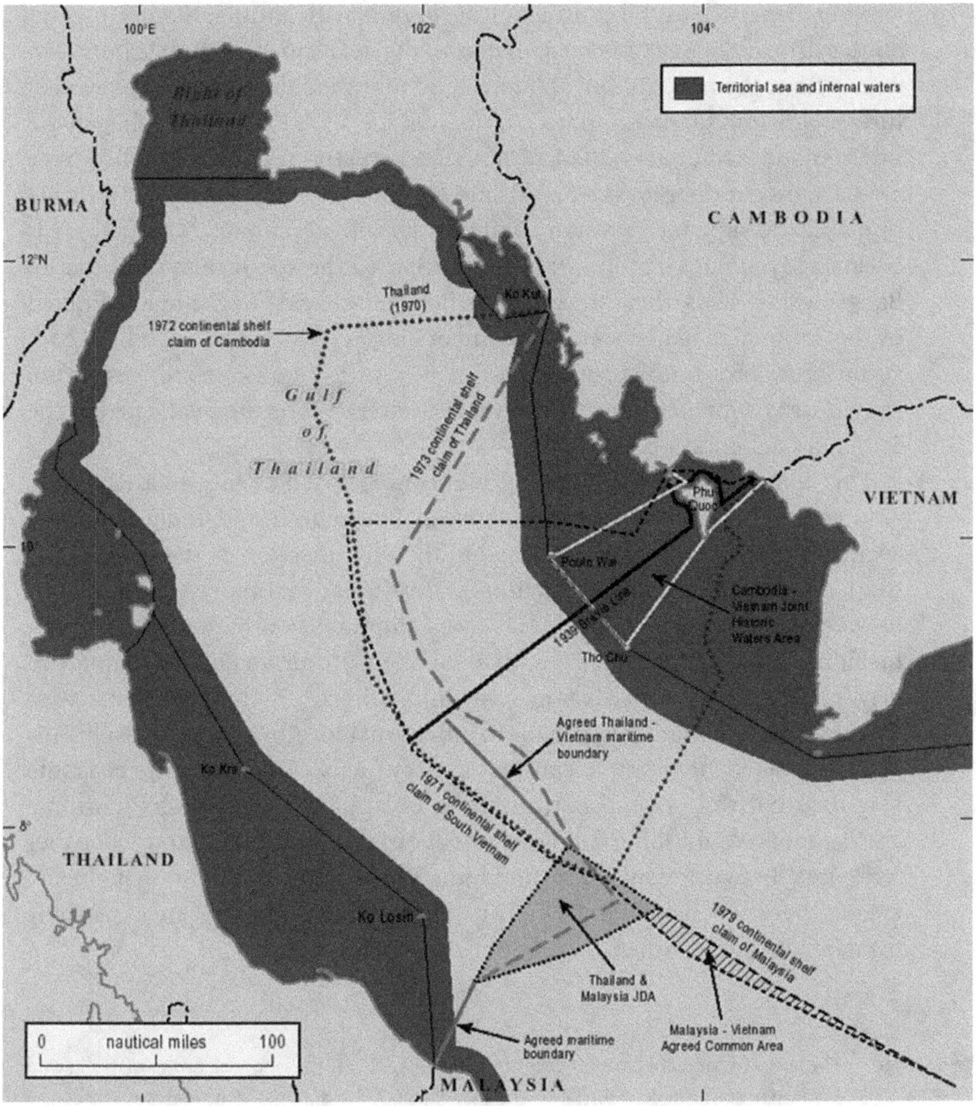

FIGURE 17.4 *Maritime Claims and Joint Development Zones in the Gulf of Thailand.*
SOURCE: SCHOFIELD AND ARSANA, 2016.

Shades of Grey

An intriguing parallel to joint zones is provided by maritime spaces where sovereign rights and jurisdiction over one maritime zone, such as the water column, rests with one coastal State while sovereign rights and jurisdiction over another area, such as the continental shelf, rests with another State. This situation exists in the Bay of Bengal, thanks to the decision of arbitral tribunals to depart from the median line and to extend maritime delimitation seaward of the 200 nm limit to the extended continental shelf. For example, a ruling of the International Tribunal on the Law of the Sea (ITLOS) creates a so-called "grey area" where continental shelf rights are accorded to Bangladesh, but jurisdiction over the superjacent water column rests with Myanmar. This outcome is because Myanmar's 200 nm limit lies farther to the south than does that of Bangladesh.[78] Thus, those areas on the Bangladeshi side of the line delimited by the Tribunal that are located seaward of Bangladesh's 200 nm EEZ limit constitute areas of extended continental shelf, yet this same area still lies within 200 nm of Myanmar, and its rights over the water column are unaffected by the Tribunal's delimitation.[79]

This complexity of the situation was compounded in July 2014 when the Arbitration Tribunal in the case between Bangladesh and India issued its award.[80] In addition to issues related to the identification of the terminus of the land boundary and delimitation of territorial sea and EEZ boundaries between the parties, the Tribunal also delimited continental shelf areas, including areas of extended continental shelf. In an analogous manner to ITLOS in the Bangladesh-Myanmar scenario, a departure from equidistance as a method of delimitation was made in order to achieve an equitable result. This decision was in large part because of the concavity of Bangladesh's coastline meaning that were equidistance lines to be applied, Bangladesh's maritime entitlements would suffer from severe 'cut off' effects. Consequently, a further "grey area" between Bangladesh and India was defined, similar to that already created between Bangladesh and Myanmar. These two grey areas partially overlap with one another.

78 The entirety of Bangladesh's 200 nm limit lies within the 200 nm limits of India and Myanmar combined by virtue of the fact that both India and Myanmar possess coasts located to the south of Bangladesh's coast.

79 Dispute Concerning Delimitation of the Maritime Boundary between Bangladesh and Myanmar in the Bay of Bengal (Bangladesh/Myanmar), International Tribunal for the Law of the Sea (ITLOS), Case no.16, Judgment, 14 March 2012, available at, http://www.itlos.org/fileadmin/itlos/documents/cases/case_no_16/1-C16_Judgment_14_02_2012.pdf, [hereinafter *Bay of Bengal Case*], para. 474.

80 In the matter of the Bay of Bengal Maritime Boundary Arbitration, The People's Republic of Bangladesh and the Republic of India, Award of the Arbitration Tribunal, 7 July 2014, paras 498–508, available at, http://www.pca-cpa.org/BD-IN%2020140707%20Award2890.pdf?fil_id=2705.

Such arrangements of different jurisdictional lines for different maritime zones are not entirely unique. In the Timor Sea, for example, Australia and Indonesia have delimited non-coincident continental shelf and EEZ boundaries, meaning that areas of Australian continental shelf are overlain by water column under Indonesian jurisdiction.[81] Similarly, the Torres Strait treaty also splits the vesting of rights in this case between Australia and Papua New Guinea.[82] Moreover, the Tribunal observed that there are many ways for the parties to address their responsibilities, including 'appropriate cooperative arrangements.'[83] It can also be suggested that the delimitation of a maritime boundary only represents the starting point for collaborative oceans management. It can be anticipated that such novel arrangements will become more common in the future, as any departure from the equidistance methodology relating to the delimitation of extended continental shelf areas between adjacent States will inevitably yield a "grey area" situation.

A Problematic Absence of Lines: Overlapping Claims Areas

A lack of marine jurisdictional lines, and particularly an absence of maritime delimitation or even the clear definition off the extent of maritime claims is likely to impede oceans governance. The extensive overlapping claims to maritime jurisdiction that result will tend to exacerbate management problems and impair access to valuable marine resources. While a distinction can be drawn between merely undelimited maritime boundaries and active maritime boundary disputes, the existence of broad areas of competing claims to the same maritime space is problematic.

Where overlapping maritime claims exist, the consequent uncertainty over jurisdiction complicates ocean resource management. With regard to living resources, the oceans remain an important source of living resources, with fisheries representing a major industry and playing a key food security role for many coastal states (despite increasing rates of stock depletion). The sustainable management of, for example, fish stocks, can be severely hampered

81 Treaty between Australia and Indonesia establishing an Exclusive Economic Zone Boundary and Certain Seabed Boundaries, 14 March 1997, available at www.austlii.edu.au/au/other/dfat/treaties/notinforce/1997/4.html (not in force).

82 Treaty between Australia and Papua New Guinea concerning Sovereignty and Maritime Boundaries in the area between the Two Countries, including the area known as Torres Strait and Related Matters, 18 December 1978, (1985) 4 *Australian Treaty Series*.

83 *Bay of Bengal Case*, para. 476.

through, at the least, uncoordinated policies and, at the more severe end of the spectrum, potentially destructive and unsustainable competition for access to the resources in question. Such activities can lead to confrontation between rival fishing fleets and such friction can lead to the involvement of the armed forces of the coastal States concerned with the attendant potential for incidents, clashes and ultimately escalation towards conflict. Further, the jurisdictional uncertainty inherent in areas of overlapping claims has the potential to undermine maritime security as, in cases where jurisdiction is contested, coastal State rights with regard to surveillance and enforcement will remain similarly uncertain. Additionally, such a scenario tends to place rival naval vessels in close proximity to one another providing the potential for incidents and even confrontation between them as each side exerts its enforcement rights in what they regard as rightfully 'their' maritime space. In short, rival maritime claims can act as a major irritant in bilateral, and indeed multilateral, relations.

Overlapping offshore claims are broad, yet frequently ill-defined. This lack of definition results from the fact that coastal States often simply make ambit claims without specifying the exact spatial limits to their claimed jurisdictional rights. Further, the rights and responsibilities of claimants and other States within such contested maritime spaces is often unclear. Such lack of maritime jurisdictional clarity is anathema to the proper development and management of marine resource. With regard to seabed energy resources for instance, which could prove crucial to the well-being and political stability of coastal States, extensive overlapping claims forestall access to these valuable resources. Indeed, overlapping claims essentially represent no-go areas for the international oil and gas industry because of the absence of fiscal and legal certainty that is a fundamental to their decision to commit the tens of millions of dollars necessary to undertake offshore exploration and development projects. Numerous examples exist where the issuing of overlapping petroleum exploration acreages on the part of rival neighbouring States has led to diplomatic (and not so diplomatic) tensions. Similarly, in a form of modern-day gun-boat diplomacy, should one claimant State seek to initiate exploration let alone drilling activities in disputed waters, confrontation often results, with the naval forces of the "other" side ordering the cessation and removal of survey or exploration vessels from the contested zone.

Fragmentation and Complexity in Oceans Governance

Traditionally marine resources and activities have predominantly been managed on a sector-specific basis. This approach is evident from ocean zonation to address specific marine resources, such as fishing zones and aquaculture sites or activities such as routeing measures including traffic separation schemes,[84] roadsteads[85] and anchoring zones (or, alternatively, no anchoring zones).[86] The offshore oil and gas industry is also familiar with the designation of blocks of within which exploration and exploitation for seabed hydrocarbons can proceed. Analogous approaches are increasingly being adopted with respect to potential deep sea mining activities as demonstrated by the above-mentioned lease areas defined by the ISA. Additionally, marine areas have been allocated for dredging, disposal or access restricted for security purposes.

The significant growth in the marine renewable energy sector, that is, wave, wind or tidal power, is also leading to new spatial impacts at sea. In particular, offshore wind has expanded substantially in recent years with over 3,000 offshore turbines have been installed and connected to the grid in 11 European countries with a cumulative total of 10,393.6 megawatts in 82 wind farms.[87] Ambitious future forecasts include 40 GW of European offshore wind power generation by 2020, with a further 110GW to be installed between 2020 and 2030. The uptake of marine renewable energy is also growing in Asia, especially on the part of China and Korea.[88] Just as is the case for oil and gas installations such as oil rigs and floating production platforms have safety zones designated around them, so too do offshore wind turbines.[89]

Moreover, as alluded to above, growing realization that the oceans and vulnerable marine environments, habitats and biodiversity are under threat has led to major advances in terms of ocean zoning for the sake of marine conservation. An excellent, if now rather venerable, example of this practice

---

84 As approved by the International Maritime Organisation (IMO) under Chapter V of the International Convention for the Safety of Life at Sea (SOLAS). See, http://www.imo.org/en/OurWork/Safety/Navigation/Pages/ShipsRouteing.aspx.

85 That is, marine areas used for the loading or unloading of ships or for anchoring. Under Article 12 of UNCLOS if such areas exist beyond territorial sea limits, they may be included in the territorial sea.

86 For example, Indonesia, Malaysia and Singapore, through the IMO, have jointly sought to prohibit anchoring in the Straits of Malacca and Singapore to ensure safety of navigation and to prevent damage to submarine cables and pipelines. See, IMO, 'Safety of Navigation: Information Concerning Anchoring in the Traffic Separation Scheme in the Straits of Malacca and Singapore,' SN.1/Circ.282, 27November 2009, available at, Ent damage to submarine cables and pipelines.

87 The European Wind Energy Association (EWEA), The European offshore wind industry—key trends and statistics, 1st half 2015, available at, http://www.ewea.org/fileadmin/files/library/publications/statistics/EWEA-European-Offshore-Statistics-H1-2015.pdf.

88 See, for example, Clive H. Schofield and Tavis Potts, "If You Don't Like Looking at Wind Farms, Why Not Build Them at Sea?," *The Conversation*, 23 June 2015 <https://theconversation.com/if-you-dont-like-looking-at-wind-farms-why-not-build-them-at-sea-43614>.

89 UNCLOS, Articles 60(5) and 80.

is provided by the Great Barrier Reef Marine Park.[90] The park, established over four decades ago in 1975, encompasses an area of 344,400 and is therefore among the oldest and largest marine parks in the world.[91] Indeed, the Great Barrier Reef Marine Park was the largest marine park on the planet but no longer holds that record thanks to a proliferation in the establishment of large-scale marine parks including the Northwestern Hawaiian Islands (NWHI) Marine Sanctuary, subsequently renamed the Papahānaumokuākea Marine National Monument,[92] was declared June 2006 which covers 362,580 km^2, the Chagos Marine Reserve[93] declared around the Chagos Islands group in the British Indian Ocean Territory (BIOT) in April 2010 as the world's largest fully no-take marine reserve at 640,000 km^2. The establishment of the latter MPA was, however, successfully challenged by Mauritius in a UNCLOS Annex VII arbitration case concluded in March 2015.[94] Additionally, the Coral Sea Commonwealth Marine Reserve,[95] declared November 2012, which covers 989,842 km^2 located to the north and east of the Great Barrier Reef Marine Park.

The advent of marine protected areas (MPAs) as well as the designation of broad particularly sensitive sea areas (PSSAs),[96] on such a grand scale has been prompted partially by the realisation that the oceans are so under-protected yet increasingly under threat. The Convention on Biological Diversity (CBD)[97] of 1992 also appears to have proven to be influential. Under the CBD agreement was reached to conserve 10% of the marine environment through

---

90 The Great Barrier Reef Marine Park Authority's webpage is available at, http://www.gbrmpa.gov.au/.

91 The Great Barrier Reef Marine Park was established through the Great Barrier Reef Marine Park Act, Act no.75 of 1975 which is available at, https://www.comlaw.gov.au/Details/C2011C00149.

92 See, http://www.papahanaumokuakea.gov/.

93 See, http://chagos-trust.org/about/chagos-marine-reserve.

94 The Arbitration Tribunal found that in declaring the MPA the UK had failed to give due regard to the pre-existing fishing rights of Mauritius around the Chagos Islands. See, In the Matter of the Chagos Marine Protected Area Arbitration, between the Republic of Mauritius and the United Kingdom of Great Britain and Northern Ireland, Award, 18 March 2015, available at, http://www.pca-cpa.org/MU-UK%2020150318%20Awardd4b1.pdf?fil_id=2899.

95 See, http://www.environment.gov.au/topics/marine/marine-reserves/coral-sea.

96 See, the International Maritime Organization, "Particularly Sensitive Sea Areas," available at, http://www.imo.org/en/OurWork/Environment/PSSAs/Pages/Default.aspx.

97 The text of the *Convention on Biological Diversity* (CBD) is available at, https://www.cbd.int/doc/legal/cbd-en.pdf.

MPAs with parties to the CBD to meet this target by 2020.[98] Here, however, it should be noted that the term marine protected area covers a broad range of mechanisms and oceans management arrangements from closed or 'no take' areas such as the above-mentioned Chagos Marine Reserve to multiple-use areas such as the Great Barrier Reef Marine Park where the 'general use' zone within the park covers fully 34% of its overall area.[99]

There has also been strong take up of MPAs on a regional basis. For example, the Framework for a Pacific Oceanscape[100] for oceans governance and conservation was endorsed by Pacific Island Forum in 2009. The Pacific Oceanscape initiative is designed to encourage action in support of the Pacific Islands Regional Oceans Policy and has acted as the catalyst for establishment of or proposals for a network of large MPAs on the part of Pacific Island States. Examples include Kiribati's 408,250 km^2 Phoenix Islands Protected Area (PIPA)[101] declared in 2006 (yet only 'no take' from 2015),[102] proposals for a Cook Islands Marine Park[103] potentially covering around 1.1 km^2 million which was announced in 2012 but has yet to be realized, and Palau's Marine Sanctuary,[104] proposed in 2014 with the intention of encompassing fully 80 percent of Palau's EEZ, or an area of 492,923 km^2 (see Figure 17.5). It remains to be seen how a number of these proposals develop and are implemented in practice and whether this nascent network of marine parks evolves towards a more integrated regional approach to oceans management.

98 CBD Aichi target 11 states that: "[b]y 2020, at least ... 10 per cent of coastal and marine areas, especially areas of particular importance for biodiversity and ecosystem services, are conserved through effectively and equitably managed, ecologically representative and well-connected systems of protected areas and other effective area-based conservation measures, and integrated into the wider landscape and seascape (CBD)."

99 For more on the complex zoning arrangements that exist within the Great Barrier Reef Marine Park see, http://www.gbrmpa.gov.au/zoning-permits-and-plans/zoning.

100 See, Pacific Islands Forum Secretariat, 'Framework for the Pacific Oceansacpe, available at, http://www.forumsec.org/pages.cfm/strategic-partnerships-coordination/pacific-oceanscape/pacific-oceanscape-framework.html.

101 See, http://www.phoenixislands.org/.

102 See, Christopher Pala, 'Kiribati commits to fishery-free reserve,' *Science Insider*, available at, http://news.sciencemag.org/asiapacific/2014/06/kiribati-commits-fishery-free-reserve. See also, http://www.phoenixislands.org/shownews.php?newsId=88.

103 See, http://www.conservation.org/projects/Pages/marae-moana-cook-islands-marine-park.aspx.

104 See, http://www.palauoceans.org/marinesanctuary/.

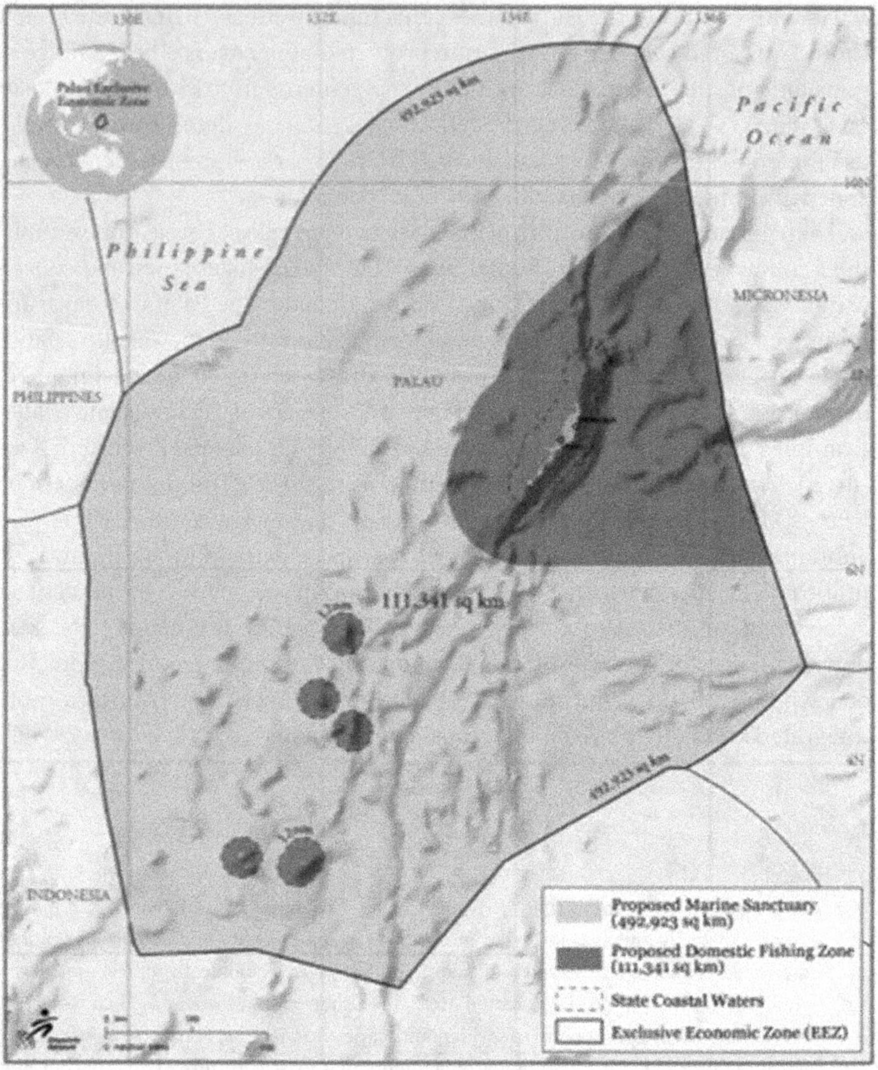

FIGURE 17.5 *Palau's Marine Sanctuary.*
SOURCE: BY KIND PERMISSION OF THE GOVERNMENT OF PALAU.

Weaving the Lines Together

The substantial extension of national claims to maritime jurisdiction seawards has generated a large number of overlapping maritime claims and boundaries in keeping with the layered, overlapping maritime zones provided for under UNCLOS. Considerable progress has been achieved in the delimitation of maritime boundaries and settlement of maritime disputes over time, albeit somewhat gradually and incrementally. That said, almost half of potential maritime boundaries remain undelimited, meaning that there remains substantial areas of overlapping claims within which maritime jurisdiction is uncertain and contested. Moreover, there are ever increasing demands for spatial management of maritime spaces across scales, as illustrated by growing requirements for sub-national maritime boundaries and transboundary arrangements or, indeed, cooperative mechanisms such as maritime joint development zones. In short, the number and types of different boundaries and lines in the sea to define, delineate and manage has evolved in character and complexity.

The remarkable proliferation of lines and zones offshore outlined in this chapter is indicative of increasingly crowded and intensively utilized marine spaces. The concern in this context is that potentially incompatible marine resource uses and activities interact and collide, leading to tensions, competition for access, disputes and conflict as well as cumulatively adverse impacts on the marine environment in particular.

Accommodating and reconciling increasingly diverse and conflicting marine uses and activities, especially where relatively 'new' uses, such as marine renewable energy developments, MPAs and marine ecotourism activities impinge on or conflict with longstanding traditional uses such as fishing or shipping activities, presents enormous challenges.[105] For example, the potential benefits to be derived from the definition of areas of the oceans for the purpose of protecting or conserving valuable marine biodiversity and habitats is often acknowledged and welcomed in principle and proven in practice. Indeed, it has been calculated that tourism expenditure in the Great Barrier Reef Marine Park in 2011–2012 was over A$6.4 billion, directly or indirectly supporting almost 70,000 jobs.[106] Nonetheless, their designation has frequently proved to be highly contentious and contested, as their introduction tends to equate to constraints on the activities of existing users and interests.[107]

---

105 See, for example, Tavis Potts, "MSP and Various Uses and Interests Relating to the Marine Environment," in Daud Hassan, Tuomas Kuokkanen and Niko Soininen (eds.), *Marine Spatial Planning and International Law: A Transboundary Perspective*, (Oxford: Routledge/ Earthscan, 2015), 42–60.

106 68,978 full time equivalent jobs according to Delotte Access Economics, *Economic contribution of the Great Barrier Reef Marine Park 2013*, Report for the Great Barrier Reef Marine Park Authority, 2013, cited in, AIMS, *The AIMS Index of Marine Industry 2014*, 10.

107 See, for example, Marcus Haward and Joanna Vince, *Oceans Governance in the Twenty-first Century: Managing the Blue Planet* (Cheltenham: Edward Elgar, 2008), 20.

A particular theme that has emerged over recent decades is dissatisfaction with traditional, generally sector-specific approaches to oceans governance, giving rise to sustained calls for more cross-cutting, coordinated and ideally integrated oceans management frameworks. Advocates of more comprehensive and integrated approaches invariably lament the manifest deficiencies of existing sectoral oceans management resulting in fragmented, disjointed, piecemeal management efforts featuring spatial and/or temporal mismatches, overlaps or lacunae. Such approaches tend to consider the requirements of a particular sector or activity in isolation from other uses of the ocean, often allowing for the uncoordinated expansion of existing ocean activities and resource exploitation coupled with the threats posed by conflicting uses where activities in one sector impact on another, ultimately leading to compounded and cumulative impacts derived from multiple activities—something aptly described as "the tyranny of small decisions."[108] However, despite general agreement regarding the merits of shifting from sectoral to more integrated approaches to coastal and oceans governance, progress in this direction has been summed up as "slow, tortuous and patchy."[109]

The increasing spatialization of the oceans in recent times represents an important response to this challenge. In the context of the enduring quest for integrated oceans governance, a key mechanism or tool that has emerged is marine spatial planning (MSP). Here it can be emphasised that the objective of MSP is not necessarily to replace well established sector-oriented management approaches but rather to offer strategic planning with a view to improving decision-making, thereby delivering more rational and harmonised uses of marine space, ideally eliminating overlaps, clashes and conflicts and identifying co-development opportunities.[110] This type of approach is by no means new.

Indeed, the concept of MSP builds on decades of debate on how to improve oceans governance and consequent moves towards enhanced coastal and marine environmental protection emerging since at least the 1970s such as integrated coastal zone management (ICZM) initiatives.[111] Further, it can be argued that the sophisticated zoning and management plans established for the multiple-use Great Barrier Reef Marine Park and particularly the extensive rezoning of the park that occurred in the mid-1990s, represents an

108 *Ibid.*, 21.

109 *Ibid.*, 4.

110 See, Potts, "MSP and Various Uses and Interests Relating to the Marine Environment."

111 Haward and Vince, *Oceans Governance in the Twenty-first Century: Managing the Blue Planet*, 21–23.

early example of MSP in practice. Indeed, the rezoning process involved a lengthy and frequently intense process of engagement with users, scientists and environmentalists and the general public.[112] Consequently, conservation-oriented spatial conservation measures have been viewed as 'one of the precursors' to MSP.[113]

A substantial literature continues to develop on MSP. These contributions frequently emphasize the ecological principles or concepts that underpin the concept, seeing MSP as "a framework for implementing an ecosystem-based coordinated governance structure in the world oceans."[114] It has been argued that MSP is especially well attuned ecosystem based management approaches as "ecosystems are places" such that "ecosystem-based management is... inherently place-based."[115] Area or place-based analysis is inherently bounded, and spatial management can therefore be ideal since it can be oriented towards a particular ecosystem. It has also been observed that "management always occurs in a delimited space," emphasising the importance of defining lines in the sea.[116] Moreover, the MSP approach has gaining considerable traction and is increasingly embedded in oceans governance policy-making.[117]

MSP has been characterised as "an idea whose time has come."[118] However, for all the promise seemingly offered by this approach, substantial challenges exist to the successful implementation of MSP. This is especially the case when existing resource uses or socio-economic, cultural and political priorities or objectives conflict with new uses or environmental concerns. For example, the fishing industry tends to view MSP with suspicion, regarding it as a means whereby marine spaces that were traditionally available to be fished are taken away as a result of "new" activities such as the construction of offshore windfarms or the designation of marine protected areas. Consequently, it

---

112 See, for example, Melissa M. Foley *et al.*, "Guiding Ecological Principles for Marine Spatial Planning," *Marine Policy* 34 (2010): 955–966, at 957.

113 Potts, "MSP and Various Uses and Interests Relating to the Marine Environment."

114 Foley *et al.*, "Guiding ecological principles for marine spatial planning," 963.

115 Larry Crowder and Elliott Norse, "Essential Ecological Insights for Marine Ecosystem-based Management and Marine Spatial Planning," *Marine Policy*, 32 (2008): 772–778.

116 *Ibid.*

117 The European Parliament and Council of the European Union (2014). Directive of the European Parliament and of the Council 2014/89/EU Establishing a framework for maritime spatial planning, available at, http://eur-lex.europa.eu/legal-content/EN/TXT/?uri=celex:32014L0089.

118 Paul M. Gilliland and Dan Laffoley, "Key Elements and Steps in the Process of Developing Ecosystem-based Marine Spatial Planning," *Marine Policy* 32 (2008): 787–796, at 787.

is generally well acknowledged that stakeholder engagement such as that illustrated through the Great Barrier Reef Marine Park rezoning process noted above, is a crucial ingredient to the success of MSP.[119] There are, however, limitations to how far MSP can reconcile competing and conflicting interests and activities and trade-offs are therefore inevitable.

Overall, the definition of ever more lines in the sea and the advent of MSP can be viewed as a significant positive step or even step-change in approach to the management of marine spaces. Indeed, spatial approaches offer huge potential to yield significantly enhanced and even integrated oceans governance. Nonetheless, it can also be recognized that lines on their own are not enough. Not only does the location and significance of such lines need to be properly communicated to users to be useful, but they need to be underpinned by appropriate regulation and legislation and backed up by adequate maritime management, surveillance and enforcement assets and systems. For all the challenges involved in implementing MSP, this approach offers the tantalizing promise of integrated oceans governance becoming closer to a practical reality.

---

119 Foley *et al.*, "Guiding Ecological Principles for Marine Spatial Planning," 963. See also, Gilliland and Laffoley, "Key Elements and Steps in the Process of Developing Ecosystem-Based Marine Spatial Planning," 795.

CHAPTER 18

Responding to Changing Coasts: The Need for Fixed and Flexible Limits and Boundaries in the Face of Sea-Level Rise

Kerrylee Rogers and Clive Schofield***

Introduction

The most recent report from the Intergovernmental Panel on Climate Change (IPCC) is unequivocal in identifying warming of the Earth's system.[1] Observations indicate elevated atmospheric and oceanic temperature, diminishing snow and ice extent and increased concentrations of carbon dioxide and other greenhouse gases in the atmosphere since the 1950s.[2] Further, and of especial importance to coastal States, are observations of sea-level rise including significant increases in the rate of sea-level rise into the 21st century. In particular, while the mean rise in sea level of 0.19 m over the course of the 20th century was relatively low, there exist strong indications that the rate of sea-level rise has accelerated. Tide-gauge and satellite altimetry data correspond and indicate a mean rate of global average sea-level rise of 3.2 mm y^{-1} between 1993 and 2010, which is progressively greater than the mean rate of global averaged sea-level rise for the period 1971 to 2010 (2.0 mm y^{-1}), and 1901 to 2010 (1.7 mm y^{-1}).[3]

* Australian Research Council (ARC) Future Fellow in the School of Earth and Environmental Science and GeoQuest Research Centre at the University of Wollongong. Dr. Rogers is the recipient of an Australian Research Council Future Fellowship (FT130100532). Email: <kerrylee@uow.edu.au>.

** Director of Research, Australian National Centre for Ocean Resources and Security (ANCORS) and Leader of the University of Wollongong's Global Challenge on Sustaining Coastal and Marine Zones. E-mail: <clives@uow.edu.au>.

1 IPCC, ed. *Climate Change 2013: The Physical Science Basis. Contribution of Working Group I to the Fifth Assessment Report of the Intergovernmental Panel on Climate Change* (Cambridge, United Kingdom and New York, NY, USA: Cambridge University Press, 2013), 4.

2 *Ibid.*

3 IPCC, ed. *Climate Change 2013: The Physical Science Basis. Contribution of Working Group I to the Fifth Assessment Report of the Intergovernmental Panel on Climate Change* (Cambridge, United Kingdom and New York, NY, USA: Cambridge University Press, 2013), 11.

 | DOI 10.1163/9789004311442_020

Sea-level rise will have global implications. The primary focus for concern among policy-makers and commentators in this context has generally been threats to low-lying coastal land territory and the populations inhabiting these areas, together with the property and infrastructure located there. This has become an increasing concern, since as a consequence of longstanding trends in the movement of populations from predominantly interior and rural settings to coastal and urban ones, global populations are increasingly concentrated in the coastal zone. It has been estimated that 44% of the global population are located within 150 km of the coast.[4] As a result of increasing populations and associated development in coastal contexts, coastal landscapes are highly contested. In particular, there is increasing pressure for urban expansion, industrial development, and agriculture. Especially parlous is the situation of significant populations living in especially low-lying coastal areas, such as the mega-deltas of the world, that are increasingly vulnerable to permanent or at least periodic inundation owing to sea-level rise.[5] Associated concerns relating to increased salt-water intrusion and disturbance to valuable habitats have also been raised.[6]

It is clear that the effects of sea-level rise extend beyond environmental impacts to social and economic impacts, with the costs associated with minimising the impacts of sea-level rise likely to be substantial.[7] These costs will

---

4 This is more than populated the globe in 1950. Source: UN Atlas of the Oceans, "Human Settlements on the Coast," United Nations, available at, http://www.oceansatlas.org/servlet/CDSServlet?status=NDoxODc3JjY9ZW4mMzM9KiYzNz1rb3M~.

5 For example, Vietnam's Mekong and Red Rivers in the Asia-Pacific are likely to be particularly susceptible to the threat of inundation, as large areas are below mean sea level at present sea levels. Consequently it has been suggested that over 7 million people living in the Mekong delta area alone could be displaced by sea-level rise of 1 metre Thomas W. Doyle, Richard H. Day and Thomas C. Michot, *Development of Sea Level Rise Scenarios for Climate Change Assessments of the Mekong Delta, Vietnam* (Reston, Virginia, US Geological Survey Open-File Report, 1165, 2010), available at http://pubs.usgs.gov/of/2010/1165/. See also, UNDP/United Nations Development Programme, 2011, "Vietnam: Sea Level Rise Could 'Displace Millions'," 20 May 2011, available at http://www.undp.org.vn/digitalAssets/25/25093_VIETNAM_sea_level_risex.pdf.

6 David Freestone, "International Law and Sea Level Rise," in Robin R. Churchill and David Freestone (eds.) *International Law and Global Climate Change*, (London/Dordrecht: Graham and Trotman/Martinus Nijhoff, 1991), 119–122. See also, David Freestone and John Pethick, "Sea Level Rise and Maritime Boundaries: International Implications of Impacts and Responses," in Gerald Blake (ed.) *International Boundaries; Fresh Perspectives*, Volume 5, (London: Routledge, 1994), 73–90.

7 IPCC, ed. *Climate Change 2014: Impacts, Adaptation, and Vulnerability. Part A: Global and Sectoral Aspects. Contribution of Working Group I to the Fifth Assessment Report of the*

vary depending upon the action taken to mitigate and adapt to the impacts of climate change. Consequently, there is an increasing need to plan for sea-level rise so as to minimize the environmental, socio-economic and jurisdictional effects of projected sea-level rise in the 21st century.

This chapter starts by exploring the meaning of the term *sea level* and noting some of the complexities and uncertainties of projecting this phenomenon. Two particular challenges arising from rising sea levels are then highlighted. First, the implications of sea-level rise for the management of coastal wetlands against the context of the coastal zone as increasingly contested space; and, second, the potential impact of sea-level rise with regard to maritime jurisdictional issues, especially as concerns the baselines, maritime limits and boundaries that are defined, delineated and delimited under the international law of the sea. The discussion continues by exploring the response of coastal wetlands (primarily mangrove and salt marsh areas) to relative sea-level rise. This section emphasizes the complexity inherent in projecting the response of coastal ecosystems to sea-level rise by presenting a case study of what is termed "ecosystem response modelling." Then, our discussion turns to planning for coastal ecosystem services, using the above-mentioned case study to highlight the potential failings of existing approaches to coastal ecosystem conservation and protection when ecosystems are impacted by sea-level rise.

The potential threat to the spatial extent of the maritime jurisdiction of coastal States resulting from the landward retreat of territorial sea baselines as a consequence of sea-level rise is then discussed. The likelihood of significant diminution in the spatial scope of national maritime claims needs to be kept in perspective. Nonetheless, the traditional linkage between baselines tied to the location of low-water lines along the coast that are susceptible to change means that the critical basepoints from which maritime zones are measured are potentially vulnerable to sea-level rise. In this context, a range of potential response options are noted.

Finally, the concluding part of the Chapter draws out the similarities, contrasts, and analogies between legal and policy frameworks for coastal ecosystems and associated maritime jurisdictional rights. A key commonality is that both coastal and marine planning are complex undertakings; both of them feature increasing and often conflicting yet coincident activities and pressures; and both involve multiple competing actors and interests. We acknowledge this complexity and recognize that there is not an all-purpose approach that suffices in planning for sea-level rise in the 21st century. Consequently, while

Intergovernmental Panel on Climate Change (Cambridge, United Kingdom and New York, NY, USA: Cambridge University Press, 2014).

planning in the context of increasingly contested coastal and marine spaces and ecosystems is likely to demand adaptive approaches featuring flexible rather than static management boundaries, there is simultaneously a desire for the clarity and certainty associated with fixed and stable baselines, limits, and boundaries in the maritime jurisdictional context. Reconciling these apparently conflicting approaches, with their divergent objectives, may demand a paradigm shift in approaches to managing coastal and marine zones in the face of sea-level rise.

Relative Sea-Level Rise

There is broad agreement in the scientific community that sea levels are rising and doing so at an increasing rate, as exemplified by the IPCC's conclusions noted above. Nonetheless, changes in sea level exhibit substantial spatial and temporal variability. Uncertainties remain, in large part because of the multitude of factors involved and thus the complexity of the issue.[8] One particularly salient point with respect to these issues is that the coastline and its associated ecosystems, and indeed the low-water line critical to the determination of maritime limits and boundaries under the international law of the sea, is defined by both elevation of the land and the sea. This phenomenon is called "relative sea level." Changes in relative sea level may be attributed to glacio-eustatic processes that influence the global volume of water in the oceans; to processes that influence the shape of the oceans or the shape of the land's surface, such as tectonism and isostasy; and also to subsidence that may be attributed to anthropogenic impacts.[9] Additionally, seasonal, regional, and temporal tidal variability and the influence of complex atmospheric-oceanic interactions such as the El Niño Southern Oscillation (ENSO) can lead to substantial, periodic and geographically variable, changes in sea level.[10] Importantly, eustatic, that is, global, sea-level rise also has feedbacks on the shape of the land surface by causing coastal retreat and enhanced storm erosion—the two processes do not act independently. Defining relative sea level changes is therefore complex

8 See, for example, Clive Schofield, "Holding Back the Waves: Sea Level Rise and Maritime Claims," in Ruppel, O.C., Roschmann, C. and Ruppel-Schlichting (eds), *Climate Change: Legal Responses and Global Responsibility* (Nomos Publishing, Germany, 2013), 593–614.

9 Colin V. Murray-Wallace and Colin D. Woodroffe, *Quaternary Sea-Level Changes* (Cambridge, United Kingdom and New York, NY, USA: Cambridge University Press, 2014), 41–78.

10 *Ibid.*, 72.

and dependent upon projecting changes in eustatic sea level, the shape of the land and seabed, and feedback between these processes that causes changes in coastal geomorphology.

Despite the considerable challenges involved, considerable scientific research effort is being directed towards projecting relative sea level at timescales relevant for coastal planning and management, and to facilitate adaptation to sea-level rise. However, as illustrated by the case study we present in this paper, accurate projections of relative sea-level rise for a particular coastline is a complex undertaking and would require consideration of the response of the full range of coastal ecosystems along that coastline. The case study presented is for one coastal ecosystem in isolation, so that extrapolating findings from this example to other coasts and ecosystems is problematic. Nonetheless, the case study illustrates the complex interactions between sea level change and coastal systems and ecosystems, highlighting the need for flexible, nuanced management responses, and illustrating the need to factor into those responses the problems of defining properly associated management and jurisdictional limits and boundaries.

Coastal Wetlands and Modelling Ecosystem Responses to Sea-Level Rise

As coastlines change in response to sea-level rise, so will the ecosystems associated with and dependent on them. Coastal ecosystems occur at the zone where the land and sea meet; that is, where relative sea level is changing. They will be among the first ecosystems impacted by sea-level rise, due primarily to their position within the intertidal zone. These coastal ecosystems include mangroves, salt marsh, coral reefs, seagrass beds, beaches and rocky shorelines; and they provide a range of ecosystem services including coastal protection, fisheries and wildlife habitat, and carbon sequestration.[11] Due to the variable interaction of marine and terrestrial processes on these ecosystems, their geomorphic response to eustatic sea-level rise, is complex. Accordingly, high resolution projections of relative sea-level rise are typically ecosystem-focussed and are not readily transferable to other coastal ecosystems, or even to similar ecosystems elsewhere. There is, however, an opportunity to use these projections as an analogy or indicator of the likely effect that sea-level rise will

11 Edward B. Barbier *et al.*, "The Value of Estuarine and Coastal Ecosystem Services," *Ecological Monographs* 81, no. 2 (2011): 169–193.

have on baselines and other distance related boundaries, such as those delineated in keeping with LOSC.

The impact of sea-level rise on coastal wetlands, in particular, has been the focus of considerable research effort as the combined effect of erosion due to coastal retreat, inundation due to sea-level rise, and land use conversion due to urban expansion threatens these ecosystems and their services. 'Coastal squeeze'[12] is a term increasingly used to describe the collective effect of sea-level rise and impediments to tidal exchange related to development of coastal landscapes, for example through the construction of roads, buildings and other infrastructure, on coastal ecosystems. Adaptation planning that accounts for relative sea level change is essential for limiting coastal squeeze and maintaining coastal ecosystem services. This approach requires realistic projections of the response of coastal ecosystems that accounts not only for the effect of eustatic sea-level rise, but also accounts for the geomorphic response of the coastal ecosystem to sea-level rise, and other factors that may influence relative sea level, such as groundwater extraction.[13]

As sea levels rise, coastal wetlands may autonomously adapt by also increasing their elevation and thereby counteract some of the risk of being submerged. This dynamic change may be achieved through two mechanisms: translating their position to a higher elevation within the coastal landscape by horizontal migration; or by increasing their elevation *in situ*. There is now evidence that coastal wetlands, responding to sea-level rise during the 20th century, have been translating to higher elevations by migration. This progressive translation has been observed for some time and was documented as a widespread trend occurring at many independent sites and representing a range of geomorphic settings in Australia[14] and across five continents.[15] Additionally, there is also substantial evidence indicating that feedbacks between sea-level rise, sediment deposition, and plant productivity may facilitate *in situ* elevation gains.[16]

---

12 Nigel Pontee, "Defining Coastal Squeeze: A Discussion," *Ocean & Coastal Management* 84 (2013): 204–207.

13 Kerrylee Rogers and Neil Saintilan, "Relationships between Surface Elevation and Groundwater in Mangrove Forests of Southeast Australia," *Journal of Coastal Research* 24 (2009): 63–69.

14 Neil Saintilan and Robert J. Williams, "Mangrove Transgression into Saltmarsh Environments in South-East Australia," *Global Ecology and Biogeography* 8, no. 2 (1999): 117–124.

15 N. Saintilan *et al.*, "Mangrove Expansion and Salt Marsh Decline at Mangrove Poleward Limits," *Global Change Biology* 20 (2014): 147–157.

16 See for example Ken W. Krauss *et al.*, "How Mangrove Forests Adjust to Rising Sea Level," *New Phytologist* 202, no. 1 (2013), 19–34; Simon M. Mudd, Susan M. Howell, and

In highly contested coastal landscapes where there is increasing pressure for precision and accuracy in projecting the distribution of ecosystems, consideration must be given to the aforementioned adaptation mechanisms when projecting the response of coastal wetlands. In addition, these projections should be adequately communicated with coastal planners and natural resource managers to ensure they are adequately incorporated into coastal zone management. Against this backdrop, "ecosystem response modelling" (ERM) thus can be a useful tool for projecting the response of coastal wetlands to sea-level rise and can aid sustainable coastal management in the 21st century.

Projecting horizontal migration of coastal ecosystems, in the absence of *in situ* vertical elevation gains, is relatively simple. Termed "bathtub modelling," these approaches assume that there is no geomorphic change as sea-level rises, that land elevation remains constant over time, and that the coastline will fill like a bathtub as water levels increase.[17] The new coastline is then delineated on the basis of the degree of sea-level rise at one point in time, and coastal ecosystems can be projected across the landscape based upon their position with respect to the coastline. The major drawback of such approaches is that they fail to account for the feedbacks between sea-level rise and coastal wetland elevations. They act only as a first order assessment of the likely gross impact of projected eustatic sea-level rise on inundation of the land surface. The failure of this approach to adequately indicate the response of coastal wetlands to sea-level rise has been demonstrated by a comparison made with an empirically based process-driven ERM that was developed for the Hunter River, Australia,[18] (Figure 18.1a–b) that accounted for feedbacks between wetland elevations and sea-level rise.[19] This ERM included annual time-steps that accounted for the spatially variable effect of sea-level rise on geomorphology, which was mediated by feedbacks between sea-level rise, sedimentation, plant production, and root decomposition (Figure 18.1c). In particular, the ERM demonstrated that "bathtub" modelling approaches (Figure 18.1d) overestimated the area lost to sea-level rise by 6 percent and at

---

James T. Morris, "Impact of Dynamic Feedbacks between Sedimentation, Sea-Level Rise, and Biomass Production on near-Surface Marsh Stratigraphy and Carbon Accumulation," *Estuarine, Coastal and Shelf Science* 82, no. 3 (2009): 377–389.

17 K. Rogers, N. Saintilan, and C. Copeland, "Modelling Wetland Surface Elevation and Its Application to Forecasting the Effects of Sea-Level Rise on Estuarine Wetlands," *Ecological Modelling* 244 (2012): 148–157.

18 The Hunter River is a wave-dominated barrier estuary that drains a catchment of approximately 21,500 km^2. The estuary of the Hunter River is located near the city of Newcastle, on the central coast of NSW; approximately 140 km north of Sydney.

19 Rogers, Saintilan, and Copeland, "Modelling Wetland Surface Elevation," 148–157.

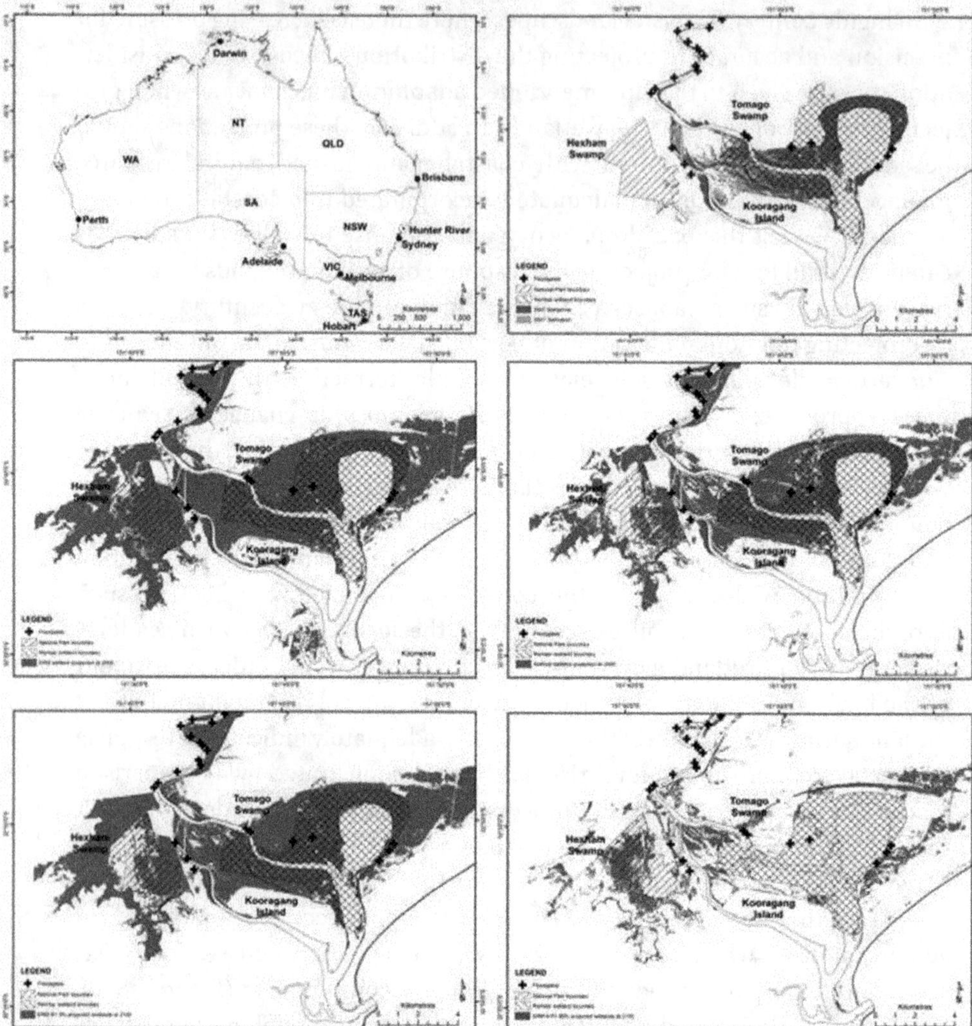

FIGURE 18.1 *Projections of Wetland Distributions Using a Range of Modelling Approaches and Sea-level Rise Scenarios. (a) Location of the Hunter River,* NSW, *Australia. (b) Mapped distribution of coastal wetlands in 2007 and the management boundaries, including Ramsar wetland boundaries and* NSW *National Park boundaries. (c–d) Projection of coastal wetland distribution at 2050 based on current rates of sea-level rise and ecosystem response model (*ERM*), and 'bath-tub' modelling approach, respectively. (e–f) Projection of coastal wetland distribution at 2100 using* ERM *and based on a low sea-level rise and high sea-level rise scenario, respectively.*

SOURCE: ADAPTED FROM KERRYLEE ROGERS, NEIL SAINTILAN, AND CRAIG COPELAND, "MANAGED RETREAT OF SALINE COASTAL WETLANDS: CHALLENGES AND OPPORTUNITIES IDENTIFIED FROM THE HUNTER RIVER ESTUARY, AUSTRALIA," *ESTUARIES AND COASTS* 37, NO. 1 (2013): 67–78.

the same time *underestimated* the projected area of horizontal expansion by 16 percent under observed rates of sea-level rise.

This approach was extended further to demonstrate the influence of management actions that cause coastal squeeze on the extent of coastal ecosystems and the services they provide.[20] The Hunter River ERM incorporated both horizontal and vertical elevation adjustment and modelled two sea-level rise projections of wetland extent that were derived from extremes on the spectrum of possible sea-level rise scenarios—one low (Figure 18.1e) and one high (Figure 18.1f). The low sea-level rise projection used a very low probability, low sea-level rise estimate derived from a projection of a convergent future world with increasing emphasis on economic, social and environmental sustainability.[21] The high sea-level rise projection used a very low probability, high sea-level rise estimate derived from a projection of a future world with rapid economic growth and intensive use of fossil resources.[22] In addition, two different management scenarios were applied to the ERM: one where a network of existing floodgates and levees was maintained and continued to limit tidal exchange and horizontal migration across the coastal landscape, and the other where the levees and floodgates were managed to facilitate tidal exchange and wetland migration across the landscape. These sea-level rise estimates and management scenarios were used to consider the resilience of coastal wetlands to sea-level rise and to demonstrate the effect that coastal management may have on coastal wetland extent and some of their ecosystem services.

The ERM projection indicated that even when the full range of autonomous pathways were incorporated (that is, horizontal migration and vertical elevation adjustment), the extent of coastal ecosystems and the value of ecosystem services provided by them are dependent not only on the degree of sea-level

---

20 Kerrylee Rogers, Neil Saintilan, and Craig Copeland, "Managed Retreat of Saline Coastal Wetlands: Challenges and Opportunities Identified from the Hunter River Estuary, Australia," *Estuaries and Coasts* 37, no. 1 (2013): 67–78.

21 That is, 5% minimum derived from B1 sea-level rise projections in *Climate Change 2007: The Physical Science Basis. Contribution of Working Group I to the Fifth Assessment Report of the Intergovernmental Panel on Climate Change* (Cambridge, United Kingdom and New York, NY, USA: Cambridge University Press, 2007), 820–822. Estimates derived from J. Hunter, "Estimating Sea-Level Extremes under Conditions of Uncertain Sea-Level Rise," *Climatic Change*, 99(3) (2010): 331–350.

22 That is, 95% maxima estimates derived from A1FI sea-level rise projections in *Climate Change 2007: The Physical Science Basis. Contribution of Working Group I to the Fifth Assessment Report of the Intergovernmental Panel on Climate Change* (Cambridge, United Kingdom and New York, NY, USA: Cambridge University Press, 2007), 820–822. Estimates derived from Hunter, "Estimating Sea-Level Extremes," 331–350.

rise that will occur in future, but also upon the management actions taken to facilitate autonomous adaptation and thus limit coastal squeeze.[23] Based on the empirically established relationship between current rates of sea-level rise and elevation adjustment, a "resilience threshold" was identified approximately at the year 2060. This threshold to resilience occurs when the rate of sea-level rise exceeds approximately 9 mm y^{-1}, thereafter surpassing the capacity of the ecosystem to build elevation by migration or *in situ* elevation gain. In such a case, a rapid submergence of the wetland was projected to occur. The result of this process includes a loss of ecosystem services. Importantly, the ERM demonstrated that management actions taken to facilitate adaptation of coastal wetlands, even under high rates of sea-level rise, may drastically limit the loss of ecosystem services. In the case of the Hunter River, opening a network of levees and floodgates that currently limit tidal exchange across the coastal floodplain would facilitate the horizontal migration of coastal ecosystems to higher elevations, and would enable continued delivery of sediments across the entire floodplain. The gain in carbon sequestered and fisheries benefit by 2100 was estimated to be in the order of $2–13 million (AUD) in a carbon sensitive economy (based on current value).[24]

Planning for Coastal Wetlands and the Ecosystem Services in the 21st Century

The ERM for the Hunter River coastal wetlands demonstrates the capacity of ecosystems for self-conservation, particularly when external pressures are low (for example, in the context of a low sea-level rise scenario) and autonomous mechanisms for self-conservation are not limited (for example, where tidal exchange is facilitated). Importantly, the ERM also demonstrates that coastal wetlands are four-dimensional ecosystems with a location in space (that is with a length and width), an elevation (height relative to sea level), and a trajectory through time. Projecting the distribution of coastal wetlands is complex and the ecosystem response is influenced by a range of processes, including sea-level rise and the management frameworks imposed on them. That coastal wetlands, and other coastal ecosystems, are four-dimensional entities has been well understood by earth and environmental scientists who

23 Rogers, Saintilan, and Copeland, "Managed Retreat of Saline Coastal Wetlands," 67–78.

24 K. Rogers *et al.*, "How Wetlands Can Help Us Adapt to Rising Seas," *The Conversation*, 1 April 2014, available at, https://theconversation.com/how-wetlands-can-help-us-adapt-to-rising-seas-23962.

have plotted their movement as sea level rose over the Holocene.[25] Similarly, the concept that coastlines (and low-water lines) are stationary has long been discredited,[26] yet this fact is yet to be adequately captured in policy where jurisdictional limits and boundaries are related to sea level.

This understanding that coastlines and the ecosystems associated with them are dynamic contrasts with legal and policy frameworks that are largely preservationist in character and tend to assume that natural assets (or ecosystems) are stationary and that their spatial position can be easily defined. Defining boundaries, either for jurisdictional or management purposes, is a basic paradigm in environmental and natural resource regulation. Perhaps unsurprisingly legal, policy, and to a large extent management frameworks traditionally favor defined and fixed boundaries. Problematically, such approaches are founded on the premise that "change" is unnatural and should be managed or reversed through preservation or restoration.[27]

Characterization of ecosystems or management units on the basis of defined, static boundaries will fail for natural coastal assets exposed to sea-level rise, as they do not adequately capture the complexity of their distribution in space (length, width, elevation) and time. This scenario will be exacerbated should sea-level rise prove to be both significant and occur rapidly. These challenges have important implications for the future protection and management of coastal wetlands and other natural coastal assets that necessitates a planned approach to protection and conservation, including the application of more flexible approaches to the definition or application of management boundaries.

The Hunter River ERM case study illustrates the potential for failure of existing legal and policy frameworks aimed at protection and conservation of coastal wetlands. The coastal wetlands on the Hunter River comprise a complex matrix of mangrove, salt marsh and tidal flats, which provide important roosting habitat for migratory wader birds; and as such, are afforded significant protection under a number of legal instruments that operate at a range of spatial scales. Specifically:

---

25 C.D. Woodroffe, "The Impact of Sea-Level Rise on Mangrove Shorelines," *Progress in Physical Geography* 14, no. 4 (1990): 483–520.

26 Colin V. Murray-Wallace and Colin D. Woodroffe, *Quaternary Sea-level Changes* (Cambridge, United Kingdom and New York, NY, USA: Cambridge University Press, 2014), 72.

27 R.K. Craig, "'Stationary Is Dead'—Long Live Transformation: Five Principles for Climate Change Adaptation Law," *Harvard Law Review* 34 (2010): 9–73.

- 2,962 hectares is listed as internationally important under the International Convention on Wetlands (Ramsar Convention) in 1984,[28] which is formally administered in the Australian legal context through the Commonwealth Environmental Protection and Biodiversity Conservation (EPBC) Act (1999).[29]
- 4,257 hectares is formally protected within the Hunter Wetlands National Park, which is administered by the New South Wales (NSW) National Parks and Wildlife Service under the auspices of the NSW National Parks and Wildlife Act (1974)[30] and also includes a significant portion that is listed as internationally important under the Ramsar Convention in 1984 (2,962 hectares).
- Coastal wetlands are broadly protected under the NSW State Environmental Planning Policy no. 14 (SEPP 14),[31] within the NSW Environmental Planning and Assessment Act (1979).[32]

Management of these wetlands has largely been deferred to the provincial (state) or district (shire or municipality) scale, with the Ramsar wetland management deferred to the state government and SEPP 14 coastal wetland protection deferred to local government. Deferring of management obligations and splitting of management between environmental agencies is conventional for natural resource management and has been linked with poor ecosystem outcomes due to a mismatch between resource allocation, jurisdictional arrangements, and the scale at which ecosystem services operate.[33]

28 *Convention on Wetlands of International Importance especially as Waterfowl Habitat 1971* (Ramsar Convention), opened for signature 2 February 1971, 996 UNTS 245 (entered into force 21 December 1975), available at, http://sedac.ciesin.org/entri/texts/ramsar.wetlands.waterfowl.habitat.1971.html.

29 *Environmental Protection and Biodiversity Conservation Act 1999* (Cth), available at, http://www.comlaw.gov.au/Details/C2015C00063.

30 *National Parks and Wildlife Act 1974* (NSW), available at, http://www.austlii.edu.au/au/legis/nsw/consol_act/npawa1974247/.

31 Available at, http://www5.austlii.edu.au/au/legis/nsw/consol_reg/seppn 14w543/.

32 *Environmental Planning and Assessment Act 1979* (NSW), available at, http://www.austlii.edu.au/au/legis/nsw/consol_act/epaaa1979389/.

33 See for example David W. Cash and Susanne C. Moser, "Linking Global and Local Scales: Designing Dynamic Assessment and Management Processes," *Global Environmental Change* 10, no. 2 (2000), 109–120; Megan K. La Peyre, Margaret A. Reams, and Irving A. Mendelssohn, "Linking Actions to Outcomes in Wetland Management: An Overview of U.S. State Wetland Management," *Wetlands* 21, no. 1 (2001), 66–74.

More generally, management of these wetlands and natural resources are largely undertaken on the basis of a defined boundary. For example, SEPP 14 boundaries were delineated on the basis of mapping of wetland extent from aerial photography dating to 1981,[34] while the Hunter Wetland National Park and Ramsar wetland boundaries are delineated on the basis of cadastral boundaries of land that has been acquired by the state of NSW for conservation purposes. However, these static preservation and conservation boundaries do not confine the distribution of natural coastal assets, which has been demonstrated elsewhere by mapping the changing distribution of SEPP 14 wetlands between 1981 and 1997.[35] Indeed, this aim would be contrary to the purposes of conservation and preservation legislation. Rather, management boundaries facilitate the limitation of threatening activities and processes that are likely to have an effect within the areas defined. As natural coastal assets respond to sea-level rise and migrate beyond existing preservation and conservation boundaries, the capacity of legislation to protect against threatening activities and processes is curtailed. Due to shifting shoreline position associated with sea-level rise, the Hunter River ERM demonstrated an 87 percent decline in coastal wetland extent within current Ramsar boundaries at 2100 under a high sea-level rise scenario; with the wetland distribution largely shifting outside the current Ramsar boundaries.

Climate change, temperature increases and sea-level rise cannot be adequately mitigated; and accordingly, ecosystems will be transformed and will not be maintained in their current form within the confines of existing management boundaries. A strong case can be made for the potential failure of current legal and policy frameworks based on management and jurisdictional boundaries as presently defined. This is evident through scrutiny of guidance provided by Ramsar for contracting parties of the Ramsar Convention, which is an intergovernmental treaty providing the framework for national action and international cooperation aimed at the conservation and wise use of wetlands and their resources. Ramsar COP11 Resolution XI.14[36] on climate change and

---

34 P. Adam *et al.*, *Coastal Wetlands of New South Wales* (Sydney: Coastal Council of NSW, 1985).

35 K. Rogers and N. Saintilan, "Remapping of the Sepp 14 Wetlands in the Shoalhaven District," *Wetlands (Australia)* 20, no. 2 (2002): 55–65.

36 Ramsar, "Resolution Xi.14. Climate Change and Wetlands: Implications for the Ramsar Convention on Wetlands," in *Resolutions of the 11th meeting of the Conference of Parties to the Convention on Wetlands* (Romania: Ramsar, 2012).

wetlands confirms the need for parties to "stem the progressive encroachment and loss of wetlands now and in the future";[37] and specifically:

- expresses concern 'that mechanisms may not be in place for determining specified limits of change in ecological character of wetlands, or adequate baselines or reference conditions available against which change can be assessed or for reporting under Article 3.2 of the Convention;[38]
- encourages contracting parties to undertake studies of the role of conservation and/or restoration of both forested and non-forested wetlands in relation to adaptation to climate change, including water regulation at local and regional scales, such as flood risk reduction, water supply and storage, and reducing the impacts of sea-level rise and extreme weather events, including rainfall situations';[39] and,
- requests the Scientific and Technical Review Panel (STRP), to continue to 'prepare advice on the implications of climate change for maintaining the ecological character of wetlands'.[40]

The STRP has been preparing advice for assessing the biophysical vulnerability of wetlands to climate change[41] and undertaken a first pass evaluation of the risk to Ramsar sites from climate change induced sea-level rise,[42] among other activities. However, stemming the loss of wetlands to sea-level rise is best achieved by facilitating autonomous adaptation; this requires action to be taken early so as to enable horizontal migration and limit coastal squeeze. To this end the STRP has yet to provide any guidance on management and legal action that could be taken to facilitate autonomous adaptation by coastal ecosystems to areas beyond the current management boundaries.

This lack of guidance or mechanism for adjusting management boundaries as ecosystems migrate is not limited to the Ramsar convention. The prevailing majority of coastal wetland conservation and protection areas are

---

37 Item 5 in *ibid.*

38 Item 24 in *ibid.*

39 Item 32.2 in *ibid.*

40 Item 35.1 in *ibid.*

41 Habiba Gitay, C. Max Finlayson, and N.C. Davidson, "A Framework for Assessing the Vulnerability of Wetlands to Climate Change," in *Ramsar Technical Report No. 5/CBD Technical Series No. 57.* (Gland, Switzerland and Montreal, Canada: Ramsar Convention Secretariat and Secretaraiat for the Convention on Biological Diversity, 2011): 1–26.

42 Ramsar STRP, "Evaluating the Risk to Ramsar Sites from Climate Change Induced Sea Level Rise," Ramsar STRP, http://www.ramsar.org/sites/default/files/documents/bn/bn5.pdf.

managed on the basis of static boundaries. Buffers around wetlands based on distance are increasingly advocated as corridors for migration,[43] though these buffers are also likely to fail as they do not capture all four dimensions, and more specifically the elevation component of coastal wetland distribution. For coastal developments, rolling easements are increasingly advocated as a sustainable approach to coastal adaptation;[44] and achieve this by progressively preventing land adjoining shorelines or coastal ecosystems from being developed and shifting existing land use beyond the easements so as to enable unimpeded increase in shoreline elevation and ecosystem migration, and limit impacts of sea-level rise on developments. Rolling boundaries or baselines are yet to be effectively applied as a sustainable approach to coastal ecosystem management. Fortunately, due to political circumstances that provided opportunities and resources for land acquisitions adjacent to existing mangrove and salt marsh on the Hunter River and management of levees and floodgates that limited migration of ecosystems to higher elevations, there are now fewer impediments to coastal wetland adaptation in the 21st century. This new land is now managed as a national park, and is yet to be fully incorporated into the Ramsar management framework.

Potential Impacts of Changing Coasts on Baselines, Limits and Boundaries

The guiding framework governing claims to maritime jurisdiction is provided by the United Nations Convention on the Law of the Sea (UNCLOS),[45] which has achieved near universal acceptance with, at the time of writing, 166 parties to it comprising 165 States plus the European Union.[46] A key achievement of UNCLOS was the realization of a clear spatial framework for the limits to national claims to maritime jurisdiction. The Convention provides for a series

43 Kevin L. Erwin, "Wetlands and Global Climate Change: The Role of Wetland Restoration in a Changing World," *Wetlands Ecology and Management* 17, no. 1 (2009): 71–84.

44 James G. Titus, *Rolling Easements* (Washington D.C., USA: Climate Ready Estuaries Program, US Environemntal Protection Agency, 2011), pp. 1–175.

45 United Nations Convention on the Law of the Sea, adopted 10 December 1982, 1833 UNTS 3 (entered into force 16 November 1994) (UNCLOS).

46 See, United Nations, Status of the United Nations Convention on the Law of the Sea, of the Agreement relating to the implementation of Part XI of the Convention and of the Agreement for the implementation of the Convention relating to the conservation and management of straddling fish stocks and highly migratory fish stocks, New York, updated to 10 October 2014, at <http://www.un.org/Depts/los/reference_files/status2010.pdf>.

of national zones of maritime jurisdiction measured offshore from baselines along the coast. These zones include a territorial sea of 12 nautical miles (nm)[47] (UNCLOS, Article 3), a contiguous zone out to 24 nm from baselines (or 12 nm from territorial sea limits) (Article 33), an exclusive economic zone (EEZ) out to 200 nm from baselines (Article 57) and continental shelf rights that may extend up to 350 nm or even further in certain circumstances (Article 76).[48]

Rights over such maritime zones are founded on sovereignty over land territory in keeping with the principle that "the land dominates the sea." As a result of this formula there is a direct link between the limits or "edge" of the land, that is, the coast, and maritime jurisdictional claims, or as Weil put it: "... the land dominates the sea and it dominates it by the intermediary of the coastal front."[49] Accordingly, the jurisdictional limits of these maritime zones are dependent on measurements from baselines along the coast. While the Law of the Sea Convention provides for several different types of baseline,[50] the fundamental type consist of "normal" baselines coincident with the "low-water line along the coast as marked on large scale charts officially recognized by the coastal state."[51] Here it is important to note that the various straight-line alternatives to normal baselines that may be claimed by a coastal State still depend to some extent on the location of normal baselines, because in order to be valid they need to be linked or anchored back to low-water lines representing the maximum extent of land territory. Also of note here is that the selection of the low-water line as the "starting line" for measuring maritime claims has the dual benefit of maximizing the land territory of coastal States

47 It is acknowledged that 'M' is the correct abbreviation for nautical miles. However, 'nm' is frequently used to denote nautical miles in the law of the sea literature and is employed for this purpose in this chapter.

48 The outer limits of the continental shelf where it extends beyond the 200 nm exclusive economic zone (EEZ) limits, unlike the limits of other zones that are based solely on a distance measurement, depend also on the geology and geomorphology of the seabed. That is why determining the outer limit of the continental shelf is a more complex task than that for other zones of maritime jurisdiction. Nonetheless, distance measurements from baselines remain critical to this task.

49 Proper Weil, *The Law of Maritime Delimitation—Reflections*, (Cambridge: Grotius, 1989): 50.

50 These include straight baselines (UNCLOS, Article 7), river closing lines (Article 9), bay closing lines (Article 10), lines related to ports and roadsteads (Articles 11 and 12), and in respect of archipelagic States (Article 47).

51 UNCLOS, Article 5.

whilst simultaneously advancing the baselines from which maritime zones are measured further, as it were, "down the beach."[52]

As noted above, it has long been recognized that coasts are dynamic and may change location and configuration in relatively short periods of time.[53] Inevitably, this phenomenon has implications for maritime jurisdictional limits and boundaries since as the low-water line moves or "ambulates" over time in accordance with changes in the coast,[54] so the normal baselines and the maritime claims measured from them will shift. The longstanding, generally accepted implication of this phenomenon is that as the coast/normal baselines change, so will the maritime jurisdictional limits measured from them. Thus where the baseline advances (for example by the deposition of material along the coast or isostatic uplift of shorelines) the outer limits of the maritime claims measured from that baseline will expand seawards. Conversely, where the normal baseline recedes (as through coastal erosion or submergence from rising sea levels) the coastal State may lose maritime areas as their maritime limits are pulled back.

Baselines are also critical to the construction of equidistance lines between coastal States in the delimitation of maritime boundaries. Importantly, equidistance lines have proved to be the basis for the delimitation of the majority of maritime boundaries delimited thus far.[55] In recent cases on maritime delimitation international courts and tribunals have adopted a three-stage approach whereby a provisional boundary line is defined on the basis of equidistance in the first instance. Subsequently, at the second stage, any factors that may justify a modification of the provisional line in order to

---

52 See, Clive Schofield, "Shifting Limits?: Sea Level Rise and Options to Secure Maritime Jurisdictional Claims," *Carbon and Climate Law Review* 4 (2009): 405–416, at 408. For a history of sea level and baselines under international law see also, David D. Caron, "Climate Change, Sea Level Rise and the Coming Uncertainty in Oceanic Boundaries: A Proposal to Avoid Conflict," in Seoung-Yong Hong and Jon M. Van Dyke, eds., *Maritime Boundary Disputes, Settlement Processes and the Law of the Sea* (Law of the Sea Institute, Martinus Nijhoff Publishers, 2009), 1–17.

53 See, for example, Bill Hirst and David Robertson, "Geographic Information Systems, Charts and UNCLOS—Can They Live Together?" *Maritime Studies* 136 (2004): 1–6. See also, Schofield, "Shifting Limits?" 408.

54 Aaron Shalowitz and Michael W. Reed, *Shore and Sea Boundaries: The Development of International Maritime Boundary Principles Through United States Practice* (3 vols., Washington D.C.: US Department of Commerce, 2000): Vol. I, p. 185. See also J.R. Victor Prescott and Clive Schofield, *The Maritime Political Boundaries of the World*, 2nd ed. (The Hague: Martinus Nijhoff Publishers, 2005), 100–101.

55 See, for example, Prescott and Schofield, *ibid.*, 239–240.

achieve an equitable result are considered; and then at the third stage, the resulting line is tested for any disproportionality.[56] An important distinction does, however, exist, between unilaterally defined maritime limits such as the 12nm limits to the territorial sea and 200nm limits to the EEZ, as well as undelimited maritime boundaries, and maritime boundaries that have been delimited through an international treaty or agreement. While unilateral limits and undelimited potential maritime boundaries may be subject to change as a result of changes in the location of baselines along the coast, once a maritime boundary has been agreed it is not subject to change except through agreement among the parties concerned.[57]

While there has been a longstanding recognition that as coastlines, and therefore baselines, change to the outer limits of maritime zones will shift, this approach was grounded in a perception that coastlines were generally relatively stable and unlikely to experience radical changes. Thus such baselines, limits and boundaries tended to be viewed as merely subject to moderate and gradual change over time whilst otherwise being relatively stable overall. The advent of sustained, significant and increasingly rapid sea-level rise has served to undermine the illusion of permanency with respect to coastal location. Such expectations of coastal stability were at least partly derived from the ability to observe fluctuations in the sea level over geologically short periods of time (years to decades). Problematically, however, the apparent stability of sea levels in the recent past is owed partly to an anomalous period of deceleration in the rate of sea-level rise over the late Holocene period.[58]

56 *Ibid.*, 240–241. See also, Clive Schofield, "One Step Forwards, Two Steps Back? Progress and Challenges in the Delimitation of Maritime Boundaries since the drafting of the United Nations Convention on the Law of the Sea," in J. Xue, J. and A. White (eds) *30 Years of UNCLOS: Progress and Prospects* (Beijing: China University of Political Science and Law Press, 2013), 217–239, at 228–237.

57 This outcome is because boundary treaties have a privileged status in the international law of treaties and are not subject to change even in the case of "subsequent fundamental change of circumstances." See, *Vienna Convention on the Law of Treaties*, 23 May 1969, UNTS 1155, 331(VCLT). Article 62(2)(a). See also, Alfred Soons, "The Effects of a Rising Sea Level on Maritime Limits and Boundaries," *Netherlands International Law Review* 32(2) (1990): 207–232, at 222; also, Clive Schofield, "Rising Waters, Shrinking States: The Potential Impacts of Sea Level Rise on Claims to Maritime Jurisdiction," *German Yearbook of International Law* 53 (2010): 189–231, at 211.

58 Kurt Lambeck *et al.*, "Sea Level and Global Ice Volumes from the Last Glacial Maximum to the Holocene," *Proceedings of the National Academy of Sciences* 111, no. 43 (2014): 15296–15303.

By the time the Law of the Sea Convention was negotiated, it was well known that substantially different sea levels had existed in the past, and a number of legal scholars have highlighted this issue subsequently.[59] Glacial theory and its influence on sea level was first documented by MacLaren in 1842[60] using previously recorded evidence from marine deposits at higher elevations than present sea level;[61] and the theory became widely accepted in the earth sciences by the end of the nineteenth century.[62] Since that time numerous techniques have been applied by scientists to establish that sea levels have been variable over geological time scales. Nonetheless, sea-level rise was largely not taken into consideration when the presently applicable law of the sea rules were drafted. With recent increases in our capacity to accurately and precisely measure sea levels over the past century, thanks to the use of satellite altimetry and global navigation satellite systems (GNSS) receivers, we now know that sea levels are not stable, even over relatively short time periods. Rather, the sea is rarely level.[63] This finding has important legal implications when sea level metrics are applied as a baseline from which maritime jurisdictional limits and boundaries are measured.

In this context, concerns have been raised that sea-level rise may lead to receded low-water lines, and thus the normal baseline can retreat inland to a substantial extent, leading to potentially significant "knock-on" effects on the limits of maritime jurisdictional claims because the base-points on which the limits of such claims depend have migrated landward. This threat to the extent of national maritime jurisdictional claims is especially significant for coastal States such as Bangladesh and parts of India, as well as Vietnam in Southeast Asia, which have large stretches of low-lying coasts. The maritime claims of States in possession (or even entirely composed) of low-elevation islands, such as Kiribati, the Marshall Islands, and Tuvalu in the Pacific Ocean

59 See, for example, Freestone, "International Law and Sea Level Rise"; F. Bird and J.R. Victor Prescott, "Rising Global Sea Levels and National Maritime Claims," *Marine Policy Reports* 177 (1989): 177–96; Soons, "The Effects of Sea Level Rise"; David D. Caron, "When Law makes Climate Change Worse: Rethinking the Law of Baselines in Light of Rising Sea Level," *Ecology Law Quarterly*, 17 (1990): 621.

60 Charles MacLaren, "The Glacial Theory of Professor Agassiz of Neuchatel," *American Journal of Science* 42 (1842): 346–365.

61 Colin V. Murray-Wallace and Colin D. Woodroffe, *Quaternary Sea-Level Changes* (Cambridge, United Kingdom and New York, NY, USA: Cambridge University Press, 2013): 1–38.

62 *Ibid.*

63 *Ibid.*

are also under threat from this phenomenon.[64] Small, remote, and low-lying islands can give rise to significant maritime jurisdictional entitlements. However, sea-level rise could change the legal status of such insular features. For example, an island that is currently always above the water surface, even during high tide, and thus (under existing ocean law) has entitlements to adjoining ocean space, may eventually disappear during high tide as a consequence of sea-level rise. This dynamic could lead to reclassification of the feature from being an island, from which claims to the full range of maritime zones may be made under Article 121(2) of UNCLOS, to one of the categories of insular formations from which only restricted maritime claims can be made, such as a "rock"[65] or a low-tide elevation (LTE) formation (whose features that are exposed at low tide but are submerged at high tide),[66] or even a fully submerged feature that cannot be used to generate maritime claims.

Such dire anticipated consequences of sea-level rise affecting the spatial scope of the maritime claims of coastal States may overstate the case, however. As noted above, such overly-simplistic "bathtub" type approaches fail to take into account complex interactions between sea level and coastlines. These interactions are highlighted by the ERM case study in relation to coastal wetlands in particular; such interactions are by no means exclusive, however, to such environments. For example, there is evidence that coral atolls have proved to be remarkably robust over long periods, including periods when sea levels were considerably higher than they are now. This fact suggests that some insular features may similarly be able to adapt naturally to climate change and sea-level rise. The analysis is underpinned by observations in and analysis of 27 coral atoll islands in the central Pacific Ocean, which found that the majority (86 percent) of these features had either remained stable or increased in area over a 20–60 year period despite reported sea-level rise in the central Pacific region.[67] Indeed it has been argued that uninterrupted sediment flows are necessary to sustain the island-building processes that maintain the integrity of

---

64 Clive Schofield and David Freestone, "Options to Protect Coastlines and Secure Maritime Jurisdictional Claims in the Face of Global Sea Level Rise," in Gerrard, M. and Wannier, G. (eds.), *Threatened Island Nations: Legal Implications of Rising Seas and a Changing Climate* (Cambridge: Cambridge University Press, 2013): 141–165, at 144–145. See also David D. Caron, "Time and the Public Trust Doctrine: Law's Knowledge of Climate Change," *University of Hawaii Law Review*, 35 (2013):441–458.

65 UNCLOS, Article 121(3).

66 UNCLOS, Article 13.

67 Arthur P. Webb and Paul S. Kench, "The Dynamic Response of Reef Islands to Sea-Level Rise: Evidence from Multi-Decadal Analysis of Island Change in the Central Pacific," *Global and Planetary Change*, 72 (2010) (3): 234–246.

coral reef islands. As Kenchington has written, "The physical dynamics of sediment supply and transport are critical factors in the context of management of rocky and sedimentary oceanic islands."[68] Accordingly, population pressures, land use and development issues, and therefore planning and management concerns in the context of an increasingly contested coastal zone, can once again be considered to be salient issues.

Furthermore, not all of a coastal State's baselines contribute toward the construction of the outer limits of its maritime claims. Maritime limits are commonly constructed through the "envelope of arcs" method (see Figure 18.2).[69] Consequently only certain base-points along the normal baseline—essentially the outermost points along the baseline such as headlands and offshore islands—will be relevant to the limits of the maritime zones. In contrast, those parts of the baseline that are located on the inner portion of a bay, for example, are unlikely to contribute to the outer limit of maritime zones. Indeed, most of the normal baseline is, in fact, irrelevant to the construction of the outer limits to maritime jurisdictional zones. A salient consideration to take into account with respect to potential losses of claimed maritime zones is that the broader the maritime zone in question, such as the EEZ out to a limit of 200nm, the fewer basepoints are required to measure that zone. Consequently, the number of basepoints that can be considered as critical, where sea-level rise will significantly impact the scope of maritime claims, is often limited.

The foregoing does not suggest baselines and the maritime limits and boundaries dependent upon them as well as coastal ecosystems will necessarily be immune to the impacts of sea-level rise. However, it does suggest that it is by no means straightforward to predict the scope, scale and pace of such impacts. A more nuanced understanding of the consequences of sea-level rise therefore is crucial, together with an expectation that these impacts are likely to be experienced in a temporally and spatially complex manner.

Much will also depend, analogously for the management of coastal ecosystems, on the response options adopted by coastal States to address potential threats to their maritime jurisdictional limits and boundaries. One of a suite of responses to coastlines/baselines under threat from inundation

68 Richard Kenchington, "Maintaining Coastal and Lagoonal Ecosystem and Productivity," in Hiroshi Terashima (ed.), *Proceedings of International Symposium on Islands and Oceans* (Tokyo, Ocean Policy Research Foundation January 2009), 22–23, available at https://www.sof.or.jp/en/report/pdf/200903_ISBN978-4-88404-217-2.pdf.

69 Christopher Carleton and Clive Schofield, *Developments in the Technical Determination of Maritime Space: Charts, Datums, Baselines, Maritime Zones and Limits*, Maritime Briefings 3 (3) (Durham, International Boundaries Research Unit, 2001), 62.

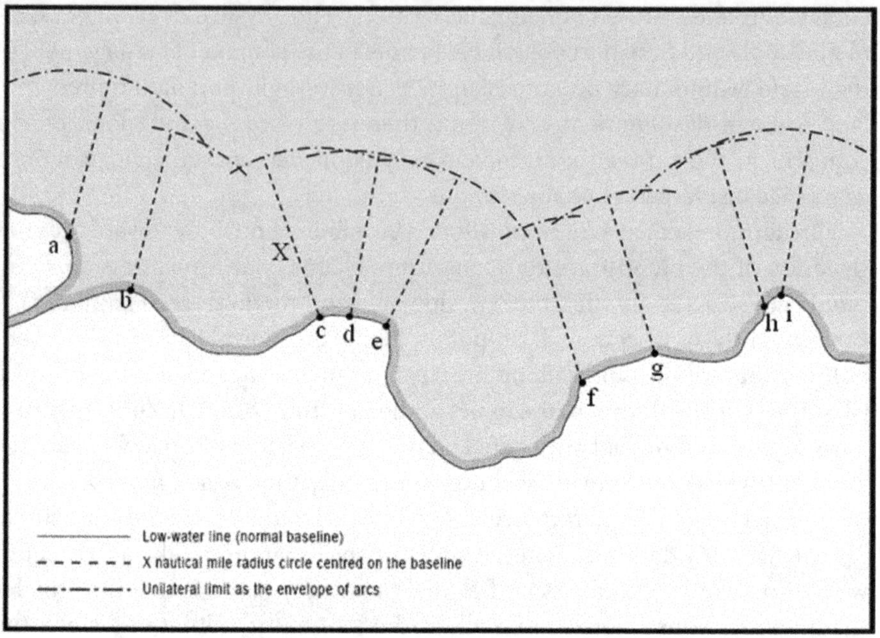

FIGURE 18.2 *Defining the Outer Limits of Maritime Claims through an "Envelope of Arcs".* SOURCE: ADAPTED FROM INTERNATIONAL HYDROGRAPHIC ORGANIZATION (IHO) TECHNICAL ASPECTS OF THE LAW OF THE SEA (TALOS) MANUAL, 5TH EDITION, SPECIAL PUBLICATION NO.51, (MONACO: INTERNATIONAL HYDROGRAPHIC BUREAU, 2014), PREPARED BY THE INTERNATIONAL HYDROGRAPHIC ORGANISATION (IHO) AND THE INTERNATIONAL ASSOCIATION OF GEODESY (IAG).

or erosion has been to attempt to protect the coastline concerned through the application of what has been termed a "bulkhead" policy,[70] often featuring the construction of hard structures such as sea walls, groynes, and wave reduction structures.[71] While being costly to implement, such interventions have significant drawbacks. In particular, sea defences and structures of this type that are intended to protect and stabilise the location of the coast can disrupt sediment supply and alter hydrodynamic energy, resulting in erosion elsewhere along the coast,[72] and leading to "unfortunate and sometimes very serious unintended

70 Freestone, "International Law and Sea Level Rise," 117–119.

71 Schofield, "Shifting Limits?" 411.

72 Schofield and Freestone, "Options to Protect Coastlines and Secure Maritime Jurisdictional Claims," 152–154. See also, John Pethick, "Waves of Change: Coastal Response to Sea Level Rise," *Geographical Analysis* 19 (1989): 1–4.

consequences."[73] Consequently, the focus has shifted more towards "soft engineering" shoreline protection approaches, and more ecosystem-based and sustainable management approaches that include principles of accepting retreat or accommodating shifting shorelines, and incorporating staged plans of action that are triggered by changes to shorelines as they occur. Accordingly, adaptation actions for sea-level rise are likely to increasingly feature retreat from the hazards associated with rising seas and extreme weather events, or accommodate the rising sea by adjusting use of the coastal zone (e.g. increasing floor heights).[74]

Adopting such approaches, essentially featuring acceptance that sea-level rise will likely lead to retreating coastlines, and therefore the retreat of normal baselines, implies that the extent of maritime zones will also be under threat. This likelihood has prompted some coastal States to seek to break the traditional link between changing and migrating normal baselines and shifting outer limits to maritime claims. For example, having received recommendations from the UN Commission on the Limits of the Continental Shelf (CLCS)[75] concerning its submissions relating to the delineation of its extended or outer continental shelf limits,[76] that is, with respect to continental shelf areas seaward of its 200nm EEZ limits, Australia has sought to declare and fix the outer limits of its continental shelf rights. Significantly, for present purposes, Australia did so not only for extended continental shelf limits but continental shelf limits consistent with 200nm arcs from baselines along the

---

73 Kenchington, "Maintaining Coastal and Lagoonal Ecosystem and Productivity," 4.

74 R.J. Nicholls, "Planning for the Impacts of Sea Level Rise," *Oceanography* 24, no. 2 (2011): 144–157.

75 For information on the United Nations Commission on the Limits of the Continental Shelf (CLCS) see, http://www.un.org/Depts/los/clcs_new/clcs_home.htm. For details of the Commission's recommendations regarding Australia's submission see, "Recommendations of the Commission on the Limits of the Continental Shelf (CLCS) in regard to the Submission made by Australia on 15 November 2004," adopted 9 April 2008, available at, http://www.un.org/Depts/los/clcs_new/submissions_files/aus04/Aus_Recommendations_FINAL.pdf.

76 "United Nations Convention on the Law of the Sea: Submission to the Commission on the Limits of the Continental Shelf on the Outer Limits of Australia's Continental Shelf Extending Beyond 200 Nautical Miles from the Territorial Sea Baseline—Executive Summary," (Canberra: Commonwealth of Australia, AUS-DOC-ES, 2004), available at, www.un.org/Depts/los/clcs_new/submissions_files/aus04/Documents/aus_doc_es_web_delivery.pdf.

coast.[77] It can be anticipated that this sort of State practice will increase into the future although there may well also be resistance to this practice.[78]

Conclusions: A Paradigm Shift for the 21st Century

The sea is rising and there will be increasing difficulties in the future concerning the defining both of appropriate management boundaries and baselines and of maritime jurisdictional limits on the basis of a relationship between the land and the sea. The Hunter River ERM case study demonstrated the complexity in projecting relative sea-level rise, and yet many questions still remain. Defining mean sea level, or low water level in the case of the UNCLOS provisions, is an ongoing issue due to the spatial and temporal variability in oceanic water levels. Indeed, this is a challenge that has plagued scientists since the emergence and acceptance of glacial theory.[79] As noted by Chambers writing in 1848:

> Has the change of the relative level of land and sea been accomplished by an upward movement of the land, or by the recession of the sea? Has the shift been slow and equable with regard to time, or by fits and starts with long pauses between, or by a slow movement interrupted by pauses? Has the time embraced by the whole series of phenomena been long or short, geologically speaking? What have been the general and particular circumstances and results of the whole movement?[80]

However, there are more unknowns about the future, many of which are not dissimilar to the queries of the past: how efficiently will climate change be mitigated; how high and fast will the sea rise; how will coastal landforms change in the future due to coastal developments, climatic perturbations such as storms,

---

77 See, Australia, *Seas and Submerged Lands (limits of the Continental Shelf) Proclamation 2012*, 24 May 2012, at, <www.comlaw.gov.au/Details/F2012L01081>. See also, Sam Bateman and Clive Schofield, "Australia's Outer Continental Shelf Regime," *Australian Journal of Marine and Ocean Affairs*, 4(4) (2012): 131–140.

78 See, for example, International Law Association (ILA), "Baselines Under the International Law of the Sea," Conference Report of the ILA Committee, Sofia 2012, 30–31, available at, http://www.ila-hq.org/en/committees/index.cfm/cid/1028.

79 See, for example, Robert Chambers, *Ancient Sea-Margins: As Memorials of Changes in the Relative Level of Sea and Land* (W. & R. Chambers, 1848): 3–4.

80 *Ibid.*

and planned coastal adaptation; and how will the feedbacks between the land and the sea alter landform shape?

Time likely will be the best resolution of these unknowns; and although there is now considerable scientific effort directed towards resolving these challenging questions, there is little time available to wait for resolutions when coastal landscapes and marine spaces are increasingly contested. As illustrated by the ERM for the Hunter River, existing legal and policy frameworks are likely to be found wanting with current management boundaries projected to be inadequate to the task of protecting or conserving coastal wetlands in the future with urgent action required to facilitate horizontal migration.

One abiding issue here relates to the emergence of new scientific knowledge that is not adequately captured in existing policy or legislation. Lack of awareness of such scientific knowledge on the part of policy makers may be one factor here, coupled with longstanding difficulties associated with capturing the intricacies of scientific knowledge and translating this into policy. Whatever the cause, there remains an evident need for greater communication between scientists, other knowledge brokers, and policy makers. The current impact of biophysical science on the political process can be regarded as somewhat linear, with scientific research, assessment and policy development leading to decision making; with only minimal revision and feedbacks between the political process of policy development and decision making, and the scientific process of research and assessment.[81] In the context of a changing climate and significant uncertainty there is a patent need to abandon the currently linear approach to incorporation of scientific knowledge into policy and law. Craig correctly highlights that ecosystem and natural resource regulation and management requires a paradigm shift in the 21st century, stating that:

> ... climate change means that regulatory objectives based on pre-climate change characteristics of particular places can and will become obsolete. Climate change adaptation law must be able to accommodate the transforming ecological realities of particular places and not attempt to freeze ecosystems and their components into some prior state of being.[82]

---

81 Laura Stoker *et al.*, "Sustainable Coastal Management?," in Richard Kenchington, Laura Stoker, and David Wood, eds., *Sustainable Coastal Management and Climate Adaptation: Global Lessons from Regional Approaches in Australia* (Collingwood, Victoria, Australia: CSIRO Publishing, 2012), 29–55.

82 Craig, "'Stationary Is Dead'—Long Live Transformation," 31.

The way forward for the 21st century will not be simple; there is no all-purpose approach to sustainable coastal and marine management. Static juristic boundaries may limit the State's capacity to adequately manage coastal resources. Consequently, more flexible, rolling boundaries may increasingly be required. The counterpoint with regard to maritime jurisdictional limits and zones are efforts to realign the existing direct connection between ambulatory baselines and shifting maritime limits to declared and fixed outer limits to claimed maritime zones. The latter trend addresses the desire of coastal States for stable maritime jurisdictional limits in order to provide clarity and certainty to marine stakeholders, to limit conflicts arising from the uncertainty in boundaries based on relationships between the land and sea[83] and, critically, to ensure that coastal States do not "lose" maritime jurisdictional entitlements as a result of the threat of sea-level rise.

The benefit in fixing baselines, and/or the maritime jurisdictional limits derived from them, under UNCLOS, ensures that under conditions of sea-level rise, a State's maritime entitlements and rights over the valuable marine resources within them will be safeguarded. However, there remain parallel risks for coastal wetlands and for UNCLOS when seas are rising, as fixed boundaries or baselines may limit the management of migrating resources, such as fisheries and ecosystems.

The complexity associated with contested landscapes requires approaches that are fit-for-purpose and that reflect the latest scientific and socio-economic knowledge in order to reconcile the apparently conflicting requirements for both flexibility in boundaries for coastal ecosystem management and stability and certainty in maritime jurisdictional limits. Adaptive pathways to management and political decision making, that integrate political decision making in a co-ordinated manner with scientific and socio-economic knowledge, and incorporates continual assessment and revision, is increasingly advocated as a way forward for sustainable management in a changing climate.[84] The trigger for revision of the status quo should be substantiated by the scientific and socio-economic outcomes of assessments; and while uncertainty will remain in scientific assessments, to forge forward with sound decisions based on the best available information there is a need to accept the inherent complexity and uncertainty in natural processes.

---

83 Caron, "Climate Change, Sea Level Rise and the Coming Uncertainty in Oceanic Boundaries: A Proposal to Avoid Conflict," 1.

84 See for example: Craig, "'Stationary Is Dead'—Long Live Transformation," 53–63; Stoker *et al.*, "Sustainable Coastal Management?," 43–49.

Adaptive approaches that are dynamic and include knowledge gathering, review and revision will be costly; but even more so are climate change impacts for the unprepared. In addition, in contested landscapes there is scarce time to wait for scientific consensus on the effect of sea-level rise on baselines and boundaries. Rather, there is a patent need to alter legal and policy frameworks early to achieve the best outcomes for sustainable coastal and marine management. In summary, five principles are proposed for coastal and marine legal and policy frameworks that will improve sustainable outcomes for coastal and marine management:

1) The sea is not level! Baselines and maritime limits founded on metrics of sea level will be dynamic in character;
2) Climate change science is complex; climate change impacts projection is complex; and uncertainty will remain;
3) Policy and law should also be dynamic and flexible, recognising that uncertainty will remain;
4) Revision of legal and policy frameworks should not be limited by scientific or socio-economic complexity and uncertainty; rather reviews and revisions should be triggered by current scientific and socio-economic knowledge; and, finally,
5) Monitoring and assessment is essential for scientific and socio-economic knowledge gathering.

CHAPTER 19

Arctic Navigation: Reflections on the Northern Sea Route

*Said Mahmoudi**

Introduction

International interest in the Arctic Ocean has increased since the United Nations Convention on the Law of the Sea (UNCLOS) entered into force in 1994. The reason is the greater attention now being paid to the consequences of global warming[1] and the new possibilities opening up due to climate change. Access to resources, including previously unknown biological ones, and increased use of navigational routes, are just two of such possibilities.

One example of this increased attention to the Arctic waters is the meeting of the ministers of transport of Japan, China and South Korea in Yokohama on 25 August 2014. Their purpose was to agree on an arrangement for the safety of ship operations in Arctic waters and to establish an efficient logistic system in that area. This was the fifth recent meeting of these ministers about Arctic navigation.

The two main sea routes in the Arctic Ocean today are the Northwest Passage, which runs along the northern coast of North America, and the Northeast Passage, which follows the coasts of Norway, Russia and Alaska. The major part of its Russian section, the Northern Sea Route, is the focus of this paper. A third route, the Central Arctic Ocean Route, which is in international waters, is considered as a potential trans-Arctic transport seaway.

The Northern Sea Route is the most attractive option for transport due to its more favorable ice conditions and available infrastructure, including nuclear-powered ice breakers. This may be compared with the Northwest Passage with relatively scant infrastructure[2] for shipping and perhaps lack of clear

---

* Professor of International Law, Stockholm University.

1 According to one recent estimate, 4.57 million square miles of Arctic sea ice melted between March and September 2012. See *Arctic Strategy*, United States Coast Guard, 2013, p. 5.

2 The Arctic Council's important report—Arctic Marine Shipping Assessment—from 2009 addresses the problem of insufficiency of shipping infrastructure in general. Mention is made of *i.a.* lack of timely information for navigation, lack of search and rescue assets, pollution

 | DOI 10.1163/9789004311442_021

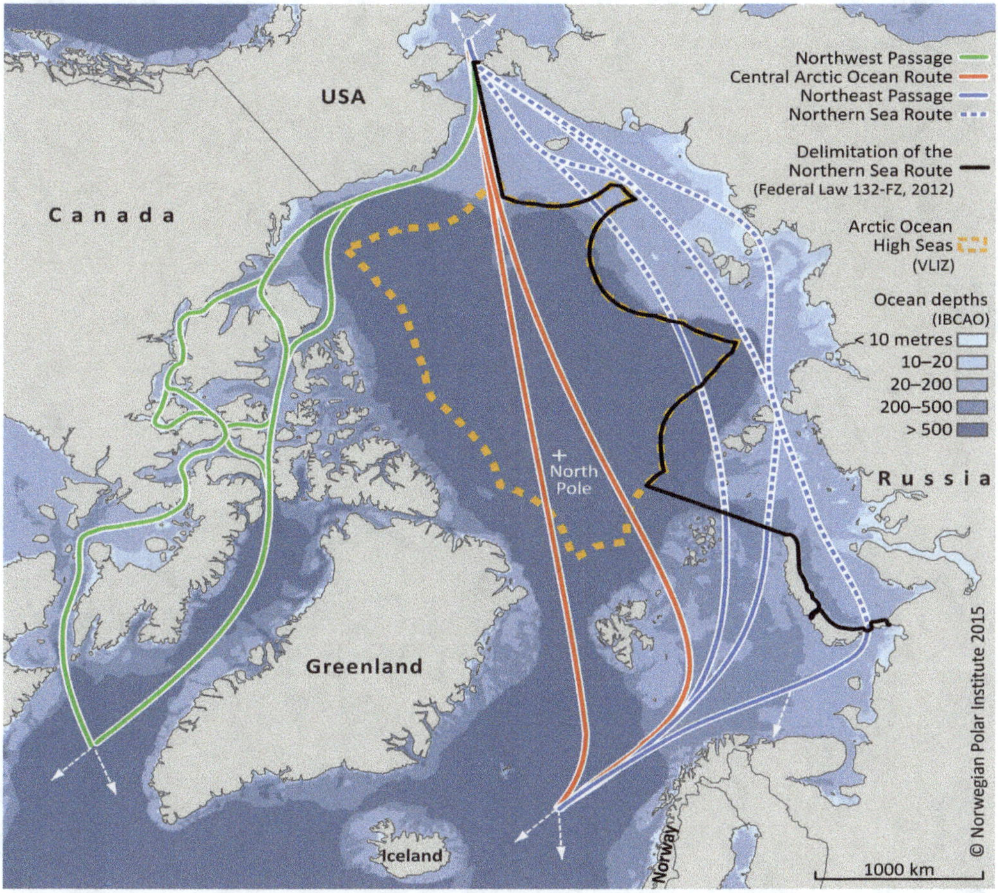

FIGURE 19.1 *Navigational Routes in the Arctic.*
SOURCE: NORWEGIAN POLAR INSTITUTE.

political commitment from the Canadians to develop the navigational routes in the Passage.

The Northern Sea Route is not one fixed route. Unlike most other sea routes, there is no single, set channel for ships to follow, but a series of channels through icy waters. The Route includes the Vilkitsky and Shokalskiy Straits connecting the Kara and Laptev Seas, the Dimtry Laptev and Sannikov Straits connecting the Laptev and East Siberian Seas, and the Proliv Long Strait, which joins the East Siberian and Chukchi Seas. The passage of

response assets, port reception facilities, and salvage resources for vessels in distress. http://www.arctic.noaa.gov/detect/documents/AMSA_2009_Report_2nd_print.pd, pp. 154–155.

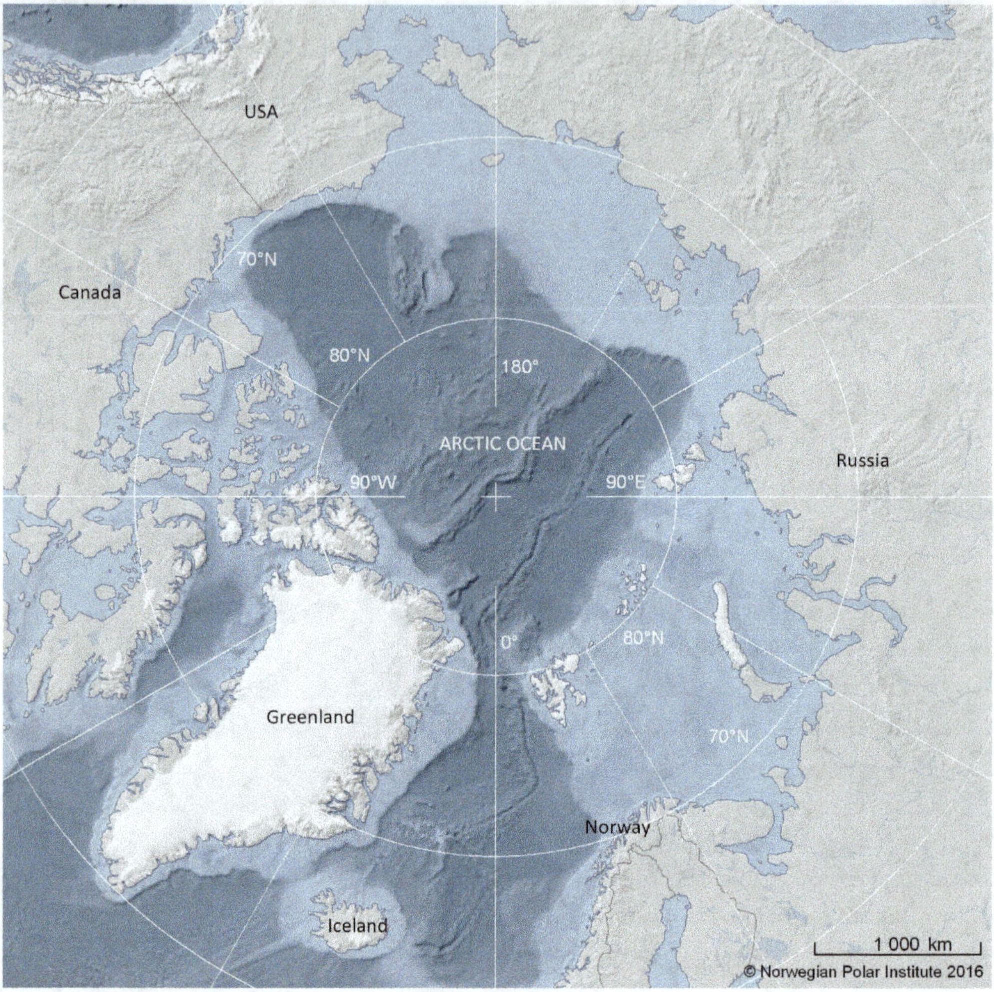

FIGURE 19.2 *Map of the Circumpolar North.*
SOURCE: NORWEGIAN POLAR INSTITUTE.

vessels through this route shortens travel from Asia to Europe by some 5700 nautical miles, saving transportation costs of between $ 60 and 120 billion annually.

The number of ships using the Northern Sea Route as a transport seaway between Europe and Asia is increasing. The first non-Russian navigation of the Route occurred in summer 1997 when two ships with Finnish and Latvian flags separately completed a full transit. In September 2009, two German ships became the first Western commercial vessels to navigate the Northeast

Passage, starting from South Korea towards the Russian North. In 2010 the number increased to four, in 2011 it was 34, in 2012 it became 46 and in 2013 it was 71. Compared with some 20,000 ships that pass through the Suez Canal every year, the number is still very low.[3]

However, despite the obvious increase in traffic in Arctic waters in general and in the Northern Sea Route in particular, the constraints of Arctic navigation are deemed to outweigh those of the routes through the Suez and the Panama Canals. One informed estimate is that it will take another 15 years before Arctic navigation becomes significant enough for the shipping industry.[4]

In August 2013, the 19,000-ton cargo vessel Yong Sheng became the first Chinese commercial vessel to reach Europe via the Northern Sea Route, sailing from the northeastern China port of Dalian to Rotterdam. In July 2014, a 21,000-ton Chinese cargo ship deployed for Europe through the Arctic reached Rotterdam in 35 days, a trip that normally takes 45 days through the Suez Canal.

Regulatory Framework

Generally speaking, navigation in the Arctic is governed by several global legal instruments, one specific regional agreement and the national legislation of the coastal States. Internationally, UNCLOS and several IMO conventions are relevant. UNCLOS applies generally to various maritime zones including those in the Arctic.[5] In addition, Article 234 of this convention specifically grants coastal Arctic States unilateral rights to adopt non-discriminatory laws and regulations for the prevention, reduction and control of marine pollution in the Exclusive Economic Zone (EEZ). Compare this with coastal State legisla-

3 Arild Moe, "The Northern Sea Route: Smooth Sailing Ahead?" *Strategic Analysis* 38 (2014): pp. 784–802, at p. 787. For more recent statistics, see www.arctic-lio.com/nsr_transits.

4 Aldo Chircop, "International Shipping in the Arctic Region: Confronting Legal Uncertainty," in Johan Schelin (ed.), *Shipping & Environment—A Legal Perspective* (Institutet för sjörätt och annan transporträtt, Stockholm, 2014), p. 2; *Strategic Assessment of Development of the Arctic, Changes in Arctic Maritime Transport* (The Arctic Centre, University of Lapland, 2014), p. 5.

5 According to Jenisch: "Throughout the LOS Conference [1973–1982], there were no objections to the full application of the new law to Arctic waters. The Conference avoided any tendencies to establish a 'special' or 'regional' regime for polar areas, with the sole exception of art. 234 with 'ice-covered waters.'" Uwe Jenisch, "The Arctic Ocean and the New Law of the Sea," in Budislav Vukas (ed.), *Essays on the New Law of the Sea* (Sveucillisnanklada Liber, 1985), p. 484. In a recent article, Aldo Chircop argues that it is not clear whether "instruments developed with navigation through open waters in mind and fundamentally different marine environments will function as well as they do in other marine regions." Chircop, *ibid.*

tion and enforcement jurisdiction in other EEZ areas where the exercise of such jurisdiction is mainly to give effect to generally accepted international rules and standards established through the competent international organization, primarily the IMO.[6]

Canada and Russia have developed comprehensive regulations based on the power entrusted to them by Article 234.[7] Even the establishment of their straight baselines in the Arctic may partly be explained by this additional power. Article 234 was negotiated almost exclusively with the exceptional environmental conditions of the Arctic in mind. Although navigation and its safety were not explicitly excluded while drafting this article during the 1970s, the possibility of international passage through ice-covered straits was considered remote. There is therefore no record of State negotiations at the conference

---

6 For a detailed study of Article 234, see Kristin Bartenstein, "The 'Arctic Exception' in the Law of the Sea Convention: A Contributor to Safe Navigation in the Northwest Passage," *Ocean Development & International Law* 42 (2011): 22–52. For the relation of legislative and enforcement jurisdiction of the coastal State in the EEZ, see Myron H. Nordquist (ed.), *United Nations Convention on the Law of the Sea 1982: A Commentary*, Vol. IV (Martinus Nijhoff Publishers, 1990), p. 393 and UNCLOS, Part XII, particularly Articles 207, 2011 and 220.

7 The application of Article 234 by Canada has, according to McDorman, given rise to questions relating to the extent of the power of a coastal State to enact and enforce a specific national law. An example is the Canadian requirement for all commercial vessels to report to Canadian authorities when entering Canada's 200-nautical-mile EEZ with a potential penalty for not doing so. Ted L. McDorman, "The Safety of Navigation in the Arctic Ocean and the Role of Coast Guards: The International Legal and Institutional Context," *Korean Journal of International and Comparative Law* 2 (2014): 27–45, at p. 34. Chircop also refers to the Canadian and Russian regulations and says that both these countries have regulated international navigation within the Arctic straits without prior resort to the IMO and despite the claim of the US and other States about the international status of these straits and the applicability of the right of transit passage. Chircop, *supra*, note 4, p. 8. Molenaar too comments similarly on vessel traffic reporting in the Arctic Canada traffic zone (NORDREG). This was voluntary since 1977 and became mandatory as from 1 July 2010. NORDREG applies not only to the Canadian EEZ in the Arctic, but also to the waters of Canada's Arctic archipelago, including the routes of the Northwest Passage. The adoption of NORDREG without IMO approval and *de facto* requirement of prior authorization have been contested by the United States, France, Germany and some other States. See Erik Molenaar, "Options for Regional Regulation of Merchant Shipping Outside IMO, with Particular Reference to the Arctic Region," *Ocean Development & International Law* 45 (2014): 272–298, at p. 275; see also Ashley Roach and Robert Smith, *Excessive Maritime Claims*, 3rd. ed. (Martinus Nijhoff Publishers, 2012), p. 493. An important and relevant legal question is the status and fate of Canadian and Russian legislation with respect to their respective exclusive economic zones in the Arctic once the ice has melted enough not to qualify these areas for being regulated according to the powers of the coastal State according to UNCLOS Article 234.

about the relation of this provision with the possible character of some Arctic straits as international.[8]

IMO conventions are generally relevant for navigation along the Northern Sea Route. They include the International Convention for the Prevention of Pollution from Ships 73/78 (MARPOL) and its annexes on vessel-source pollution,[9] the 1974 International Convention for the Safety of Life at Sea (SOLAS),[10] the 1978 International Convention on Standards of Training, Certification and Watch-keeping for Seafarers (STCW),[11] and the 1972 Convention on the Prevention of Marine Pollution by Dumping of Wastes and Other Matter[12] with its 1996 Protocol.

Parallel to the melting of ice and the increased possibility of navigation, particularly for commercial ships, a need has been felt for a uniform set of rules for the Arctic. This need can be met partly by harmonization of the Canadian and Russian rules that to a large extent set the standards for Arctic navigation. It can also be achieved through a new global initiative—the Polar Code—that has been under preparation within the IMO since 2009.

The Arctic Council[13] has also played an important role in responding to the need for a harmonized set of navigational rules and standards. Its perhaps most important contribution so far has been the Arctic Marine Shipping Assessment of 2009[14] with follow-up activities and non-binding guidance for its Member States. This Assessment was reviewed in 2013, and recommendations were issued in response to the new experience and increasing shipping.[15] In addition, under the Council's auspices, eight Arctic States adopted the Agreement on Cooperation on Aeronautical and Maritime Search and Rescue in the Arctic in May 2011.[16] The Agreement entered into force in January 2013.

8 Nordquist, *supra*, note 6, pp. 392–398.

9 Adopted November 2, 1973, amended on February 17, 1978, 1340 *United Nations Treaty Series*, 1983, p. 61.

10 Adopted November 1, 1974, 1148 *United Nations Treaty Series*, 1980, p. 2.

11 Adopted July 7, 1978, 1361 *United Nations Treaty Series*, 1984, p. 2.

12 Adopted 1972, 1046 *United Nations Treaty Series*, 1977, p. 138.

13 On the role of the Arctic Council in this respect, see Timo Koivurova, "Governing Arctic Shipping: Finding a Role for the Arctic Council," *The Yearbook of Polar Law* 2 (2010): 115–138.

14 *Supra*, note 2.

15 Protection of the Arctic Marine Environment Working Group, *Arctic Ocean Review—Final Report*, Arctic Council, 2013.

16 *International Legal Materials* 50 (2011): 1110–1112.

They also adopted the Agreement on Cooperation on Marine Oil Pollution Preparedness and Response in the Arctic, which is not yet in force.[17]

Polar Code

International awareness of the increasing possibility of shipping in Arctic waters during the 1990s led to the IMO's adoption in 2002 of the Guidelines for Ships Operating in Arctic Ice-covered Waters.[18] This was followed by the 2009 Guidelines for Ships Operating in Polar Waters.[19] However, the need was felt for a mandatory instrument regulating navigation in Polar waters. In 2009 the IMO started preparations for the adoption of a mandatory polar code. The drafters sought to create a new document different from the Guidelines as regards both approach and legal status. Thus, a risk/goal approach was adopted and it was decided that the document should consist of both mandatory and recommendatory parts.[20] A comprehensive draft was submitted to the IMO by Norway (Sub-Committee on Ship Design and Construction) on 15 November 2013. Other countries also, particularly Canada, Argentina and the US, had contributed to this draft, entitled Development of a mandatory code for ships operating in polar waters—Draft text of a polar code. The International Code for Ships Operating in Polar Waters (the Polar Code), and related amendments to (SOLAS) were adopted in November 2014. Even certain environmental provisions and amendments of MARPOL were adopted in 2015. They are all expected to enter into force in January 2017.

The code addresses in some 20 chapters most issues regarding the safety of navigation and pollution prevention according to SOLAS, STCW and the International Convention for the Prevention of Pollution from Ships (MARPOL) with special regard to the particular requirements of polar waters. Each of its two parts is subdivided into mandatory provisions (A) and recommendations (B). The Code's provisions regarding vessel safety measures constitute a new chapter of the SOLAS Convention—Chapter XIV—whereas provisions relating to pollution prevention are to be adopted in 2015 through amendments to the relevant annexes to the MARPOL Convention. Incorporation of the provisions

---

17 https://oaarchive.arctic-council.org/bitstream/handle/11374/529/MM08_agreement_on_oil_pollution_preparedness_and_response_%20in_the_arctic_formatted%20%282%29.pdf?sequence=1&isAllowed=y.

18 IMO Document MSC/Circ. 1056, MEPC/Circ. 399, December 23, 2002.

19 IMO Document A 26/Res. 1024, January 18, 2010.

20 IMO Document DE 53/26, March 5, 2010.

of the Code into the existing IMO Conventions instead of adopting a separate new convention has speeded up the entry into force of the Polar Code, since amendments to the named conventions can come into force through tacit acceptance by State Parties. Compare this with the time-consuming procedure of ratification of a new convention.[21]

Although the Code is intended to apply as a whole to both the Arctic and the Antarctic, the legal and geographical differences between the two areas have been taken into account.

The Code is to cover all possible sources of hazard to navigation: ice, very low temperature, extended periods of darkness, high latitudes that affect navigation and communication systems, remoteness and possible lack of accurate information, limited infrastructure, and lack of experience in polar operations.[22]

Given the prospect of increased commercial traffic in Arctic waters, particularly along the Northern Sea Route, and the exceptional environmental conditions of this marine area, it is fully understandable that specific protective measures shall be taken at all levels. All these measures seem to assume that the legal status of these waters and the navigational routes is a more or less settled issue.

However, a quick look at the positions of the three major regional powers, Canada, the Russian Federation, and the USA, shows that the issue of the legal status of the navigational routes is still unsettled. This can have implications for the passage of foreign vessels through these routes.

Legal Status of the Navigational Routes[23]

The Northwest Passage and the Position of Canada

The basic position of Canada concerning the Northwest Passage and the Russian Federation regarding Northern Sea Route is that these two Routes constitute internal waters of the respective countries.[24]

Canada invokes two legal bases in support of this position: the waters are internal by virtue of the historic title that Canada enjoys; and also because they

21 McDorman, *supra*, note 7, p. 44.

22 For a list of Arctic navigational challenges, see Chircop, *supra*, note 4, p. 4.

23 For a thorough study of the legal status of Arctic navigational routes and the positions of Canada and Russia, see Erik Franckx, Maritime Claims in the Arctic—Canadian and Russian Perspectives, Martinus Naijhof, 1993.

24 For a brief account of the main Canadian arguments, see Robert Dufresne, *Controversial Canadian Claims over Arctic Waters and Maritime Zone*, PRB-07–47E, Library of Parliament, Law and Government Division, January 10, 2008.

are on the landward side of the straight baselines that Canada drew around the Arctic Archipelago in 1985.[25] Other States, primarily the US and some EU countries, reject these arguments and consider that all States enjoy navigational rights according to the law of the sea since the waters are *not* internal. The main argument in support of this position is that States enjoy the right of navigation—innocent passage—that existed prior to enclosure of the Passage with the straight baselines.[26] Moreover, it is argued that States could enjoy a right of transit passage if the Route qualified as an international strait. Finally, it is maintained that if the Passage is not yet a strait used for international navigation, it could become one, and a right of transit passage could in that case develop over time.[27]

The difference of opinion between the United States and Canada as regards the legal status of the Northwest Passage became evident first after the passage of the US tanker *S.S. Manhattan* in 1969–1970 through a part of the Canadian Arctic waters then considered as high seas. This was followed by navigation of the US icebreaker *CGS Polar Sea* in 1985 through the Northwest Passage without requesting permission. Canada strongly protested the latter transit.[28] The US argument in response to the Canadian objection to both passages was that permission from Canada was unnecessary since the vessels were navigating through waters which constituted a strait used for international navigation. The US invoked the ICJ decision in the *Corfu Channel Case*[29] that the volume of international traffic is not relevant for a body of water to qualify as a strait used for international navigation.[30] Moreover, later statements of position by the US seem to imply that the *potential* for the Northwest Passage to be used for international maritime traffic, as opposed to its *actual* use, was sufficient to make it a strait used for international navigation.[31] This is challenged by Canada and the Russian Federation, supported by commentators who opine

25 Donat Pharand, "The Arctic Waters and the Northwest Passage: A Final Revisit," *Ocean Development & International Law* 38 (2007): 3- 69, at 5–28. This position was formally stated by the Secretary of State for External Affairs before the House of Commons on 10 September 1985. Statement No. 85/49, pp. 6462–6464.

26 UNCLOS, Article 8(2).

27 Dufresne, *supra*, note 24, p. 2.

28 Pharand, *supra*, note 25, p. 42.

29 Judgment of 9 April 1949, *ICJ Reports 1949*, p. 4.

30 The Court nevertheless referred to the limited volume of navigation through the Corfu Channel between 1936 and 1937, which assisted it in determining that the said channel had been "a useful route for international maritime traffic." *Ibid.*, p. 28.

31 The European Union has also had the same position and opposed Canada's use of straight baselines around the Arctic Archipelago. Dufresne, *supra*, note 24, p. 6.

that it is the actual use which is decisive for the establishment of a strait's status as international.

Following the voyage of the CGS *Polar Sea* in 1985, Canada established straight baselines around the Arctic Archipelago, effective as from 1 January 1986. Canada's historic title argument, which was originally claimed in 1973,[32] is generally considered weak. Were this claim well-founded, other States would have an obligation under international law to obtain permission for entry into these waters. The argument of internal waters arising from straight baselines may have more support in current international law. According to Article 8 of UNCLOS, in such internal waters all foreign States have a right to innocent passage.[33] Note that a possible increase in the number of vessels passing through the Northwest Passage without seeking Canadian authorization can strengthen the US position on the international-strait status of the Passage and the applicability of the right of transit passage for all foreign vessels.[34]

As regards the passage of American icebreakers, a settlement was achieved in January 1988 through the conclusion of the Agreement on Arctic Cooperation between the US and Canada.[35] According to this agreement, the passage of American icebreakers shall take place with the consent of the Government of Canada. However, this does not seem to have changed the formal position of the US with respect to the legal status of the Passage as a strait used for international navigation.[36] The real reason for US insistence on the international-strait argument in this case may be its desire to invoke the same argument with respect to similar national claims by other States, primarily by the Russian Federation regarding the Northern Sea Route. Like Canada, the Russian Federation claims that the Northern Sea Route constitutes a part of that country's internal waters, a claim which is rejected by some other States including the US.

---

32 Pharand cites a statement by the Bureau of Legal Affairs, which reads "Canada also claims that the waters of the Canadian Arctic Archipelago are internal waters of Canada, on a historical basis, although they have not been declared as such in any treaty or by any legislation." *Supra*, note 25, p. 11.

33 Some authors argue that since the straight baselines were established in 1985 when Canada was not a party to UNCLOS, the rule of innocent passage as enshrined in the Convention's Article 8 is not applicable here. The counter-argument is that by ratifying the LOS Convention in 2004, Canada has become bound by the provision in Article 8, and innocent passage of foreign ships in waters on the landside of the baselines will be applicable. For a discussion, see Dufresne, *supra*, note 24, p. 5.

34 *Ibid.*, p. 9.

35 Adopted on January 11, 1988, 1852 *United Nations Treaty Series*, p. 60.

36 Roach and Smith, *supra*, note 7, p. 318.

The Northern Sea Route and the Position of the Russian Federation

Russian authorities claim that the integral nature of the Northern Sea Route as a national transport route is unaffected by the fact that individual portions of it may pass outside its internal waters, territorial sea and exclusive economic zone. Russia claims historic title to these waters.[37]

Russia argues that the Northern Sea Route is a coastal transport route, similar to Norway's Indreleia,[38] which the International Court of Justice treated in the *Anglo-Norwegian Fisheries Case*[39] as a navigational route under the complete control and administration of Norway, regardless of whether parts of it fell within internal and territorial waters. In that case, Norway employed 47 straight baselines to measure its territorial sea, prompting a complaint from the United Kingdom regarding that practice and the Norwegian claim that sea space on the landward side of these straight lines was part of its internal waters. The Court accepted Norway's argument and considered the Indreleia not as a strait, but rather a navigational route prepared as such by means of artificial aids to navigation. The Court further accepted that Norway had exercised effective jurisdiction over the said route for a long time without opposition from other States. Consequently this uncontested long practice had, according to the Court, given legal validity to the Norwegian claim.

As regards the Northern Sea Route, in 1965 the Soviet Union enacted internal regulations requiring compulsory pilotage for vessels navigating the Vilklitsky and Skokalsky Straits. Similar regulations followed in 1972 for passage through the Dimitry Laptev and Sannikov Straits. In 1991 and 1993, Russia extended domestic escort rules to warships, requiring "icebreaker-assisted pilotage" and a host of other regulatory controls for all surface vessels transiting Russian Arctic straits. These regulations affected only the scant Russian traffic that existed in these waters. However, the decision in the early 1990s to actively promote international navigation in the Route led to the adoption of new regulations in the coming years.

Mention should particularly be made of the Regulations for Navigation on the Seaways of the Northern Sea Route (1990), Regulations for Icebreaker-Assisted Pilotage of Vessels on the Northern Sea Route (1996), and the Federal Act on the Internal Maritime Waters, Territorial Sea and Contiguous Zone of the Russian Federation (1998). A revision of the existing laws resulted in the

---

37 Alexander Kovalev, *Contemporary Issues of the Law of the Law of Sea: Modern Russian Approaches*, translated by William Butler (Eleven International Publications, 2004), p. 189. See also Moe, *supra*, note 4, p. 786.

38 *Ibid.*, p. 184.

39 *ICJ Reports 1951*, p. 116.

adoption in 2012 of the Federal Law on Amendments to Specific Legislative Acts of the Russian Federation Related to Governmental Regulation of Merchant Shipping in the Water Area of the Northern Sea Route. The act sets forth centralized state management of the Northern Sea Route, providing equal access to all interested carriers subject to prior authorization.[40]

Navigation of State Vessels

Public information about navigation by foreign State vessels along the Northern Sea Route is limited to the voyages of some US icebreakers during the 1960s. In the summer of 1963, the USCGC *Northwind* visited the Laptev Sea and engaged in activities formally characterized by the US as oceanographic survey. This was followed in 1964 by a visit of the USS *Burton Island* to the East Siberian Sea for the same purpose. Further to the latter visit, the Soviet Union reacted in July the same year by presenting an *aide mémoire* regarding the voyage of *Burton Island.* This document underscored that the Route "has been used and is used only by ships belonging to the Soviet Union ..." It was further underlined:

> the northern seaway route at some points goes through Soviet territorial and internal waters. Specifically, this concerns all straits running west and east in the Karsky Sea. Inasmuch as they are overlapped two-fold by Soviet territorial waters, as well as the Dmitry, Laptev and Sannikov Straits, which unite the Laptev and Eastern Siberian Seas and belong historically to the Soviet Union. Not one of these stated straits, as is known, serves for international navigation. Thus over the waters of these straits the statute for the protection of the state borders of the USSR fully applies, in accordance with which foreign military ships will pass through territorial and enter internal sea waters of the USSR after advance permission of the Government of the USSR ...[41]

40 Roach & Smith, *supra*, note 7, p. 495. For the text of the relevant regulations and particularly the 2012 Federal Law, see www.arctic-lio.com/nsr_legislation. For a brief and useful comment on the content of the Federal Law, see Konstantin Krasnokutskiy, *The Northern Sea Route—A Major Change in Regulation*, http://navicus.ru/files/The%20Northern%20Sea%20Route%20%E2%80%93%20A%20major%20change%20in%20regulation.pdf.

41 Roach & Smith, *supra*, note 7, p. 312.

In its answer dated 22 June 1965, the United States rejected the Soviet claim and emphasized the right of innocent passage of all ships through straits used for international navigation between two parts of the high seas (including the straits of the Karsky Sea), and that this right cannot be suspended.[42]

The episode was repeated in 1967 when the United States commissioned an Arctic circumnavigation by two Coast Guard icebreakers *Edisto* and *East Wind*. They entered the Karsky Sea but encountered severe ice conditions blocking passage north of the Severnaya Zemlya archipelago. The US informed the Soviet Union of the situation and notified the latter that to continue circumnavigation the icebreakers had to transit Vilkitsky Straits. The Soviet Union responded that "the proposed passage of the 'Edisto' and 'Eastwind' would be in violation of Soviet regulations", and raised "the possibility of action by the Soviet Government to detain the vessels or otherwise interfere with their movement."[43]

The US cancelled the planned circumnavigation and strongly protested the Soviet position considering the Vilkitsky Straits as territorial waters subject to the regime of prior authorization. The US, referring to the fact that both countries were parties to the 1958 Convention on Territorial Sea and Contiguous Zone, pointed out in its protest note that "Soviet law cannot have the effect of changing the status of international waters and the rights of foreign ships with respect to them."[44] The US has since periodically asserted its right to free transit through the disputed straits, but has not dispatched icebreakers there again.

Appraisal

The main difference of views about the legal status of the Northern Sea Route revolves around the fact that the Route should be considered either as a single long strait used for international navigation in the sense of the LOS Convention—the view that seems to be held by the US—or mainly internal and historical waters as claimed by the Russian Federation.

The position of the United States seems to reflect a matter of principle rather than an issue of absolute practical needs. Freedom of navigation is a pivotal, general principle in the global naval strategy of the United States. Any restriction to this freedom is an exception and not easily permitted. The expedition of several US icebreakers to Soviet-claimed areas in the Arctic Ocean

42 *Ibid.*, p. 313.
43 *Ibid.*, p. 316.
44 *Ibid.*, pp. 318–319.

during the 1960s was a clear manifestation of this position as described in the American responses to the Soviet protests. However, all later voyages of US research vessels to these waters seem to have been made in compliance with the Russian requirements.[45]

The security-military value of the Northern Sea Route for the US surface and submarine fleet is considered to be low.[46] That is why the possibility of deploying US icebreakers to challenge the Russian legal regime of prior authorization may be remote even if the United States continues to formally consider that the existing Russian regulations contravene international law.

The Russian Federation has indeed effectively exercised its sovereignty over the Northern Sea Route. This has been done not only by imposing a license system for all foreign vessels and mandatory pilotage for passage through certain straits, but also by providing necessary support services without which the passage of foreign vessels through the Route would have been very difficult, if not impossible. The available public information testifies that there have been no unauthorized passages by non-Russian ships in the Vilkitsky, Shokalskiy, Dimtry Laptev, and Sannikov Straits. This is also evidenced by the effort of the Greenpeace ship *Arctic Sunrise* to enter the Northern Sea Route in August 2013.[47] This is certainly a manifestation of the Russian sovereign claim of control over the Route.

The increased attention of major economic powers, China, Japan and South Korea, to the Northern Sea Route and their more frequent use of the Route would not have been possible if there had been serious doubts about

---

45 R. Douglas Brubaker and Willy Østreng, "The Northern Sea Route Regime: Exquisite Superpower Subterfuge?" *Ocean Development and International Law* 30 (1999): 299–331, p. 318.

46 *Ibid.*, pp. 305–311. Brubaker and Østreng argue that "The military-strategic value of the NSR will be shown to be low as a link between the Northern and Pacific Fleets, as a station area for nuclear ballistic missile submarines (SSBNs), and as a militarized surveillance area. In terms of perceptions, it has, however, a long history of consistent overrating, which also affects the legal regimes." Their conclusion is that "most of the reasons behind the overrated military security perceptions and the expansive legal positions relate to global macro-strategic interests in regions far removed from the NSR itself" pp. 299 and 324.

47 For the attempt of the *Arctic Sunrise* to enter the Northern Sea Route and the alleged reasons of the Russian authorities' refusal to issue a permit, see http://www.ft.com/cms/s/0/1136d6da-0a4f-11e3-9cec-00144feabdc0.html#axzz3VxjLya7U. For a detailed account of the *Arctic Sunrise* mission and the ensuing dispute between the Netherlands and the Russian Federation, See *The Arctic Sunrise Arbitration* (*Kingdom of the Netherlands v. Russian Federation*), the Award of the Permanent Court of Arbitration, Case No. 2014-2, 14 August 2015, http://www.pcacases.com/web/sendAttach/1438.

the certainty of complete Russian control over this body of water. Despite the formal positions of these countries, which like the US continue to emphasize the international status of these waters, their *de facto* acceptance of the applicability of the Russian legislation shows that the Russian Federation has in fact consolidated its legal claims in a definitive manner.

In summary, the fragmentation of regimes, varying degrees of development in the consolidation of legal claims, divergences in the specific postures and objectives of both the States with jurisdictional claims in the Arctic and the States aspiring to become major players, and the often-problematic prospects of new territorial determinations and regulatory rules coming into the picture, all suggest that the future of the Arctic's navigational regime holds out the prospect of new challenges—certainly interesting, possibly formidable, in their nature—for the international oceans order.

Index

www.ingramcontent.com/pod-product-compliance
Lightning Source LLC
LaVergne TN
LVHW010623300926
849162LV00001B/40
* 9 7 8 9 0 0 4 3 1 1 4 3 5 *